Springer Proceedings in Physics

Volume 270

Indexed by Scopus

The series Springer Proceedings in Physics, founded in 1984, is devoted to timely reports of state-of-the-art developments in physics and related sciences. Typically based on material presented at conferences, workshops and similar scientific meetings, volumes published in this series will constitute a comprehensive up-to-date source of reference on a field or subfield of relevance in contemporary physics. Proposals must include the following:

- name, place and date of the scientific meeting
- a link to the committees (local organization, international advisors etc.)
- scientific description of the meeting
- list of invited/plenary speakers
- an estimate of the planned proceedings book parameters (number of pages/articles, requested number of bulk copies, submission deadline).

Please contact:

For Americas and Europe: Dr. Zachary Evenson; zachary.evenson@springer.com
For Asia, Australia and New Zealand: Dr. Loyola DSilva; loyola.dsilva@springer.com

More information about this series at https://link.springer.com/bookseries/361

Guixiong Liu · Fengjie Cen
Editors

Advances in Precision Instruments and Optical Engineering

Proceedings of the International Conference on Precision Instruments and Optical Engineering, 2021

Springer

Editors
Guixiong Liu
South China University of Technology
Guangzhou, China

Fengjie Cen
Guangzhou Keo Information Technology
Co., Ltd.
Guangzhou, China

ISSN 0930-8989 ISSN 1867-4941 (electronic)
Springer Proceedings in Physics
ISBN 978-981-16-7257-6 ISBN 978-981-16-7258-3 (eBook)
https://doi.org/10.1007/978-981-16-7258-3

This Springer imprint is published by the registered company Springer Nature Singapore Pte Ltd.
The registered company address is: 152 Beach Road, #21-01/04 Gateway East, Singapore 189721, Singapore

Committee Members

Committee Chair

Prof. Guixiong Liu, South China University of Technology, China

Program Committees

Prof. (Hub) Minin, Igor, Tomsk Polytechnic University, Russia
Prof. SAEED ISLAM, Abdul Wali Khan University Mardan (AWKUM), Pakistan
Prof. Zhong-Cheng Liang, Nanjing University of Posts and Telecommunications, China
Prof. Xiangdong Xu, University of Electronic Science and Technology of China, China
Prof. Jianfeng Li, University of Electronic Science and Technology of China, China
Prof. Morteza Sasani Ghamsari, Photonics and Quantum Technologies Research School of Iranian Nuclear Science and Technology Research Institute, Iran
Assoc. Prof. Xianguang Yang, Jinan University, China
Assoc. Prof. Jiahao Yan, Jinan University, China

Organizing Committees

Prof. V. Vijayakumar, VIT University, India
Prof. Behzad Niroumand, Isfahan University of Technology (IUT), Iran
Dr. Zhe Zhu, Duke University, USA

Preface

The 2021 International Conference on Precision Instruments and Optical Engineering (PIOE 2021) was planned to be held during August 20–22, 2021, in Guangzhou, China. Currently, the entire world is struggling against the virulent pandemic COVID-19. Unfortunately, each of us is affected, either overtly or covertly. Our conference, the 2021 International Conference on Precision Instruments and Optical Engineering (PIOE 2021), was not an exception.

Nowadays, mass gatherings are not permitted by the government. It is uncertain when the COVID-19 would end, so it remains unclear for postponement time, while many scholars and researchers wanted to attend this long-awaited conference and have academic exchanges with their peers. Therefore, in order to actively respond to the call of the government, and meet the author's request, the PIOE 2021, which was planned to be held in Guangzhou, China, during August 20–22, 2021, was changed to be held online through Zoom software. This approach not only reduces people gathering but also meets their communication needs.

The conference conducted in-depth exchanges and discussions on related topics such as "precision instruments", "intelligent instruments", "optical engineering" and "laser technology", aiming to provide a platform for scientific research scholars, technicians and related personnel engaged in related topics such as precision instruments and optical engineering to share scientific research achievements and cutting-edge technologies, understand academic development trends, broaden research ideas, strengthen academic research and discussion and promote cooperation in industrialization of academic achievements. Experts, scholars, business people and other relevant personnel from universities and research institutions at home and abroad are cordially invited to attend and exchange.

The conference model was divided into three parts: oral presentations, keynote speeches and online Q&A discussion. There were over 70 participants who attended the meeting. In the first part, we were honored to have invited three professors as our keynote speakers. The first one, Assoc. Prof. Xianguang Yang, Guest Editor of Frontiers in Materials, Institute of Nanophotonics, Jinan University, China. He performed a speech: *Optical Properties and Lasing Physics in Flexible Fibers Doped with Nanocrystals.* And then we had Prof. Aleksey I. Mis'kevich, from National

Researches Nuclear University (Moscow Engineering Physics Institute), Russia. He performed a speech: *Nuclear Pumped Excimer Lasers: Achievements, Prospects and Problems.* The last keynote speakers, Assoc. Prof. Jiahao Yan, is from Institute of Nanophotonics, Jinan University, China. He shared a speech: *All-dielectric nanostructures for tunable light-matter interactions.* Their insightful speeches had triggered a heated discussion in the third session of the conference. The WeChat discussion lasted for about 30 minutes. Every participant praised this conference for disseminating useful and insightful knowledge.

The proceedings of PIOE 2021 span over 2 topical tracks that include Optical Engineering, Precision Instruments and other related fields. All the papers have been through rigorous review and process to meet the requirements of International publication standards.

We would like to acknowledge all of those who have supported PIOE 2021. Each individual and institutional help were very important for the success of this conference. Especially, we would like to thank the organizing committee for their valuable advices in the organization and helpful peer review of the papers. We hope that PIOE 2021 will be a forum for excellent discussions that will put forward new ideas and promote collaborative researches. We are sure that the proceedings will serve as an important research source of references and knowledge, which will lead to not only scientific and engineering progress but also other new products and processes.

Committee of PIOE 2021

Guangzhou, China

Guixiong Liu
Fengjie Cen

Contents

Contributors

Yu Ai Tianjin Boomscience Technology Co., Ltd., Tianjin, China

Chenxing Bao Beijing Aerospace Institute for Metrology and Measurement, Beijing, China

Jiahan Bao Engineering Technology Research Center of Hydraulic Vibration Technology, Anhui University of Technology, Maanshan, Anhui Province, China

Fushen Cai Guangdong Jiangmen Supervision Testing Institute of Quality & Metrology, Jiangmen, China

Jiafu Cai South China University of Technology Guangzhou, Guangdong, China

Jingwen Cai State Grid Dalian Electric Power Supply Company, Dalian, China

Jing Cao School of Optoelectronic Science and Engineering, University of Electronic Science and Technology of China, Chengdu, China

Yuwan Cen Engineering Technology Research Center of Hydraulic Vibration Technology, Anhui University of Technology, Maanshan, Anhui Province, China

Feng Chen Key Laboratory of New Energy Automotive Drive Systems for Zhejiang Market Regulation, College of Mechanical Engineering, Zhejiang University of TechnologyZhejiang Fangyuan Testing Group Co., Ltd, Hangzhou, China

Haiqing Chen China Tobacco Jiangsu Industrial Co., Ltd., Nanjing, Jiangsu, China

Hang Chen College of Military Basic Education, National University of Defense Technology, Changsha, Hunan, P. R. China

Jia-Hui Chen School of Physics and Optoelectronic Technology, South China University of Technology, Guangzhou, China

Jiaxuan Chen Hubei Engineering Technology Research Center of Energy Photoelectric Device and System, Hubei University of Technology, Wuhan, China;
School of Science, Hubei University of Technology, Wuhan, China

Jie Chen Elevator Inspection Department, Shanghai Institute of Special Equipment Inspection and Technical Research, Shanghai, China

Jin-Zhong Chen Pressure Pipeline Department, China Special Equipment Inspection and Research Institute, Beijing, China

Ke Chen Inner Mongolia First Machinery Group Co.,Ltd, Inner Mongolia, China

Wen Chen Innovation Academy for Microsatellites of CAS, Shanghai, China

Wenjia Chen Hubei Engineering Technology Research Center of Energy Photoelectric Device and System, Hubei University of Technology, Wuhan, China;
School of Science, Hubei University of Technology, Wuhan, China

Yang Chen CATARC New Energy Vehicle Test Center (Tianjin) Co., Ltd., Tianjin, China

Zhijun Chen Institute of Army Aviation, Beijing, China

Chunfu Cheng Hubei Engineering Technology Research Center of Energy Photoelectric Device and System, Hubei University of Technology, Wuhan, China;
School of Science, Hubei University of Technology, Wuhan, China

Guangxin Cheng College of Advanced Interdisciplinary Studies, National University of Defense Technology, Changsha, China

Xiaomeng Cheng State Key Laboratory of Electronic Thin Films and Integrated Devices, School of Optoelectronic Science and Engineering, University of Electronic Science and Technology of China (UESTC), Chengdu, People's Republic of China

Jin Cui Research Institute for Frontier Science, Beihang University, Beijing, China;
Ningbo Institute of Technology, Beihang University, Ningbo, China

Yulong Dai College of Military Basic Education, National University of Defense Technology, Changsha, Hunan, P. R. China

Chenchen Dong Key Laboratory of New Energy Automotive Drive Systems for Zhejiang Market Regulation, Zhejiang Fangyuan Testing Group Co., Ltd, Hangzhou, China

Mingming Dong Beijing Institute of Technology, Beijing, China

Zhuangzhuang Dong College of Advanced Interdisciplinary Studies, National University of Defense Technology, Changsha, China

Shaopeng Dong School of Automation Science and Electrical Engineering, Beihang University, Beijing, China;
Ningbo Institute of Technology, Beihang University, Ningbo, China

Kang Du School of Precision Instrument and Optoelectronics Engineering, Tianjin University, Tianjin, 300072 China

Jiashun Duan College of Military Basic Education, National University of Defense Technology, Changsha, Hunan, P. R. China

Yifei Duan Key Laboratory of Quantum Optics, Shanghai Institute of Optics and Fine Mechanics, Chinese Academy of Sciences, Shanghai, China;
Center of Materials Science and Optoelectronics Engineering, University of Chinese Academy of Sciences, Beijing, China

Huiying Fan College of Advanced Interdisciplinary Studies, National University of Defense Technology, Changsha, China

Li Fang Hubei Engineering Technology Research Center of Energy Photoelectric Device and System, Hubei University of Technology, Wuhan, China;
School of Science, Hubei University of Technology, Wuhan, China

Chun Feng College of Electrical Engineering and Control Science, NanJing Tech University, NanJing, China

Shuangchang Feng Elevator Inspection Department, Shanghai Institute of Special Equipment Inspection and Technical Research, Shanghai, China

Weidong Feng Jilin Business and Technology College, Changchun, China

Yuanting Feng State Key Laboratory of Electronic Thin Films and Integrated Devices, School of Optoelectronic Science and Engineering, University of Electronic Science and Technology of China (UESTC), Chengdu, People's Republic of China

Shao Fengbo Henan University of Science and Technology, Luoyang, Henan, China

Renxuan Fu School of Electrical and Mechanical Engineering, Guangdong Polytechnic of Industry and Commerce, Guangzhou, China

Jiyu Gao Mechanical and Electrical Engineering Department, Changchun University of Science and Technology, Changchun, Jilin, China

Zhiqiang Gao Department of Basic Sciences, Air Force Engineering University, Xi'an, China

Nizhi Ge Chuxiong Redrying Factory of Yunnan tobacco redrying Co., Ltd., Chuxiong, Yunnan, China

Qiuwen Gong School of Electronics and Materials Engineering, Leshan Normal University, Leshan, China

Tianmao Guo Beijing Aerospace Institute for Metrology and Measurement, Beijing, China

Wentong Guo School of Water Conservancy Engineering, Zhengzhou University, Zhengzhou, China

Xuebing Han Jilin Business and Technology College, Changchun, China

Hongyu He School of Mechanical and Power Engineering, Chongqing University of Science and Technology, Chongqing, China

Jing He Xi'an Technological University, Xi'an, Shanxi, China

Ren-Yang He Pressure Pipeline Department, China Special Equipment Inspection and Research Institute, Beijing, China

Xiaofei He National University of Defense Technology, Changsha, China;
Beijing Aerospace Times Optical-Electronic Co., Ltd, Beijing, China

Chang'an Hu Institute of Optics and Electronics, Chinese Academy of Sciences, Chengdu, China;
University of Chinese Academy of Sciences, Beijing, China;
National Institute of Measurement and Testing Technology, Chengdu, China

Guichuan Hu School of Mechanical and Power Engineering, Chongqing University of Science and Technology, Chongqing, China

Jing Hu School of Intelligent Manufacturing, Chongqing College of Architecture and Technology, Chongqing, China

Junjie Hu State Key Laboratory of Electronic Thin Films and Integrated Devices, School of Optoelectronic Science and Engineering, University of Electronic Science and Technology of China (UESTC), Chengdu, People's Republic of China

Song Hu Institute of Optics and Electronics, Chinese Academy of Sciences, Chengdu, China

Zongyu Hu China Tobacco Jiangsu Industrial Co., Ltd., Nanjing, Jiangsu, China

Jian Huang Beijing Institute of Tracking and Telecommunications Technology, Beijing, China

Jianlu Huang Rocket Force University of Engineering, Xi'an, Shanxi, China

Jianzhong Huang School of Mechanical Engineering, Hefei University of Technology, Hefei, China

Yafeng Huang Key Laboratory of Quantum Optics, Shanghai Institute of Optics and Fine Mechanics, Chinese Academy of Sciences, Shanghai, China;
Center of Materials Science and Optoelectronics Engineering, University of Chinese Academy of Sciences, Beijing, China

Guokai Jiang CATARC Software Testing (Tianjin) Co., Ltd., Tianjin, China

Jinfeng Jiang AVIC Manufacturing Technology Institute, Beijing, China

Shu-Bo Jiang College of Electrical Engineering and Control Science, NanJing Tech University, NanJing, China

Boyang Jin Inner Mongolia University of Science & Technology, Inner Mongolia Autonomous Region, Baotou, China

Niu Jin Departments of Optoelectronic Engineering, University of Changchun Science and Technology, Jilin, China

Yan Jin AVIC Manufacturing Technology Institute, Beijing, China

Yiwen Ju Elevator Inspection Department, Shanghai Institute of Special Equipment Inspection and Technical Research, Shanghai, China

Qing Jv School of Mechanical and Electrical Engineering, Jiangxi University of Science and Technology, Ganzhou, China

Cui Kai Macau University of Science and Technology, Macau, China

Xiao-Wei Kang Pressure Pipeline Department, China Special Equipment Inspection and Research Institute, Beijing, China

Chao Li School of Physics and Optoelectronic Technology, South China University of Technology, Guangzhou, China

Chunxu Li Intelligent Shipping Center, China, Waterborne Transport Research Institute, Beijing, China

Dong Li Innovation Academy for Microsatellites of CAS, Shanghai, China

Guangchang Li School of Mechanical and Electrical Engineering, Jiangxi University of Science and Technology, Ganzhou, China

Hong Pei Li School of Mechanical and Automotive Engineering, South China University of Technology, Tianhe, Guangzhou, Guangdong, China

Hongda Li Mechanical and Electrical Engineering Department, Changchun University of Science and Technology, Changchun, Jilin, China

Jinsong Li Innovation Academy for Microsatellites of CAS, Shanghai, China

Mengmeng Li Hubei Engineering Technology Research Center of Energy Photoelectric Device and System, Hubei University of Technology, Wuhan, China; School of Science, Hubei University of Technology, Wuhan, China

Pengyuan Li Beijing Institute of Tracking and Telecommunications Technology, Beijing, China

Shaopeng Li China Tobacco Jiangsu Industrial Co., Ltd., Nanjing, Jiangsu, China

Tang Li Key Laboratory of Quantum Optics, Shanghai Institute of Optics and Fine Mechanics, Chinese Academy of Sciences, Shanghai, China

YanDe Li Zhuzhou CRRC Times Electric Co., Ltd., Zhuzhou, Hunan, China

Youliang Li Mechanical and Electrical Engineering Department, Changchun University of Science and Technology, Changchun, Jilin, China

Yue Li National Key Laboratory of Science and Technology On ATR, National University of Defense Technology, Changsha, Hunan, China

Yanli Li Key Laboratory of Quantum Optics, Shanghai Institute of Optics and Fine Mechanics, Chinese Academy of Sciences, Shanghai, China;
Center of Materials Science and Optoelectronics Engineering, University of Chinese Academy of Sciences, Beijing, China

Mingbo Liang Shenzhen Han's Scanner S&T Co., Ltd., Shenzhen, China

Yanchun Liang Elevator Inspection Department, Shanghai Institute of Special Equipment Inspection and Technical Research, Shanghai, China

Guomin Liao School of Digital Engineering, Chongqing College of Architecture and Technology, Chongqing, China

Chunqing Lin State Grid Dalian Electric Power Supply Company, Dalian, China

Xuesong Lin State Grid Dalian Electric Power Supply Company, Dalian, China

Feifei Liu School of Mechanical and Electrical Engineering, Jiangxi University of Science and Technology, Ganzhou, China

Guixiong Liu School of Mechanical and Automotive Engineering, South China University of Technology, Tianhe, Guangzhou, Guangdong, China

Haiming Liu China Automotive Technology and Research Center Co., Ltd., Tianjin, China

Jinrong Liu State Key Laboratory of Electronic Thin Films and Integrated Devices, School of Optoelectronic Science and Engineering, University of Electronic Science and Technology of China (UESTC), Chengdu, People's Republic of China

Junbo Liu Institute of Optics and Electronics, Chinese Academy of Sciences, Chengdu, China

Ke Liu Beijing Aerospace Institute for Metrology and Measurement, Beijing, China

Liang Liu Key Laboratory of Quantum Optics, Shanghai Institute of Optics and Fine Mechanics, Chinese Academy of Sciences, Shanghai, China

Qian Liu China Academy of Engineering Physics, Mianyang, Sichuan, China

Shuang Liu School of Physics and Optoelectronic Technology, South China University of Technology, Guangzhou, China;
Innovation Academy for Microsatellites of CAS, Shanghai, China

Ting Liu Key Laboratory of New Energy Automotive Drive Systems for Zhejiang Market Regulation, Zhejiang Fangyuan Testing Group Co., Ltd, Hangzhou, China

Tonghua Liu Inner Mongolia University of Science & Technology, Inner Mongolia Autonomous Region, Baotou, China

Xinquan Liu School of Precision Instrument and Optoelectronics Engineering, Tianjin University, Tianjin, China

Xu Liyou Henan University of Science and Technology, Luoyang, Henan, China

Yufeng Long Naval Aviation University, Yantai, China

Guowei Lu Hangzhou Weiheng Technology Co., Ltd, Hangzhou, China

Jianan Luo Intelligent Shipping Center, China, Waterborne Transport Research Institute, Beijing, China

XiaoHui Luo Equipment Procurement & Supply Division, China Nuclear Power Engineering Co. Ltd., Shenzhen, Guangdong, China

Jiapeng Lv Naval Aviation University, Yantai, China

Shuaijie Mei School of Automation Science and Electrical Engineering, Beihang University, Beijing, China

Siwei Meng Mechanical and Electrical Engineering Department, Changchun University of Science and Technology, Changchun, Jilin, China

Tao Meng Pressure Pipeline Department, China Special Equipment Inspection and Research Institute, Beijing, China

Chenlin Miao China Tobacco Jiangsu Industrial Co., Ltd., Nanjing, Jiangsu, China

Changsong Ni State Grid Dalian Electric Power Supply Company, Dalian, China

Zhaodong Niu National Key Laboratory of Science and Technology On ATR, National University of Defense Technology, Changsha, Hunan, China

Yiwen Ou Hubei Engineering Technology Research Center of Energy Photoelectric Device and System, Hubei University of Technology, Wuhan, China;
School of Science, Hubei University of Technology, Wuhan, China

Baitian Ouyang Guangdong Jiangmen Supervision Testing Institute of Quality & Metrology, Jiangmen, China

Yufeng Qin Naval Aviation University, Yantai, China

Haiyin Qing School of Electronics and Materials Engineering, Leshan Normal University, Leshan, China

Liwei Qiu Beijing Aerospace Times Optical-Electronic Co., Ltd, Beijing, China

Qiuzhi Qu Key Laboratory of Quantum Optics, Shanghai Institute of Optics and Fine Mechanics, Chinese Academy of Sciences, Shanghai, China

Tao Shen Rocket Force University of Engineering, Xi'an, Shanxi, China

Guangfeng Shi Mechanical and Electrical Engineering Department, Changchun University of Science and Technology, Changchun, Jilin, China

Lili Shi Beijing Aerospace Times Optical-Electronic Co., Ltd, Beijing, China

Xianjun Shi Naval Aviation University, Yantai, China

Haoyu Song School of Electronics and Materials Engineering, Leshan Normal University, Leshan, China

Kai Song Key Laboratory of Nondestructive Testing, Ministry of Education, Nanchang Hangkong University, Nanchang, Jiangxi, China

Li Song Equipment Procurement & Supply Division, China Nuclear Power Engineering Co. Ltd., Shenzhen, Guangdong, China

Zhuolun Song Qingdao Foreign Language School, Qingdao, Shandong, China

Xuemiao Su Chuxiong Redrying Factory of Yunnan tobacco redrying Co., Ltd., Chuxiong, Yunnan, China

Haifeng Sun Institute of Optics and Electronics, Chinese Academy of Sciences, Chengdu, China

Changhui Tian Department of Basic Sciences, Air Force Engineering University, Xi'an, China

Kaiwen Tian College of Advanced Interdisciplinary Studies, National University of Defense Technology, Changsha, China

Baochang Wang College of Science, China Three Gorges University, Yichang, China

Changyuan Wang China Automotive Technology and Research Center Co., Ltd., Tianjin, China

Guopeng Wang College of Computer and Information, China Three Gorges University, Yichang, China

Hengtao Wang College of Computer and Information, China Three Gorges University, Yichang, China

Lingke Wang Key Laboratory of Quantum Optics, Shanghai Institute of Optics and Fine Mechanics, Chinese Academy of Sciences, Shanghai, China

Pinghe Wang School of Optoelectronic Science and Engineering, University of Electronic Science and Technology of China, Chengdu, China

Qinghua Wang School of Electrical and Mechanical Engineering, Guangdong Polytechnic of Industry and Commerce, Guangzhou, China

Xiaoguang Wang Beijing Aerospace Institute for Metrology and Measurement, Beijing, China

Xidong Wang College of Computer and Information, China Three Gorges University, Yichang, China

Yitong Wang Beijing Institute of Technology, Beijing, China

Yupeng Wang National Key Laboratory of Science and Technology On ATR, National University of Defense Technology, Changsha, Hunan, China

Lijun Wei Engineering College, Yunnan University Of Bussiness Management, KunMing, YunNan, China

Jie Wen Intelligent Shipping Center, China, Waterborne Transport Research Institute, Beijing, China

Changjun Wu School of Electronics and Materials Engineering, Leshan Normal University, Leshan, China

Jun-Fang Wu School of Physics and Optoelectronic Technology, South China University of Technology, Guangzhou, China

Yanbao Wu School of Electronics and Materials Engineering, Leshan Normal University, Leshan, China

Zhiyi Xiang College of Advanced Interdisciplinary Studies, National University of Defense Technology, Changsha, China

Tong Xiao Department of Basic Sciences, Air Force Engineering University, Xi'an, China

Chen Xiaolong School of Frontier and Interdisciplinary Sciences, National University of Defense Technology, Changsha, China

Wang Xingshu School of Frontier and Interdisciplinary Sciences, National University of Defense Technology, Changsha, China

Ke Xiong State Key Laboratory of Electronic Thin Films and Integrated Devices, School of Optoelectronic Science and Engineering, University of Electronic Science and Technology of China (UESTC), Chengdu, People's Republic of China

Xu Xiping Departments of Optoelectronic Engineering, University of Changchun Science and Technology, Jilin, China

Minghui Xu State Key Laboratory of Electronic Thin Films and Integrated Devices, School of Optoelectronic Science and Engineering, University of Electronic Science and Technology of China (UESTC), Chengdu, People's Republic of China

Qiang Xu China Tobacco Jiangsu Industrial Co., Ltd., Nanjing, Jiangsu, China

Ruyan Xu China Tobacco Jiangsu Industrial Co., Ltd., Nanjing, Jiangsu, China

Xiangdong Xu State Key Laboratory of Electronic Thin Films and Integrated Devices, School of Optoelectronic Science and Engineering, University of Electronic Science and Technology of China (UESTC), Chengdu, People's Republic of China

Xiangqian Xu College of Systems Engineering,, National University of Defense Technology, Changsha, Hunan, P. R. China

Yu Xu State Grid Dalian Electric Power Supply Company, Dalian, China

Zhang Xu Zhuhai College of Science and Technology, Zhuhai, China

Liang Xue Guangdong University of Education, Guangzhou, China;
Institute of Data Science, City University of Macau, Macau, China

Pan Ya Southwest University of Science and Technology, Mianyang, Sichuan, China

Fang Yan Inner Mongolia University of Science & Technology, Inner Mongolia Autonomous Region, Baotou, China

Jiabin Yan Engineering College, Yunnan University of Business Management, Wuhua, Kunming, Yunnan, China

Jun Yang College of Systems Engineering,, National University of Defense Technology, Changsha, Hunan, P. R. China

Liu Yang Southwest University of Science and Technology, Mianyang, Sichuan, China

Yingquan Yang Innovation Academy for Microsatellites of CAS, Shanghai, China

Zhibo Yao School of Mechanical and Electrical Engineering, Jiangxi University of Science and Technology, Ganzhou, China

Meifeng Ye Key Laboratory of Quantum Optics, Shanghai Institute of Optics and Fine Mechanics, Chinese Academy of Sciences, Shanghai, China

Zhao Yingwei School of Frontier and Interdisciplinary Sciences, National University of Defense Technology, Changsha, China

Jinpei Yu Innovation Academy for Microsatellites of CAS, Shanghai, China

Kehan Yu School of Electronics and Materials Engineering, Leshan Normal University, Leshan, China

Xudong Yu College of Advanced Interdisciplinary Studies, National University of Defense Technology, Changsha, China

Di Yu State Grid Dalian Electric Power Supply Company, Dalian, China

Daocheng Yuan China Academy of Engineering Physics, Mianyang, Sichuan, China

Mei Yuan School of Automation Science and Electrical Engineering, Beihang University, Beijing, China;
Ningbo Institute of Technology, Beihang University, Ningbo, China

Jie Zhang CATARC New Energy Vehicle Test Center (Tianjin) Co., Ltd., Tianjin, China

Junlin Zhang Inner Mongolia University of Science & Technology, Inner Mongolia Autonomous Region, Baotou, China

Lei Zhang State Grid Dalian Electric Power Supply Company, Dalian, China

Rongzhuo Zhang Beijing Aerospace Institute for Metrology and Measurement, Beijing, China

Wangxing Zhang Fujian Sanming Jinye redrying Co., Ltd., Sanming, Fujian, China

Xu Zhang China Automotive Technology and Research Center Co., Ltd., Tianjin, China

Xue-Wei Zhang Nanchang Hangkong University, Nanchang, China

XueZhi Zhang Zhuzhou CRRC Times Electric Co., Ltd., Zhuzhou, Hunan, China

Yuan-Bao Zhang School of Physics and Optoelectronic Technology, South China University of Technology, Guangzhou, China

Yuhai Zhang Zhengzhou Tobacco Research Institute of China National Tobacco Corporation, Zhengzhou, Henan, China

Zhongjie Zhang Key Laboratory of New Energy Automotive Drive Systems for Zhejiang Market Regulation, Zhejiang Fangyuan Testing Group Co., Ltd, Hangzhou, China

Can Zhao Equipment Procurement & Supply Division, China Nuclear Power Engineering Co. Ltd., Shenzhen, Guangdong, China

Juanru Zhao School of Automation Science and Electrical Engineering, Beihang University, Beijing, China

Kunyang Zhao China Airport Construction Group Co., Ltd., Beijing, China; Beijing Super-Creative Technology Co., Ltd., Beijing, China

Likun Zhao Changchun Medical College, Changchun, China

Panfei Zheng China Airport Construction Group Co., Ltd., Beijing, China; Beijing Super-Creative Technology Co., Ltd., Beijing, China

Xiaorui Zheng Beijing Aerospace Times Optical-Electronic Co., Ltd, Beijing, China

Duan Zhenhao Departments of Optoelectronic Engineering, University of Changchun Science and Technology, Jilin, China

Wang Zhisheng School of Frontier and Interdisciplinary Sciences, National University of Defense Technology, Changsha, China

Jian Zhou College of Advanced Interdisciplinary Studies, National University of Defense Technology, Changsha, China

Weiwei Zhu Hangzhou Xiaoshan Technician College, Hangzhou, China

Yimo Zong School of Mechanical Engineering, Anhui University of Technology, Maanshan, China

Part I
Sensor and Radar Signal Monitoring Technology

Chapter 1
High-Speed Short-Stroke Displacement Calibration Method Based on Pull-Wire Sensor

Rongzhuo Zhang, Ke Liu, Xiaoguang Wang, Tianmao Guo, and Chenxing Bao

Abstract In order to solve the problem where high-speed motion measurement is needed due to the fixed length of cable in the dynamic calibration process of pull-wire displacement sensor, a high-speed short-stroke calibration method was proposed. This paper introduces the various mechanisms and innovative features of this calibration method in detail. The pull-wire displacement sensor can solve the contradiction between speed and displacement through this method, and realize high-speed and short-stroke dynamic and accurate measurement.

1.1 Preface

The pull-wire displacement sensor is a common instrument in engineering testing. It is generally used to measure relative displacement, strain, and distance. The accuracy of the measurement process is directly related to the reliability of the measurement results, so the measuring instrument is calibrated regularly.

At present, pull-wire sensors are calibrated in accordance with the national metrology technical specification JJF1305-2011. This technical regulation only describes the calibration method in the static process. In actual engineering applications, many of them are dynamic high-speed measurements, such as partial separation testing on rockets and missiles, and engine hydraulic cylinder testing. Therefore, dynamic calibration should be added in the calibration process. When calibrating at different speeds, because the length of the cable is fixed, the low-speed displacement can be realized directly, but the high-speed displacement is difficult to be calibrated, and the problems such as overshoot and the break of the cable are easy to occur. In order to cope with these difficulties, this paper proposes a high-speed short-stroke displacement calibration method based on a cable sensor.

R. Zhang (✉) · K. Liu · X. Wang · T. Guo · C. Bao
Beijing Aerospace Institute for Metrology and Measurement, Beijing 100076, China

G. Liu and F. Cen (eds.), *Advances in Precision Instruments and Optical Engineering*, Springer Proceedings in Physics 270, https://doi.org/10.1007/978-981-16-7258-3_1

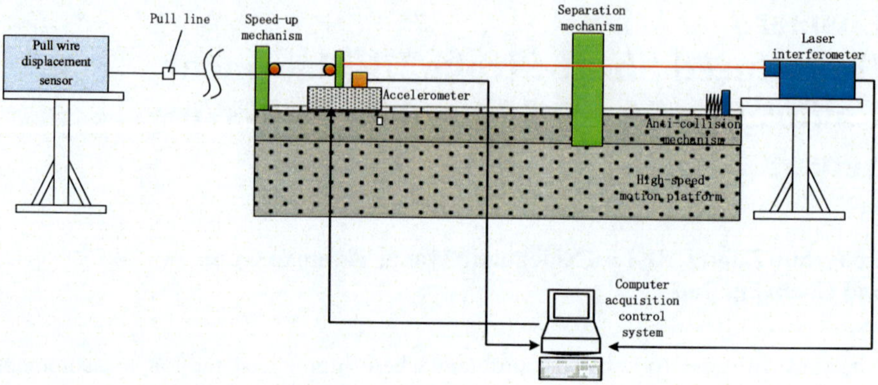

Fig. 1.1 Dynamic calibration overall scheme diagram

1.2 The Overall Scheme of Dynamic Calibration

According to the national measurement technical specification JJF1305-2011, taking the wavelength of 633 nm frequence-stabilized He–Ne laser as the reference for dynamic calibration, through the principle of interference fringe technology and with the high-speed short-stroke displacement calibration device proposed in this paper, the dynamic measurement of the cable type sensor is carried out [1, 2].

Fix the pull-wire displacement sensor on one end of the high-speed motion platform. The top of the cable is tied to the high-speed motion mechanism. When installing, it should meet the Abbe principle as much as possible, and the linear motion of the cable should be aligned with the axis of the high-speed motion platform, thereby reducing the cosine error [3, 4]. A laser interferometer is placed on the other end of the high-speed moving platform for calibration [5]. Taking into account that the distance of the cable is fixed, high-speed movement cannot be achieved directly by the cable, so an acceleration mechanism is built on the motion platform using the variable-speed pulley block. In dynamic measurement, if the motion platform performs variable deceleration motion, the cable will overshoot and the calibration result cannot be correctly reflected. In order to avoid the occurrence of overshoot, a mechanical and electronic compound separation mechanism is designed at the end of the motion platform. Figure 1.1 shows the overall scheme of dynamic calibration.

1.3 Implementation Method of High-Speed Short-Stroke Displacement

The pull wire of the pull-wire sensor is fixed. The entire calibration process must be accelerated to the maximum speed within a certain displacement and then naturally decelerated to a stop, so as to ensure the safety of the pull-wire sensor and avoid the wire break. This calibration method builds a high-speed motion actuator, a motion speed-increasing mechanism, and a mechanical-electronic composite separation mechanism through moving guides, dovetail air-floating platforms, pulleys, etc., and solves the contradiction between speed and displacement, high speed and safety.

1.3.1 High-speed Motion Actuator

In the entire calibration method, the cable end of the displacement sensor is moved quickly through a high-speed motion actuator. The pull-wire displacement sensor in the industrial application of most of the pull for linear high-speed stretching state, so the use of linear guide as a motion simulation mechanism. The high-speed motion actuator mainly includes two parts: linear aerostatic guide and linear motor. The linear aerostatic guide has the characteristics of low friction, which can reduce the resistance of the system movement and increase the movement speed. For ordinary air-floating platforms, the phenomenon of "floating" is prone to occur during high-speed movement, and the direction of the pull line movement is shifted up and down, which affects the calibration result. The linear guide rail adopting a closed air-floating structure can improve the binding force of the platform by increasing the air flow, and avoid the phenomenon of "floating" under high-speed conditions. The guide rail adopts a dovetail air-floating platform, and a high-stiffness air foot is formed between the air-floating platform and the granite base to form a high-precision air-floating rail, which has the characteristics of small friction coefficient and good straightness error uniformity effect. The working range of the entire linear guide is (0 ~ 5000) mm, and a certain margin is left for the slide to safely decelerate.

The measurement speed of the pull-wire sensor can generally reach 10 m/s ~ 50 m/s. During the calibration process, it is necessary to ensure that the speed is fast enough and the control accuracy is high enough. Ball screw as a traditional transmission mode, its motion mode cannot reach the required speed. Although the speed of the belt transmission can meet the requirements, the control accuracy is somewhat poor, so the method of motor rotation winding wire drawing is used as the transmission mode of the high-speed motion actuator. This method can meet the requirements of calibration speed and control accuracy. In the high-speed motion state, the linear motor platform realizes high-precision motion control through closed-loop control, as shown in Fig. 1.2. The motion controller collects the position data of the grating ruler, analyzes the motion controller, and sends a servo signal to the driver, and then the motion controller collects the position data of the grating ruler, thereby forming a closed-loop structure to achieve precise control of linear motion.

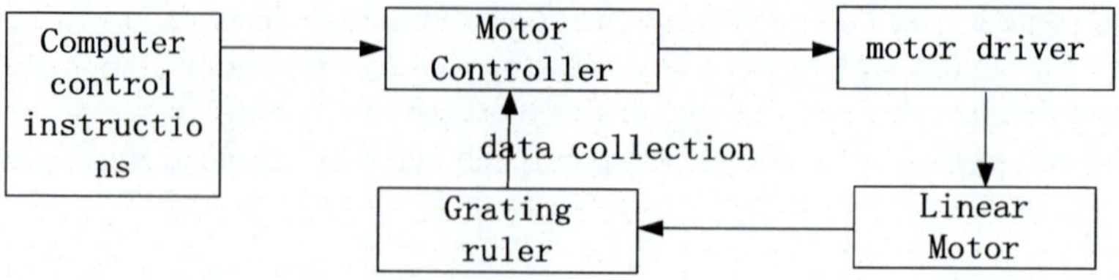

Fig. 1.2 Linear motor control structure diagram

1.3.2 Movement Acceleration Mechanism

If only a linear motor is used as a transmission device on a short-stroke guide rail, it is difficult to achieve a sufficiently high-speed state. In order to solve this problem, it is necessary to design a speed increasing device to cooperate with the linear motor for transmission. The speed-increasing device is composed of multiple movable pulleys and fixed pulleys. The fixed pulley acts as an equal arm lever to change the direction of the wire. The movable pulley can increase the distance and acceleration of the wire. The combination of the movable pulley and the fixed pulley can realize short-distance high-speed wire drawing. Figure 1.3 is a schematic diagram of the structure of the speed increasing device.

In the speed-increasing structure, the use of different numbers of movable pulleys will change the total stroke of the movable sliding table, the maximum speed of the movable platform, acceleration, and other parameters. This article takes the pull-wire sensor with the 2500 mm range as an example, the measurement speed is 15 m/s, and the relationship between the number of movable pulleys and the parameters of the mobile platform is shown in Table 1.1. It can be seen from the table that the larger the number of movable pulleys, the smaller the movement range of the sliding table, and the faster the falling speed. It is necessary to select a suitable number of movable pulleys according to the working range of the pull-wire sensor.

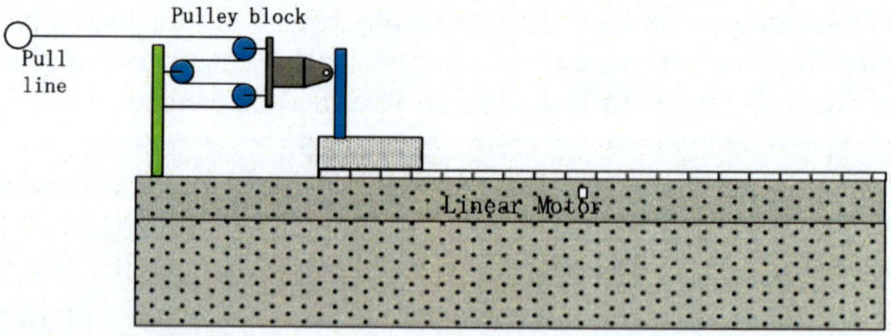

Fig. 1.3 Structure diagram of speed increasing device

Table 1.1 The relationship between the number of movable pulleys and the change of motion parameter

Number of moving pulleys	Stroke(mm)	Maximum speed(m/s)	Acceleration(m/s$^{2)}$
0	2500	15	45
1	1250	7.5	22.5
2	625	3.75	11.25
4	312.5	1.875	5.625

1.3.3 Mechatronics Composite Separation Mechanism

The above-mentioned two mechanism devices can realize the short-distance high-speed movement of the pull-wire sensor, but there are still some problems with only relying on them to complete the entire calibration process. During the calibration process, the movement process of the mobile platform includes four states: uniform acceleration, uniform speed, uniform deceleration, and stop. When the platform changes from a constant speed to a deceleration state, the cable end of the sensor will overshoot due to inertia, and the data at this time is invalid for the sensor's characteristic calibration. Only by avoiding the overshoot phenomenon of the cable end, the data collected under the state of uniform acceleration and uniform speed can truly reflect the stretching state of the sensor. In order to obtain more effective data within the limited stroke range of the displacement sensor, it is necessary to ensure that within the measurement range, the wire end is in an acceleration or uniform motion state and reaches the required speed index. Therefore, it is necessary to design a separation mechanism: the cable end is in a state of uniform acceleration and uniform speed before separation, and when the cable end is about to reach the full scale and there is still a certain safe distance, the movable pulley group, and the mobile platform are separated. Make sure that the cable end of the sensor is not stretched, and the moving platform slows down until it stops to prevent disconnection and ensure the safety of the device. The separation mechanism of the entire mechatronics compound is shown in Fig. 1.4. The pulley block bracket is a steel structure, and an electromagnet is installed on the moving platform. After power on, the pulley bracket is attracted by electromagnetic gravity, so that the moving pulley and the moving platform move together. A proximity switch is installed on the moving platform. When the platform moves to the limit of the separation bracket, the proximity switch is triggered to generate a pulse signal, and the signal triggers the electromagnet to power off, and the moving pulley block and the electromagnet of the moving platform lose electromagnetic attraction and separate. After the movable pulley group is separated, it enters the slideway of the separation mechanism, and its roller mechanism ensures that the pulley group moves linearly in the slideway until it is decelerated by the spring and stops and falls on the slideway. In order to prevent the rebound of the movable pulley group, a limit reed is designed. At this time, the motion platform decelerates to a stop by itself through the separation mechanism.

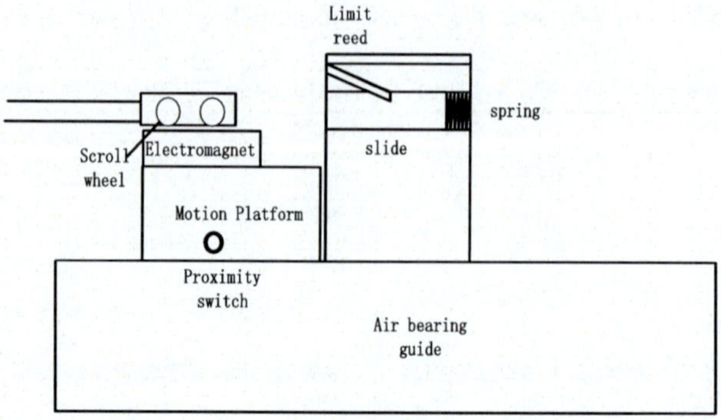

a) Power on.

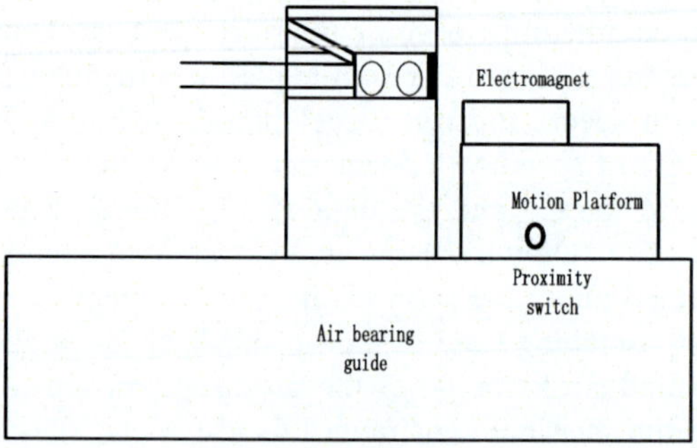

b) Power off.

Fig. 1.4 Schematic diagram of the structure and principle of the separation device

According to the sensor parameters in the above example, the cable length limits the movement stroke of the entire platform to 625 mm. The platform accelerates first and then moves at a constant speed within the entire stroke range to ensure that the movement speed is not less than 3.75 m/s. In this way, the movement speed of the cable end of the sensor is not less than 15 m/s. When the maximum range of the cable is reached, the variable speed pulley group is separated from the mobile platform, and the platform continues to decelerate to a stop. According to formula (1.1) and formula (1.2), the movement time t before platform separation can be obtained as 0.33 s.

$$s = \frac{1}{2}at^2 \tag{1.1}$$

$$v = at \tag{1.2}$$

According to the calculation results, the relationship between platform stroke s, speed v, acceleration a, and time t can be obtained as shown in Figs. 1.5, 1.6 and 1.7.

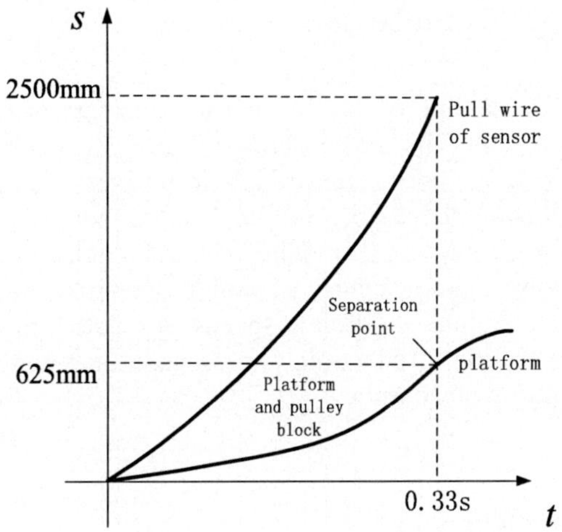

Fig. 1.5 Platform travel and time

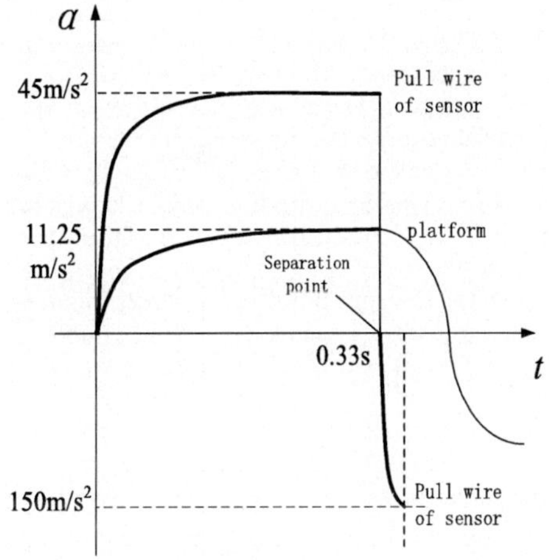

Fig. 1.6 Platform acceleration and time

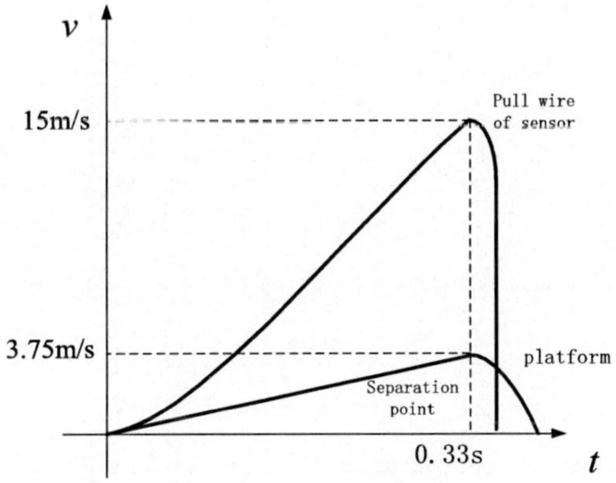

Fig. 1.7 Platform speed and time

1.4 Conclusion

The calibration device proposed in this paper solves the problem of the fixed length of the pull-wire sensor during the dynamic calibration process of the pull-wire sensor, which cannot achieve high-speed motion measurement. The measuring device in this article can be used to measure the displacement at higher speed and meet the requirements of dynamic calibration of the pull-wire sensor. The device and method have the advantages of simple operation, low cost, etc., and can realize dynamic calibration at different speeds, simulate high-speed displacement states of various industrial production, and propose new method ideas for dynamic calibration of pull-wire sensors.

References

1. Y. Zhang, P. Liang, *JJF 1305–2011 Linear displacement sensor calibration specificatio*n (China Quality Inspection Press, Beijing, 2011)
2. Y. Ding, Y. Deng, *JJF 1033–2008 Measurement standard assessment specification* (China Quality Inspection Press, Beijing)
3. S. Changyan, Y. Depei, *JJF 1305–2011 Evaluation of Measuring Instrument Characteristics* (China Quality Inspection Press, Beijing, 2011)
4. Y. Dou, W. Guan, JJF 1305–2011 Displacement sensor measurement method. Metrol. Technol. 16–18 (2013)
5. D. Li, The application of laser interferometer in calibrating the rope type displacement sensor. Metrol. Test. Technol. **34**(3), 29–32 (2007)

Chapter 2
Fine Guidance Sensor Attitude Determination System of the High-Accuracy Satellite Mission

Jinsong Li, Dong Li, Shuang Liu, Yingquan Yang, Wen Chen, and Jinpei Yu

Abstract Fine guidance sensor is becoming a norm for future satellite missions. Missions requiring attitude stability better than sub-arcsec to milli-arcsec employ satellite specific Fine Guidance Sensor (FGS), which is an interferential instrument with complex research, as it forms extended field of view that focuses the optic signal from the telescope primary dish to provide high pointing stability information to the satellite for control. High precision and high stability pointing are hot issues in attitude determination research both in the past and now. First, this work introduces the FGS of foreign astronomical telescopes mission (Hubble and Kepler), cameras of TESS mission, and the FGS of the domestic mission astronomical telescope mission (SVOM). Secondly, the attitude estimation algorithms of SVOM and foreign high-precision satellites are compared. Finally, a closed-loop test of the SVOM system achieves a stability of 0.8 arcsec /100 s.

2.1 Introduction

Satellite attitude determination is the basis of satellite attitude control and the premise of satellite function realization. The Space Variable Objects Monitor (SVOM), a joint Space telescope project between China and France, is an astronomical satellite in which the main goal is to observe the characters of the most energetic phenomena in the universe, called gamma-ray bursts. The pointing stability of SVOM satellite is required to be less than 0.8 arcsec per 100 s [1]. As an important part of attitude determination system, the accuracy synthesis of attitude sensor and attitude determination algorithm determines the pointing accuracy of attitude determination system [2, 3]. Pointing accuracy of more than sub-arcsec has become a common requirement for future astronomical observation satellites.

Due to the relative motion of the spacecraft and the observation object, uneven gravity of orbit, the sun's radiation pressure caused by the spacecraft orbit perturbation, spacecraft attitude, such as pitch, yaw, and roll change, operation of spacecraft

J. Li (✉) · D. Li · S. Liu · Y. Yang · W. Chen · J. Yu
Innovation Academy for Microsatellites of CAS, Shanghai 201210, China
e-mail: lijs@microsate.com

G. Liu and F. Cen (eds.), *Advances in Precision Instruments and Optical Engineering*, Springer Proceedings in Physics 270, https://doi.org/10.1007/978-981-16-7258-3_2

moving parts and some space inside the platform operation and walking astronauts are likely to cause the vibration of a spacecraft platform. The image blur caused by the relative motion of the object and the detector during the exposure time of the imaging detector is called image shift. Image shift will blur the outline of the observed object. There is a certain gap between the target and the background. The transition region expands with the increase of image shift. When the transition region reaches a certain extent, the images of two adjacent targets overlap and even cannot be distinguished. Hence, the existence of image shift will seriously affect the resolution and imaging quality of the telescope.

This paper introduces the development and analysis of a high-precision attitude determination system which consists of a precise sensor and an optimal EKF. The next section briefly describes the FGS of well-known telescopes at home and abroad, and introduces the attitude pointing accuracy and pointing stability required by different space missions. Further, the selection of FGS, gyro and star sensors and the design of EKF estimation algorithms are described. Finally, the FGS-star tracker—Gyro-fusion was verified by the closed-loop test of the satellite system.

2.2 FGS Overview and Attitude Accuracy Requirements

2.2.1 Hubble Mission

Hubble's FGS measurement accuracy has been reached to 0.0028 arcsec, with a frequency of 40 Hz for feedback, the boresight axis can be kept within the range of 0.007arcsec within 10 min [4]. Three sets of FGS were applied to the Hubble in a 90-degree alignment around the main focal plane array. Two of them are used to point and lock the observational target of the telescope, and the other is used as an astrometric instrument to measure the position of a specific celestial body. Each FGS consists of two moveable Koester prism interferometers with a 60 $arcmin^2$ field of view to search and track stars and a 5.0 $arcsec^2$ instantaneous field of view to accurately locate stars. The two sets of FGs work together to determine the pointing position [5]. First one FGS searches for and locks on one guide star, then another FGS searches for and locks on the other. Once the two guides are locked, the image of the observed object is kept in the selected focal plane of the scientific instrument for a long period of time. The attitude pointing accuracy and pointing stability of the Hubble satellite is undoubtedly the best among launched and studied satellites (Fig. 2.1).

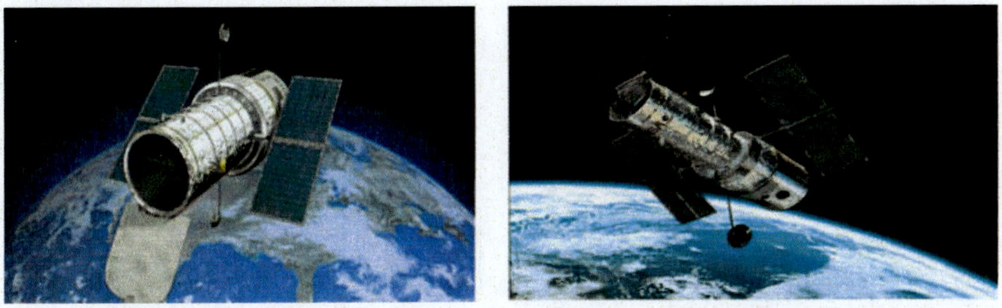

Fig. 2.1 Hubble Space Telescope

2.2.2 Kepler Mission

The Kepler mission is a space telescope designed by NASA to find planets in the habitable zone [6]. A 0.95 m Schmitt telescope provides the 96 megapixel Kepler focal plane array with a field of about 13°, allowing continuous monitoring of more than 100,000 stars in the field. Four FGS CCD modules are mounted on the corners of invar substrate to collect additional pointing information for the attitude control system to achieve the required 2.5 milli-pixel (0.01") pointing accuracy. The FGS module contains an E2V-made backlight, 3-phase, 3 MHz pixel readout rate operation FGS module, allowing a 10 Hz frame rate. The FGS module is located on the same mechanical surface as the science module and is subject to the same opto-thermo mechanical effects. The pointing precision required of the attitude determination and control system is 0.009 arcsec at 3σ on timescales of 15 min and longer.

The focal plane array assembly consists of 21 Science CCD and 4 FGS CCD modules, all located on a curved invar substrate located at the telescope image surface [7] (Fig. 2.2).

Fig. 2.2 Science CCD and FGS CCD modules

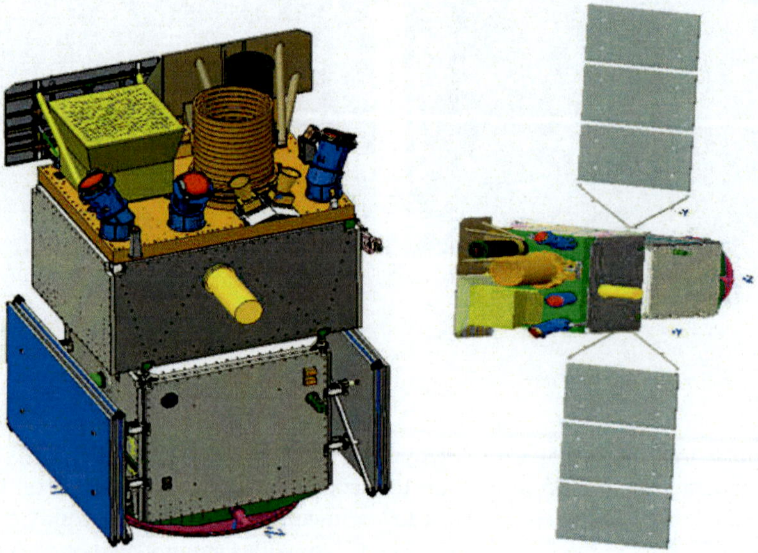

Fig. 2.3 TESS Cameras and spacecraft

2.2.3 *TESS Mission*

The Transiting Exoplanet Survey Satellite (TESS) is a two-year all-sky survey mission designed to look for transiting exoplanets around bright and nearby stars [8]. The TESS instrument consists of four superposed wide-angle cameras that provide a combined field of view of approximately 24 degrees by 96 degrees from the ecliptic plane to the ecliptic pole. The attitude determination errors of the four cameras are less than 5.2 arcsec. Equivalent Angle requirements of precision noise of instrument camera under nominal precision condition < 0.6 arcsec (3σ) in the cross-boresight axes and < 4.2 arcsec (3σ) in the roll axis. Pointing stability of the system during scientific operation is less than 0.06 (3σ) arcsec per hour and less than 2(3σ) arcsec per minute [9] (Fig. 2.3).

2.2.4 *SVOM Mission*

The attitude and orbit control system (AOCS) of SVOM satellite is responsible for the required spacecraft attitude during all mission phases and in particular for high pointing accuracy during science observation [10]. The major tasks are stabilization of the spacecraft after separation from the launcher, autonomous attitude acquisition and determination, stabilization and control of the attitude in 3 axes with very demanding fine pointing requirements, and attitude maneuvers to allow for scanning

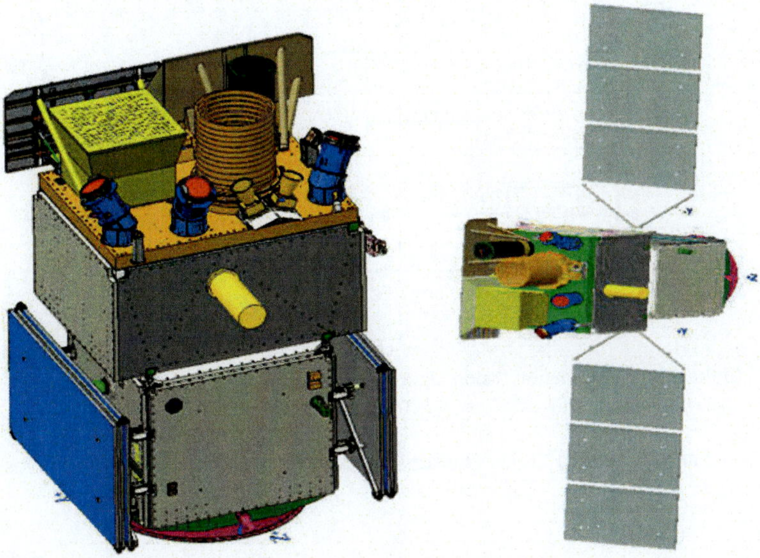

Fig. 2.4 SVOM satellite schematic

the extragalactic sky according to the gamma ray burst(GRB)observation strategy (step and stare).The stability requirement based on the FGS as considered for the following AOCS analysis is 0.8 arcsec/100 s (Ys, Zs axis, 0-peak) (Fig. 2.4).

2.3 Mission High Stability Pointing Mode

2.3.1 Sensor Requirements

The Sun Sensor part includes six analog sun sensors (ASS), two 01 sun sensors (01-SS), and one electronic box. The quantity of the earth magnetic field vector in orbit, which determines the attitude information of the spacecraft (S/C), is provided by Magnetometer. The absolute inertial attitude reference is provided by star trackers (STRs). There are three STR on S/C, in which vertical optical axis measurement accuracy is better than 3.5 arcsec. The Fine Guidance Sensor is used for SVOM to provide accurate attitude measurements during the scientific observations. Two following requirements on FGS are derived from the S/C system design. First is to provide the signal in a frequency of not less than 1 Hz. Second to provide the relative attitude measurement with a precision of not bigger than 0.2 arcsec in the whole FOV of visible telescope (VT). The Gyro Unit is used as rate measurement unit. Two fiber optic gyroscopes (FOGs) are foreseen for redundancy. Absolute error of output

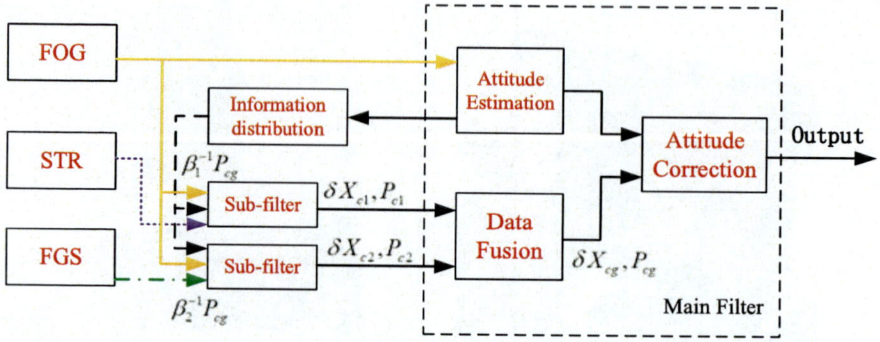

Fig. 2.5 Multi-sensor information fusion algorithm

precision of angle rate of FOGs is better than 3×10^{-5}°/s when angle rate is less than 0.1°/s.

2.3.2 *Data fusion*

The federated Kalman filter is used as the basic information fusion structure of the satellite attitude determination system. FOG is used as the common reference system, and the other sensors are combined with the gyro to form two subsystems: FOG/star tracker and FOG/FGS. The measurement information is fed into the corresponding sub-filter and processed. The attitude parameter error of gyro output is taken as the common state of each sub-filter. Each sub-filter uses the measurement information of its own sensor to give local estimation of the public state and input it to the main filter. The local estimation of the sub-filter is fused by the main filter to obtain the global estimation of the gyro attitude error, and then the attitude parameters obtained by the gyro are corrected to obtain the satellite three-axis attitude information. In the figure below, $\delta \vec{X}$ is the estimate of the error state variable, $\vec{P}$ is the estimated mean square error matrix, and β is the information distribution coefficient. The subscript g means the filtering result based on all measurements, that is the global estimate. i(i = 1,2) is the sub-filter and c is the common reference system gyro (Fig. 2.5).

2.3.3 *System Test Result*

The SVOM satellite is currently in phase C and is about to enter phase D. It is scheduled to launch in 2023. Joint tests of the satellite system, ground system, and application system are under way. Under the working mode of high precision and high stability, the stability of Y-axis and Z-axis is mainly assessed through FGS

Fig. 2.6 Y-axis stability

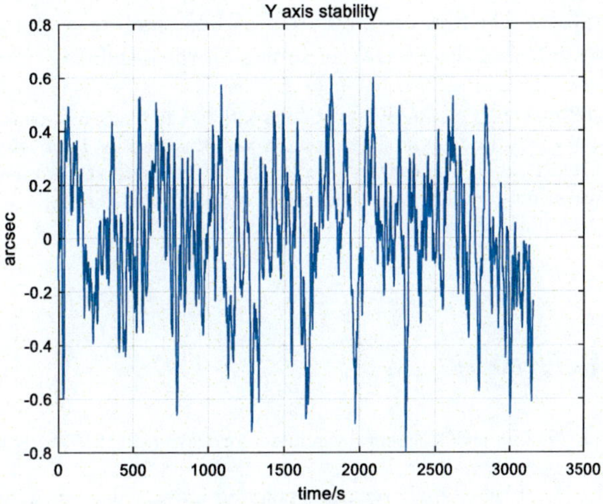

Fig. 2.7 Z-axis stability

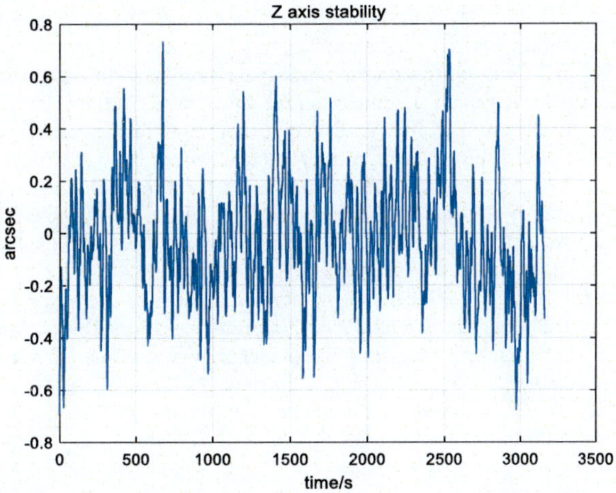

attitude determination and PID wheel control. The test data shows that the index of pointing stability is better than 0.8 arcsec/100 s in this mode (Figs. 2.6 and 2.7).

2.4 Conclusions

In this paper, we design and develop a high-precision attitude determination system, which includes an EKF algorithm, three high-precision star tracker, and two gyros. Accurate pointing also requires sensor calibration prior to scientific observation to counteract the effects of launch shocks, in-orbit thermal fluctuations, and exhaust effects. The filtering method of SVOM satellite is extended Kalman filtering. Closed-loop test of the SVOM system achieves a stability of 0.8 arcsec/100 s.

Acknowledgements First author would like to thank our team members for algorithm processing and hardware testing. The research was financed by Innovation Academy for Microsatellites of CAS. I would like to thank the leaders of the Institute for their support of the research work. At the same time, the contributions of domestic and foreign researchers in high-precision attitude determination and high stability control are also appreciated.

References

1. S. Liu, AOCS general Architecture Design for SVOM Satellite, in *IWAACE.Zhuhai* (2020), pp. 238–244
2. D. Bhatia, High accuracy pointing attitude determination estimator system of the future infrared astronomy satellite swarm mission, in *International ESA Conference on Guidance, Navigation & Control Systems* (Salzburg, 2017), pp. 2–18
3. F.L. Markley, *Fundamentals of Spacecraft Attitude Determination and Control Space Technology Library* (Springer, 2014)
4. G.S. Nurre, Preservicing mission, on-orbit modifications to hubble space telescope pointing control system. J. Guid. Control. Dyn. **18**, 222–228 (1995)
5. E. Nelan, The fine guidance sensors aboard the hubble space telescope, in The Scientific Capabilities of these Interferometers Proceedings of SPIE.L.A (1998), pp. 237–247
6. J. Li, Photometer performance assessment in Kepler science data processing, in *International Society for Optical Engineering* (San Diego, 2010), pp. 70–81
7. W.J. Chaplin, Asteroseismic fundamental properties of solar-type stars observed by the NASA kepler mission. Astrophys. J. Suppl. Series.L.A 1–22
8. T. Nguyen, Fine-pointing performance and corresponding photometric precision of the Transiting Exoplanet Survey Satellite. J. Astron. Telesc. Inst. Syst. **4**(4), 47–60 (2018)
9. G.R. Ricker, Transiting exoplanet survey satellite. J. Astron. Telesc. Inst. Syst. **1**(1), 3–13 (2015)
10. J.S. Li, A physical simulation method for SVOM satellite platform high stability control algorithm based on FGS simulator. Sci. Technol. Eng. **17**(30), 92–97 (2017)

Chapter 3
Research on OFDR Pressure Sensor Based on PDMS

Zhang Xu and Cui Kai

Abstract OFDR, as a new type of optical fiber sensor, can be sensitive to a variety of physical quantities, such as pressure and temperature. Physical parameters can be measured by using different materials to wrap the optical fiber. To improve the accuracy of OFDR sensing, researchers focus on processing the fiber structure, such as adding fiber gratings. These solutions increase sensor manufacturing cost and manufacturing complexity while improving accuracy. This paper proposes a low-cost OFDR sensor and pressure measurement method, using PDMS flexible material to wrap ordinary single-mode quartz fiber, and using a self-made pressure test platform for pressure measurement. Experimental measurement shows that this optical fiber pressure sensor has good linearity and repeatability. The results show that the linear fit of the sensor reaches 0.99, and the measured pressure can reach 20 N (24 kPa). At the same time, the sensor has good stability, so it has engineering application value.

3.1 Introduction

As a new type of sensing technology, optical fiber sensing technology has been greatly developed in recent decades [1]. Fiber optic sensors have many advantages over traditional sensors, including anti-electromagnetic interference, light weight, small size, high sensitivity, and easy implementation of multiplexing or distributed sensing [2]. The research on fiber optic sensors mainly focuses on strain, temperature, and vibration [3]. Among them, fiber grating sensors [4], fiber optic gyroscopes, and fiber current sensors have been widely used.

Optical Frequency Domain Reflectance (OFDR) is an optical fiber sensing technology based on the principle of Rayleigh backscattering [5]. It can measure temperature and strain in a distributed manner by using the inherent structural defects and

Z. Xu
Zhuhai College of Science and Technology, Zhuhai 519000, China

C. Kai (✉)
Macau University of Science and Technology, Macau 518063, China
e-mail: nucck1995@163.com

G. Liu and F. Cen (eds.), *Advances in Precision Instruments and Optical Engineering*, Springer Proceedings in Physics 270, https://doi.org/10.1007/978-981-16-7258-3_3

weak backscattering caused by damage in the optical fiber. OFDR has the advantages of high spatial resolution and high accuracy. It only needs one scan to obtain all the data in the measurement fiber. The current research on OFDR focuses on improving the strength of back reflection by changing the internal structure of the fiber, such as fiber grating and fiber tapering, which increases the accuracy and greatly increase the cost.

This paper proposes a low-cost distributed flexible sensor structure, using PDMS material to wrap ordinary single-mode fiber to establish the relationship between pressure and fiber strain. PDMS is a flexible material with good stability, and is often used as a pressure material in the laboratory. By changing the ratio of PDMS stock solution and curing agent to obtain flexible substrates of different hardness, the relationship between fiber strain and pressure can be changed to obtain different pressure ranges, accuracy, and resolution.

3.2 OFDR Optical Fiber Sensing System

3.2.1 Measurement System Structure

The sensing system of the fiber sensor is composed of the optical fiber sensing unit, OFDR fiber demodulation equipment, a computer, and an optical fiber jumper. The sensing system with a double fiber junction is shown in Fig. 3.1, which is composed of two circular structures with a diameter of 2 mm. The test signal can directly escape from the circular junction that exceeds the maximum bending curvature of the optical

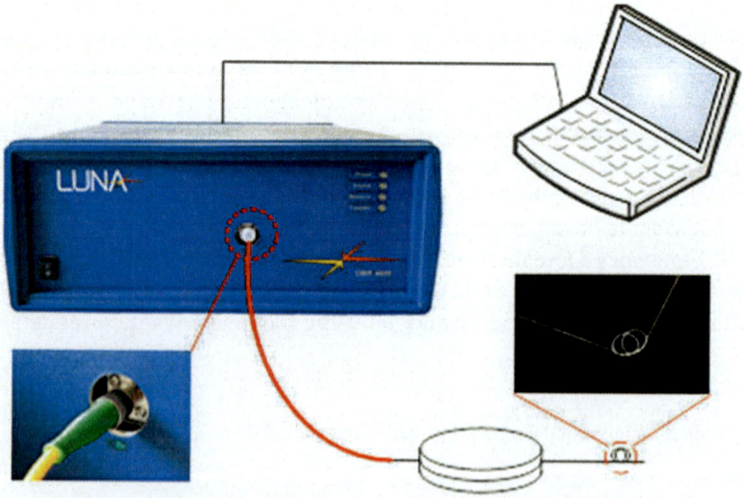

Fig. 3.1 The sensing system with a double fiber junction

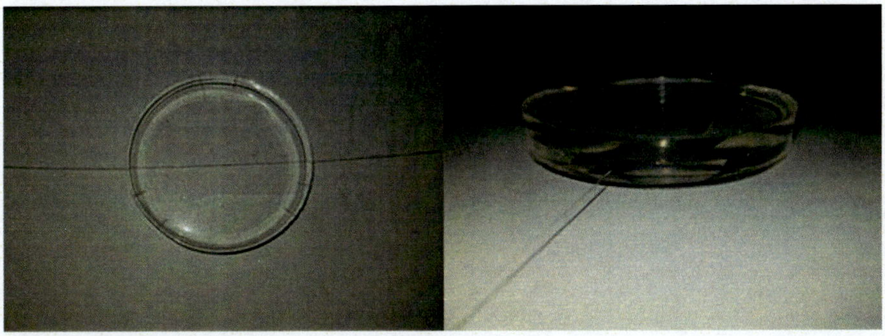

Fig. 3.2 Pressure sensor

fiber. This structure can prevent the end face reflected signal from covering the weak back Rayleigh scattered signal, which is generated at the end of the fiber, which is a low-cost solution.

The optical fiber demodulation equipment was LUNA's OBR4600, its spatial resolution can reach 0.01 cm, and strain measurement error is less than 1 $\mu\varepsilon$. This experiment uses 1 cm resolution, 1 m sensing distance, and 1 cm sensing interval. Use FC-APC fiber optic jumper to splice with the sensing unit.

3.2.2 Pressure sensor

First, the mixture is uniformly stirred, wherein the mass ratio of the PDMS liquid to the curing agent in the mixture is 9:1. Slowly pour the liquid mixture into two containers and place it horizontally in a vacuum machine for 30 min. The vessel was then placed in an incubator, set at 60 °C curing, the curing time 2 h. After curing, the PDMS is peeled and taken out of the container, one of which is the lower surface and the other is the upper surface. The single mode optical fiber having a diameter of 125 μm straightened passes between the upper and lower surfaces, fixed with glue to form a pressure sensor. The pressure sensor is shown in Fig. 3.2.

3.2.3 Pressure Test Bench

The test system uses a pressure test bench as shown in Fig. 3.3 to apply pressure.

When measuring, place the weight on the bearing platform, and insert the cylindrical structure under the bearing platform into the center of the bearing platform. There is a cylindrical structure with a smaller radius under the bearing platform, which forms a solid-shaped, fixed-size contact surface with the sensor placed under it. The pressure can be changed by changing the mass of the weight, and the pressure

Fig. 3.3 Pressure test bench

applied to the surface of the pressure sensor can be changed by changing the contact surface.

3.3 Experimental Results

The experiment uses linear pressurization. Select the step length of the weight to be 100 g for continuous pressure. The experimental results are shown in Table 3.1. When the mass of the weight is 1900 g, continue to increase the 100 g weight, the optical fiber strain value obtained by the demodulator does not change, and the strain value fluctuates after the mass of the weight continues to be increased. Therefore, set the end-point of pressure measurement to 2000 g for repeatability test. The first set of measurement results are linearly fitted, and the results obtained are shown in Fig. 3.4. The experimental results show that the sensor has good linearity, and the degree of fit reaches 0.99.

Because the experimental results show that when the pressure value of the sensor reaches 2000 g, there will be residual strain inside. As a result, the strain value measured by the interrogator will not return to zero. After all the weights are removed, the strain will continue for a while. So, after the last set of measurements, first remove all the weights, let the measurement system stand for 1 min, and then repeat the application of weights after resetting the interrogator to zero for repeatable measurement. Displayed equations are centered and set on a separate line. The repeatability measurement results are shown in Table 3.1. The repeatability measurement

Table 3.1 The repeatability measurement results

Mass (g)	Group 1	Group 2	Group 3	Group 4	Pressure (MPa)
100	18	19	24	20	0.001209
200	40	48	50	50	0.002418
300	66	67	71	74	0.003627
400	91	92	100	100	0.004836
500	112	113	121	125	0.006045
600	135	138	130	145	0.007254
700	158	161	166	164	0.008463
800	174	184	178	186	0.009672
900	193	205	209	210	0.010881
1000	211	217	227	229	0.01209
1100	227	235	246	244	0.013299
1200	250	256	272	271	0.014508
1300	262	280	289	288	0.015717
1400	280	298	311	306	0.016926
1500	301	315	329	324	0.018135
1600	319	338	348	347	0.019344
1700	335	348	366	364	0.020553
1800	351	372	384	380	0.021762
1900	369	390	404	401	0.022971
2000	369	400	420	416	0.02418

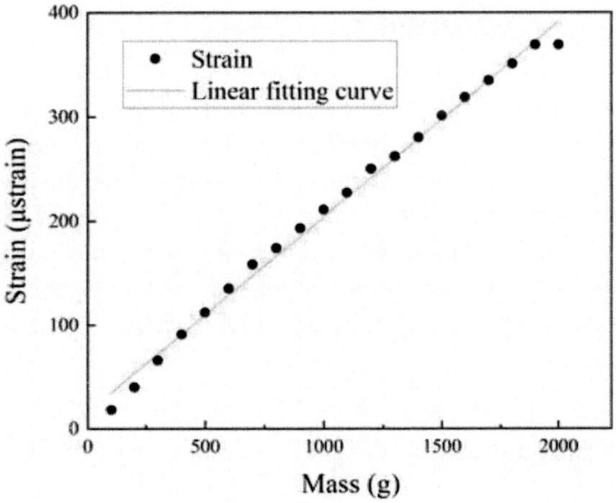

Fig. 3.4 Single pressure measurement result and linear fitting curve graph

Fig. 3.5 Repeatability measurement result graph

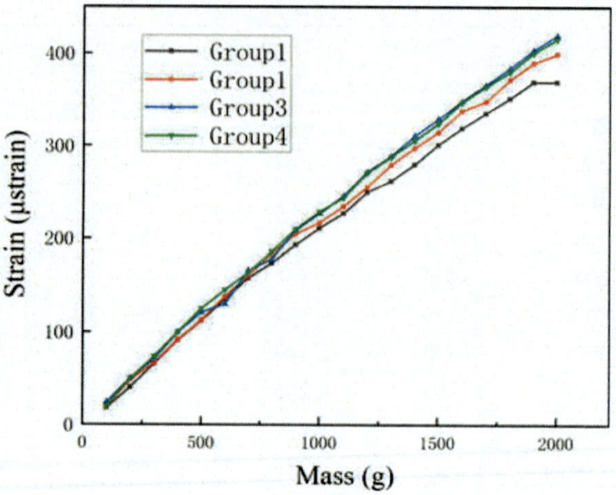

results are shown in Fig. 3.5. The measurement results show that the sensor has good repeatability.

After the third measurement, the stability of the sensor gradually increased, and the repeatability was better. The reason for the analysis may be, after the manufacturing was completed, the inside of the sensor was not fully contacted, and the glue material was too hard. After the first two measurements, the internal structure of the sensor was stable, so the stability was slightly improved.

3.4 Conclusions

The optical fiber pressure sensor based on PDMS material drives the optical fiber to produce strain through the deformation produced by the PDMS material and establishes a stable connection between pressure and optical fiber deformation when pressure is applied. A stable pressure response characteristic can be achieved with ordinary single-mode fiber, and the linearity of measurement can be as high as 0.99 under the constant temperature and stable environment of the laboratory. The pressure measurement can be done at a lower cost.

References

1. Z. Zhou, Z. Li, N. Tang, J. Sun, K. Han, Z. Wang, On-line temperature measurement of fiber Bragg gratings inside a fiber laser. Opt. Fiber Technol. **45**, 137–140 (2018). https://doi.org/10.1016/j.yofte.2018.07.009
2. B. Zhou et al., Highly sensitive small pressure monitoring using hyperelastic Silicone-Cladding/Silica-Core composite optical fiber. IEEE Photonics J. **9**(6), 1–8 (2017). https://doi.org/10.1109/JPHOT.2017.2779605

3. X. Li, H. Zhang, C. Qian, Y. Ou, R. Shen, H. Xiao, A new type of structure of optical fiber pressure sensor based on polarization modulation. Opt. Lasers Eng. **130**, 106095 (2020). https://doi.org/10.1016/j.optlaseng.2020.106095
4. H. Ahmad et al., High-sensitivity pressure sensor using a polymer-embedded FBG. Microw. Opt. Technol. Lett. **50**, 60–61 (2008). https://doi.org/10.1002/mop.23021
5. M. Wegmuller, J.P. von der Weid, P. Oberson, N. Gisin, High resolution fiber distributed measurements with coherent OFDR (2000)

Chapter 4
Space-Domain Fiber Cavity Ring-Down Magnetic Field Sensor Using D-shaped Fiber Coated with Magnetic Fluid

Wenjia Chen, Chunfu Cheng, Jiaxuan Chen, Yiwen Ou, Li Fang, and Mengmeng Li

Abstract A space-domain fiber cavity ring-down magnetic field sensor is proposed and experimentally demonstrated. The enhanced evanescent field effect and stability was achieved by using D-shaped fiber coated with magnetic fluid as the sensor head, and a temperature controller was introduced to improve the measurement accuracy. With this novel technique, magnetic field sensing has been demonstrated by measuring ring-down distance and a sensitivity of 8.4×10^{-4} dB/Gs in the linear region was achieved.

4.1 Introduction

Time-domain fiber cavity ring-down (FCRD) technique has attracted much attention recently due to its high sensitivity [1]. In this scheme, through measuring the ring-down time [2], the loss induced by the tested sample can be deduced. As this technique does not measure the absolute intensity of light source, it is insensitive to light intensity fluctuations of the light source [3]. In addition, it has the advantages of high sensitivity due to multiple effective interactions with the sample; therefore, it was utilized by many researchers to monitor different parameters. Combined with different sensor heads, different parameters including refractive index of liquids [4–6], macro-bend loss of single-mode fiber (SMF) have been measured using this scheme [7]. However, the traditional time-domain FCRD technique uses expensive equipment, such as pulsed light sources, pulse modulators, fast detectors, and high-speed oscilloscopes, to observe ring-down signals in time-domain, thus resulting in high costs. Ye and Qian et al. [8], based on frequency-shifted interferometry (FSI), propose a new FCRD technique which adopts continuous light, low-speed differential detector and a low-speed, low-cost acquisition card device to reduce instrument

W. Chen · C. Cheng (✉) · J. Chen · Y. Ou · L. Fang · M. Li
Hubei Engineering Technology Research Center of Energy Photoelectric Device and System, Hubei University of Technology, Wuhan 430068, China
e-mail: chengchunfu@hbut.edu.cn

School of Science, Hubei University of Technology, Wuhan 430068, China

G. Liu and F. Cen (eds.), *Advances in Precision Instruments and Optical Engineering*, Springer Proceedings in Physics 270, https://doi.org/10.1007/978-981-16-7258-3_4

cost. In this technique, through Fast Fourier transform (FFT), a ring-down curve in the space domain can be obtained. Therefore, this new scheme can be called space-domain FCRD sensing technique. Compared with traditional time-domain FCRD technique, Space-domain FCRD crucially reduces technical costs while maintaining its original advantages, and the employed differential detection method can effectively eliminate DC noise which can increase signal-to-noise ratio and thus improving the system stability. Due to the excellent characteristics of space-domain FCRD, it has attracted wide attention from researchers and has been successfully applied to gas sensing [9], pressure sensing, [10] and stress sensing [11]. The magnetic field sensing plays an important role in a wide range of areas. Combining space-domain FCRD and D-shaped fiber to acquire highly sensitive optical fiber magnetic field sensor with high stability is of great significance and has vast application prospects.

In the paper, we propose and demonstrate a magnetic field sensor based on space-domain FCRD technique, and design a sensor head using D-shaped fiber coated with magnetic fluid. The technical principle and manufacturing method of the sensor head were demonstrated and the magnetic field has been monitored by measuring the ring-down distance. The experimental results have proved that a highly sensitive magnetic field sensor can be realized by using space-domain FCRD technique combined with D-shaped fiber.

4.2 Operation Principle and Experimental Setup

The proposed highly sensitive magnetic field sensing experimental setup based on space-domain FCRD is shown in Fig. 4.1. It is essentially composed of a ring-down cavity embedded in a Sagnac loop. An acousto-optic modulator is inserted

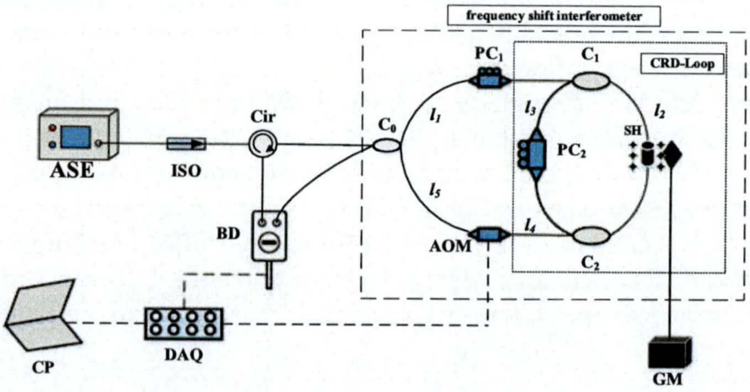

Fig. 4.1 Magnetic field sensing experimental setup based on space-domain FCRD technology. ASE: amplified spontaneous emission source; ISO: isolator; Cir: circulator; C_1:3 dB fiber coupler; C_2 and C_3: 99.5/0.5 fiber couplers; PC_1 and PC_2: polarization controller; SH: sensor head; AOM: acousto-optic modulator; BD: balanced detector; DAQ: data acquisition card; CP: Computer; GM: gaussmeter

asymmetrically in the Sagnac loop as a frequency shifter to form a frequency-shifted Sagnac interferometer. The principle of the frequency-shifted Sagnac interferometer is detailly explained in [12]. ASE emits continuous light of a frequency v and enters the Sagnac loop through the fiber coupler C_0, and thus it is divided into two light beams, one clockwise and another counterclockwise. The clockwise light directly enters the ring-down cavity, and the light leaks out of the ring-down cavity from the fiber coupler C_2 in each propagation and returns to the fiber coupler C_0 through the frequency shifter AOM. Meanwhile, the counterclockwise light firstly passes the frequency shifter AOM, then enters the ring-down cavity, next leaks out of the ring-down cavity through the fiber coupler C_1, and eventually returns to the fiber coupler C_0. The two light beams have the same frequency and a phase difference. When the coherence length of the light source is less than the ring-down cavity length d, the two light beams will interfere at the fiber coupler C_0. The differential interference signal detected by the balanced detector can be expressed as

$$\Delta I \infty \sum_{m=0}^{\infty} I_m \cos[2\pi \frac{n(mL + L_0)}{c} f] = \sum_{m=0}^{\infty} I_m \cos(2\pi F_m f) \tag{4.1}$$

where I_m is the light intensity after the light propagates m times in the ring-down cavity, n is the effective refractive index of the single-mode fiber, L_O is the length of the ring-down cavity, $L = l_1 + l_2 + l_4\text{-}l_5$ is a constant, c is the speed of light, f is the frequency shift produced by the acousto-optic modulator, $F_m = n(mL + L_O)/c$ is the oscillation frequency. F_m is unique for each round-trip number m and is equally spaced by nd/c. Due to the inherent loss of the optical device in the cavity, the light intensity decays exponentially in the following relationship. It can be expressed as

$$I_m = I_0 \exp(-m\alpha_c/4.34) = I_0 \exp(-\frac{l}{4.34L}\alpha_c) \tag{4.2}$$

where I_0 is the initial light intensity, $l = mL$ is the propagation distance of light in the ring-down cavity, and α_c is the inherent loss in the cavity. In traditional time-domain FLRD technique, the time that it takes for the light intensity to decay to 1/e of the initial light intensity is called the ring-down time. Correspondingly, in the space-domain FCRD technique, the distance of the light intensity attenuation to 1/e of the initial light intensity is called the ring-down distance. It can be expressed as

$$d_0 = 4.34\frac{L}{\alpha_c} \tag{4.3}$$

When different magnetic field intensity is applied, the refractive index of the magnetic fluid will change, and the loss in the cavity will also change, thus resulting in an additional loss. The ring-down distance becomes

$$d = 4.34\frac{L}{\alpha_c + \alpha_m} \tag{4.4}$$

where $\alpha_m = \gamma CHl_s$, γ is the magnetic-field-induced absorption coefficient, C is the concentration of magnetic fluid, H is the magnetic field strength, l_s is the length of the D-shaped fiber coated with magnetic fluid. Based on (4.3) and (4.4), the additional loss α_m can be expressed as

$$\alpha_m = 4.34L\left(\frac{1}{d} - \frac{1}{d_0}\right) = kH \tag{4.5}$$

For a given space-domain FCRD sensing system, the additional loss has a linear relationship with the magnetic field strength (H). The slop k represents the sensitivity of the sensor to the magnetic field.

4.3 Experimental Results and Discussion

Combining space-domain FCRD scheme with magnetic field sensor head, the magnetic intensity can be measured. For magnetic field sensing, the key is to make a highly sensitive sensor head. The schematic diagram of the proposed sensor head is shown in Fig. 4.2a. The sensor head is composed of a D-shaped fiber (NIR-SPF-W1550-2, Micro photons Technology Co.) and magnetic fluid (EMG Series Water-based Ferrofluid, EMG-603P, Ferrotec, USA) depicted in Fig. 4.2b, which is packed by plexiglass as shown in Fig. 4.2c. There are two small holes on the surface of the upper plexiglass to fill the magnetic fluid. When the optical fiber is completely soaked in the magnetic fluid and there is not any bubble in the groove, the small hole is sealed with paraffin. In order to avoid the influence of temperature on the performance of the sensor head, we place the sensor head on a temperature controller based on Peltier effect. Due to the special structure of the D-shaped fiber, a strong evanescent wave into the magnetic fluid was generated. It is worth mentioning that the excessively strong evanescent wave and the excessively high concentration ($C_0 = 18.7\%$) of the water-based magnetic fluid make the light greatly absorbed by the magnetic fluid and thus the cavity loss is too large. Therefore, we use deionized water to dilute the magnetic fluid to an appropriate concentration ($C = 4.39\%$) for reducing the cavity loss.

In the experiment, we wrapped the copper wire around the iron core to generate the magnetic intensity which is controlled by changing the current load on the copper coils. A Peltier-effect-based temperature controller was used to keep temperature unchanged. For magnetic fluid, when the external magnetic field H increases, the effective permittivity of the magnetic fluid system will change, thus resulting in changes in the refractive index of the magnetic fluid, and also it is worth noting that when the external magnetic field intensity H reaches a certain threshold, the refractive index is basically saturated without any change because higher magnetic field intensity can agglomerate more magnetic nanoparticles(MNPs) to form more longer chains, leading to the constant change of RI of the magnetic fluid until it

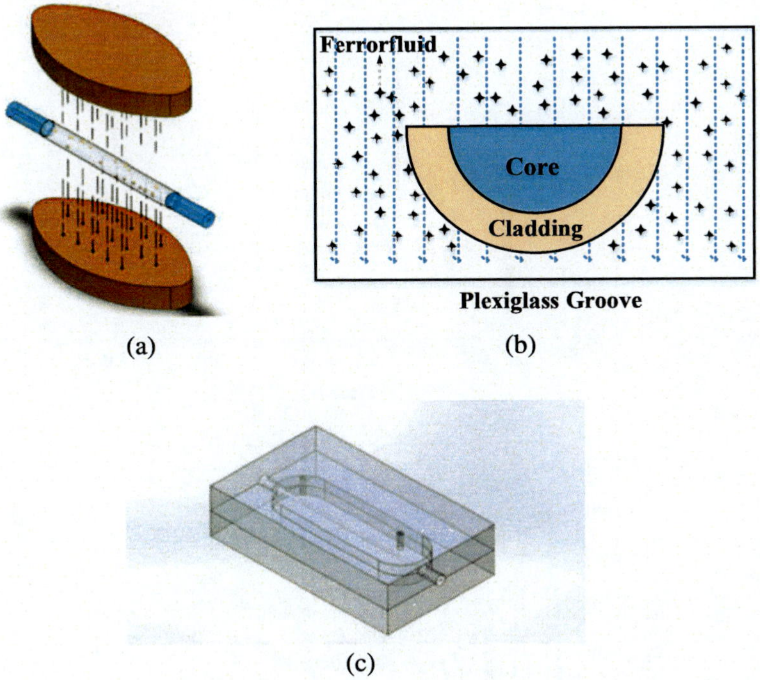

Fig. 4.2 **a** Schematic diagram of sensor head based on D-shaped fiber, **b** the cross section of the sensor head, **c** encapsulation of the sensor head

reaches saturation. When no external magnetic field was applied, the typical time-domain differential interference signal is collected by DAQ as shown in Fig. 4.3a. A series of ring-down signals in the space domain were obtained by performing FFT on the time-domain differential interference signal as shown in Fig. 4.3b. Then the peak value is extracted and fitted with an exponential function. The difference of the abscissas of the adjacent two peaks is 85.14 m (i.e., the cavity length) and the cavity loss is 1.138 dB, corresponding the ring-down distance which is 292.19 m according to (4.3).

Applying a magnetic field intensity of 0 Gs-400 Gs at increments of 50 Gs to the sensor head, we measure a series of ring-down signals in the space domain as shown in Fig. 4.4a and the corresponding exponential fit ring-down curves were shown in Fig. 4.4b. With the increase of the applied magnetic field intensity, the ring-down distance keeps decreasing,

which mainly resulted from the magnetic field deformations that occur as the magnetic field intensity varies [13]. To improve the measurement accuracy, the cavity loss under different magnetic field intensity is measured 30 times as shown in Figs. 4.4 and 4.5a. The increase of the magnetic field intensity leads to the increase of the ring-down cavity loss. On the other hand, it can be observed that when the magnetic field intensity is increased from 350 to 400 Gs, the cavity loss of the sensing

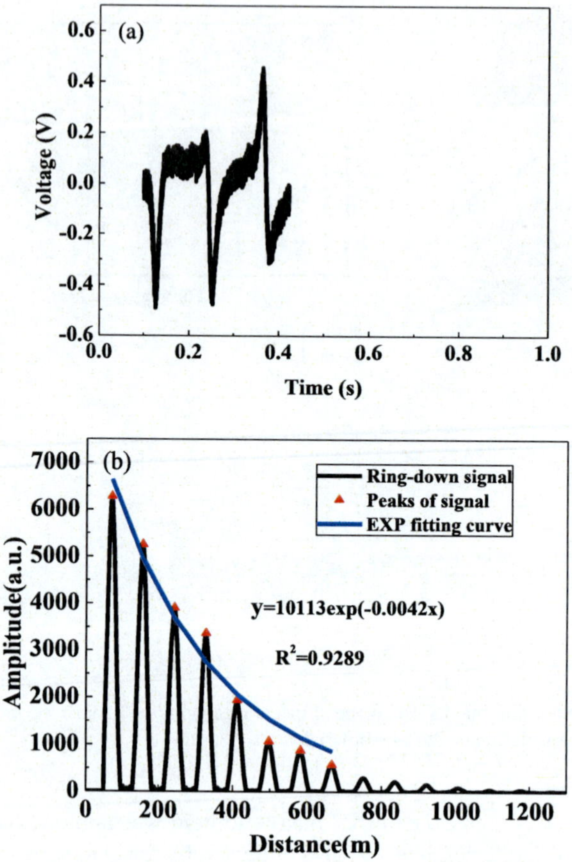

Fig. 4.3 **a** Detected ring-down signals, **b** Exponential fitted decay curve under 0–400 Gs magnetic field intensity

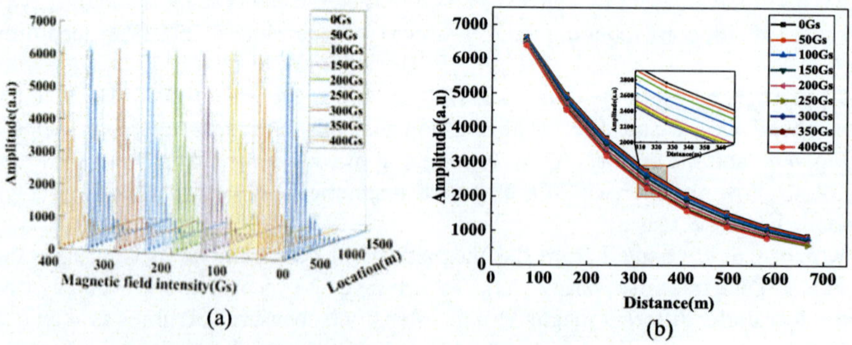

Fig. 4.4 **a** The Ring-down cavity loss responses to magnetic field applied on the sensor. **b** The relationship between the additional cavity loss and the magnetic field intensity

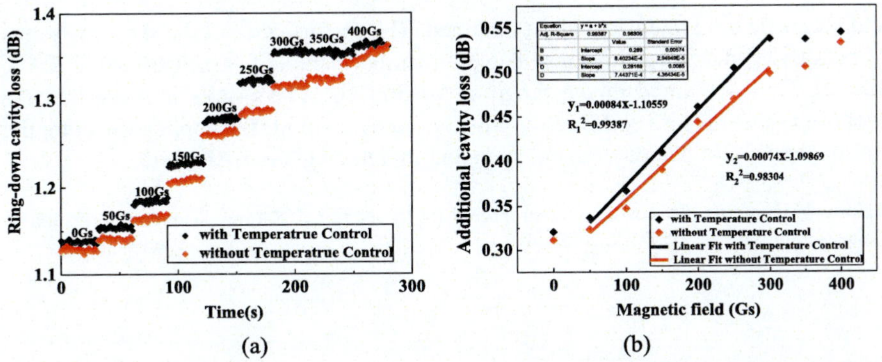

Fig. 4.5 Typical differential interference signal ΔI when the magnetic field intensity is 0 Gs. **a** the time-domain signal sampled by the DAQ; **b** the obtained space-domain FCRD signal after performing fast Fourier transform (FFT) on (**a**)

system is almost the same as that of the 300 Gs, thatis to say, the saturation effect of magnetic fluid occurs. MNPs determine the saturation magnetic field intensity of the magnetic fluid. When MNPs can no longer form longer and more chains under the impact of an external magnetic field, the magnetic field intensity at this time is regarded as the saturation magnetic field intensity. The saturation intensity of the magnetic fluid (EMG Series Water-based Ferrofluid, Ferrotec, USA) used in the experiment is 330 Gs, which is consistent with the experimental result. The basic carrier fluid of the magnetic fluid is usually deionized water or various oil-based magnetic field; therefore, the temperature has a significant impact on the magnetic field [14]. Figures 4.4 and 4.5a also shows the influence of magnetic field intensity on cavity loss with or without temperature control. Figures 4.4 and 4.5b illustrates the relationship between additional cavity loss and magnetic field intensity. Additional cavity loss linear increase with magnetic field intensity in range from 50 to 300 Gs, the R-square of 0.99 indicates that the sensing system have a good linear response, which agrees well with the theoretical result described in (4.5). The slope of the fitted straight line suggests that the measurement sensitivity is 8.4×10^{-4} dB/Gs. It also can be observed that sensitivity of sensor without temperature control is 7.4×10^{-4} dB/Gs which is lower than the sensitivity of sensor with temperature control. Therefore, the sensitivity is increased using the sensor head with temperature controller.

4.4 Conclusion

In conclusion, a novel magnetic sensing method based on space-domain FCRD is demonstrated. A highly sensitive sensor head is proposed by combining D-shaped fiber and magnetic fluid. By measuring the ring-down distance under different magnetic field intensities, the relationship between additional cavity loss and magnetic field intensity can be obtained. The performance of the sensor in terms of sensitivity and linear response is experimentally studied. A sensitivity of 8.4×10^{-4} dB/Gs is obtained in the range of 50–300 Gs. The results indicate that the proposed sensor is low-cost, highly sensitive, and it exhibits the application potential for magnetic field detection in environment monitoring, aerospace, etc.

Acknowledgements This work was supported by the National Natural Science Foundation of China (Grant Nos. 61805075, 61475044).

References

1. G. Berden, R. Engeln (eds.), Cavity ring-down spectroscopy: techniques and applications (2010)
2. F. Ye, C.M. Zhou, B. Qi, L. Qian, Continuous-wave cavity ring-down evanescent-field sensing with a broadband source based on frequency-shifted interferometry. Sensor Actuators B Chem. **184**, 150–155 (2013)
3. C. Wang, Fiber loop ringdown—a time-domain sensing technique for multi-function fiber optic sensor platforms: current status and design perspectives. Sensors **9**, 7595–7621 (2009)
4. N. Ni, C. Chan, L. Xia, P. Shum, Fiber cavity ring-down refractive index sensor. IEEE Photon. Technol. Lett. **20**, 1351–1353 (2008)
5. M. Jiang, W. Zhang, Q. Zhang, Y. Liu, B. Liu, Investigation on an evanescent wave fiber-optic absorption sensor based on fiber loop cavity ring-down spectroscopy, Opt. Commun. **283**, 249–253 (2010)
6. C. Wang, C. Herath, High-sensitivity fiber-loop ringdown evanescent-field index sensors using single-mode fiber. Opt. Lett. **35**, 1629–1631 (2010)
7. H. Berberoglu, H. Altan, A simple single-mode fiber loss measurement scheme in the C-band based on fiber loop-cavity ringdown spectroscopy. Opt. Commun. **317**, 29–33 (2014)
8. F. Ye, B. Qi, L. Qian, Continuous-wave fiber cavity ring-down measurements using frequency-shifted interferometry. Opt. Lett. **36**(11), 2080–2082 (2011)
9. C. Cheng, Z. Yang, Y. Ou, Z. Chen, J. Chen, H. Lv, Simultaneous measurement of gas composition and concentration combined fiber cavtiy ringdown and frequency-shifted interferometry. Optical Fiber Technol. **48**, 303–307 (2019)
10. Y. Ou, C. Cheng, Z. Chen, Z. Yang, H. Lv, L. Qian, Continuous-wave fiber cavity ringdown pressure sensing based on frequency-shifted interferometry. Sensors **18**(4), 1207 (2018)
11. C. Cheng, Z. Chen, Y. Ou, J. Chen, Multipoint fiber loop ringdown sensor for large strain measurement using frequency-shifted interferometry. Sensor **19** (13), 2907 (2019)
12. Y. Zhao, R. Lv, H. Li, Q. Wang, Simulation and experimental measurement of magnetic fluid transmission characteristics subjected to the magnetic field. IEEE Trans. Magn. **50**, 4600107 (2014)
13. E. Petrova, D. Kotsikau, V. Pankov, Structural characterization and magnetic properties of sol-gel derived $Zn_x Fe_{3-x}O_4$ nanoparticles. J. Magn. Magn. Mater. **378**, 429–435 (2015)
14. B. Sun, T. Shen, Y, Feng, Fiber-loop ring-down magnetic field and temperature sensing system based on the principle of time-division multiplexing. Optik **147**, 170–179 (2017)

Chapter 5
Research on Unsaturated Magnetization MFL Detection of Gouge in Oil and Gas Pipeline

Jin-Zhong Chen, Xiao-Wei Kang, Xue-Wei Zhang, Ren-Yang He, and Tao Meng

Abstract Oil and gas pipeline online detection technology can effectively detect metal loss defects, but the ability to identify mechanical damage gouge is limited. Given the shortcomings of the existing pipeline inspection technology, a DC magnetic yoke unsaturated magnetic flux leakage detection technology is proposed. The transient mechanics three-dimensional finite element model of the gouge site of the pipeline mechanical damage is established and the simulation results show that the obvious compressive stress is generated in the gouge part of the pipeline. A test device for unsaturated magnetic flux leakage detection is set up, and an experiment of gouge of oil and gas pipelines under unsaturated magnetization is carried out. The results show that the amplitude and peak-to-peak value of the MFL signal are affected by the stress at the defect region, which is very different from the leakage magnetic field signal of gouge without stress. Unsaturated magnetic flux leakage detection technology can not only effectively detect gouge defects, but also realize the identification and differentiation of them. The research results provide technical support for the application of unsaturated magnetic flux leakage detection technology in the detection and identification of oil and gas pipeline mechanical damage gouge.

5.1 Introduction

Oil and gas pipelines are exposed in the field or buried underground for a long time and are affected by various climatic conditions or human factors. Pipeline accidents occur frequently, posing a serious threat to oil and gas transportation and personnel

J.-Z. Chen · X.-W. Kang (✉) · R.-Y. He · T. Meng
Pressure Pipeline Department, China Special Equipment Inspection and Research Institute, Beijing, China
e-mail: kxw816@126.com

J.-Z. Chen
e-mail: cchenbeiter@aliyun.com

X.-W. Zhang
Nanchang Hangkong University, Nanchang, China

G. Liu and F. Cen (eds.), *Advances in Precision Instruments and Optical Engineering*, Springer Proceedings in Physics 270, https://doi.org/10.1007/978-981-16-7258-3_5

safety [1–3]. Among the reasons for pipeline failure, mechanical damage is one of the most common reasons. Mechanical damage can be divided into dent and gouge, which are mainly caused by external forces, ground disturbance, third-party excavation, and construction equipment [4]. In general, dents can be detected online by a high-precision geometric deformation detector [5]. However, gouges usually occur at the point of contact between the wall and the external force, resulting in metal loss or the pipe wall compression and thinning. The gouge is accompanied by stress concentration and cold working, but it is very similar to the corrosion defect in appearance [6]. At present, most pipeline online inspection tools detect and quantify corrosion and other metal loss defects through axial magnetization to near magnetic saturation [7]. Under a strong magnetic field, residual stress and cold working have little effect on the magnetic flux leakage signal, so it is impossible to identify the defect type accurately. Therefore, this paper proposes a method that can not only detect gouges but also identify them effectively, which can provide technical support for pipeline companies to make scientific and effective maintenance plans.

Under the condition of high magnetic field strength, the defect geometry shape has different degrees of influence on the MFL signal, and the residual stress caused by the gouge in the mechanical damage significantly changes the microstructure and mechanical properties of the pipeline but has little influence on the MFL signal [8]. Foster [9] proposed that under low field strength, the MFL signal would gradually increase with the increase of the external magnetic field, but as the magnetization intensity of the pipe wall approached saturation value, the influence of stress on MFL signal would almost disappear. To further study the influence of stress on MFL signal at low magnetization intensity, Kvasnica [10] and Jlies [11] et al. proposed that both elastic strain and plastic strain would significantly change the magnetic permeability of carbon steel. At low field intensity, the stress perpendicular to the magnetic field size of 120 MPa would reduce the magnetic leakage flux to 1/6 of the original value. Bubenik [12] designed an online detection tool for dual magnetization fields and found that the final decoupling signal was mainly affected by changes in residual stress and magnetic properties. However, the above researches on magnetic force mechanism and the relationship between pipeline defect types and detection signals are still not comprehensive enough, and the detection of pipeline gouge is not reliable.

In this paper, the transient mechanics simulation model of mechanical damage gouge in pipelines is established to analyze the geometric deformation and residual stress distribution in the process of the gouge. Based on the principle of magnetic flux leakage detection, an unsaturated magnetic flux leakage detection test device is built, and the experiment under unsaturated magnetization is carried out, which can detect and identify the gouges accurately.

5.2 Principle of MFL

In the process of MFL online detection, a complete magnetic loop is formed between the magnetized pipe wall, the yoke, and the iron core. The air permeability is much less than that of the pipeline. The lines parallel to the inner surface of the pipe will happen internal magnetic field deformation, resulting in the magnetic field lines in the defect leak due to the geometric discontinuity. Ultimately, the signal of leakage magnetic field is collected by the magnetic sensor. The detection principle is shown in Figs. 5.1 and 5.2.

In magnetic flux leakage testing, the selection of magnetization intensity is the most important step. The intensity of magnetization affects the characteristics of magnetic flux leakage signal, and the structure of the detection device determines whether the signal of mechanical damage gouge can be effectively detected and identified. In the B-H curve of the material, the magnetic permeability changes with the magnetic field intensity, as shown in Fig. 5.3. In the "A-B" stage, the magnetic permeability increases rapidly with the increase of magnetization intensity, which is not conducive to the detection of magnetic flux leakage. In the "B-C" phase, the magnetic permeability first rises to the highest point and then decreases gradually. In practical situations, other factors need to be considered, such as SNR and the design of detection equipment.

Important factors affecting MFL signals include MFL testing tool design, defect geometry, magnetism, and operating conditions, such as speed and stress. To study the magnetic flux leakage signal characteristics of mechanical damage gouge, it is very important to understand the pipe wall magnetism under stress. Most pipeline steels are magnetic anisotropic, and their easily magnetized axes are the result of

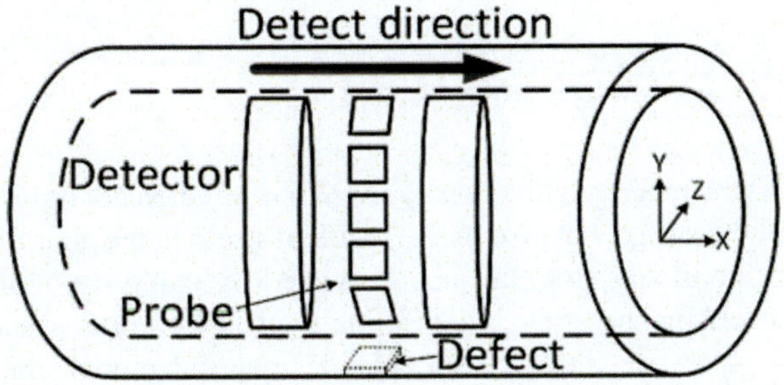

Fig. 5.1 Schematic diagram of MFL internal detection model

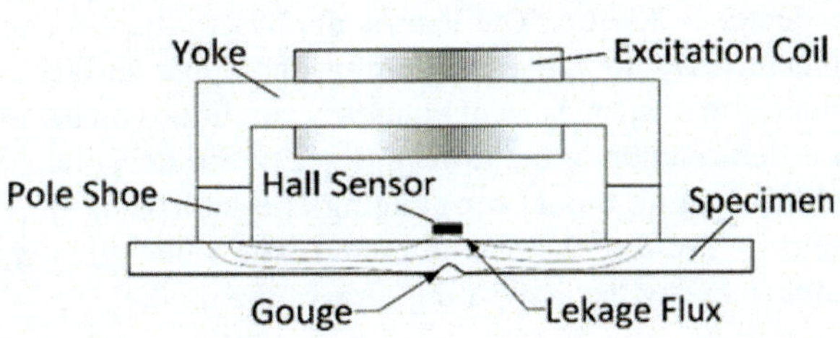

Fig. 5.2 Principle diagram of magnetic leakage detection

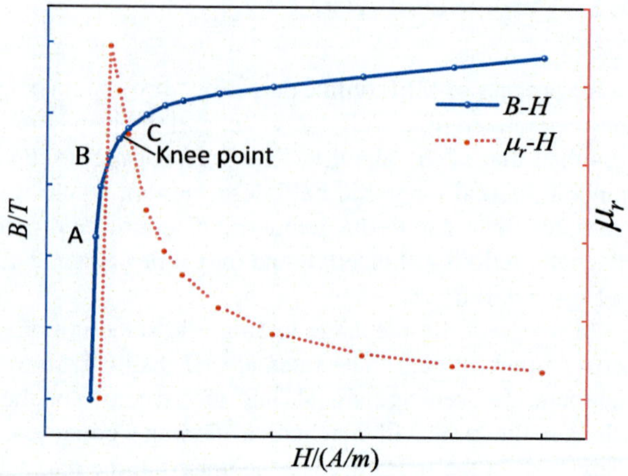

Fig. 5.3 Magnetic characteristic curve

the interaction between stress and domain orientations, usually along the pipeline axis. For materials with positive magnetostriction, such as pipeline steel, the easily magnetized axis tends to deflect in the direction of the applied force when a load is applied [13].

The relationship between the initial permeability and the residual stress of the material is first determined by coupling the magnetism with the applied stress. This relationship is based on the minimization of the total energy of the magnetic field and the material residual stress when it reaches the most stable state.

5.3 Simulation

5.3.1 Simulation Model

As the magnetic flux leakage signal is affected not only by the geometric deformation of the gouge, but also by the residual stress in the gouge area, the finite element model of nonlinear structure is established by using the finite element software ANSYS to obtain the stress distribution caused by gouge. Assuming that the pipe wall is a piecewise linear elastic–plastic material model, the steel plate with the size of 40 mm*40 mm*8 mm is used as a part of the pipe wall, and the diameter of the indenter is 20 mm. The load is applied to the circular section of the center of the indenter. By applying pressure on the upper surface of the vertical steel plate and imposing fixed constraint boundary conditions on the lower surface of the steel plate, the deformation of the lower surface of the steel plate is prevented. Finally, a simple elastic–plastic model for gouging is established. For the sake of model symmetry, Fig. 5.4 is 1/4 of the overall model. The model greatly reduces the number of grids, thus reducing the computing time of the model.

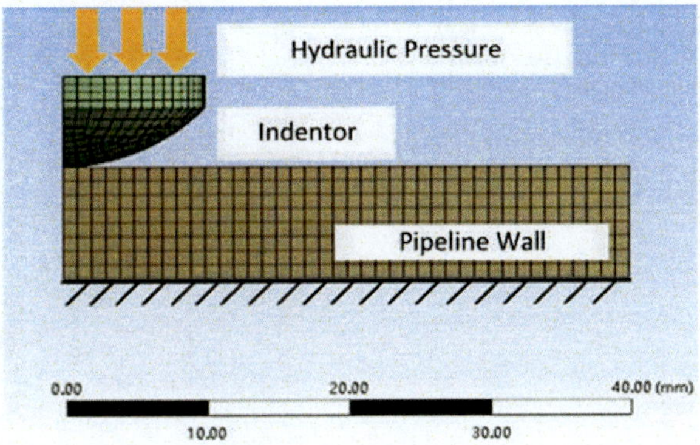

Fig. 5.4 Three-dimensional model

5.3.2 Model Parameters

The steel plate is made of 20# steel, with a density of 7850 kg/m^3, Young's modulus of 206 GPa, Poisson's ratio of 0.3, yield strength of 245 MPa, and tangential modulus of 0.113 GPa. The stress–strain behavior of the material is described by a bilinear total stress–strain curve, which starts from the origin and has positive stress and strain values.

According to the Coulomb friction law, the friction coefficient of the contact surface is set to 0.15. The automatic surface contact algorithm is used to model the contact surface between the steel plate and the indentation head. The algorithm uses the material characteristics of the two contact surfaces to calculate the stiffness of the contact element.

5.3.3 Simulation Results

In the process of gouge formation, the load pressure increases linearly from 0 to 120 MPa and then decreases linearly to 0 by applying instantaneous pressure on the surface of the indentation head to simulate the loading conditions. When the load is divided, the residual stress distribution of the steel plate is an approximate value of the residual stress representing the gouge. Figure 5.5 shows the stress and strain result after unloading.

Considering Fig. 5.5, it is seen that the load is perpendicular to the outer surface, and geometric deformation (wall thinning) occurs in the gouge area, while the stress is not uniform over the whole thickness of the pipe wall.

Figure 5.6 shows the stress distribution curve in the gouge area. It can be seen that the stress begins to increase at the initial position of the gouge, reaches the maximum at the center of the gouge, and then decreases sharply.

According to the stress distribution of the defect under load, the external magnetization direction is generally along the pipeline axis and perpendicular to the

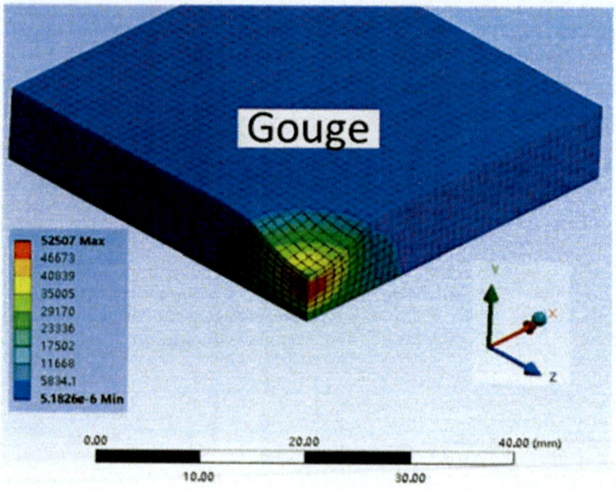

Fig. 5.5 The stress and strain result after load unloading

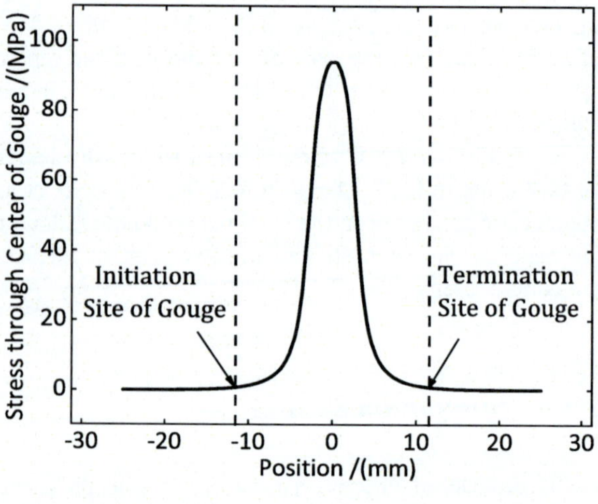

Fig. 5.6 The stress distribution curve in the gouge area

maximum component of stress vector. Therefore, the compressive residual stress of vertical magnetization increases the permeability.

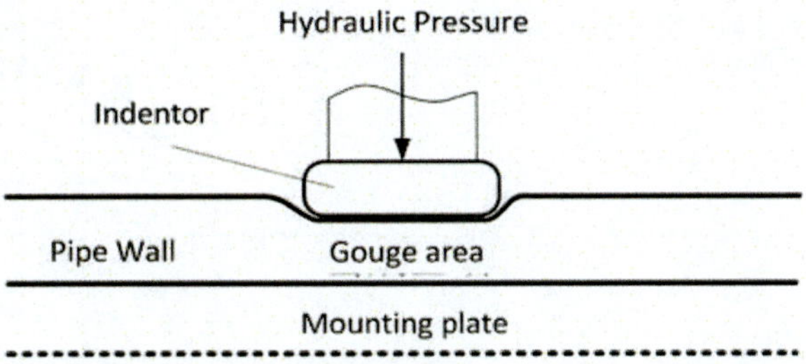

Fig 5.7 Schematic diagram of specimen processing method

Fig 5.8 Gouge specimen of pipeline

5.4 Experiment

5.4.1 The Pipeline Specimen of Gouge

To perform the experiments, the tested pipeline specimen has an outer diameter of 323.9 mm, a wall thickness of 8 mm, a length of 1200 mm, and a material of 20# steel.

To ensure the applicability of unsaturated magnetic flux leakage detection technology, it is particularly important to analyze whether the magnetic flux leakage signal of the gouge has obvious differences in some characteristic values, and verify whether the extracted characteristic values can effectively detect and identify the gouge defect.

During gouging, a pressure load is applied to the indentation head and placed on the supporting plate on the other side of the pipe wall to prevent radial displacement of the inner wall of the pipe. After the load is removed, gouge only contains residual stress and strain and pipe wall thinning without metal loss. The gouging schematic diagram is shown in Fig. 5.7. The maximum depth of gouge measured is 3.0 mm, and the maximum outer diameter is 19.1 mm. The gouge specimen is shown in Fig. 5.8.

5.4.2 Magnetic Flux Leakage Probe and Device

The magnetic flux leakage signal of defects can be measured by a magnetic sensor. The Hall sensor based on the Hall effect is applied to magnetic flux leakage detection for measuring signal. The principle of Hall effect is shown in Fig. 5.9. After the current is input, the Hall voltage is generated at both ends of Hall sensor by a magnetic field

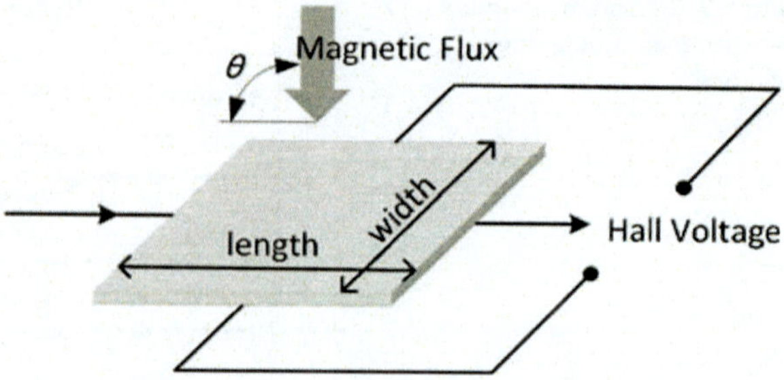

Fig. 5.9 Schematic of the Hall effect

perpendicular to the current direction. There is a linear relationship between the value of voltage and the intensity of magnetic field. Finally, the Hall voltage indicating the magnetic flux signal can be measured through the data acquisition device. In this study, the characteristic parameters of the Hall sensor selected, SS49E, are shown in Table 5.1.

In the experiment, an electromagnetic MFL setup is established, which is mainly composed of a magnetization, an amplification processing module, an A/D conversion circuit, a scanning system, and a sample pipe to be tested, as shown in Fig. 5.10. The pipe wall is magnetized by using an assembly of electromagnet. Yoke, having the same curvature as the pipe, is used to couple the flux into the pipe wall, thus completing the magnetic circuit from through the electromagnet to the pipe wall and back again. To detect the gouge under the low magnetic field, the direct current applied is 4A and the ampere-turns of the circuit is 4100AT. The applied magnetic field is 2850A/m.

Moreover, the scanning system consisted of an SS94E Hall sensor that is controlled by computer software and moved smoothly in a two-dimensional grid with an increment of 1 mm. It is connected to an XY digital plotter, which is controlled by a LabVIEW program. To simulate the working condition of the inner detector in the pipeline, the Hall sensor is near to the surface of the pipeline.

Result Analysis

In the experiments, the magnetic flux leakage signals in the state of no stress and the state with stress 100 MPa are studied by using the MFL analysis model. The

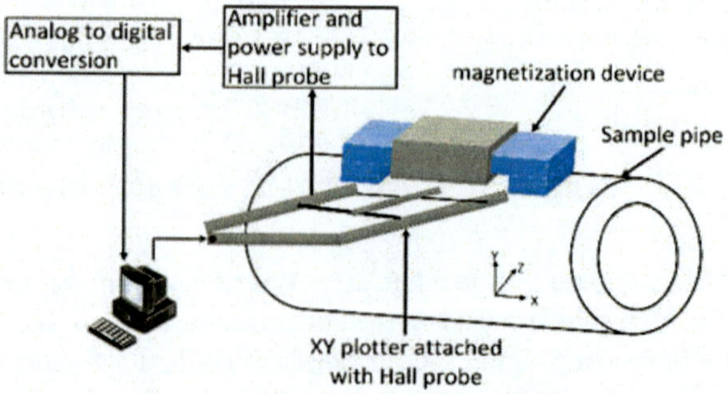

Fig. 5.10 Experimental setup of magnetic flux leakage detection

axial component and radial component of MFL signal are obtained experimentally and the defect leakage signals are measured at 2850A/m, respectively.

The signals are shown in Figs. 5.11 and 5.12. It can be seen that the horizontal coordinate of the signal is the distance to the center of the detect and the vertical coordinate is the intensity of the detected magnetic leakage signal. The axial amplitude of the leakage signal without stress is approximately 80.3Gs with a base value of approximately 11.2Gs and the peak-to-peak value of radial MFL signal is approximately 10Gs with a base value of approximately 2.0Gs. When the specimen is applied at 100 MPa, the axial signal amplitude is approximately 77.6Gs with a base value of approximately 1Gs and the shoulder peak value decreased by 13Gs due to residual stress. Moreover, the radial signal peak-to-peak value is approximately 8.5Gs with a

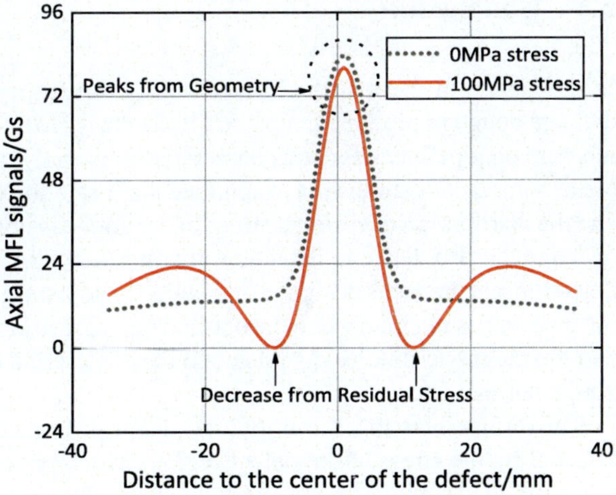

Fig. 5.11 The magnetic flux leakage axial signal of the gouge

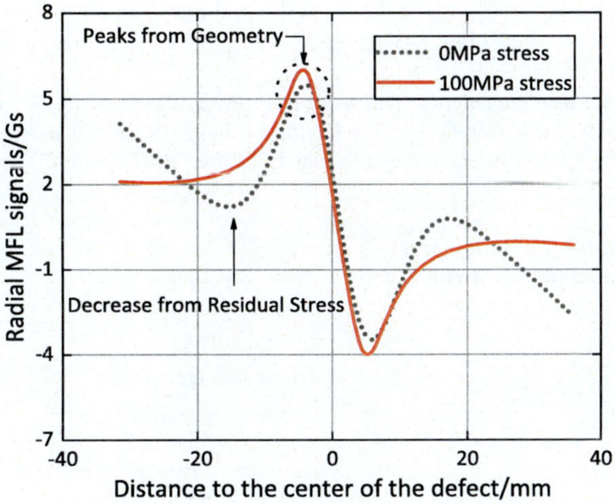

Fig. 5.12 The magnetic flux leakage radial signal of the gouge

base value of approximately 0.5Gs and the shoulder peak value decreased by 1.8Gs due to residual stress.

Based on data analysis, the contrast shows that the characteristic intensity of the defect leakage signal under stress is weakened. Due to the stress concentration phenomenon at the gouge, the amplitude and peak-to-peak value of the MFL signal are affected by the stress at the defect region. By comparing the measured data with the calculated results of the model, the shoulder peak value of the leakage signals above the stress region is less than that in the unstressed region. The axial and radial peaks from the geometry of the leakage signals without stress are greater than those with stress, which is consistent with the simulation analysis.

5.5 Conclusion

In this paper, in view of the limited ability to identify and distinguish mechanical damage gouge in pipeline magnetic flux leakage (MFL) online inspection, the effect of stress on pipeline magnetic characteristics is analyzed. A three-dimensional finite element model is established to simulate the formation of mechanical damage gouge, and the distribution of residual stress in the gouge is obtained. Based on the principle of magnetic flux leakage detection, the unsaturated magnetic flux leakage internal detection test device is designed and built. By adjusting the excitation source current, the pipe wall is magnetized to the optimal magnetization intensity, which is 2850A/m, and the magnetic flux leakage detection test is carried out for gouge with stress and gouge without stress.

The results show that the amplitude and peak-to-peak value of the MFL signal are affected by the stress at the defect region, which is very different from the leakage magnetic field signal of gouge without stress. The test results conform to the theory of magnetic field leakage and have high reliability and SNR, which proves that the unsaturated magnetic field leakage detection technology is feasible to detect mechanical damage gouge.

Acknowledgements This work is supported by the Science and Technology Plan Project of the State Administration for Market Regulation 2019MK138 and Youth Fund Project of China Special Equipment Inspection and Research Institute 2019 youth 07.

Table 5.1 Hall sensor characteristic parameter

Name	Parameter
Model	SS49E
Input voltage	8 V
Linear range	−1500G to + 1500G
Operating temperature	−40 to + 150 °C
Sensitivity	2.0 to 3.0 mV/G
Static output voltage	2.25 to 2.75 V

References

1. Y. Shi, C. Zhang, R. Li et al., Theory and application of magnetic flux leakage pipeline detection[J]. Sensors **15**(12), 31036–31055 (2015)
2. H.M. Kim, Y.W. Rho, H.R. Yoo, et al., A study on the measurement of axial cracks in the magnetic flux leakage NDT system[C], in *2012 IEEE International Conference on Automation Science and Engineering (CASE), 2012*, pp. 624–629
3. R. Wagner, O. Goncalves, A. Demma et al., Guided wave testing performance studies: comparison with ultrasonic and magnetic flux leakage pigs[J]. Insight-Non-Destruct. Test. Condit. Monit. **55**(4), 187–196 (2013)
4. M. Rosenfeld, J.W. Pepper, K. Leewis, Basis of the new criteria in ASME B31. 8 for prioritization and repair of mechanical damage[C]. Int. Pipel. Conf. **2002**, 647–658
5. S. Paeper, B. Brown, T. Beuker, Inline inspection of dents and corrosion using "high quality" multi-purpose smart-pig inspection data[C]. Int. Pipel. Conf. **2006**, 243–248
6. S. Razvarz, R. Jafari, A. Gegov, Basic concepts of neural networks and deep learning and their applications for pipeline damage detection, flow modelling and control in pipeline systems, 2021, pp. 101–119
7. G. Kopp, H. Willems, Sizing limits of metal loss anomalies using tri-axial MFL measurements: a model study[J]. NDT and E Int. **55**, 75–81 (2013)
8. P. Ivanov, Y. Sun, L. Udpa, et al., Stress characterization by local magnetic measurements, review of progress in quantitative nondestructive evaluation (Springer, 1999), pp. 1895–1902
9. F. Forster, On the way from the «Know-haw» to the «Know-why» in the magnetic leakage field method of nondestructive testing. I[J] Mater. Evaluat. **43**(10), 1154–1162 (1985)
10. B. Kvasnica, P. Fabo, Highly precise non-contact instrumentation for magnetic measurement of mechanical stress in low-carbon steel wires[J]. Meas. Sci. Technol. **7**(5), 763 (1996). S. Weidong, H. Ye, Z. Liang, Analysis of ultimate strength of depressed pipeline based on reverse engineering technology. J. Xian Shiyou Univ. (Natural Science Edition), vol. 34, pp. 97–103, May 2019
11. D. Jiles, The effect of compressive plastic deformation on the magnetic properties of AISI 4130 steels with various microstructures[J]. J. Phys. D: Appl. Phys. **21**(7), 1196 (1988)
12. T. Bubenik, J. Nestleroth, R. Davis, et al., Multiple magnetization level MFL for pipeline mechanical damage characterization[C]. Int. Pipel. Conf. **2000**, V002T06A026 (2000)
13. C. Coughlin, L. Clapham, D. Atherton, Effects of stress on MFL responses from elongated corrosion pits in pipeline steel[J]. NDT & E Int. **33**(3), 181–188 (2000)
14. B. Liu, N. Luo, G. Feng, Quantitative study on MFL signal of pipeline composite defect based on improved magnetic charge model[J]. Sensors **21**(10), 3412 (2021)

Chapter 6
The Infusion Monitor System Based on Single-Chip Micyoco

Haiyin Qing, Changjun Wu, Yanbao Wu, Haoyu Song, Qiuwen Gong, and Kehan Yu

Abstract This system uses the Single-Chip Micyoco as the main control chip, uses the capacitance sensor of the liquid level measurement as the detecting element, and uses the port wireless receiving and dispatching module (LC12S) as the signal transmission unit, which can realize the information collection of the patients and unified management. When the infusion is over, the alarm signal is transmitted wirelessly to the nurse station. The hospital central control system will realize sound and light alarm according to the test results, and display the specific alarm position for those hospitals with lower condition; for the hospital with best condition, every working nurse could be equipped with an alarm bracelet which can display and alarm need to change droplet bed ID. This system can reduce the medical negligence and improve the efficiency of health care.

6.1 Introduction

With the rapid development of automation equipment in recent years, the progress of intelligent medical equipment will become the focus of people. At present, in clinical medicine, the most commonly used treatment method is intravenous infusion; due to the need for human monitoring, it not only leads to the decline of medical staff's work efficiency, but also causes medical accidents due to human negligence; how to realize the liquid-level monitoring, sound and light alarm and display without affecting the normal infusion of patients? [1–4]. Therefore, it is necessary to improve the existing man-made monitoring mode and study the intelligent infusion monitoring system, so as to reduce the probability of medical accidents and improve the efficiency of medical care.

The concept of portable intelligent infusion monitoring system is introduced: the infusion information of patients in the inpatient building is collected at multiple points and managed in a unified way. When the infusion is about to end, the alarm signal is transmitted to the nurse desk through wireless transmission. The nurse desk

H. Qing (✉) · C. Wu · Y. Wu · H. Song · Q. Gong · K. Yu
School of Electronics and Materials Engineering, Leshan Normal University, Leshan, China
e-mail: qinghaiyin123@163.com

G. Liu and F. Cen (eds.), *Advances in Precision Instruments and Optical Engineering*, Springer Proceedings in Physics 270, https://doi.org/10.1007/978-981-16-7258-3_6

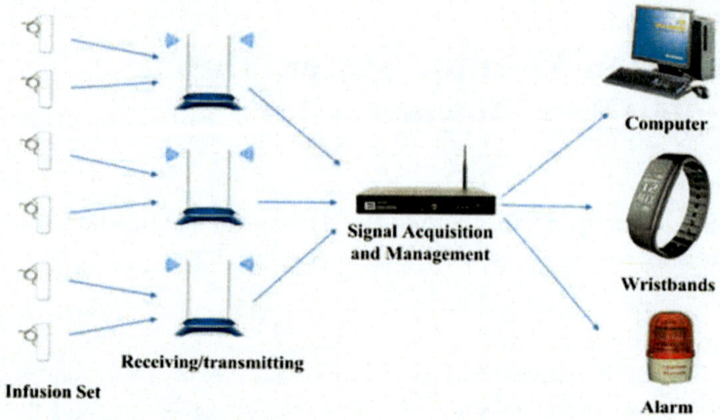

Fig. 6.1 System work flow chart

can clearly see the infusion process of patients through the display screen, and realize the sound and light alarm. If conditions permit, each nurse is equipped with an alarm bracelet, which can realize the real-time monitoring. The workflow of the system is shown in Fig. 6.1.

6.2 Product Design

The software part of the system consists of sending end and receiving end. The transmitter is mainly used to receive the output signal of the liquid-level detection sensor. The signal is analyzed and processed by the single-chip microcomputer to judge the infusion situation. If the infusion is about to end, the transmitter module will send an alarm signal. The receiving end is mainly used to receive the alarm signal sent by the sending end. If the alarm signal is received, the display module will display the bed information and send out the sound and light alarm [5, 6]. The alarm Bracelet processing method is similar.

The hardware part of the system consists of transmitter and receiver. The transmitter is mainly composed of liquid-level detection module and wireless transmission module, and STC89C52 is the main control chip. The receiving end is mainly composed of wireless receiving module, OLED display module, and sound light alarm module. At the same time, STC89C52 MCU is used as the main control chip [7].

6.3 System Composition

Now, SCM is not only a powerful assistant for developers, but also an indispensable tool in practical application, which is widely used in various fields [8]; as an important representative of embedded system, it plays an important role in the development of electronic technology [9]. STC89C52 has a maximum clock frequency of 80 MHz. The chip contains 4 K bytes Rom. The device is compatible with the standard MCS-51 instruction system and 80C51 pin structure. The chip integrates a general 8-bit CPU and ISP flash memory unit. With the PC control program, the program can be downloaded into the MCU. In this way, there is no need to buy a general programmer, and the speed is faster than the editor. The chip pin diagram of STC89C52 is shown in Fig. 6.2.

Sensors are the eyes of the system. In practical application, there are many monitoring points and the environment is more complex, so only manual duty cannot meet the requirements [10]. At present, most of the infusion monitors in the market use photoelectric sensors to detect the flow of liquid. Its disadvantage is that the photoelectric sensor is greatly affected by the light, especially in the strong light environment; the photoelectric sensor almost completely fails. In addition, the transparency of the liquid is very high; sometimes it cannot block the infrared signal from the infrared emitting tube of the photoelectric sensor, resulting in the inaccurate signal transmitted by the sensor; the medical equipment must be of high precision and high stability; in the above situation, we use the capacitive liquid-level sensor based on the change of capacitance, that is, when the liquid flows through, according

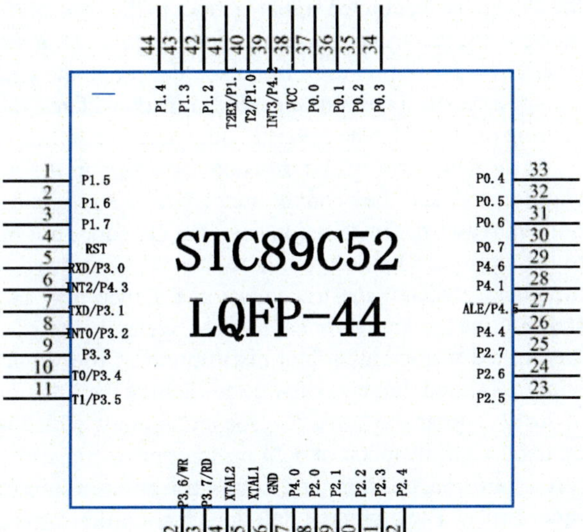

Fig. 6.2 Chip pin diagram of STC89C52 single-chip microcomputer

Table 6.1 Product parameters of xkc-y26a

Index	Parameter
Input volage	DD 5 ~ 24 V
Current consumption	3 ~ 8 mA
Output voltage (high)	Vin
Output volage (low)	0 V
Output current	1 ~ 100 mA
Response time	500mS
Working environment temperature	0 ~ 75 °C
Applicable diameter range	3 ~ 10 mm
Liquid level error	±2 mm
Humidity	5% ~ 100%
Texture of material	ABS
Waterproof performance	IP65

to the difference between the two plates of the capacitor the flow of liquid can be judged by the change of medium [11].

The intelligent non-contact liquid-level sensor uses the inductive capacitance of water to detect whether there is liquid. When there is no liquid approaching the sensor, there is a certain static capacitance between the sensor and the ground due to the distributed capacitance. When the liquid level rises slowly and approaches the sensor, the parasitic capacitance of the liquid will be coupled to the static capacitance to make the final electric resistance of the sensor. When the capacitance value becomes larger, the changed capacitance signal is input to the control IC for signal conversion, which converts the changed capacitance into the change of some electrical signal, and then a certain algorithm is used to detect and judge the degree of the change. When the change exceeds a certain threshold, it is considered that the liquid level reaches the sensing point [12]. The specific indicators are shown in Table 6.1.

At present, most of the infusion monitors on the market use wired transmission alarm signal, and they cannot understand the infusion situation centrally, which not only increases the installation cost, but also does not improve the medical efficiency to the extreme. Our solution is to use MCU to drive wireless module to send real-time data. Because the transmission distance may be far away, in order to ensure stability, we use signal repeater to receive and amplify the signal. Finally, we realize the unified management and distribution of the signal at the nurse end, and make windows-based display software to display the infusion information of each bed.

Lc12s adopts the latest 2.4gsoc technology, which is characterized by free development, sight distance of 120 m, integrated transceiver, no switching, transparent serial port transmission, and communication protocol. It can be quickly debugged successfully. The module is half duplex communication, with strong anti-interference ability, where working frequency can be set, multi-module frequency's complementary interference, small volume MSD package, no antenna, convenient installation,

Table 6.2 Product parameters of lc12s serial wireless transceiver module

Index	Parameter
Working frequency	2.4 Ghz
Receiving sensitivity	95 dBm
Receiving working current	27 mA
Interface	UART Serial port
RF chip	BK2461
Emission working current	40 mA@12 dBm
Working volage	2.89–3.6 V typical 3.3 V
Sight distance	120 m

and medical personnel can monitor each infusion point that carry out the whole process monitoring [13]. The specific index parameters are shown in Table 6.2.

Organic light emitting diode (OLED): organic light emitting diode (OLED) is also known as organic laser display. OLED display technology has the characteristics of self-illumination. It uses very thin organic material coating and glass substrate. When current passes through, these organic materials will emit light. Moreover, OLED display screen has large viewing angle and low power consumption [14]. OLED has the advantages of high contrast, thin thickness, wide viewing angle, fast reaction speed, flexible panel, wide temperature range, simple structure, and process. The first 12,864 screens are LCD, which need backlight and high power consumption, while OLED has low power consumption, which is more suitable for small systems; due to the different light-emitting materials, OLED displays better in different environments. The module power supply can be 3.3 V or 5 V without modifying the module circuit.

6.4 Physical Debugging

According to the hardware design, we use Altium designer software to complete the schematic drawing. The figure below is to draw the wireless sending and receiving module together. The physical schematic diagram is shown in Fig. 6.3.

According to the hardware design, using Altium designer software to complete the drawing of PCB diagram, drawing the wireless sending and receiving module together can save cost; most of the components of PCB adopt chip packaging, pursue small volume and are portable; the specific figure is as follows: the physical PCB diagram is as shown in Fig. 6.4.

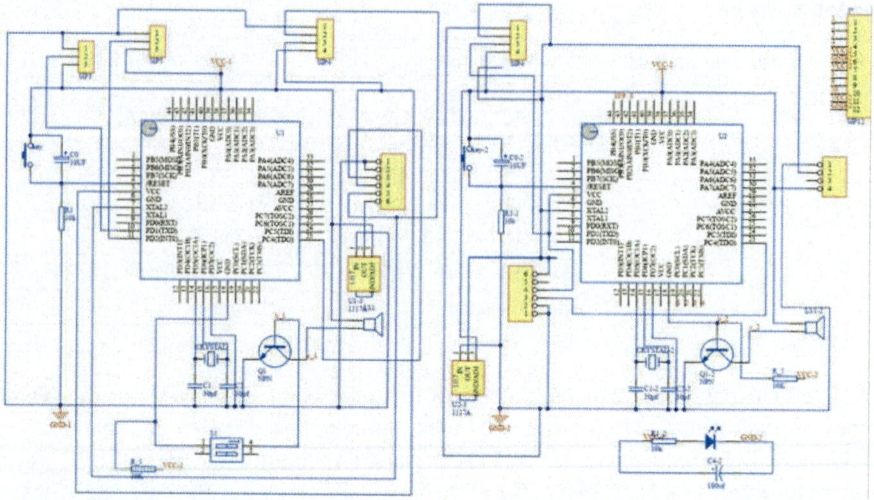

Fig. 6.3 Physical schematic diagram

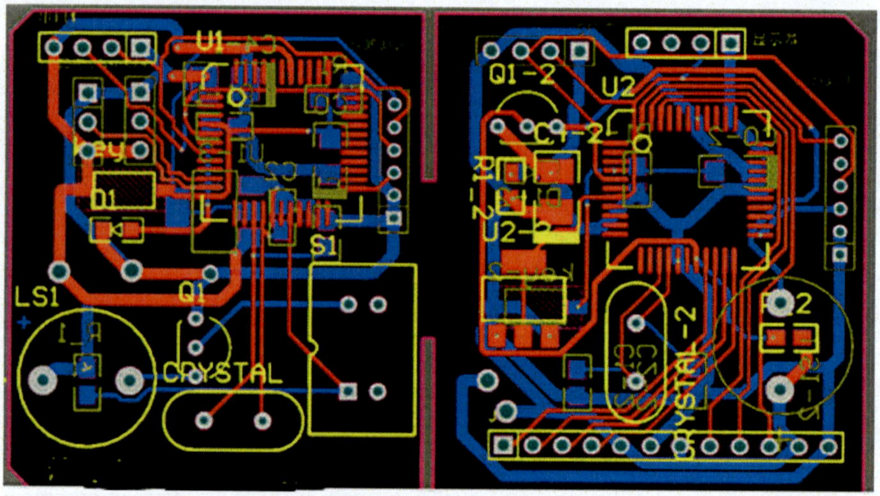

Fig. 6.4 Physical PCB

6.5 Conclusion

At present, the equipment that can realize transfusion monitoring can be seen everywhere on the Internet, but it is rarely used in the market. The reason is that most of the equipment is complex in technology, expensive in cost, and large in volume, or it needs to change the national standard structure of transfusion tube or the original alarm system of the hospital, which greatly reduces the feasibility of the practical application of the product; moreover, the high cost makes small hospitals. It is difficult for target customers such as general clinics to afford.

At present, the system has the reliability and stability to participate in the practical application, which is superior to other similar products in the market in terms of operation difficulty and cost. However, due to the problems of manufacturing technology and medical equipment licensing, it is difficult to be put into the market at present. It is expected that if it can be widely used, it will be an indispensable part of hospitals and animal husbandry. Therefore, the system has a broad market prospect and certain social significance.

Acknowledgements The authors would like to thank the Chinese Meridian Project for the use of data. This work was supported by the Sichuan provincial University Key Laboratory of Detection and Application of Space Effect in Southwest Sichuan. This work was supported in part by the National Natural Science Foundation of China under Grant 41804148, in part by the Application Foundation of Science and Technology Department of Sichuan Province under Grant 2019YJ0302, and in part by the Natural Science Research Foundation of Leshan Normal University under Grant LZDP013, Grant ZZ201803, and Grant DGZZ202001.

References

1. L. Zhiying, C. Mingxia, L. Hongfu, L. Shunyan, Intelligent infusion monitoring system based on STC89C52. China Sci. Technol. Inform. **10**, 64–66 (2019)
2. Z. Yue, Y. Songli, X. Hui, Y. Tianhao, L. Yifan, Z. Xiaochen, Software design of unattended infusion monitoring system based on MCU. Ind. Technol. Forum **14**, 70–71 (2018)
3. Z. Yingping, Z. hanqiang, Z. Jinpeng, L. Haitian, Design of intelligent liquid droplet monitoring system. J. Jilin Normal Univ. **3**, 69–73 (2017)
4. L. Lei, Y. Ling, L. Ping, L. Fei, W. Yanan, Development of an infusion detection and monitoring device. Med. Health Equip. **7**, 36–38 (2016)
5. W. Zhongwei, H. Xianshan, Y. Tong, Design and research of wireless loss preventer based on 51 single chip microcomputer. Ind. Control Comput. **12**, 154–156 (2018)
6. J. Xizhen, Z. Jiamin, Z. Chengming, L. Zhaolin, C. Mei, Automatic detection, monitoring and alarm device for liquid drops. Electron Technol. **11**, 3–4 (2018)
7. S. Lei, L. Zhentao, G. Ziping, H. Jian, Implementation of transfusion monitoring and management system. Chinese PLA Hosp. Manag. J. **12**, 1155–1157 (2016)
8. S. Hua, application of single chip microcomputer in electronic field. Electr. Technol. Softw. Eng. **24**, 242 (2018)
9. N. Xiaoyan, H. Xu, M. Xuchang, L. Xiwen, S. Lei, Application and development of single chip microcomputer. Hebei Agric. Mach. **12**, 53 (2019)
10. L. Ze, J. Mingshun, L. Shanshan, Z. Youfeng, S. Chenhui, Z. Lei, Z. Faye, S. Qingmei, Design of high precision ultrasonic liquid level measurement system. Automat. Instr. **11**, 56–59 (2018)
11. F. Peiyun, F. Junqing, Realization of remote water level monitoring based on STC89C52. Smart Factory **8**, 57–59 (2018)
12. Q. Zhaoyan, Design of liquid level control system. Technol. Market **12**, 102 (2016)
13. Cairns, OLED will be the next wind outlet, Yingcai, vol. 7, pp. 90 (2016)
14. W. Jinxia, W. Hao, Preliminary design of remote infusion monitoring and automatic alarm. Wireless Internet Technol **5**, 127 (2014)

Chapter 7
Density Prediction Model and Characterization Method of Density Variation Rate in Finished Tobacco Box Based on Microwave Signal Value

Zongyu Hu, Haiqing Chen, Qiang Xu, Xuemiao Su, Chenlin Miao, Yuhai Zhang, Wangxing Zhang, Nizhi Ge, Shaopeng Li, and Ruyan Xu

Abstract The detection and control of the density uniformity of finished tobacco packing can be applied in the threshing and redrying process, and in order to better popularize it, we do the following researches and practices. Through the research on the repeatability of detecting the tobacco density in the box by the microwave signal value, and the correlation between the microwave signal and the tobacco density in the box detected by the nine-point static picking-out method and the ionizing radiation method, the density prediction model based on the microwave signal value is established. According to the results, we know that the tobacco density in the box detected by microwave signal value shows good repeatability. The tobacco density detection model based on microwave signal value has a higher determination coefficient and fewer errors, and has better prediction effect. The prediction model is adopted to convert the microwave signal value in the box into the density value. And based on the results and feedback, we can guide the production and quality evaluation as well as the risk prevention and control, then the characterization methods of tobacco density variation rate in zoning and grid boxes are proposed. respectively. The results can not only be used as the adjustment basis of the homogenizing device in the tobacco packaging process, but also provide data support for the risk prevention and control of the aging process in tobacco storage.

Z. Hu · H. Chen · Q. Xu · C. Miao · S. Li · R. Xu (✉)
China Tobacco Jiangsu Industrial Co., Ltd., Nanjing, Jiangsu 210,019, China
e-mail: xy_xury@jszygs.com

X. Su · N. Ge
Chuxiong Redrying Factory of Yunnan tobacco redrying Co., Ltd., Chuxiong, Yunnan, China

Y. Zhang
Zhengzhou Tobacco Research Institute of China National Tobacco Corporation, Zhengzhou, Henan, China

W. Zhang
Fujian Sanming Jinye redrying Co., Ltd., Sanming, Fujian, China

G. Liu and F. Cen (eds.), *Advances in Precision Instruments and Optical Engineering*, Springer Proceedings in Physics 270, https://doi.org/10.1007/978-981-16-7258-3_7

7.1 Introduction

The packing density uniformity of the tobacco box in the threshing and redrying process will directly affect the aging quality in the tobacco raw material storage link and the processing quality in the primary processing. For example, the large density variation in the tobacco box in the aging process will easily lead to the appearance of oil and caking, while the loosening and conditioning in the primary processing is liable to make the leaves difficult to loosen [1–5]. Therefore, the control of packing density uniformity of tobacco has always been a hot research topic in the quality control of threshing and redrying. The nine-point static picking out method (hereinafter referred to as the nine-point method) is defined in the tobacco industry standards such as Leaf tobacco-Threshing and redrying-Technical specification (YC/T146-2001) [6] and the Threshed tobacco-Quality inspection (YC/T147-2001) [7]. Based on the definition of density [8], this method uses a drill core with a diameter of 40 mm to drill samples at 9 specified positions on the finished tobacco box, and weigh and calculate its density variation rate (DVR); However, this method has some defects in the actual operation process, mainly manifested in such aspects as the damage of tobacco, high labor intensity, long detection time, and failure to meet the needs of on-line detection [9–11];Tobacco and tobacco products-non-destructive determination of strips density deviation ratio in case-ionizing radiation method (GB/T31786-2015) [12] (hereinafter referred to as X-ray method); although this method has solved the problems of the nine-point method, it faces great limitation in its popularization and application in the tobacco industry due to the problems such as large equipment investment, cumbersome radioactive source management, and psychological obstacles of users.

In recent years, with the in-depth promotion of major special projects for the upgrading of threshing and redrying technology, the characteristic process technology of Chinese-type cigarettes has been moving forward, and the quality detection and control of the threshing and redrying process has attracted more and more attention [1–3]. As an important reference index affecting the aging and primary processing, the density uniformity in the tobacco box is particularly important for the detection and control. Studies have shown that the microwave micro winding method uses the corresponding relationship between the density of the material placed in the resonance cavity and the perturbation of the resonance cavity to detect the density [13, 14]. This method is an online, non-destructive, and safe density detection method [15, 16], and has been successfully applied in tobacco rod density detection [17–20]. In order to further strengthen the application of the detection and control of the density uniformity of tobacco packing in the threshing and redrying, after combining with the current situation that microwave moisture detector is generally set up in the preloading and packaging link in threshing and redrying enterprises, we study the relationship between the microwave signal and the density value of tobacco in the box detected by nine-point method and X-ray method, establish a prediction model of the tobacco density value based on the microwave signal value, and propose a characterization method of the density variation rate. By doing so, we hope these efforts could provide data support for the regulation basis of the homogenizing device in the leaves packaging process and the risk prevention and control of the aging process of tobacco storage.

7.2 Materials and Methods

7.2.1 Instruments, Equipment, and Materials

Materials: Middle tobacco leaves are selected in Yulong in Lijiang, Yunnan, Ninghua in Sanming, Fujian and Fei County in Linyi, Shandong, in 2020.

Instruments: MMA-4020 microwave moisture meter (MALCAM); Detector for density variation rate of tobacco in Fe3 box (Beijing Bangruida Mechanical and Electrical Equipment Co., Ltd.); Sample plate for detecting the variation rate of tobacco density in the box by nine-hole method, and the sampling drill core with diameter of 40 mm and length of 725 mm (self-made).

Equipment: Double KY161B pre-pressing packers and KY242 re-pressing packers (Baoying Yanlord Industrial Co. Ltd.), triple KY16A pre-pressing packers and KY25A re-pressing packers (Kunming Fengdong New Technology Group Development Co., Ltd), and triple KY17A pre-pressing packers and KY25A re-pressing packers (Kunming Fengdong New Technology Group Development Co., Ltd).

7.2.2 Sample Preparation and Methods

Sample preparation: In order to increase the representativeness of the data and the generalization ability of the model, nine weight gradients of 160 kg, 170 kg, 180 kg, 190 kg, 200 kg, 205 kg, 210 kg, 215 kg, and 220 kg are prepared, respectively, with 3 boxes of each weight gradient (1 box of Shandong Linyi sample with 180 kg weight gradient is missed). The net weight of packed tobacco in each gradient box is shown in Table 7.1.

Experimental methods: We use the microwave method and X-ray method to detect for 10 times, and then select the "2" sample box to implement the nine-point detection method; Among them, the detection methods of Lijiang samples are used by microwave method, nine-point method and X-ray, Sanming samples are used by microwave method and nine-point method, and Linyi samples are used by microwave method and X-ray. MMA-4020 microwave moisture meter is used for microwave method; The X-ray uses the tobacco density variation rate detector in box FE3; The nine-point method uses the sampling drill core to drill samples at the specified 9 positions on the finished tobacco box, weigh and detect its density. Among the nine points detected by the nine-point method and X-ray, the average density values of the

Table 7.1 Net weight of packed tobacco in each gradient sample box (unit: kg)

No	Gradient box number	Samples of Lijiang, Yunnan	Samples of Sanming, Fujian	Samples of Linyi, Shandong
1	160–1	160.4	160.8	157.5
2	160–2	159.8	160.3	160.2
3	160–3	160.5	160.3	162.2
4	170–1	171.8	168.8	170.1
5	170–2	168.0	170.0	170.5
6	170–3	169.9	170.2	171.6
7	180–1	179.8	179.5	180.3
8	180–2	179.5	179.7	180.9
9	180–3	179.3	180.3	/
10	190–1	189.9	191.4	187.5
11	190–2	189.2	190.5	189.7
12	190–3	188.8	191.1	190.7
13	200–1	198.3	199.6	200.7
14	200–2	200.8	200.1	200.9
15	200–3	200.4	199.7	200.9
16	205–1	205.7	203.3	203.6
17	205–2	205.3	203.8	206.2
18	205–3	205.7	203.3	206.3
19	210–1	211.5	212.0	209.8
20	210–2	208.7	211.6	210.0
21	210–3	210.1	211.7	210.8
22	215–1	213.7	213.2	214.0
23	215–2	214.1	213.7	214.8
24	215–3	215.5	214.2	216.3
25	220–1	220.9	220.5	220.4
26	220–2	221.2	221.2	220.8
27	220–3	218.2	221.5	222.0

second, fifth, and eighth points corresponding to them are selected to represent the nine-point method and X-ray detection values for analysis; The detection positions of microwave detection, X-ray, and nine-point method are shown in Fig. 7.1.

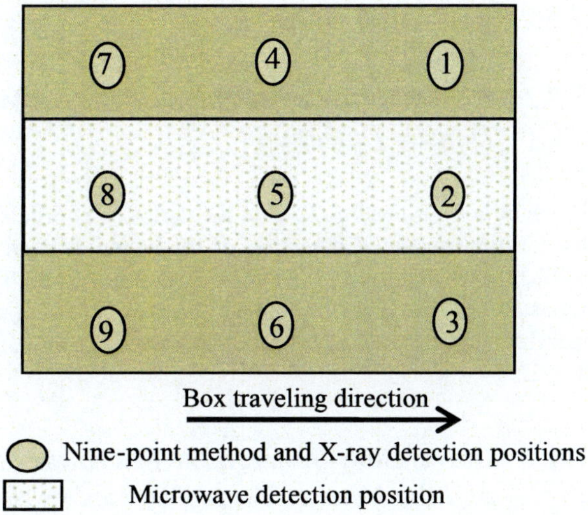

Fig. 7.1 Identification of detection area

Data statistical analysis method: The model is established by univariate linear regression method, and the prediction model of univariate linear regression analysis method is

$$Y_i = ax_i + b \tag{7.1}$$

Where, xi represents the value of phase i independent variable; Yi the value of phase i dependent variable; a and b the parameters of univariate linear regression equation;

Parameter b is obtained by the following formula:

$$b = \frac{\sum Y_i}{n} - a\frac{\sum x_i}{n} \tag{7.2}$$

Parameter a is obtained by the following formula:

$$a = \frac{n\sum x_i Y_i - \sum x_i \sum Y_i}{n\sum x_i^2 - \left(\sum x_i\right)^2} \tag{7.3}$$

7.3 Results and Analysis

7.3.1 *Repeatability of Microwave Signal Value in Tobacco Density Detection*

It can be seen from Fig. 7.2 that the variation coefficient of microwave signal value for ten-time repeated detection of the boxes of each gradient sample is less than 1.0%, except that the Linyi sample 210–2 is 9.26%. This result indicates that good repeatability of microwave signal value in detecting tobacco density can meet the stability requirements of tobacco density detection in the box.

7.3.2 *Correlation Between the Microwave Signal Value and the Density Value Detected by the Nine-point Method and X-ray Density Value*

Samples of Lijiang

From Figures 7.3 and 7.4, we can see that with the increase of the density value detected by the nine-point method and X-ray method, the microwave signal value tends to increase, and its correlation coefficients reach 0.940 and 0.897, respectively, demonstrating that there is a strong correlation between the microwave signal value and the density value detected by the X-ray method and the nine-point method.

It can be seen from Fig. 7.5 that with the increase of the density value detected by the nine-point method, the X-ray detection density value tends to increase, and its correlation coefficient reaches 0.921. It shows that there is a strong correlation between X-ray and nine-point method.

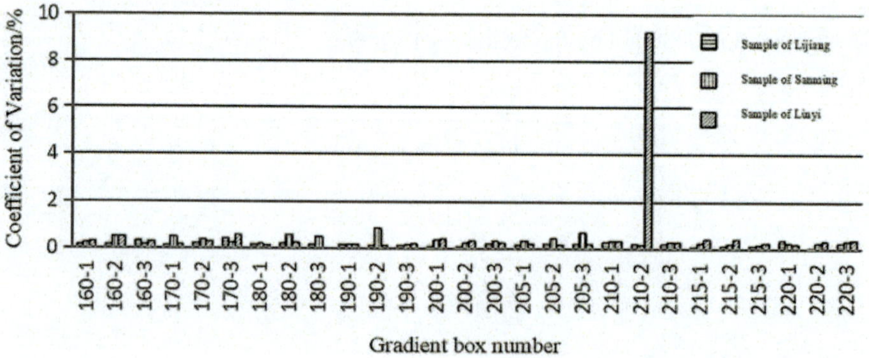

Fig. 7.2 Statistical graph of coefficient of deviation for repeated detection of microwave signal value

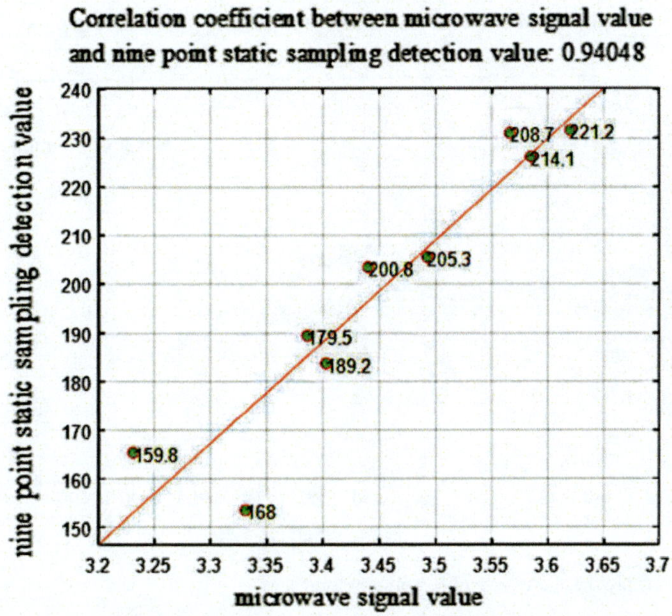

Fig. 7.3 Scatter plot of nine-point method and microwave signal value

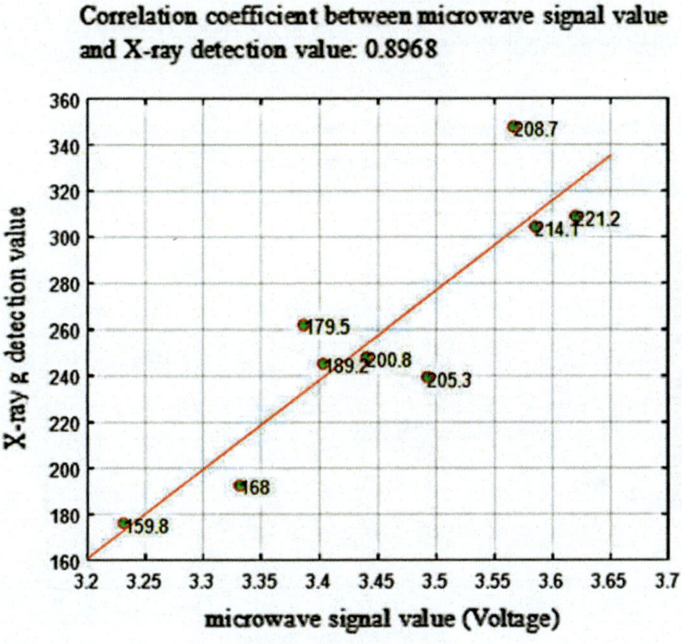

Fig. 7.4 Scatter plot of X-ray and microwave signal values

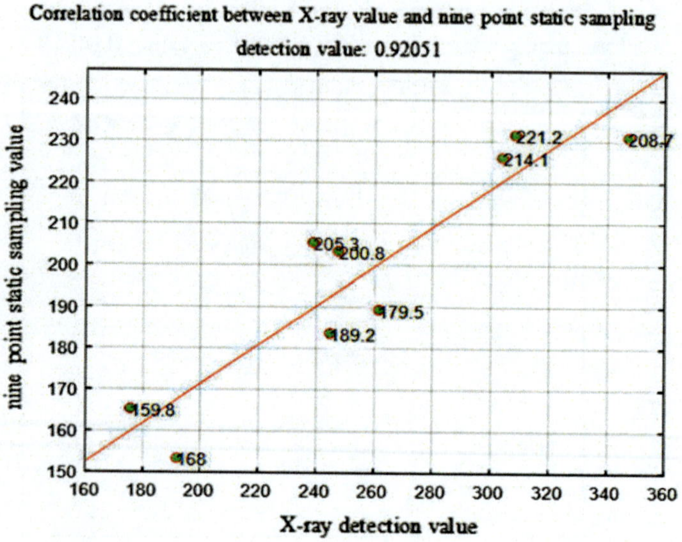

Fig. 7.5 Scatter plot of X-ray and nine-point method

Samples of Sanming

According to Fig. 7.6, we know that with the increase of the density value detected by the nine-point method, the microwave signal value tends to increase, and its

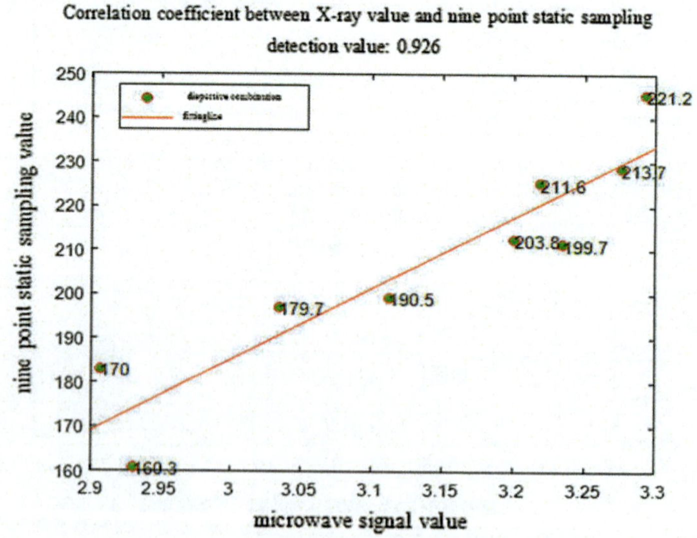

Fig. 7.6 Scatter plot of nine-point method and microwave signal value

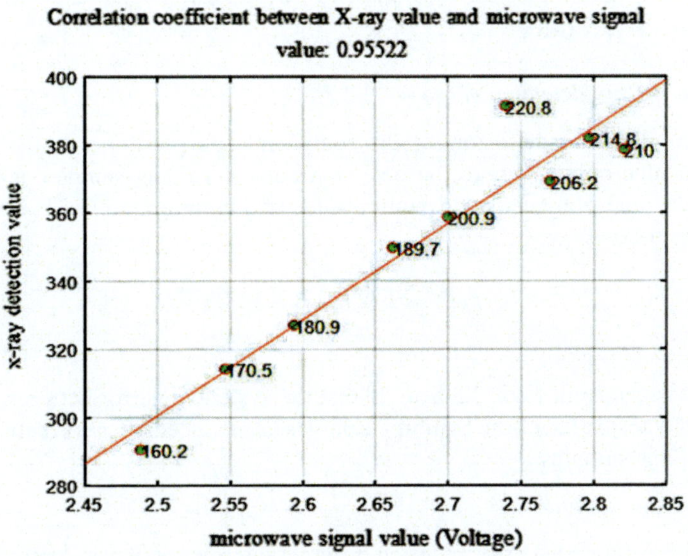

Fig. 7.7 Scatter plot of X-ray and microwave signal values

correlation coefficients reach 0.926, respectively, indicating that there is a strong correlation between the microwave signal value and the density value detected by the nine-point method.

Samples of Linyi

It can be seen from Fig. 7.7 that with the increase of X-ray detection density value, the microwave signal value tends to increase, and its correlation coefficient reaches 0.955, indicating that there is a strong correlation between microwave signal value and X-ray detection density value.

7.3.3 Establishment of Box Density Prediction Model Based on Microwave Signal Value

There is a strong correlation between the nine-point method and the X-ray detection density value. Therefore, when establishing the box density value prediction model based on the microwave signal value, we adopt the classical nine-point method detection data.

Correction of Test Data

Nine-point density detection value

In order to eliminate the errors of the equipment and personnel in the detection process of nine-point method, the detection data of Lijiang samples is corrected based on the nine-point density detection value of Sanming samples. The correction model is as follows:

$$f(x) = 1.028 * x - 14.03 \tag{7.4}$$

Where $R^2 = 0.8556$.

It can be seen from Figs. 7.8 and 7.9 that the systematic error between the nine-point density test value of the Sanming samples has reduced after correcting the test data of Lijiang samples.

Microwave signal value

In order to eliminate the error between different microwave density detector equipment, the detection data of Lijiang samples is corrected based on the microwave signal value of Sanming samples. The correction model is as follows:

$$f(x) = 0.8116 * x + 0.7468 \tag{7.5}$$

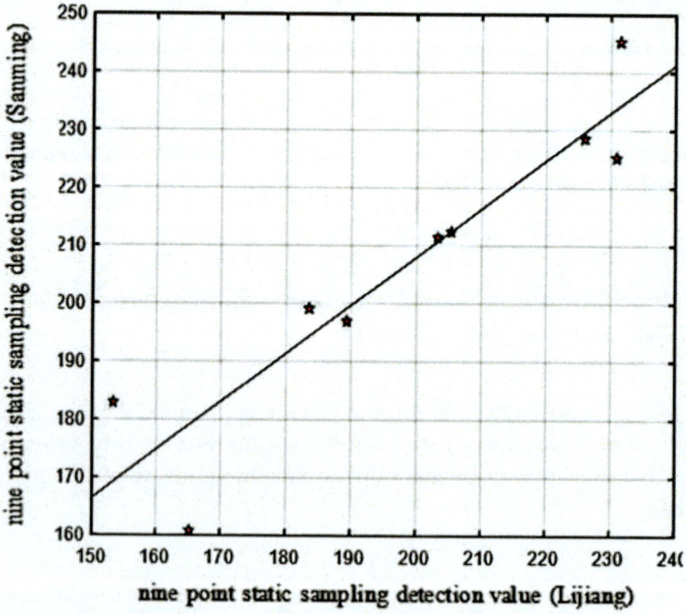

Fig. 7.8 Comparison chart of density value detected by the nine-point method

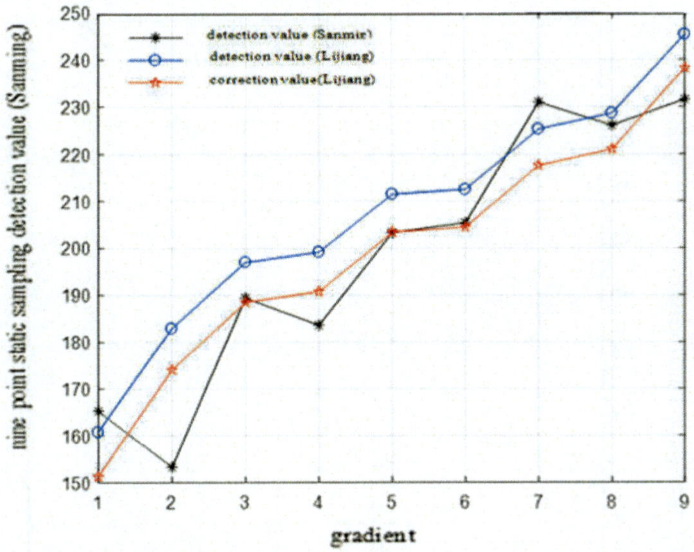

Fig. 7.9 Comparison chart of density value before and after correction by the nine-point method

Where $R^2 = 0.8341$.

From Figs. 7.10 and 7.11, we can see that after using the microwave signal value of Sanming samples to correct this value of Lijiang samples, the systematic error between the microwave signal values has effectively reduced.

The Establishment of Prediction Model

The corrected nine-point method detection density value and microwave signal value data are randomly divided into correction set (modeling set) and verification set. 80% of the samples are selected to establish the model, and 20% of the samples are used to predict the established model. The prediction model of microwave signal value and nine-point method detection density value is established as follows:

$$y = 204.9 * x - 202.4 \tag{7.6}$$

Wherein, the determination coefficient of the model is $R^2 = 0.8626$; The root mean square error of the model validation set RMSEP = 5.282, the average absolute deviation MAD = 5.0525, and the average relative deviation MRD = 2.45%.

The model has high determination coefficient and small error; Fig. 7.12 shows its good prediction effect.

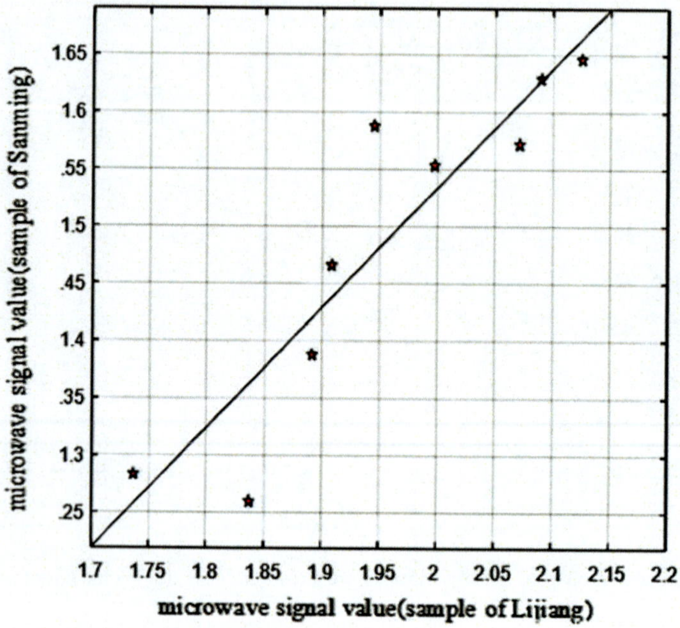

Fig. 7.10 Comparison of microwave signal value

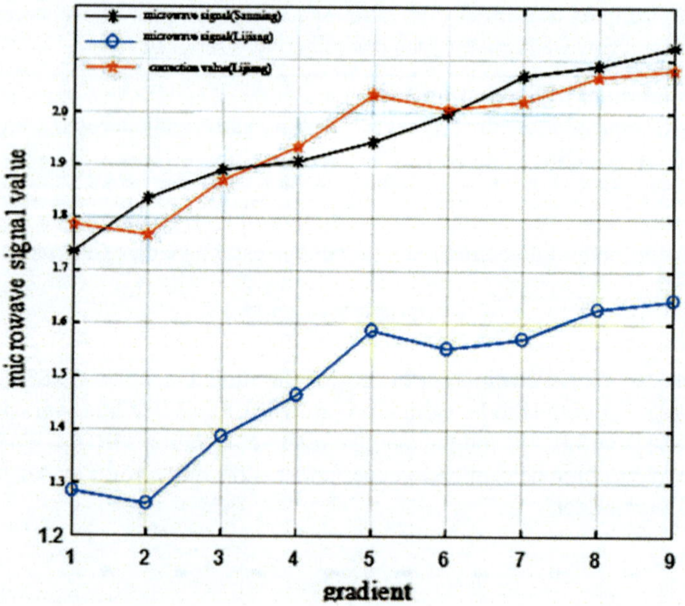

Fig. 7.11 Comparison of microwave signal values before and after correction

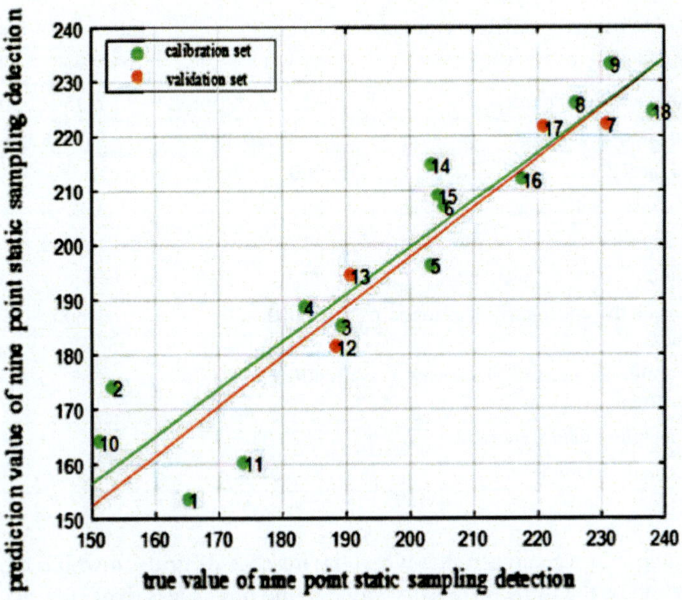

Fig. 7.12 Comparison between the real value and the predicted value of the nine-point density detection value model for microwave signal value

7.3.4 Characterization Method of Tobacco Density Variation Rate in Box Based on Microwave Signal Value

Variation Rate of Tobacco Density in Gridded Box Based on Quality Evaluation and Risk Prevention and Control

Based on the two main box models currently used in the tobacco industry, the surface area of the box is divided into grids according to the area of the nine-point sampling pipe. It can be seen from Table 7.2 that the two general-purpose boxes of model 1 and model 2 can be divided into 514.24 and 483.29 grids, respectively. In order to facilitate the application and statistics collection, the two types of boxes are uniformly divided into 500 grids.

The microwave detector is used to scan the finished box to obtain its microwave signal value. According to the prediction model of the density value in the box based on the microwave signal value, the value in each grid is transformed into the density value, and its mean value is calculated, respectively. The density variation rate DVR of the tobacco in the box is calculated by the density mean value in each grid. After that, the proportion of the number of grids with deviation of 10%, 20%, 30%, 40%, 50%, and > 50% from the absolute value of tobacco density in the box in the total grid shall be counted (see Table 7.3 for an example).

Table 7.2 Grid division of different types of tobacco box

Types of tobacco box	Type 1	Type 2
Length, width, and height/mm	1136 × 720 × 725	1115 × 690 × 725
Volume/mm3	592992000	557778750
Packing weight/Kg	200	
Theoretical density value/(kg/m3)	337.27	358.57
Surface area/mm2	817920	769350
Strap width/mm	12.5	12.5
Area of strips on the upper surface of tobacco box(four)/mm2	36000	34500
Inside and outside diameter of nine-point sampling tube/mm	40/44	
Nine-point sampling tube area/mm2	1520.53	
Number of grid division/piece	514.24	483.29

From Table 7.3, we can see that when the number of grids in which the absolute value deviation of the tobacco density value in the box is less than 10% accounts for a large proportion, the DVR value in the box is small and the density is relatively uniform; While a small proportion of this number shows large DVR value and poor uniformity. The results can finely describe the uniformity of the density of tobacco in the box, and provide data support for the prevention and control of quality risks such as mildew and carbonization in the later aging process of tobacco storage.

Density Variation Rate of Tobacco in Partitioned Box, Where the Tobacco Production is Guided by the Feedback

According to Table 7.2, the surface of two types of tobacco boxes can be divided into 9 areas on average, the average tobacco density of each area can be calculated, respectively, and then the DVR of tobacco density in the box can be calculated (see Table 7.4 for an example), which can be used to guide and optimize the homogenization device of tobacco leaves packaging process online. Therefore, the uniformity of packed tobacco density can be further improved.

Table 7.3 Density deviation rate of tobacco in gridded box (unit: %)

Samples in tobacco box	Deviation from the absolute value of the theoretical density of tobacco in the box and the proportion of the number of grids						DVR
	≤10	*10–20*	*20–30*	*30–40*	*40–50*	*>50*	
box 1	73.96	17.02	9.02	0	0	0	9.71
box 2	74.23	25.77	0	0	0	0	6.20
box 3	97.08	2.92	0	0	0	0	5.50

Table 7.4 Deviation rate of tobacco density in partitioned box

Samples in tobacco box	Tobacco density value in the area (kg/m3)									DVR
	1	**2**	**3**	**4**	**5**	**6**	**7**	**8**	**9**	
box 1	265.16	331.52	317.86	314.64	322.86	324.95	320.73	314.04	268.80	7.90
box 2	350.38	375.93	368.80	357.88	341.15	336.54	329.36	324.14	317.53	5.84
box 3	360.38	346.71	343.68	339.54	327.00	320.34	314.43	314.98	311.18	5.23

7.4 Conclusion

According to the variation coefficient of microwave signal value detected repeatedly for 10 times in each gradient sample tobacco box, we can conclude that the microwave signal value in detecting the tobacco density in the box shows good repeatability; The correlation between microwave signal value with nine-point method and X-ray detection density value can reach 0.90 or so, which has a strong correlation; And these results show that the microwave method is a feasible way to detect the density deviation rate of tobacco in the box and meets the repeatability requirements. Based on the univariate linear regression equation, the prediction model of microwave signal value and nine-point density value is established by fitting the linear relationship between microwave signal and nine-point density value; The determination coefficient of the model $R^2 = 0.8626$; Root mean square error of model validation set RMSEP = 5.282, mean absolute deviation MAD = 5.0525, mean relative deviation MRD = 2.45%; These results indicate that the model has high determination coefficient and small error, and the prediction effect is good. Based on the results and feedback, we can guide production and quality evaluation, and propose the characterization methods of density deviation rate of tobacco in partitioned and gridded boxes, respectively. Compared with the previous methods, these two characterization methods represent the overall density distribution of the box through the density values of nine points, which can comprehensively and accurately reflect the uniformity of tobacco density in the box. It can not only provide a relatively reliable basis for the on-line feedback control of the homogenizing device in the tobacco leaves packaging process, but provide data support for the prevention and control of quality risks such as mildew and carbonization in the aging process of tobacco storage.

Using the microwave method to detect the density variation rate in the finished tobacco box can not only overcome the shortcomings of nine-point method and X-ray method in practical application, but also fully combine the current situation that the threshing and redrying processing enterprises generally set up microwave moisture detectors in the preloading and packaging links, effectively reducing the investment of density variation rate detection equipment in the finished tobacco box. It provides technical support for better popularization and application of the detection and control of tobacco packing density uniformity in threshing and redrying processing. However, due to the difficulty in preparing samples with different packing weights and the heavy detection workload through the nine-point method, the paper only tested three middle tobacco leaves. In the later stage, experimental research will be carried out on the upper and lower tobacco leaves to improve the density variation rate method in the finished tobacco box. By doing so, the methods can be suitable for the research of different parts of tobacco.

Acknowledgements Fund project: Key R & D project of China National Tobacco Corporation "research and application of processing attribute classification technology based on national mechanical properties of flue-cured tobacco leaves" (No.: 110202102005).

References

1. W. Du, J. Yi, Z. Huang, etc., On-line chemical determination in threshing and redrying process and its application in quality control [J]. Acta Tabacaria Sinica **15**(01), 1–5 (2009)
2. M. Wang, Evaluation on stability of threshing and redrying lamina structure of tobacco [D]. Henan Agricultural University (2011)
3. Y. Wang, M. Long, B. He, etc., Application of chromatic aberration method in quality consistency evaluation of finished strips after threshing and redrying [J]. Tobacco Sci. Technol. **2013**(12), 9–13 (2013)
4. L. Chen, Analysis on the importance of detecting the deviation rate of smoke density inside the box [N]. East Tobacco, 2015-09-29 (002)
5. G. Guo, B. Liu, Shuai Lang, etc. The design on the hanging mechanism of offshore CPT [J]. Mech. Eng. 2009(05), 63–64 (2009)
6. YC/T146–2001 Leaf tobacco—Threshing and redrying—Technical specification [S]
7. YC/T147–2001 Threshed tobacco—Quality inspection [S]
8. L. Zhimei, C. Yingmei, Improvement of experimental method for measuring object density [J]. J. Shaoguan Univ. **31**(06), 54–57 (2010)
9. M. Song, Application of the detector of the deviation rate of the tobacco density in threshing and redrying enterprises [J]. Publ. Enterprise Technol. **31**(05), 32–33 (2012)
10. Y. Sun, Z. Wang, H. Sun, etc., Real time detection and control system of density deviation rate in redried tobacco box [J]. Plant Maint. Eng. **2014**(04), 57–59 (2014)
11. J. Zheng, Study of the detector for density deviation rate of smoke in the box [N]. East Tobacco, 2015-09-30 (002)
12. GB/T 31786–2015, Tobacco and tobacco products—Non-destructive determination of strips density variation ratio in case—ionizing radiation method [S]
13. Y. Zhou, Z. Niu, Z. Lu, etc., Method of on-line measurement of cigarette humidity and density based on microwave resonance cavity perturbation [J]. Instr. Techn. Sensor **2009**(01), 102–104 (2009)
14. Z. Lu, W. Sun, Z. Wu, etc., A study on microwave resonance cavity sensor for detecting cigarette humidity and density [J]. Chinese J. Sensor Actuat. **2007**(05), 1030–1033 (2007)
15. L. Zhu, The study of the key technology about microwave density and moisture measuring system [D]. Hefei Polytechnic University (2006)
16. S. Chen, Research and application of microwave on-line density measurement technology [D]. Chongqing University (2002)
17. M. Liu, Y. Liu, R. Wen, etc., Measurement characteristics analysis of microwave cigarette density distribution detector [J]. Food Mach. **32**(08): 35–37+62 (2016)
18. K. Li, Application of microwave technology in cigarette density detection [J]. Mod. Bus. Trade Ind. **21**(04), 301–302 (2009)

19. Y. Huang, Study on microwave detection method of cigarette density [D]. National University of Defense Technology (2009)
20. L. Deng, X. Liao, Y. Lu, etc., Application of microwave moisture meter in cigarette density measurement [J]. New Technol. New Prod. China **2010**(01), 16–17 (2010)

Chapter 8
Research on Fault Detection Technology of Primary and Secondary Circuit of Switchgear

Xuesong Lin, Yu Xu, Changsong Ni, Jingwen Cai, Chunqing Lin, Lei Zhang, and Di Yu

Abstract At present, the primary and secondary equipment integration has been realized in 10 kV switching station and ring main unit, and the functions of telesignalling, telemetering and telecontrol have been realized by assembling Data Transfer Unit (DTU). However, in actual operation and maintenance, if an equipment fault occurs, it is necessary to first determine whether it is a primary or secondary equipment failure, and then send corresponding maintenance personnel to the site. In addition, this maintenance mode also needs to be powered off. The existing primary equipment does not have a circuit breaker test position. If the test is required, the whole section of bus must be powered off. Due to the important user and reliability requirements, the power off is almost difficult to implement. If the fault identification cannot be solved at one time, the power off times need to be increased, which affects the reliability of power supply. The primary equipment is electrified, and the secondary equipment has few test methods, low safety factor and more equipment operation with faults, which is not conducive to the healthy development of distribution network (M. Adams, Effectiveness of different design solutions to control internal faults in MV switchgear, IEE Colloquium on Risk Reduction: Internal Faults in T&D Switchgear (Digest No: 1997/295), pp. 5/1–5/6, 1997). In order to solve the above problems, this paper proposes a ring network unit fault detection device that can distinguish between primary and secondary circuit failures without power failure. On the basis of the traditional fault diagnosis method, a low-voltage live detection method with high safety and easy operation is added.

8.1 Introduction

With the development of the power industry, the power distribution switchgear has evolved into a complete set of automation equipment with primary and secondary integration. Dispatch remote operation and automatic fault isolation have been widely

X. Lin (✉) · Y. Xu · C. Ni · J. Cai · C. Lin · L. Zhang · D. Yu
State Grid Dalian Electric Power Supply Company, Dalian, China

G. Liu and F. Cen (eds.), *Advances in Precision Instruments and Optical Engineering*, Springer Proceedings in Physics 270, https://doi.org/10.1007/978-981-16-7258-3_8

used in daily operation, and it has also brought great technical challenges to the operation and maintenance of power distribution.

When the switchgear refuses to operate or exits from the operation mode, it is necessary to accurately determine whether it is a primary equipment failure or a secondary circuit failure. According to different fault types of high-voltage switchgear, fault detection methods using mechanical properties, electrical properties, temperature and other parameters have been proposed at home and abroad. Various monitoring methods can be used to easily obtain the state parameters of the high-voltage switchgear and the waveforms of related signals. However, because the high-voltage switchgear is a combination of multiple electrical equipment, it involves various physical phenomena such as electricity, magnetism, and temperature, and the manifestations and mechanisms of faults are very different. It not only affects the reliability of power supply, but also causes a great waste of manpower and financial resources. Therefore, it is necessary and imminent to study the detection technology of the primary and secondary circuit failures of the distribution automation ring mains.

This paper proposes the isolation and detection of the primary and secondary circuits in the ring network unit system. It is proposed to conduct online live diagnosis for most primary and secondary circuit faults without power failure. On the basis of the traditional fault diagnosis method, a low-voltage live detection method is added, which provides high safety and easy operation.

8.2 Research on Switchgear Fault

8.2.1 Primary Faults

Investigation statistics show that the primary faults of high-voltage switchgear mainly fall into the following categories:

(1) Fault of refusing action and misoperation: This kind of fault is the main fault of high-voltage switchgear, and its causes can be divided into two categories: one is caused by mechanical faults of operating mechanism and transmission system. The other is caused by electrical control and auxiliary circuits.

(2) Insulation fault: It is manifested as external insulation to ground flashover breakdown, internal insulation to ground flashover breakdown, phase-to-phase insulation flashover breakdown, lightning overvoltage flashover breakdown, porcelain bushing, porcelain bottle fracture, etc. [3].

(3) Current-carrying fault: The main cause of current-carrying fault at voltage levels of 7.2 to 12 kV is the poor contact of the switchgear isolation plug, which causes the contacts to melt [4].

(4) External forces and other faults: including foreign body impact, natural disasters, short circuiting of small animals, etc.

8.2.2 Secondary Faults

The secondary faults of the switchgear generally fall into the following categories:

(1) Power source: AC and DC circuits; current and voltage circuits.

(2) Purpose: signal circuit, protection circuit, automatic device circuit, control circuit and measurement circuit.

(3) Equipment: transformer secondary circuits, switch secondary circuits, capacitor secondary circuits, etc.

8.3 Design of Detection System Structure

When the switchgear is put into operation, after pulling out the working plug, the detection system enters the test state. At this time, the telecontrol closing signal is given in the background, the test power indicator is on, the closing relay is pulled in, and the test closing indicator is on and maintained, which proves that the closing circuit works normally, and the test of switch-off circuit is the same (Fig. 8.1).

As shown in Fig. 8.2, there are 11 terminals in the test terminal block on the door of the ring switch cabinet, and each terminal consists of an upper end and a lower end which can be communicated or isolated. In the daily operation state, the working plug is inserted into the test terminal block of the cabinet door, so that the upper and lower ends of the terminals of the terminal block are in a communication state.

The lower ends of terminals 1, 2, 3 and 4, terminals 5 and 6, terminals 7 and 8 and terminals 9 and 10 of the cabinet door test terminal block are all equipped with a shorting ring (Table 8.1).

As shown in Fig. 8.3, the terminal of the cabinet test terminal base is composed of upper and lower parts, and the lower end of the terminal is provided with a short circuit ring. The upper and lower ends of the terminal are normally open. During daily operation, the working plug is inserted into the cabinet door test terminal block, the upper and lower ends of the terminal are connected, the short circuit ring at the lower end of the terminal is invalid, KA1, KA2 relays, HR1, HR2 and Hg indicator lights do not work, and the test circuit is separated from the control circuit.

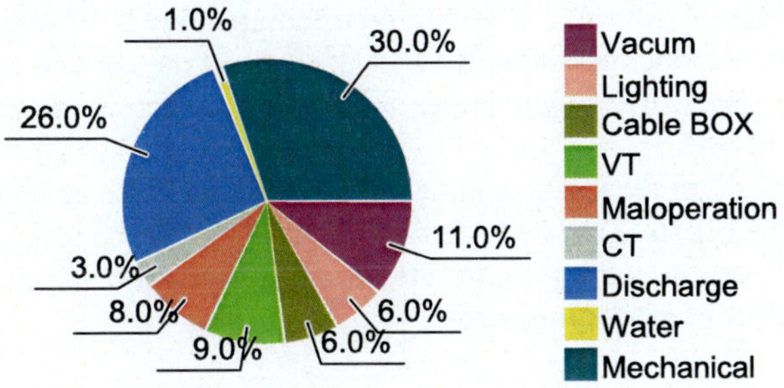

Fig. 8.1 Different kinds of faults in switchgears

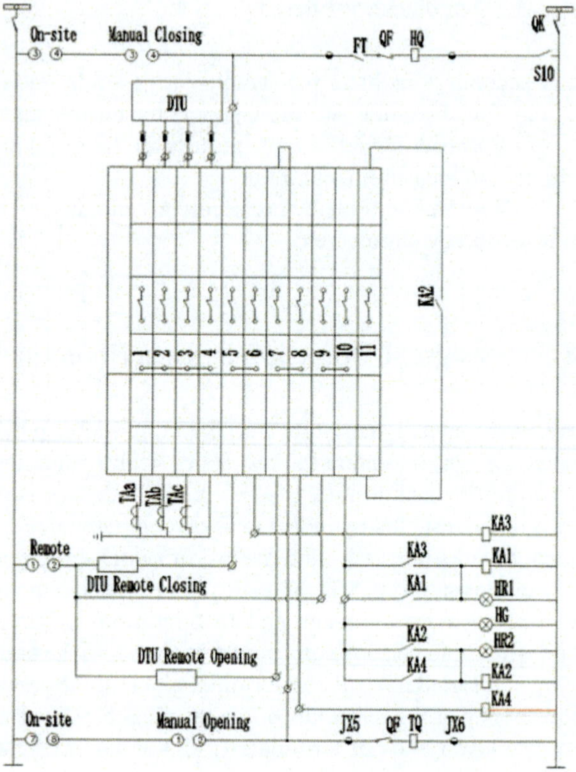

Fig. 8.2 Terminal on the door of ring switch cabinet

Figure 8.4 shows, when faults need to be detected, pull out the working plug of the cabinet test terminal block and insert the test plug. At this time, the upper and lower terminals of the cabinet test terminal block are disconnected, and the short ring at the lower end of the terminal is invalid. The short-circuited ring at the lower end of the test plug is effective, so that the lower end of terminals 1, 2, 3 and 4 is short circuited and the protection current circuit is sealed. Add an overcurrent signal to DTU at the upper end of the test plug; the lower end of terminal 7 and 8 of the test plug is short circuited, and the telecontrol trip circuit is switched from the circuit breaker trip coil to KA2 coil. Receive overcurrent signal in DTU, send overcurrent remote control trip signal, intermediate relay KA2 operation; the passive open point of intermediate relay KA2 is closed, and the closure of intermediate relay KA2 can be measured by the upper and lower conduction of terminal 11 of the plug through the test.

After the test is completed, the work switch is reset to the local position, and then the trip time can be tested again when it is turned remotely. Pull out the test plug and insert the work plug to release the self-locking and experimental circuit and return to normal working state.

Table 8.1 Terminals of ring switch cabinet

Terminal	Bottom	Top	Function
1, 2, 3, 4	Secondary winding of CT	DTU	Conduct, provide interface for external detection equipment
5	DTU output port	Circuit breaker closing circuit	Conduction and isolation
6	K1 relay	empty	Test the conduction of DTU telecontrol closing circuit
7	DTU export	Circuit breaker opening circuit	Conduction and isolation of telecontrol opening circuit
8	K1 relay	empty	Test the conduction of DTU telecontrol closing circuit
9	DTU export	empty	Detect the conduction of DTU remote switching-off circuit
10	Auxiliary contacts of KA1 and KA2 relays	empty	Detect the conduction of DTU telecontrol closing circuit and opening circuit with terminals 8 and 9
11	Auxiliary contact of KA2 relay	Auxiliary contact of KA2 relay	Backup protection tripping standby time detection point

Fig. 8.3 Cabinet test terminal block

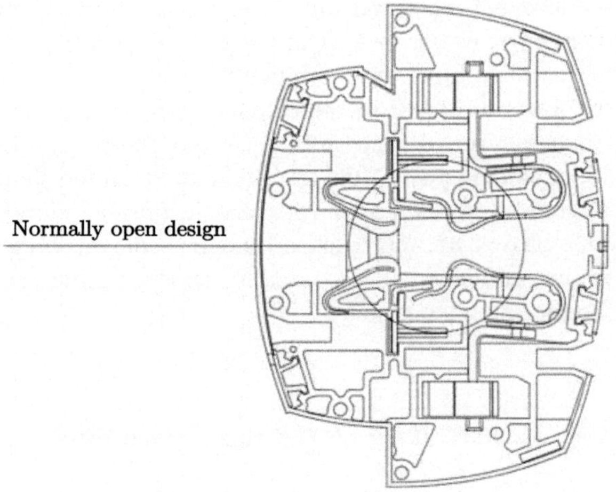

Fig. 8.4 Plugs and terminals for testing

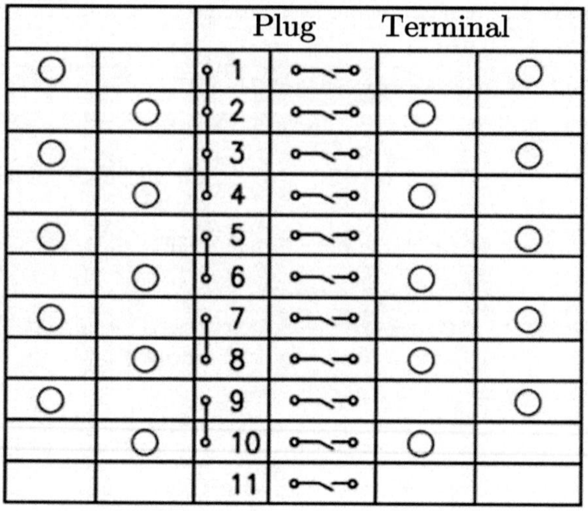

When the system is running normally, the background will monitor the current value of the current system in real time. When the three-phase current value of the power grid is found to be unbalanced, it may be the primary equipment (such as current transformer), or the secondary ammeter or detection system. At this time, the working plug of the cabinet test terminal block can be pulled out, and the short-circuit ring at the lower end of the terminal takes effect to short circuit the secondary current circuit, so as to prevent overvoltage caused by the open circuit of the secondary circuit of the current transformer. Then, the plug inserts the test plug into the cabinet test terminal block. At this time, the test terminal block at the lower end of the short ring invalid, instead of the test plug on the short ring in effect, replace the test terminal block on the function of the short ring. At the same time, the upper and lower terminals of the cabinet test terminal block are disconnected to realize separation from the measuring device of the system. The testing device is connected to the binding post of the test plug, and then the short-circuit ring on the test plug is removed to realize the role of detecting the current value of the system. If the three-phase current is balanced and the value meets the standard, it is proved that the primary equipment has no problem, and the problem lies in the secondary measuring equipment on the system.

8.4 Research on Detection Technology

At present, the primary and secondary circuit fault detection method of the switchgear mainly includes three main points: fault detection of the secondary circuit, protection setting mechanism, and the prevention of the secondary circuit CT open circuit and PT short circuit.

8.4.1 Fault Detection of the Secondary Circuit

When the equipment fails and the switch refuses to operate or the remote control fails, the current practice requires the entire section of the bus to be powered off before troubleshooting. Moreover, it is more dangerous to short circuit the CT in the current secondary circuit live test, and it is easy to misoperate and cause the switch to trip. In addition, it is necessary to remove the equipment trip connection to prevent the equipment protection from malfunctioning, which may cause the risk of malfunction [5].

By on-line monitoring and testing of relevant functions of secondary equipment, the successfully developed secondary circuit monitoring component can distinguish the specific fault section of primary and secondary equipment. When the failure reason cannot be directly determined by reading the DTU message and fault information, it can be inserted through the test terminal to enter the test state, dispatch the telecontrol command, and indicate the telecontrol state through the indicator light on site, so as to quickly distinguish the fault reason of primary and secondary equipment. In case of secondary equipment failure, troubleshooting and protection of transmission can be carried out when the primary equipment is live, which will not cause risks such as tripping at the outlet of the primary equipment and has high safety.

8.4.2 Protection Setting Mechanism

It is difficult to implement the protection setting when the relay protection is running on line, and the safety risk is high. The switch trip and personal accident caused by the disconnection error and CT short-circuit error occurs. There is an urgent need for a simple, plug-and-play test method to reduce personnel safety risks and operational difficulties.

Therefore, the protection setting mechanism of distribution automation switch under the condition of live operation is studied, and the switch action display under the test state is realized through the equipment. The protection test, set value setting, protection transmission and other functions of the electrified secondary circuit of the primary equipment are realized, and the protection set value is realized on the premise of ensuring the safety of the human body and equipment.

8.4.3 Prevent CT Open Circuit and PT Short Circuit

The equipment can automatically short circuit the CT circuit and disconnect the PT circuit, and automatically connect to the test section sub-circuit. The operator directly connects the test wiring to the test section, which is convenient and practical, and prevents personnel safety risks caused by the opening of CT and PT short circuit during the test.

8.5 Conclusion

By studying the fault monitoring method and monitoring components of secondary circuit based on primary and secondary fusion switchgear, the specific fault segments of primary and secondary equipment are distinguished, and the related functions of secondary equipment are tested through online monitoring. At the same time, it can realize the protection setting mechanism under the energized operation of the distribution automation switch, and realize the protection test, setting value setting, protection transmission and other functions of the primary equipment on the secondary circuit through the equipment. The automatic CT circuit and PT circuit can also be automatically short circuited and opened through the equipment, so as to avoid switch tripping and personal accidents caused by disconnection errors and CT short-circuit errors. Main technical advantages:

(1) During the fault detection, it can realize the isolation of the primary and secondary circuits and detect the operating state without power failure, and identify the location and section of the fault on the spot.

(2) Strong versatility, suitable for all the ring network elements of the first and second fusion system.

(3) Simple operation, easy to use, to solve the existing equipment test difficult problem, to solve the problem of live adjustment protection value, reduce the difficulty and safety risk of the primary equipment running secondary equipment live test.

References

1. M. Adams, Effectiveness of different design solutions to control internal faults in MV switchgear, IEE Colloquium on Risk Reduction: Internal Faults in T&D Switchgear (Digest No: 1997/295) (1997), pp. 5/1–5/6. https://doi.org/10.1049/ic:19970977
2. Q. Hu, M. Sahni, C. Gibune, S. Liang, W. Lee, Development of an online real time web accessible low-voltage switchgear arcing fault early warning system, in *2007 39th North American Power Symposium* (2007), pp. 20–24. https://doi.org/10.1109/NAPS.2007.4402280
3. D.W. Klaus, Internal faults in distribution switchgear-risk reduction and design considerations, IEE Colloquium on Risk Reduction: Internal Faults in T&D Switchgear (Digest No: 1997/295) (1997), pp. 2/1–2/4. https://doi.org/10.1049/ic:19970974
4. G.A. Hussain, M. Shafiq, M. Lehtonen, Predicting arc faults in distribution switchgears, in *2016 17th International Scientific Conference on Electric Power Engineering (EPE)* (2016), pp. 1–6. https://doi.org/10.1109/EPE.2016.7521722
5. Z. Hailong, J. Song, Z. Dongchen, L. Zhirui, L. Chengpu, W. Hejian, Research on detection system for 10kV distribution switcher of both Primary and secondary technology. Int. Conf. Power Syst. Technol. (POWERCON) **2018**, 3929–3935 (2018). https://doi.org/10.1109/POWERCON.2018.8601599

Chapter 9
Research on FPGA Hardware Acceleration for Real-Time Detection of Mobile Phone Lens Defects

Xidong Wang, Guopeng Wang, Baochang Wang, and Hengtao Wang

Abstract The current machine vision method for mobile phone lens defect detection has the shortcomings of high delay, high power consumption, high cost, and harsh deployment conditions. In this article, optimization strategies such as parameter reordering, dynamic quantization, loop expansion, and block segmentation are adopted to design a software and hardware collaborative processing platform based on FPGA hardware acceleration technology, and real-time detection of mobile phone lens defects is realized. The experimental process is that, firstly, the data set samples of mobile lens defect detection are sent to the YOLOv2 model for iterative training, and the optimal weight is obtained by analyzing the performance curve of the training process. Then, the YOLOv2 model is transplanted to the PYNQ-Z2 development board. Finally, the real-time detection of the validation set samples is accelerated by thc integrated neural network acceleration IP of the platform. The experimental results show that the FPGA complete real-time detection of an image takes 2 s, and the total power consumption is 2.953 W, which is 97 times higher than that of CPU. The positioning accuracy of the defect is high, and the detection accuracy is 90.80%. It meets the real-time detection requirements of low cost, low power consumption, and convenient deployment of small mobile terminals.

9.1 Introduction

With the vigorous development of convolutional neural network in the field of target detection and image recognition [1], the target detection model YOLO has become a hotspot in application research with the advantages of fast algorithm implementation, flexibility, and excellent generalization performance [2].

X. Wang (✉) · G. Wang · H. Wang
College of Computer and Information, China Three Gorges University, Yichang 443002, China
e-mail: xdwang@ctgu.edu.cn

B. Wang
College of Science, China Three Gorges University, Yichang 443002, China

G. Liu and F. Cen (eds.), *Advances in Precision Instruments and Optical Engineering*, Springer Proceedings in Physics 270, https://doi.org/10.1007/978-981-16-7258-3_9

FPGA is a semi-custom circuit, which has programmable input and output units, embedded RAM, dedicated wiring resources, configurable logic blocks. FPGA parallelizes massive data to achieve accelerated processing, and the power consumption is extremely low. Compared with CPU, FPGA has unique advantages in accelerating convolution neural network algorithm [3], which is conducive to deployment to small mobile devices.

At present, most of the domestic mobile phone lens manufacturing factories use artificial visual method to detect lens defects. This method has low efficiency and high false detection rate, which is susceptible to the subjective factors of the test workers, and the labor cost is also high [4], which does not adapt to the current automation trend. In recent years, with the development of machine vision technology, its application in lens defect detection has gradually deepened [5]. Zhu et al. used blackbody as a dark background to improve the contrast of lens defects, and used image analysis method for detection. The detection method takes 5 s per piece [6]. Zhu optimized the point source projection lighting scheme of precision microscope and other instruments to achieve the optimal imaging effect of surface defects, analyzed the shape and size of lens defects, and wrote corresponding image processing algorithms for each defect type to identify. The cost of this method was high [7]. Zhu trained a degenerated YOLO network based on deep learning, and realized the PC-controlled detection system to identify various defects in resin lens samples. The detection power consumption of this method was high [8].

On the basis of the above research on detection methods, this article adopts optimization strategies such as parameter reordering, dynamic quantization, loop expansion and block segmentation to design a software and hardware collaborative processing platform based on FPGA hardware acceleration technology, and real-time detection of mobile phone lens defects is realized. The experimental results show that the platform has low real-time detection delay, low power consumption, low cost, high accuracy, and high defect positioning accuracy, which proves the effectiveness and practicability of the accelerated detection of this platform.

9.2 Material and Methods

9.2.1 Selection of Network Model

The R-CNN series algorithm is a two-stage structure network, and its detection speed cannot meet the real-time requirements. YOLOv1 [9] is not suitable for the defect detection of mobile phone lenses due to the immature prediction frame mechanism and loss calculation, which leads to the low positioning accuracy of the target with small feature size. YOLOv2 [10] is a single-stage structural network proposed by Redmon et al. in 2017. Its core idea is to transform the target detection problem into a classification regression problem [11], and introduce the Anchor mechanism to improve the recall rate in the process of model training and enhance the classification ability of fine-grained image features. It is suitable for detecting targets with small feature size fast detection speed, which can achieve real-time detection of targets.

Due to the low cost of FPGA development board and the relatively limited computing resources, YOLOv2 with high detection accuracy for small size targets, low real-time detection delay, and lightweight network structure is selected.

9.2.2 YOLOv2 Network Structure

The overall structure of YOLOv2 network is shown in Fig. 9.1.

It can be seen from Fig. 9.1 that YOLOv2 network is composed of CBL, MCN, Route, and Reorg. CBL is the smallest component in the YOLOv2 network structure, which is composed of Conv, BN, and Leaky ReLU. The function of Conv (convolution operation) is to realize feature extraction and obtain feature map. Batch normalized BN layer is a linear transformation of feature image elements to accelerate the training convergence process [12]. The activation function is to make a nonlinear transformation of the feature image elements to enhance the nonlinear fitting ability of the network. Except that the final output layer uses Sigmoid, the rest are Leaky ReLU, as shown in (9.1).

$$f(x) = \begin{cases} x \times 0.1, & x < 0 \\ x, & x \geq 0 \end{cases} \tag{9.1}$$

MCN is a down-sampling component, which is composed of the maximum pooling layer MAX and N CBLs. The function of the maximum pooling layer is to realize the down-sampling of the feature map, reduce its size and keep its functional characteristics unchanged.

Route is the routing layer, which integrates multi-dimensional feature information.

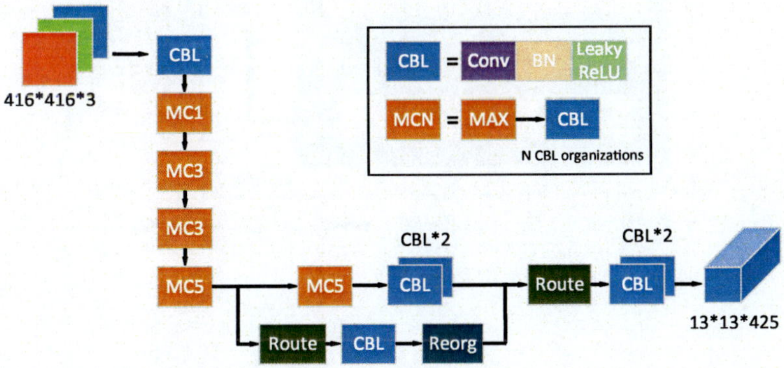

Fig. 9.1 Overall structure of YOLOv2 network

Reorg is a reordering layer that enables reordering of features. The large size feature map is divided into several small size feature maps.

9.2.3 Accelerator Framework

In the acceleration scheme, the part with large amount of calculation and redundant calculation structure in the neural network is implanted into FPGA, and the dynamically configurable neural network acceleration IP is designed. The PS terminal stores the structural parameters of the neural network, and the calculation results are obtained by calling the neural network acceleration IP. The overall framework of neural network accelerating IP design is shown in Fig. 9.2.

It can be seen from Fig. 9.2 that the input and output modules are designed in the FPGA, and the convolution calculation, pooling calculation and reorder calculation are encapsulated in the IP module, and the AXI-lite and AXI-full interfaces are encapsulated. The PS terminal configures the calculation unit through AXI-lite, and the weight parameters and sample feature data are introduced into the convolution neural network to accelerate the IP through AXI-full, and then the calculated results are received by AXI-full.

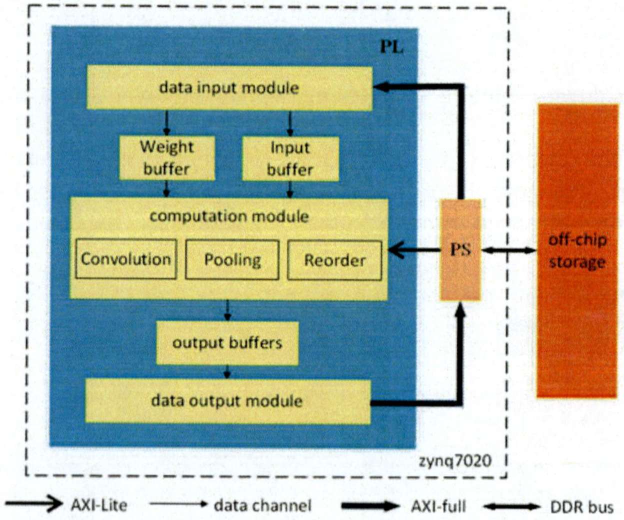

Fig. 9.2 Neural network accelerating IP framework

9.2.4 Integrated Convolutional Neural Network IP

Input and Output Module

The input and output module is used as the main device, and its interface design adopts AXI-full interface, and DDR3 at the PS terminal is used as the slave device. Considering the limited BRAM resources of FPGA, the idea of loop block segmentation is used for expansion, as shown in (9.2) and (9.3).

$$Tci = (Tco - 1) \times S + K \tag{9.2}$$

$$Tri = (Tro - 1) \times S + K \tag{9.3}$$

Among them, Tci and Tri are the size of input feature graph, Tco and Tro are the size of output feature graph, S is the step size of convolution kernel, K is the size of convolution kernel. The process of expanding the input and output modules in depth is shown in Fig. 9.3.

In Fig. 9.3, Tn is the input depth, and Tm is the output depth obtained. The buffer design of double buffers and pipeline operation are adopted to increase the throughput of data and avoid the performance bottleneck for the subsequent convolution acceleration module. The pipelined input and output structure of double buffer is shown in Fig. 9.4.

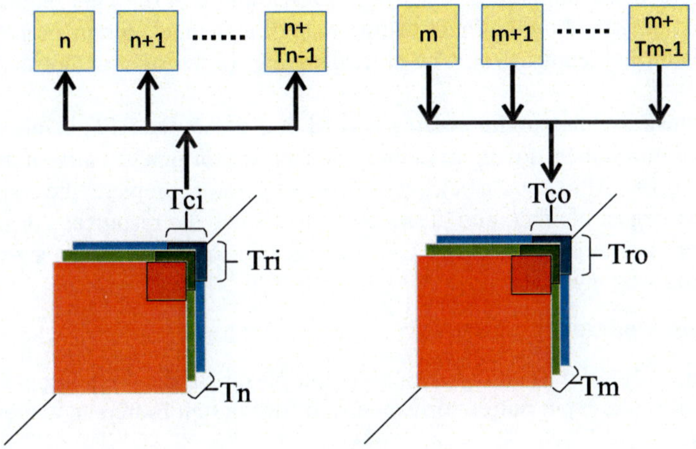

Fig. 9.3 Input and output block segmentation diagram

Fig. 9.4 Pipelined input and output structure of double buffer

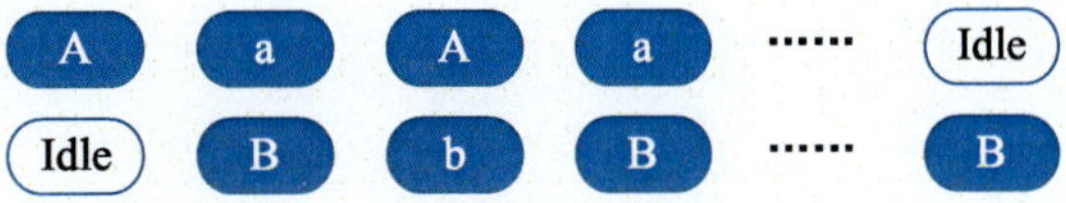

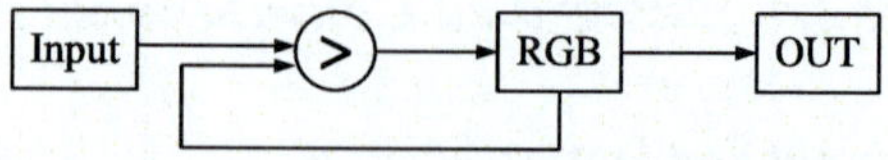

Fig. 9.5 Hardware structure corresponding to maximum pooling

Figure 9.4 for the input structure, A is to pass data from DDR3 to cache Buffer1, a is to pass data from DDR3 to cache Buffer2, B is to pass data from Buffer1 to on-chip cache, and b is to pass data from Buffer2 to on-chip cache. Figure 9.4 for the output structure, A is to pass the data from the on-chip output cache to the cache Buffer1, a is to pass the data from the on-chip output cache to the cache Buffer2, B is to pass the data from Buffer1 to DDR3, and b is to pass the data from Buffer2 to DDR3.

Convolution Module

Convolution calculation occupies the vast majority of the calculation in the neural network, and its essence is multiplication and accumulation. In order to improve the parallelism, the two optimization strategies of expansion and block segmentation are adopted. The input and output feature maps are expanded in two dimensions. Each time, Tm output feature maps and Tn input feature maps are parallelly calculated. The input feature maps are divided into blocks, and only one block is calculated each time. Reduce FPGA requirements for BRAM, reuse computing data, reduce the number of read and write data from off-chip storage.

Maximum Pooling Module

Consider the maximum pooling as a special convolution. It does not require weight parameters, only pooling the input feature map of a channel, and its corresponding operation unit is a comparator. The corresponding hardware structure is shown in Fig. 9.5.

When the 2×2 maximum pooling is used in Fig. 9.5, the REG is initially zero. When the four numbers are input, in turn, the comparator inputs a large number into the REG register. After 2×2 clock cycles, the comparator completes the comparison, outputs the largest number, and clears the REG. Since the resources consumed by comparators are mostly LUT, the pooling design adopts the block segmentation optimization strategy.

Reordering Module

Reordering within FPGA can be seen as a multiplexer, as shown in Fig. 9.6.

In Fig. 9.6, one input buffer corresponds to four output buffers to achieve 2×2 reordering.

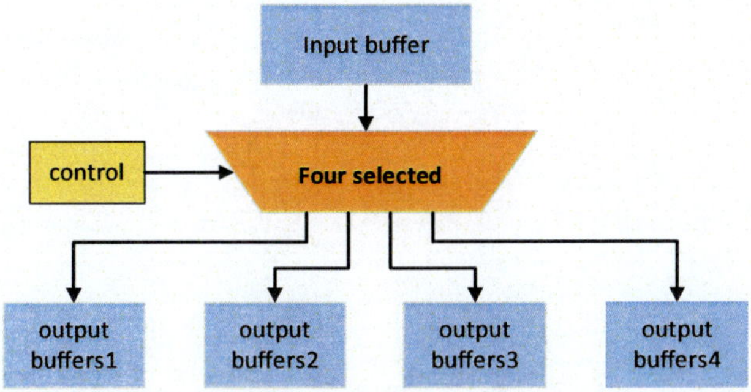

Fig. 9.6 The corresponding structure of reordering

9.2.5 Production and Training of Defect Lens Data Set

The four common defects (petal injury, glue hole, crack, membrane crack) of mobile phone lens are selected as the research objects to make data sets. The labeling software is used to mark mobile phone lens defect detection dataset. Due to the limited number of actual samples, the number of samples is expanded through data enhancement (rotation transformation, affine transformation, image enhancement, and noise addition). In this way, the purpose of increasing the number of samples is realized, the direction and size of defects are changed, and the robustness of image noise is enhanced [13].

YOLOv2 training network parameters: In order to balance the training effect and the pressure of memory occupation, the number of pictures sent to the network in batches at each iteration is 64, and the number of copies sent to the trainer in batches at each iteration is 8. In order to improve the training speed and avoid overfitting, the momentum constant is 0.9 and the weight attenuation regularization coefficient is 0.0005. With the increase of iterations, in order to make model learning more effective, the initial learning rate is 0.001, and the total number of iterations is 200. The learning rate adjustment strategy is steps. When the number of iterations reaches 160 and 180, the learning rate is reduced to 0.0001 and 0.00001, respectively. The number of categories in region layer is changed to 4, and the number of convolution kernels in the last convolution layer is changed to 45.

In the training process, the change curves of Loss and recall rate of the model are shown in Fig. 9.7a, b, respectively.

It can be seen from Fig. 9.7a that when the training model is iterative to 90 times, the loss function converges to 0.01. It can be seen from Fig. 9.7b that when the training model is iterative to 70 times, the recall rate is close to 1. Based on the analysis of training process performance curve, the weight effect of YOLOv2 network model is ideal. Use right re-results detect the samples of the validation set, and the results are shown in Table 9.1.

Table 9.1 shows that the model detection accuracy is 96.13%, which proves that the training effect of the model is ideal and can be transplanted in the next step.

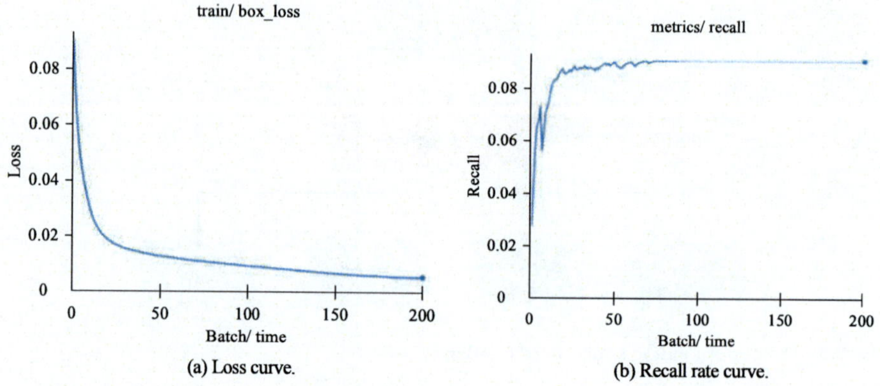

Fig. 9.7 Change curve of training process

Table 9.1 Test results of model validation set

Class	Number	Accurate number	Accuracy (%)
Petal injury	1152	1092	94.79
Glue hole	960	928	96.67
Crack	192	192	100
Membrane crack	96	95	98.96
Total	2400	2307	96.13

9.2.6 *Transplantation of YOLOv2*

Rewrite YOLOv2 feature extraction network darknet framework source code, the weight file is separated into two files of weight and offset parameters. Parameter mapping is shown in (9.4).

$$y = A \times X + B \tag{9.4}$$

In the above equation, A and B are mapping coefficients. The weight and offset parameter expressions of convolution kernel are shown in (9.5) and (9.6).

$$\text{weight}_i = \text{weight}_i \times A_i \tag{9.5}$$

$$\text{bias}_i = \text{bias}_i \times A_i + B_i \tag{9.6}$$

In the above equations, where A_i and B_i are the mapping coefficients of characteristic figure i. Store the mapped values in binary form for the separated files, reducing the amount of computation and increasing the computation speed [14].

Quantification of Data

Before quantization, each parameter in the weight and offset data occupies 32 bits, which requires high bandwidth of the transmission bus and indirectly limits the operation rate. In this article, the dynamic fixed-point 16-bit quantization [15] is used to reduce the bit width of the parameters. The optimal order of weight for each layer is searched through traversal, as shown in (9.7).

$$\exp_w = \arg\min \sum_{i=0}^{n} \left| w^i_{float} - w^i_{(bw,\exp_w)} \right| \tag{9.7}$$

In the above equation, $\exp_w$ is the order code of the minimum sum of the absolute value of the loss accuracy of all parameters in this layer after quantization, n is the number of parameters to be quantified in this layer,w^i_{float} is the original floating point value of the first parameter, $w^i_{(bw,\exp_w)}$ is the number of fixed points of the first parameter under the bit width bw and the order code $\exp_w$, and then the number of floating points is converted.

Weight Reordering

When the reordering module is designed, the idea of block segmentation is used to optimize the convolution operation. When the designed algorithm is transformed into RTL circuit by HLS, the simulation results show that the weight parameters are stored separately, resulting in the burst transmission of AXI-full, the size is only K*K, which cannot make full use of the transmission bandwidth of DRAM. In order to make full use of the transmission bandwidth, the weight parameters need to be reordered, as shown in Fig. 9.8.

PS and PL Terminal Collaborative Real-time Detection

The functional design of PS terminal is mainly to store the neural network structure, drive the USB camera, call the integrated neural network to accelerate IP, preprocess the images collected by the camera, and post-process the final output of the integrated

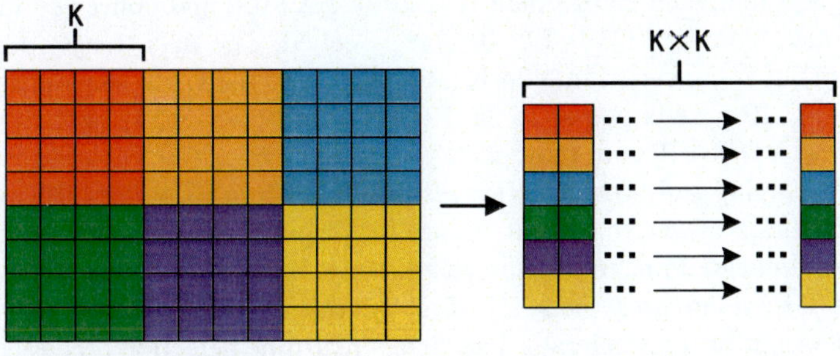

Fig. 9.8 Weight parameter reordering

Fig. 9.9 PS and PL terminal synergy framework

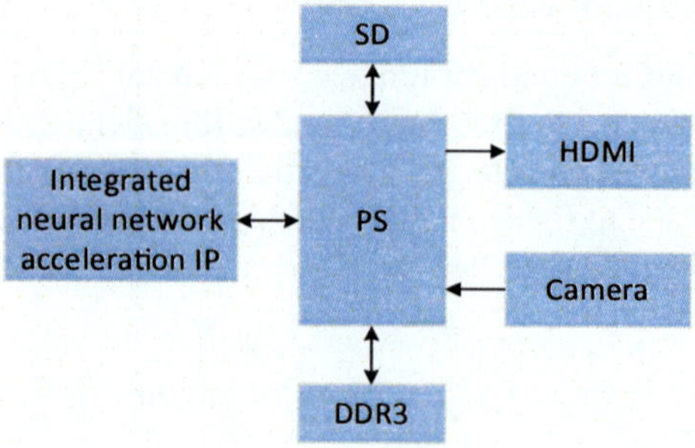

neural network to accelerate IP. The coordination design framework is shown in Fig. 9.9.

It can be seen from Fig. 9.9 that the PS terminal first reads the weight and offset files stored in the SD card and puts them to the external memory DDR3. Then, it drives the USB camera to collect real-time video frames, caches the collected images, and then preprocesses them, mainly including the shaping of the collected images and pixel normalization. Finally, it reads the neural network structure, and configures the integrated neural network to accelerate IP according to the input and output layers and order codes.

OpenCV is called to drive the camera, and its built-in function will cache 4 frames of video. In order to solve the problem that continuous acquisition will cause non-real-time video frames, set the program segment to delete the cache frame, which will affect the delay of lens detection.

In order to simplify the detection platform, HDMI is used to display, DIGILENT's RGB2DVI open-source IP core is used to complete data bit conversion, and XILINX's VTC IP core is used to control video timing. In order to solve the problem of video timing, the original data of the video should be input. After the USB camera collects the image data, the data are stored in the external memory DDR3. Since the DDR3 controller is on the PS terminal, the video data need to be transmitted to the PL terminal through the AXI-full interface. In order to improve the video transmission rate and simplify the workload of the PS terminal, the VDMA IP core is called on the PL terminal. The DDR3 data is read by AXI-full and converted into AXI-stream format. The AXIS2VIDEO IP core is called. The data transmitted by the AXI-stream bus is converted into 24-bit RGB data and the video sequence received from the VTC IP core is output to the RGB2DVI IP core. AXIS2VIDEO IP core interacts with VDMA IP core through AXI-stream bus.

When displaying real-time detection results, taking into account the PS terminal's limited computing power, in order to reduce the display time, the ipywidgets library is used. The processed image is displayed directly. If there are defects in the image, it is stored. If there are no defects, it is directly covered to the next frame.

The FPGA hardware accelerator lens defect real-time detection process is shown in Fig. 9.10.

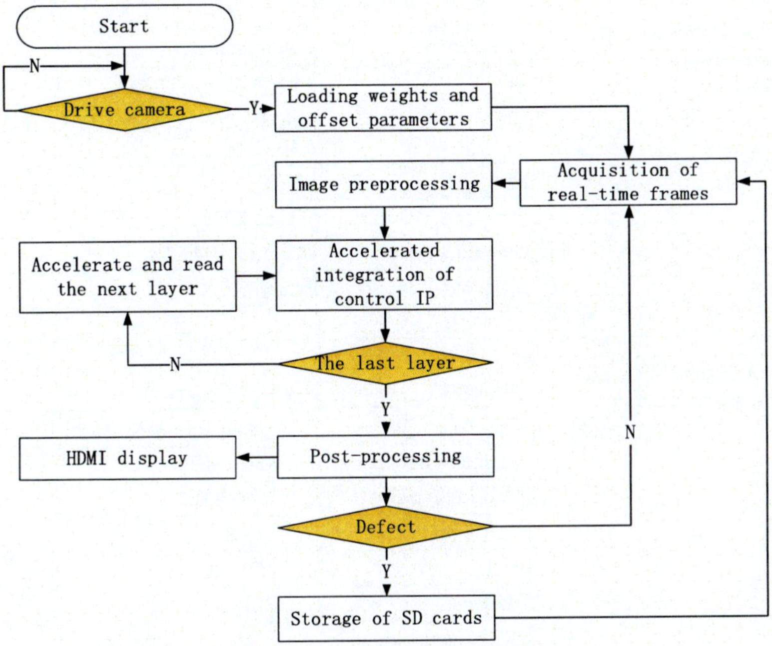

Fig. 9.10 FPGA hardware acceleration lens defect real-time detection

In Fig. 9.10, the convolution layer, the maximum pooling layer, and the reordering layer are accelerated in the integrated neural network acceleration IP according to the division of labor. The routing layer records the address information of input and output. The post-processing is mainly the final detection layer, and the PS terminal is used to realize the positioning and category calibration of the lens defect prediction box.

9.3 Results

9.3.1 Hardware and Software Platform Environment

The hardware and software platform environment used in the experiment is shown in Table 9.2.

This experiment uses TUL's PYNQ-Z2 development board based on zynq7020 chip. The main resource allocation of zynq7020 chip is shown in Table 9.3.

PYNQ-Z2 supports the open-source framework PYNQ, which is simply understood as combining Python with ZYNQ. The library module of Python language can be called under the framework to achieve efficient embedded development and provide full play to the advantages of software and hardware collaboration

Table 9.2 Hardware and software platform configuration

Platform	Configuration
FPGA	PYNQ-Z2(zynq7020)
CPU	Intel(R) Core(TM) i5-10,500 CPU @ 3.10 GHz
Frame	PYNQ
Development tools	Vivado 2020
Operating system	Ubuntu 20.04 LTS
Language	Python 3.9.5

Table 9.3 Zynq7020 partial main resource allocation

Parameter	Configuration
ARM chip	Dual-core ARM Cortex-A9
Maximum frequency	667 MHz
RAM in the SCM	256 KB
Logical unit	85 K
LUT	53.2 K
Trigger	106 K
DSP computing unit	220
BRAM36K	140

[16]. Although RTL programming cannot be directly carried out in Python environment, the PYNQ open-source framework provides convenience for transplanting the YOLOv2 model.

9.3.2 *Real-Time Detection*

Using the panel display mobile phone lens defect verification set sample for real-time detection, the effect is shown in Fig. 9.11.

Figure 9.11a shows two petal defects in the model correct box, Fig. 9.11b shows one glue hole defect in the model correct box, Fig. 9.11c shows two crack defects in the model correct box, and Fig. 9.11d shows one membrane crack defect in the model correct box. The data above the target box are labels for the prediction category. From the test results, the model can predict and distinguish four kinds of defects. At the same time, in the case of multi-defect coexistence, there is basically no defect missing.

In the process of real-time detection, the adjustment of camera focal length directly affects the image quality of real-time video frames, which indirectly affects the detection results. The real-time detection log is shown in Fig. 9.12.

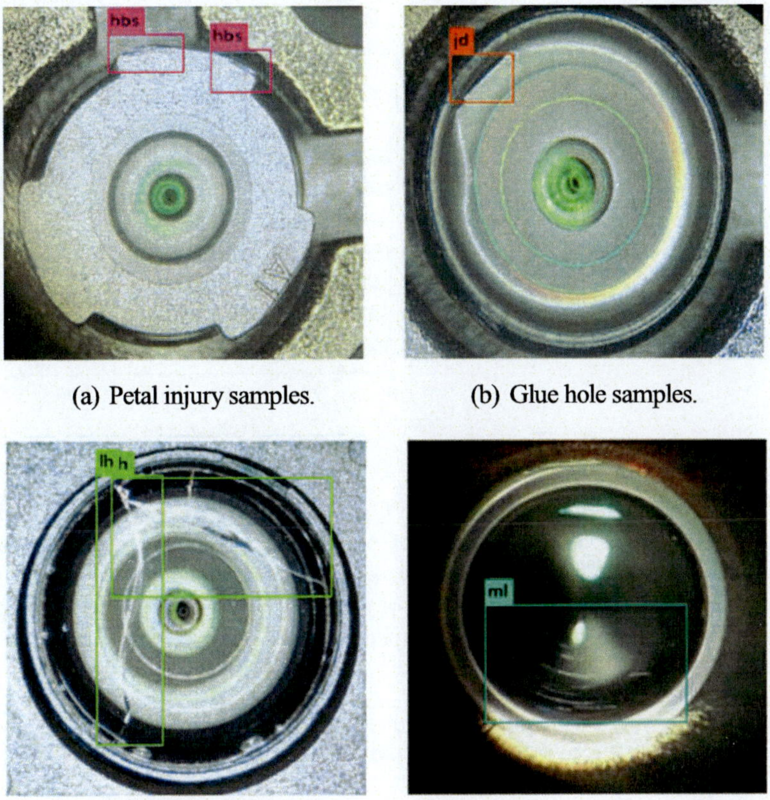

(a) Petal injury samples. (b) Glue hole samples.

(c) Crack samples. (d) Membrane crack samples.

Fig. 9.11 Real-time detection effect

```
Open pictrue success!
pictrue size: (640, 480)
pictrue mode: RGB
yolov2_image copy ok

FPGA_Accelerate_Completed!!
845
name: jiaodong 65.2170420121
fpga_process_time          : 0.997957706451416
```

Fig. 9.12 Real-time detection log

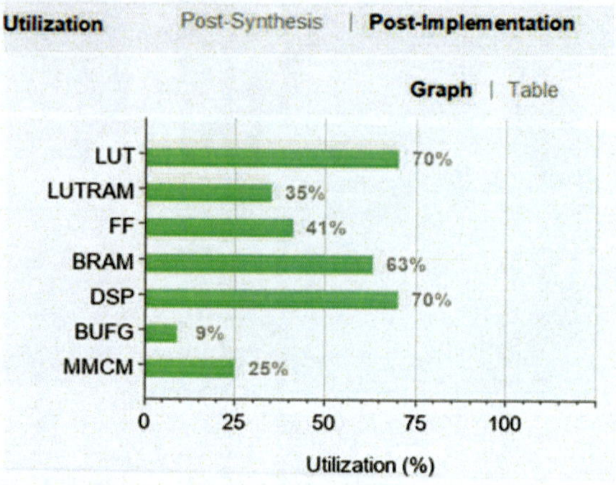

Fig. 9.13 Hardware resource consumption

It can be seen from Fig. 9.12 that the FPGA acceleration process can be completed within 1 s. With the time-consuming of pre-processing, post-processing and PS terminal real-time detection and display results, it can be concluded that it takes 2 s for FPGA to complete the detection of a real-time image by detecting multiple samples and taking the average value respectively.

9.3.3 Hardware Resource Consumption

The hardware resource consumption accelerated by porting the YOLOv2 network to the PYNQ-Z2 development board is shown in Fig. 9.13.

From Fig. 9.13, BRAM and DSP consume the most hardware resources. The reason for BRAM consumption is that the weight parameters and input–output feature maps are stored every time. The reason for DSP consumption is that in order to maximize the acceleration effect of a large number of multiplication and addition operations in the neural network, multiplication and addition operations are performed in parallel.

9.3.4 Detection Power Consumption

The power consumption comprehensive evaluation report of PYNQ-Z2 development board running YOLOv2 model on Vivado platform for lens defect detection is shown in Fig. 9.14.

It can be seen from Fig. 9.14 that the static power consumption is 0.198 W, the dynamic power consumption is 2.775 W, and the total power consumption is 2.953 W. It can be seen that the power consumption of FPGA is extremely low.

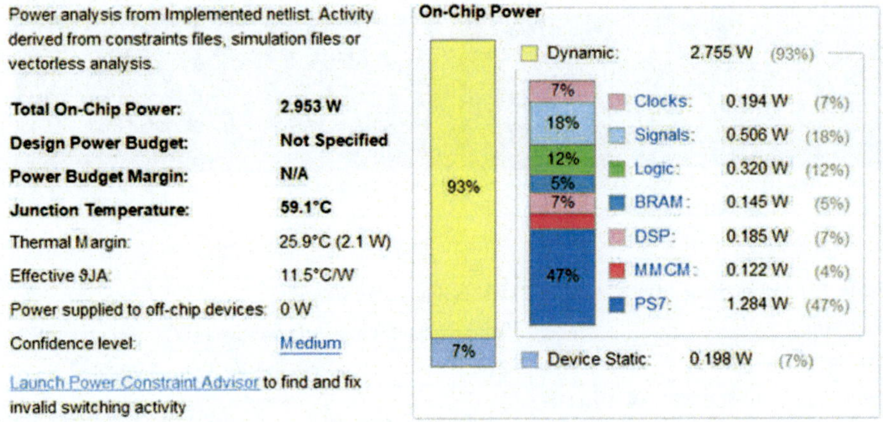

Fig. 9.14 Power consumption synthesis report

Table 9.4 The verification results of lens detection under FPGA

Class	Number	Accurate number	Accuracy (%)
Petal injury	240	208	86.7
Glue hole	200	187	93.5
Crack	40	40	100.0
Membrane crack	20	19	95.0
Total	500	454	90.8

9.3.5 Detection Accuracy

The focal length is adjusted. After the imaging is clear and stable, 500 samples of validation set are detected in real time. The statistical verification results are shown in Table 9.4.

The analysis of Table 9.4 shows that the accuracy of FPGA real-time detection is 90.80%, and the lowest accuracy is petal injury. Compared with other defects, petal injury has smaller feature size and smaller difference between feature shape and background, which leads to relatively simple feature extraction and low detection rate.

9.4 Discussion

The description of CPU detection environment in YOLO official network is shown in Fig. 9.15.

From Fig. 9.15, YOLO official CPU environment detection of an image takes 6–12 s.

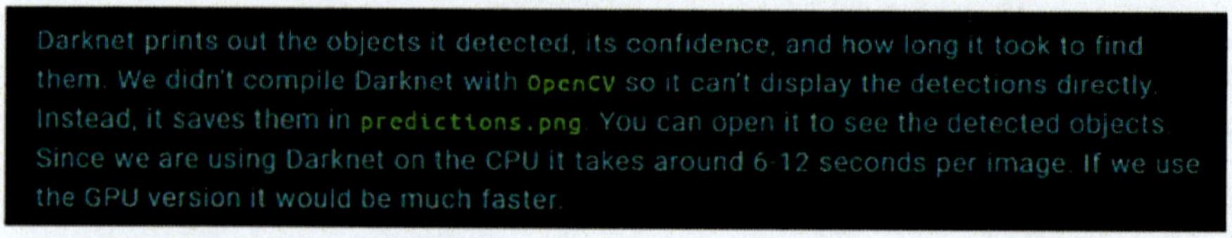
Darknet prints out the objects it detected, its confidence, and how long it took to find them. We didn't compile Darknet with OpenCV so it can't display the detections directly. Instead, it saves them in predictions.png. You can open it to see the detected objects. Since we are using Darknet on the CPU it takes around 6-12 seconds per image. If we use the GPU version it would be much faster.

Fig. 9.15 YOLO official description

Table 9.5 Comparison of FPGA and CPU detection performance

Platform	Time-consuming (s)	Power loss (W)	Accuracy (%)
PYNQ-Z2(zynq7020)	2	2.953 W	90.80
Intel Core i5-10,500 CPU @ 3.10 GHz	7	81 W	96.13

FPGA versus CPU detection performance is shown in Table 9.5.

The analysis of Table 9.5 shows that the FPGA hardware acceleration lens defect real-time detection acceleration rate is 350% of the CPU environment, FPGA acceleration effect is significant. FPGA has excellent advantages in power consumption. For long-running equipment or instruments, it is more energy-saving to choose FPGA. The detection accuracy of FPGA is slightly lower than that of CPU environment. The main reason is that the weight and offset parameters are quantified by dynamic fixed-point 16 bits, and the decrease of bit width leads to the loss of accuracy and the decrease of accuracy.

Table 9.6 compares the defect detection performance of this platform with the other three methods in references.

It can be seen from Table 9.6 that the platform in this article is superior to the other three machine vision methods in terms of detection delay, power consumption, cost and application scenarios, and meets the requirements of real-time detection of lens defects with low cost and low power consumption.

Table 9.6 Method performance comparison

Method	Time-consuming	Power loss	Cost	Application scene
[6]	5 s/piece	High	High	Real-time
[7]	High	High	High	Static state
[8]	High	High	High	Static state
This article	2 s/piece	Lower	Lower	Real-time

9.5 Conclusions

In order to solve the problems of high delay, high power consumption, high cost and harsh deployment conditions in current machine vision method for mobile phone lens defect detection, and meet the requirements of low delay, low power consumption, high accuracy and strong stability in real-time application scenarios of small mobile devices. In this article, optimization strategies such as parameter reordering, dynamic quantization, loop expansion, and block segmentation are adopted to design a soft and hard collaborative processing platform based on FPGA hardware acceleration technology, and real-time detection of lens defects is realized. The experimental results show that the FPGA complete real-time detection of an image takes 2 s, which is 3.5 times faster than the CPU performance, and the total power consumption is 2.953 W, which is equivalent to 3.6% of the CPU, and the performance power consumption ratio is increased by 97 times. The defect location accuracy of the detection method is high, and the detection accuracy is 90.80%, which proves that the detection platform has certain advantages in performance.

Limited by time and energy, this platform has the following limitations. Firstly, the current FPGA detection accuracy is 5.33% lower than that of CPU. Subsequently, the dynamic fixed-point 16-bit quantized model can be iteratively optimized by multiple trainings to reduce the loss of detection accuracy and control it within an acceptable range. Secondly, the post-processing flow in the real-time detection process of the platform is not accelerated by FPGA hardware, and the time consumption is about 0.5 s. The post-processing flow can be accelerated in FPGA, which can reduce the real-time detection delay to a certain extent.

References

1. O. Russakovsky, J. Deng, H. Su et al., ImageNet large scale visual recognition challenge. Int. J. Comput. Vision **115**(3), 211–252 (2015)
2. B. Wang, H.X. Le, W.J. Li, M.H. Zhang, Oil tank detection algorithm on remote sensing image using multi-scale parallel convolutional neural networks. Comput. Eng. Appl. **57**(08), 62–69 (2021)
3. X.F. Wang, P.L. Jiang, H. Zhou, X.B. Zhao, Design of FPGA accelerator with high parallelism for convolution neural network. Comput. Appl. **41**(03), 812–819 (2021)
4. L. Sun, C. Liu, H.B. Yao, Automatic detection of water-mask for resin glasses by machine vision. J. Jiangsu Univ. **39**(04), 425–430 (2018)
5. L. Lin, T.G. Zang, Edge detection of defect for resin lenses based on machine vision. Mach. Build. Automat. **50**(02), 230–232+240 (2021)
6. Y.D. Zhu, Y.X. Chen, Visual inspection method of detecting flaws on optical glasses surface. Appl. Opt. **41**(03), 553–558 (2020)
7. J.F. Zhu, Research on surface defect detection of resin lens based on machine vision technology. Jiangsu Univ. **2019**, 1–75 (2019)
8. H.Z. Zhu, The research on resin lens defect recognition method based on degenerate YOLO network. Harbin Inst. Technol. **2019**, 1–66 (2019)
9. J. Redmon, S. Divvala, R. Girshick et al., You only look once: unified, real-time object detection, in *Computer Vision & Pattern Recognition* (Las Vegas, 2016), pp. 779–788

10. J. Redmon, A. Farhadi, YOLO9000: better, faster, stronger, in *Computer Vision & Pattern Recognition* (Hawaii, 2017), pp. 6517–6525.
11. S.S. Xin, W.L. Zhao, Overview of UAV target detection in complex scenes based on Yolo series algorithms. Appl. Res. Comput. **37**(S2), 28–30 (2020)
12. J.W. Liu, H.D. Zhao, X.L. Luo, L. Xu, Research progress on batch normalization of deep learning and its related algorithms. Acta Automatica Sinica. **46**(06), 1090–1120 (2020)
13. B. Yu, A data augmentation method for image class imbalance problem using generative adversarial networks. South China Univ. Technol. **2018**, 1–73 (2018)
14. Y. Cui, G.H. Shu, D. Li, Research on binary neural network classification models based on the PYNQ development board. Electr. Automat. **041**(005), 53–56 (2019)
15. S. Lei, M. Zhang, D. Lin et al., A dynamic multi-precision fixed-point data quantization strategy for convolutional neural network, in *CCF National Conference on Computer Engineering and Technology* (Singapore, 2016), pp. 978–981
16. H.J. Liu, Y.J. Yue, Design of binocular vision ranging system based on pynq platform. Electron. World. **16**, 184–186 (2020)

Chapter 10
An Improved Phase Unwrapping Method Based on Geometric Constraints

Liu Yang and Pan Ya

Abstract In the field of three-dimensional measurement of structured light, the phase unwrapping method based on geometric constraints has the characteristics of high precision. This method does not need to acquire additional marking images, nor does it need to use expensive hardware equipment, and can achieve rapid and accurate phase unwrapping with the help of a reference plane, which can meet the practical requirements of fast and high precision 3D measurement. However, the 3D measurement method based on geometric constraints has a problem of limited measurement depth, and the maximum measurement depth range is within the phase domain of 2π. If the height of the measured object is beyond this range, normal phase unwrapping cannot be performed. In this paper, the target image is segmented according to the boundary discontinuous feature of the region beyond the measurement range, and then the relative phase unwrapping algorithm based on modulation intensity sorting is used for this region. Finally, the correct absolute phase information of the measured target is recovered. Experiments show that this method can not only increase the depth of measurement but also has better robustness for low phase mass region. The validity and correctness of the proposed method are verified by experiments.

10.1 Introduction

3D measurement based on structured light is one of the most important non-contact 3D measurement technologies, which has been widely used in various fields [1]. In recent years, with the rapid development of digital fringe projection technology (DFP), Fourier transformation profilometry (FTP) [2], wavelet analysis method [3] and phase shift technique (PSP) [4] have been widely used in the acquisition of 3D information of objects. However, the fringe analysis method mentioned above only produces enveloped phases containing discontinuities. To obtain a continuous phase diagram, the wrapping phase needs to be unwrapped. In general, phase unwrapping

L. Yang · P. Ya (✉)
Southwest University of Science and Technology, Mianyang 621010, Sichuan, China
e-mail: panya@swust.edu.cn

G. Liu and F. Cen (eds.), *Advances in Precision Instruments and Optical Engineering*, Springer Proceedings in Physics 270, https://doi.org/10.1007/978-981-16-7258-3_10

methods are divided into time phase unwrapping method (TPU) [5], space phase unwrapping method (SPU) [6] and the method of adding additional hardware [7]. The spatial phase unwrapping algorithm mainly confirms the discontinuity point of 2π through the inspection of adjacent pixels. And eliminate the discontinuity by adding or subtracting multiple integers of 2π. The former spatial phase unwrapping method is not robust enough to the discontinuous or highly abrupt regions on the surface of the object. Time-phase unwrapping algorithms, such as binary coding method [8] and multi-frequency phase-shifting method [9], confirm the fringe order through the recognition of additional markers. However, the projection of the auxiliary fringe needs extra time, so the traditional time phase unwrapping can not meet the requirements of fast 3D shape restoration. Researchers have proposed to solve this problem by adding a second camera or projection equipment [10]. This method combines polar geometry with stereo vision, and uses phase information as the basis for recognition of the corresponding points of the left and right cameras to achieve 3D information acquisition. However, the calculation speed of this method is slow and the second camera increases the complexity and cost of the system. In the case of no auxiliary projection mode and additional hardware, An et al. [11] proposed a per-pixel absolute phase unwrapping method based on geometric constraints, which used geometric constraints and known phase distribution of reference plane to unwrap the phase of the target. This method has the characteristics of fast and high precision. Compared with other phase unwrapping methods, the method based on geometric constraints does not need additional auxiliary images, and can effectively improve the speed of the 3D measurement system. However, the method based on geometric constraints can only play a role when the measurement range in the phase domain is smaller. It is difficult to measure objects with large depth variations.

To solve the problem of limited measurement range based on geometric constraint phase decomposition method, firstly, the target image is segmented according to the boundary discontinuous features of the region beyond the measurement range, and then the relative phase is unwrapped by using the algorithm based on modulation intensity sorting proposed in this paper, and the correct absolute phase information of the measured target is recovered by converting the relative phase to the absolute phase. Experiments show that this method can not only increase the depth of measurement, but also overcome the discontinuity problem of traditional spatial phase unwrapping. The structure of this paper is as follows. In the first section, the principle of geometric constraint method and the problem of depth limit are discussed. The second section mainly introduces the algorithm principle and process. In the third section, simulation experiments and comparative experiments are carried out to prove the effectiveness of the proposed algorithm. The fourth section is the summary of the full text.

10.2 Measuring Depth Limit Based on Geometric Constraints

The core idea of phase unwrapping method based on geometric constraints is to establish an absolute phase distribution of a known reference plane, and then to unwrap the target phase according to the geometric relationship between the reference plane and the envelopment phase. As shown in Fig. 10.1, the absolute phase of the known reference plane is represented by Φ_0, and the phase of the target to be measured is ϕ. The fringe order of the phase diagram of the target $k(x, y)$ can be obtained by the following formula

$$k(x, y) = ceil[\frac{\Phi_0(x, y) - \phi(x, y)}{2\pi}] \tag{10.1}$$

(x, y) Represents pixel coordinates, and $ceil[]$ represents rounding up operation. Therefore, the absolute unwrapping phase of the target can be obtained as follows:

$$\Phi_m(x, y) = \phi(x, y) + 2\pi \times k(x, y) \tag{10.2}$$

In real measurements, when the projected fringe pattern is projected onto the surface of the object, the phase information changes due to the height modulation of the object. As shown in Fig. 10.2, the red rectangular box represents the phase distribution of the package before the height modulation of the object, and the green rectangular box represents the phase distribution of the package after the height modulation of the object. The phases in the green box in Fig. 10.2 change in the same way as those in the red box, so the same sequence of fringes k will be obtained when unrolled using Formula 10.1. As shown in Fig. 10.2b, there will be an error between the actually unwrapped phase Φ_2 and the correctly unwrapped phase Φ_{2True}. Therefore, the per-pixel phase unwrapping algorithm based on geometric constraints has the limitation of measuring depth, and the height change of the object is limited within the phase domain 2π. Because the part beyond the measurement range will

Fig. 10.1 Target phase unwrapping process

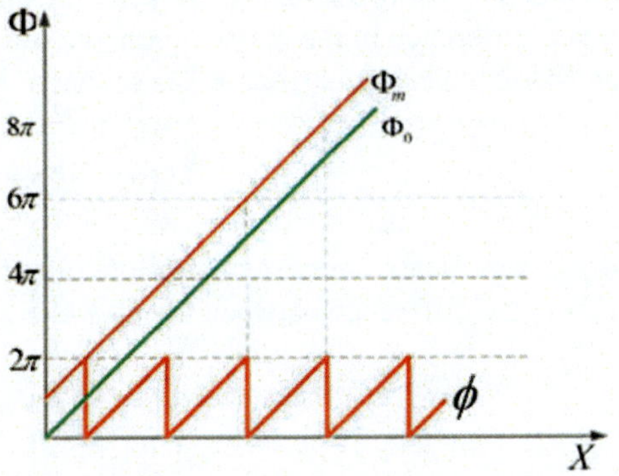

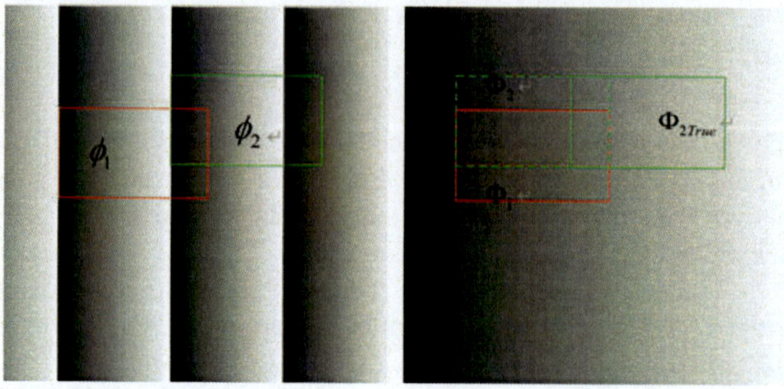

(a) (b)

Fig. 10.2 Description of depth range limits. **a** The window area represents the phase ϕ_1 and ϕ_2 of the package obtained at different depths; **b** ϕ_1 corresponds to Φ_1, ϕ_2 corresponds to Φ_2, Φ_{2True} is the correct unfolding phase of ϕ_2

have obvious discontinuous dividing line in the phase unwrapping image. This paper combines the idea of image segmentation and the method based on modulation intensity sorting to solve this problem.

10.3 Algorithm

The phase unwrapping method in this paper includes the following steps: (1) Establish a reference plane of absolute phase unwrapping. (2) The geometric constraint method is used to phase unwrap the target phase. (3) The Sobel segmentation algorithm is used to extract the correct part of the target phase unwrapping, and the binary *mask* of the region is generated. (4) The algorithm based on modulation intensity sorting proposed in this paper is used to carry out relative phase unwrapping for the region beyond the measurement range of the measured object. (5) The difference between the relative phase and the absolute phase beyond the measurement range can be obtained by using the binary mask, as shown in Formula (10.4). (6) According to the phase difference in the mask region, the relative phase distribution is converted into an absolute phase distribution, as shown in Formula (10.3).

$$\Phi_{True} = \Phi_{relative} + \Delta\Phi \tag{10.3}$$

Φ_{True} represents the correct absolute phase, $\Phi_{relative}$ represents the relative phase. $\Delta\Phi$ is the difference between the absolute and relative phases.

$$\Delta\Phi = Average((\Phi_{geom} - \Phi_{relative}) \cdot *mask) \tag{10.4}$$

$Average()$ represents the mean function, Φ_{geom} represents the phase expanded based on geometric constraints $mask$ represents the mask of the properly expanded part based on the geometric constraint method.

10.3.1 Modulation Intensity-Based Sort Method

In the phase unwrapping of the wrapped phase image, the higher the quality of the phase, the higher the precision of the phase unwrapping. In general, the quality of the phase is positively correlated with the modulation intensity of the point [12]. Based on this relation, this paper uses the idea of sorting and clustering to expand the relative phase of the region beyond the depth range. The steps are as follows: First, the target fringe pattern is scanned globally, and discontinuity detection is performed on the target phase pattern, as shown below

$$\sum_{k=1}^{4} R[w(p_k) - w(p_{k+1})] = 0 \tag{10.5}$$

$R[]$ represents rounding operation, $w(p_k)$ represents the phase of the package, p_k represents the coordinate point. Find the global complement of the set of discontinuous points P_j. The phase points $(w_1, w_2, ...w_m)$ in the set P_b are sorted by shell sorting algorithms according to the modulation intensity $M(w_1) >= M(w_2) >= ...M(w_m)$. Perform the following operation for each package phase point in order.

(1) If none of the neighborhood points surrounding the phase point w_i are processed, then a subset C_i is created for this point, the weight of this point is set as $M(w_i)$ and then the next point in the sequence list is processed.
(2) If there are neighborhood points of w_i that have been processed, they will be added to the subset C_n of neighborhood points. In this case, the weight is set as follows:

$$w_i = \sum_{w_j \in C_n} M(w_j) \tag{10.6}$$

So the weight w_i is equal to the sum of the weights of all the points in the subset. If the difference between the neighborhood point with the largest expanded phase value is greater than π:

$$|\Phi(w_i) - \Phi(w_n)| > \pi \tag{10.7}$$

The following formula is used to correct the points within the subset

$$|\Phi(w_i) - (\Phi(w_n) + 2\pi * \delta_n)| <= \pi, \delta_n \in Z \tag{10.8}$$

(3) If there are multiple neighborhood points belonging to subsets $C_{n1}, C_{n2}, \ldots, C_{n,k}$, the phase offset existing in these subsets needs to be expanded, and the subsets with lower weight need to be merged into the subsets with the highest weight. First, select the neighborhood point with the maximum weight expand as follows:

$$\forall w_{ni} |\Phi(w_i) - (\Phi(w_{ni}) + 2\pi * (\delta_{ni})| <= \pi, \delta_{ni} \in Z \tag{10.9}$$

After unwrapping the phase points in the set of P_b, phase unwrapping of the discontinuous point set P_j should be carried out. For the outermost discontinuity point in the discontinuity region, the untied phase points around the discontinuity point are used as reference to solve the phase of the discontinuity point. As shown in Fig. 10.4, a set of green dots (1, 2, ...5) represents discontinuous points, and a set of dots (6, 7, ...19) represents points that have been expanded. The discontinuity points are expanded in turn as follows:

$$w_n \in p \quad w_j = \min |w_i - w_n| \tag{10.10}$$

$$|\Phi(w_i) - (\Phi(w_j) + 2\pi * (\delta_n)| <= \pi \tag{10.11}$$

First, find the direction with the smallest gradient decline of the phase points in the neighborhood, and expand the phase of the discontinuity points along this direction. The expanded discontinuity points are added to the subset P_b until all discontinuity points are expanded (Fig. 10.3).

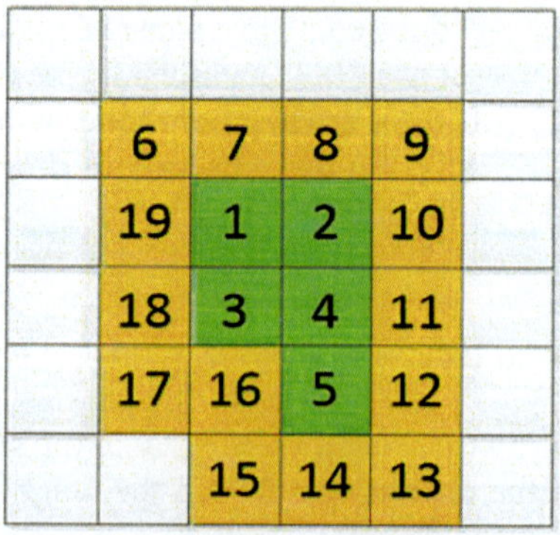

Fig. 10.3 Schematic diagram of phase solution

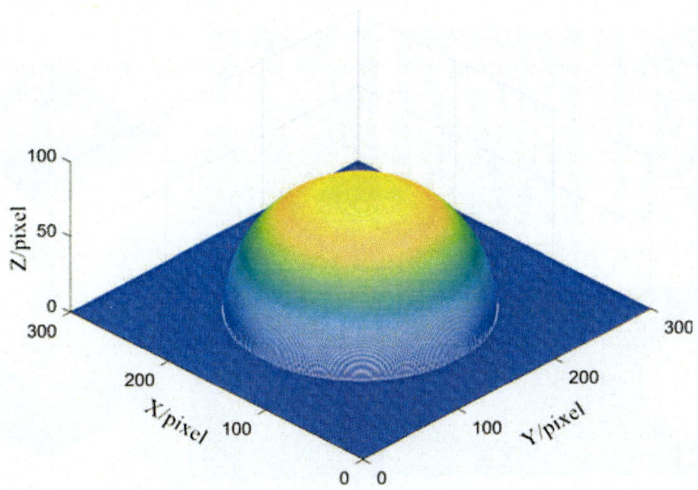

Fig. 10.4 3D view of simulated object

10.4 Experiment

10.4.1 Simulation Experiment

In order to verify the effectiveness of the proposed method, a simulation experiment was designed. Figure 10.4 is a schematic diagram of the equation below

$$Z = 0.8 * sqrt(10000 - (X - 150)^2 - (Y - 150)^2) \tag{10.12}$$

The height is 80, and the spatial frequency of the projected raster is $f = 35$ pixels. The system parameters of the projection model are set as the distance between the exit pupil center of the camera and the reference plane $Len = 20$, and the distance between the projection center and the camera center $d = 20$. According to the model parameters, the phase after modulation is shown as follows:

$$\Delta\Phi = 2 * pi * f * d \cdot * Z/len \tag{10.13}$$

According to the above equation, the maximum measured height in the 2π phase domain is 35. The simulated target is shown in Fig. 10.5. Figure 10.5a is the simulated deformed fringe with a period of 35 pixels. Figure 10.5b is the enveloped phase of Fig. 10.5a, and Fig. 10.5i is the lateral phase distribution of the truncated region in Fig. 10.5c. It can be seen that the phase distribution presents continuous truncation phenomenon due to exceeding the measurement range. Figure 10.5f is the continuous relative phase distribution of the discontinuous region solved by the modulation intensity sorting algorithm. Equation (10.4) is calculated as $\Delta\Phi = 18.8496$ rad, and the

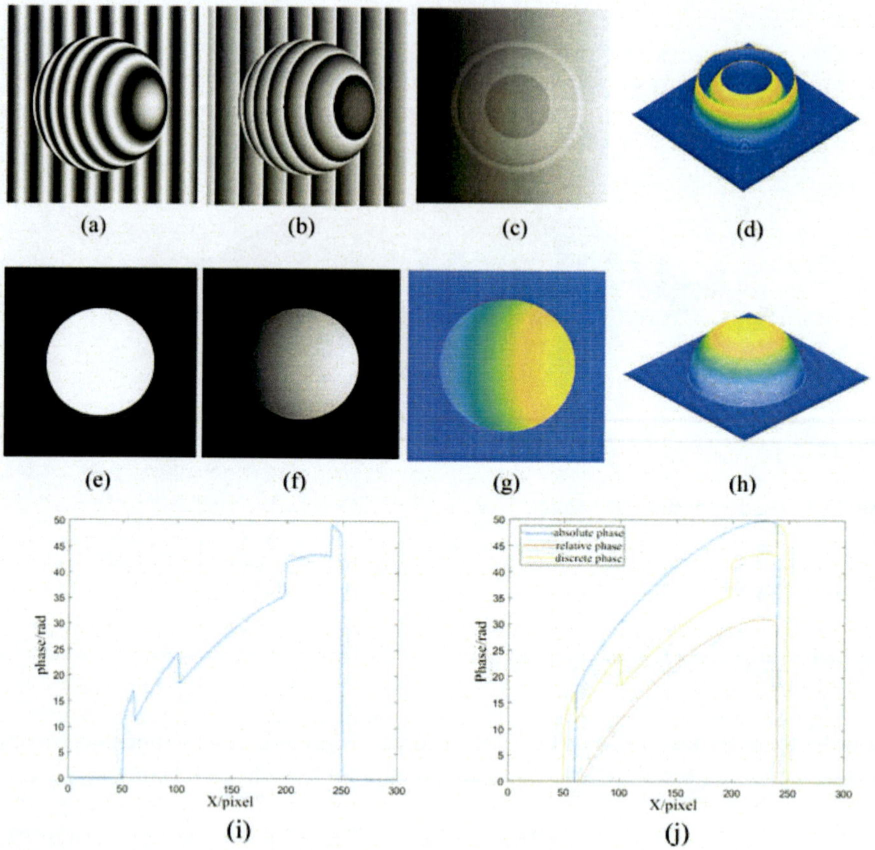

Fig. 10.5 Simulated measurement of sphericity. **a** Simulating deformed projection fringe; **b** Phase diagram of the package; **c** Phase diagrams of geometric constraint method expansion; **d** The reconstruction results were obtained by geometric constraint method; **e** segmentation results based on Sobel algorithm; **f** The relative phase distribution is obtained based on the modulation intensity sorting method; **g** The final absolute phase diagram; **h** 3D reconstruction results; **i** and **j** Phase unwrapping contrast

relative phase is expanded into absolute phase according to Eq. (10.3). Figure 10.5j shows the change of the truncation region from phase discontinuity to relative phase and from relative phase to absolute phase. Figure 10.5g is the continuous absolute phase diagram, and Fig. 10.5h is the final 3D reconstruction result. Comparing the reconstruction results with the simulated true values, the overall mean square error is 8.2×10^{-3}. Simulation experiments show the effectiveness of the proposed algorithm. The next step was to measure a model of the hemisphere and a model of the nose.

10.4.2 Measurement Experiment

In this paper, a DFP system composed of CCD camera and projector is constructed. The camera has a focal length of 15 mm and the projector has a focal length of 23 mm. The resolution of the camera is 1280 × 1920. The resolution of the projector is set to 1260 × 1366. The system adopts the method introduced by Li et al. [13] for calibration, and selects the camera lens coordinate system as the world coordinate system. The period of the fringe pattern in the following experiment is $T = 50$ pixels. Three equal-phase-shifted stripe patterns were used to generate the enveloped phase map. A standard hemisphere with a diameter of 200 mm was first measured. Figure 10.6a is a three-step phase shift stripe image collected by the camera, with a stripe period of 50 pixels. Figure 10.6b is the expanded phase diagram obtained in the first step of the algorithm in this paper. There is an obvious dividing line in Fig. 10.6b, indicating that the unwrapped phase diagram is discontinuous. Figure 10.6c is the corresponding three-dimensional measurement result in Fig. 10.6b. Then, the Sobel image segmentation algorithm is firstly used to segment Fig. 10.6b to segment the discontinuous region in Fig. 10.6c and generate a binary mask as shown in Fig. 10.6d. Figure 10.6e is the continuous relative phase distribution of the discontinuous region resolved by the modulation intensity sorting algorithm. Expand the relative phase into an absolute phase according to Eq. (10.3) in Fig. 10.6f. Figure 10.6g, h shows the final 3D reconstruction results, indicating that the three-dimensional information of the standard hemisphere is well recovered. The multi-frequency phase-shifting method is the most accurate algorithm among the current phase solving algorithms. In this paper, the dual-frequency phase-shifting method is used to measure and

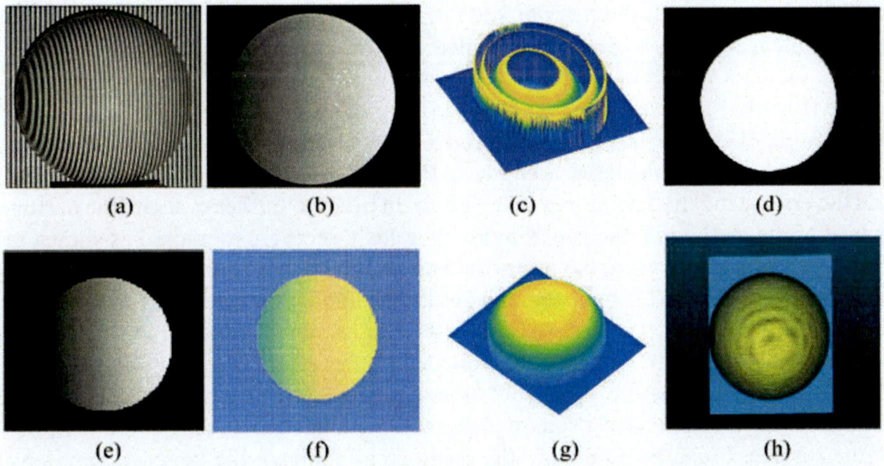

Fig. 10.6 Measurement hemispherical experiment. **a** the deformed projection fringe modulated by the measured object; **b** Phase diagram of geometric constraint method expansion; **c** The reconstruction results are obtained by geometric constraint method; **d** Segmentation results based on Sobel algorithm; **e** Relative phase diagrams; **f** Absolute phase diagrams; **g** and **h** 3D reconstruction results

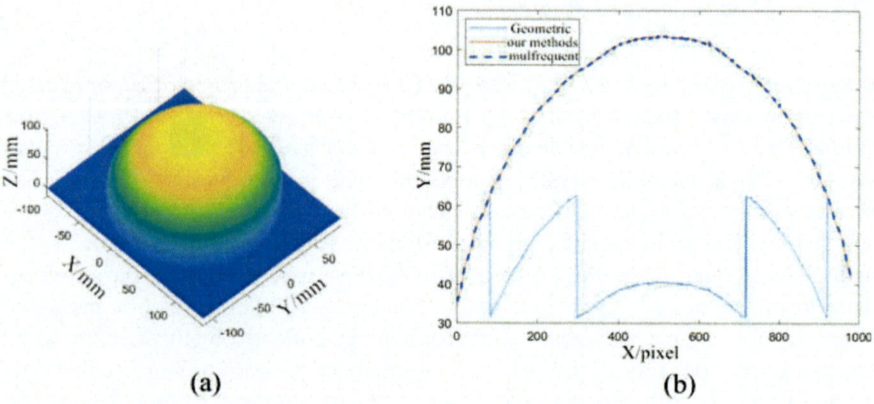

Fig. 10.7 Comparison algorithm. **a** Dual-frequency phase shift algorithm; **b** Comparison of cross section accuracy

compare the results. Three-step and five-step phase shift measurements were made using two fringe periods of 24 and 180 pixels, respectively. Figure 10.7a are the three-dimensional measurement results obtained by dual-frequency phase-shifting algorithm. Figure 10.7b shows the height distribution of the cross section of the above method. It can be seen that the height distribution represented by the method presented in this paper (solid red line) is almost identical to the results obtained by the dual-frequency phase shift (dashed blue line). However, the dual-frequency phase shift method requires at least 6 pairs of projected fringe patterns, so the measurement efficiency is low. In order to further evaluate the accuracy of the proposed method, an ideal hemisphere with a diameter of 200 mm was taken as an example to analyze the measurement accuracy of the proposed method. Since the radius of the hemisphere is known, the least square method is used to determine the position of the center of the sphere from the measured data, and then the ideal sphere model is rebuilt to calculate the difference between the ideal sphere and the measured data, as shown in Fig. 10.8a. The error diagram is calculated as shown in Fig. 10.8b. the mean error was 0.054 mm, and the error standard deviation was 0.37 mm. In order to further evaluate the performance of the method in this paper, more complex targets are measured as shown in Fig. 10.9. As a comparison, the improved least square phase solving algorithm [14], branch cutting method [15], mass graph guiding method [16] were, respectively, used to reconstruct the model in Fig. 10.9a, and the results were shown in Fig. 10.10. The phase error will occur in the shaded region, that is, the region with poor phase quality, and the error will propagate along with the expansion path of the two algorithms. The essence of the improved least square phase solving algorithm is the quadratic fitting of the surface. It can be seen from the results that although the target morphology is recovered, there is deformation and distortion in the lower left side, which affects the measurement results. Comparatively, the algorithm in this paper can recover the target's 3D appearance well.

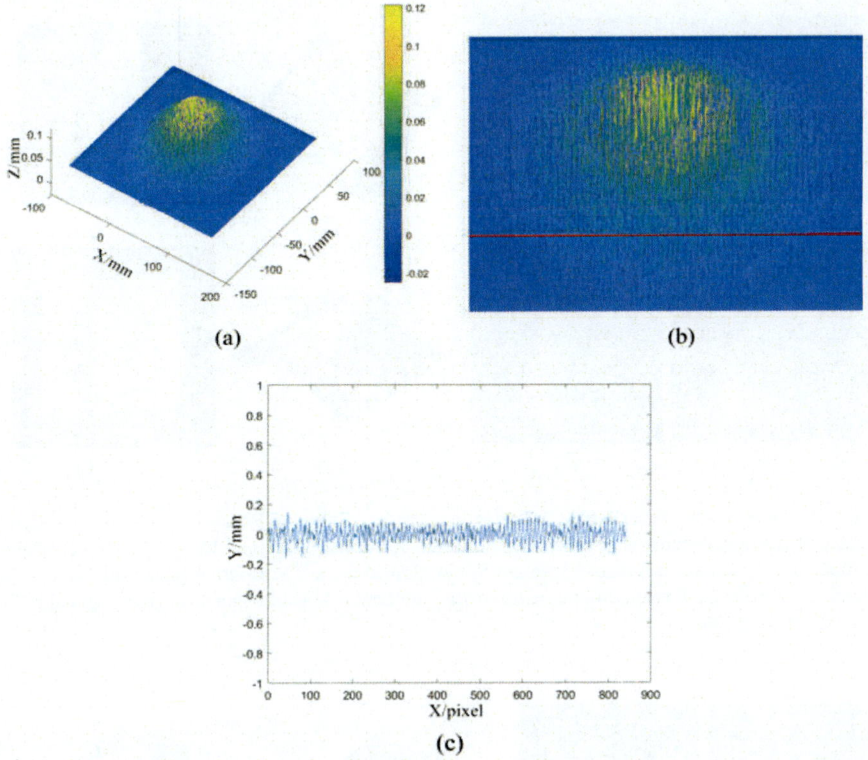

Fig. 10.8 Error analysis of hemisphere measurement. **a** Height difference between the measured sphere and the ideal sphere. **b** and **c** A cross-sectional view of the error graph. **c** Is the error distribution of the section shown in the middle red line in **b**

10.5 Conclusion

The traditional algorithm based on geometric constraint has high precision and fast speed. However, the measurement depth is limited. This problem limits the practical application of this method. In order to solve this problem, this paper proposes a depth range extension method based on discontinuous region segmentation and modulation intensity sorting, which is suitable for objects with large depth deviation. Firstly, the correct part of the discontinuous absolute phase distribution obtained by the geometric constraint method is extracted, and the corresponding binary mask is made. Then, the phase relative unwrapping is realized by the modulation intensity ordering method. Finally, the relative phase distribution is converted to the absolute phase distribution according to the obtained absolute phase value in the mask region. The validity and correctness of the method were verified by measuring standard hemisphere and nose model.

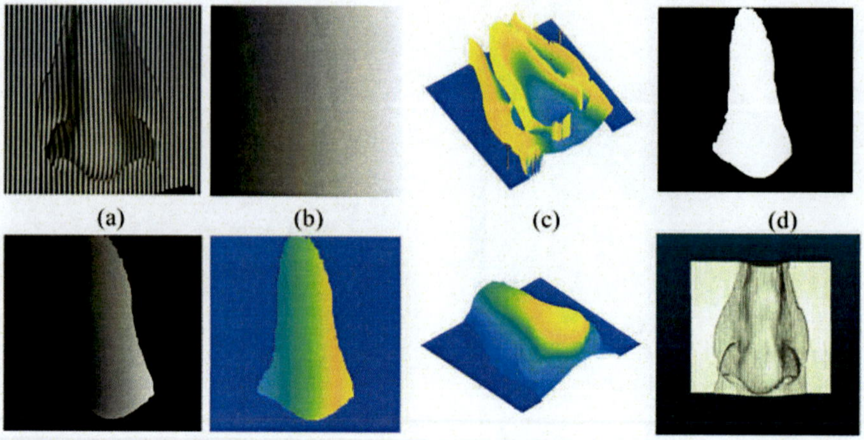

Fig. 10.9 Nose statue experiment. **a** Projection fringe pattern; **b** Phase diagram of geometric constraint method expansion; **c** The reconstruction results are obtained by geometric constraint method; **d** Segmentation results based on Sobel algorithm; **e** The relative phase distribution is obtained based on the modulation intensity sorting method; **f** Absolute phase diagrams; **g** and **h** 3D reconstruction results

Fig. 10.10 Algorithm comparison **a** branch cutting method; **b** Quality map guidance; **c** an improved least square phase solution algorithm; **d** The present methodology

References

1. S. Zhang, High-speed 3D shape measurement with structured light methods: a review. Opt. Lasers Eng. **106**, 119–131 (2018)
2. B. Li, Y. An, S. Zhang, Single-shot absolute 3D shape measurement with Fourier transform profilometry. Appl. Opt. **55**(19), 5219 (2016)
3. J. Huang, W. Chen, X. Su, Application of two-dimensional wavelet transform in the modulation measurement profilometry. Opt. Eng. **56**(3), 034105 (2017)
4. S. Ordones, M. Servin, M. Padilla et al., Shape defect measurement by fringe projection profilometry and phase-shifting algorithms. Opt. Eng. **59**(1), 1 (2020)
5. C. Zuo, L. Huan, M. Zhang, Q. Chen, A. Asundi, Temporal phase unwrapping algorithms for fringe projection profilometry: a comparative review. Opt. Laser Eng. **85**, 84–103 (2016)
6. S. Zhang, Absolute phase retrieval methods for digital fringe projection profilometry: a review. Opt. Lasers Eng. **107**(Aug), 28–37 (2018)

7. J. Chufan, L. Beatrice, Z. Song, Three-dimensional shape measurement using a structured light system with dual projectors. Appl. Opt. **57**(14), 3983 (2018)
8. Z. Dongliang, K. Qian, D. Feipeng et al., Ternary gray code-based phase unwrapping for 3D measurement using binary patterns with projector defocusing. Appl. Opt. **56**(13), 3660 (2017)
9. Y. Hu, Q. Chen, Y. Liang, S. Feng, T. Tao, C. Zuo, Microscopic 3D measurement of shiny surfaces based on a multi-frequency phase-shifting scheme. Opt. Lasers Eng. **122**(Nov), 1–7 (2019)
10. C. Jiang, S. Zhang, Absolute phase unwrapping for dual-camera system without embedding statistical features, in *Society of Photo-optical Instrumentation Engineers. Society of Photo-Optical Instrumentation Engineers (SPIE) Conference Series* (2017)
11. Y. An, J.-S. Hyun, S. Zhang, Pixel-wise absolute phase unwrapping using geometric constraints of structured light system. Opt. Express **24**, 18445–18459 (2016)
12. C. Zachiu, M. Ries, C. Moonen, B.D. de Senneville, An adaptive non-local-means filter for real-time MR-thermometry. IEEE Trans. Med. Imaging **36**(4), 904–916 (2017). https://doi.org/10.1109/TMI.2016.2627221
13. B. Li, N. Karpinsky, S. Zhang, Novel calibration method for structured light system with an out-of-focus projector. Appl. Opt. **53**(16), 3415–3426 (2014)
14. L. Yan, H. Zhang, R. Zhang, X. Xie, B. Chen, A robust phase unwrapping algorithm based on reliability mask and weighted minimum least-squares method. Opt. Lasers Eng. **112**(Jan), 39–45 (2019)
15. C. Zuo, S. Feng, L. Huang et al., Phase shifting algorithms for fringe projection profilometry: a review. Opt. Lasers Eng. **109**, 23–59 (2018)
16. J.M. Amjad, Robust and fast filtering method for enhancement of two-dimensional quality-guided path unwrapping algorithms. Appl. Opt. **59**(13), 3920–3926 (2020)

Chapter 11
Research on the Influence of Electromagnetic Interference Test Signal on Networked Vehicle Detection Equipment

Jie Zhang, Changyuan Wang, Haiming Liu, Xu Zhang, Guokai Jiang, and Yang Chen

Abstract The rapid development of intelligent and networked automobiles has driven the electronic, intelligent, and integrated industries of automobiles, but at the same time, it has also brought more complex electromagnetic environment inside and outside the automobiles. The reliable operation of vehicles in complex electromagnetic environment is related to the comfort of vehicles, and affects the safety of vehicles. Therefore, the electromagnetic reliability detection of networked vehicles needs to be improved urgently. In order to protect the reliability of the networked vehicle test equipment in the electromagnetic interference test, this paper designs a scheme to conduct a bottom-up test, and tests the signal strength received by the networked vehicle test equipment under different field strengths, different frequencies and different antenna positions.

J. Zhang (✉) · Y. Chen
CATARC New Energy Vehicle Test Center (Tianjin) Co., Ltd., Tianjin, China
e-mail: zhangjie2020@catarc.ac.cn

Y. Chen
e-mail: chenyang@catarc.ac.cn

C. Wang · H. Liu · X. Zhang
China Automotive Technology and Research Center Co., Ltd., Tianjin, China
e-mail: wangchangyuan@catarc.ac.cn

H. Liu
e-mail: liuhaiming@catarc.ac.cn

X. Zhang
e-mail: zhangxu@catarc.ac.cn

G. Jiang
CATARC Software Testing (Tianjin) Co., Ltd., Tianjin, China
e-mail: jiangguokai@catarc.ac.cn

G. Liu and F. Cen (eds.), *Advances in Precision Instruments and Optical Engineering*, Springer Proceedings in Physics 270, https://doi.org/10.1007/978-981-16-7258-3_11

11.1 Introduction

With the rapid development of intelligent and networked automobiles, automobiles are becoming more and more like a mobile large-scale "mobile phone". Intelligent networked vehicles refer to the application of electronic sensing devices on vehicles and roads to collect and process information on vehicles, roads and environments, and realize the information interaction and sharing between vehicles, vehicles and people, vehicles and roads [1], thus ensuring the safety and smoothness of vehicles during their travel. In addition to the vehicle networking function, the networked vehicles also integrate navigation, Bluetooth, cellular 4G, cellular 5G, WiFi, ecall, and other Wireless functions [2].

Networked vehicles drive the development of electronic, intelligent, and integrated vehicles, but at the same time, they also bring more complex electromagnetic environment inside and outside the vehicle. The reliable operation of vehicles in complex electromagnetic environment is related to the comfort of vehicles, but it also affects the safety of vehicles [3].

11.2 Electromagnetic Compatibility of Networked Vehicles

In the electromagnetic compatibility test of networked vehicles, the test of electromagnetic interference is particularly important. The electromagnetic interference test refers to the establishment of communication connection with vehicles through wireless signal simulation equipment and antenna in semi-anechoic darkroom, so as to set up the communication scene of networked vehicles, and then monitor their communication working state under electromagnetic interference immunity state, and judge the reliability of networked functions of networked vehicles [4]. Electromagnetic compatibility test of networked vehicles can well evaluate the reliability of vehicles running in complex electromagnetic environment and ensure people's vehicle safety [5].

In the test, because the antenna does not have the characteristics of unidirectional conduction and reverse isolation, it is inevitable to receive anti-interference signals from the laboratory, and these signals will be returned to the test equipment after being received by the antenna. However, the acceptable power of general test equipment is not high, but the price is very expensive, while the anti-interference signal in the test is large, and many enterprises have achieved 100 V/m or even 140 V/m. Therefore, considering that the withstand power of the test equipment is generally about 0dBm, when the anti-interference signal is received by the antenna and returned to the test equipment, it is very likely that the test equipment will be damaged and the test will be interrupted.

How much power is fed back to the test equipment by the anti-interference signal, will it damage the test equipment, and how should the communication antenna be arranged? In this paper, the test scheme is designed, the bottom test is carried out, and the conclusion is given.

11.3 Test Scheme of Anti-interference Signal Received by Detection Equipment

The signal size fed back to the networked test equipment is affected by frequency, field strength, and the position of communication antenna. In order to determine the signal strength fed back to the networked test equipment under different circumstances, the spectrum analyzer is used to test the signal strength fed back to the networked test equipment in this baseline test.

As shown in Fig. 11.1, the communication antenna is placed in a dark room, and the rest are placed outside the dark room. To ensure the safety of the spectrum analyzer, a 20 dB attenuator is added between the communication antenna and the spectrum analyzer to attenuate the anti-interference signal intensity from the dark room.

There is space loss when electromagnetic waves are transmitted in space [6], so different positions of communication antennas will also cause differences in signal strength fed back to network test equipment. As schematically shown in Fig. 11.2 put the communication antennas at different positions, and check the signal power intensity received by the spectrum analyzer under different frequencies and different immunity levels. Among them, the anti-interference antenna is placed according

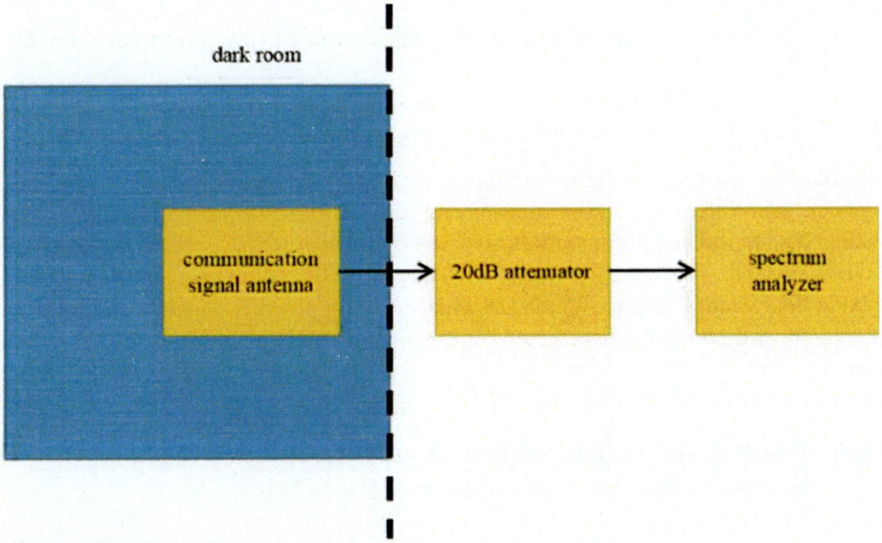

Fig. 11.1 Schematic diagram of scheme

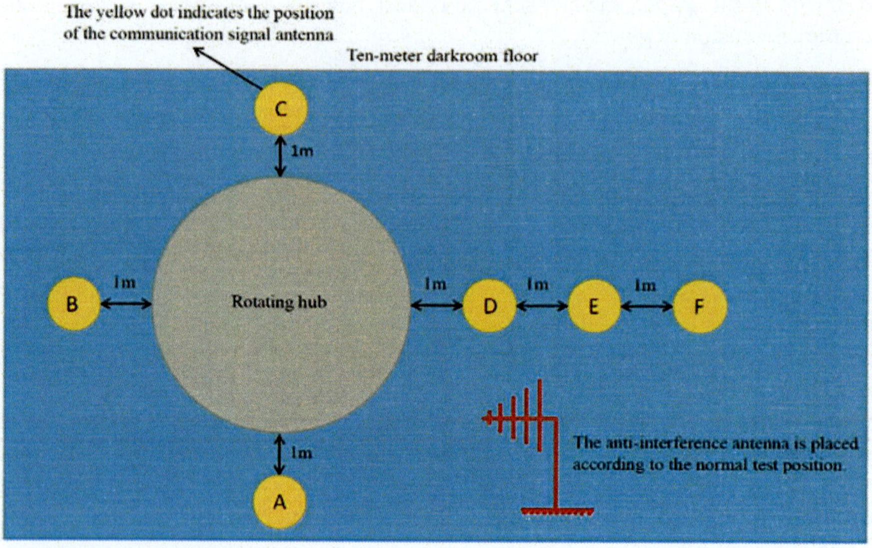

Fig. 11.2 Schematic diagram of test layout

to the position of normal anti-interference test, and the principle of selecting the position of communication antenna is to avoid the overlap of its pattern with the anti-interference antenna as much as possible, and it is convenient to select the appropriate position by uniform placement.

Schematic illustration:

(1) at points A and C, the antenna phase center is >= 1 m away from the turntable side;
(2) The distance between the antenna phase center and the ground is 1.2 m;
(3) The center of antenna is horizontally aligned with the center of turntable.

Considering that the vehicle communication frequency is now as high as 6 GHz, which is convenient for statistical analysis, the frequency points are selected for testing. The frequency points of this test are 600 MHz, 1.2 GHz, 1.8 GHz, 2.4 GHz, 3 GHz, 3.6 GHz, 4.2 GHz, 4.8 GHz, 5.4 GHz, and 6 GHz; According to the standard and the field strength grades which are widely used in practice, the field strengths of 30, 50, 80, and 100 V/m are selected for testing.

11.4 Analysis of Test Results of Anti-interference Signals Received by Test Equipment

The actual test arrangement is shown in Fig. 11.3. The parameters of communication antenna used this time are shown in Table 11.1.

Fig. 11.3 Actual test layout

Table 11.1 Communication antenna parameters

Category	Parameter
Frequency	400 MIIz–18 GHz
VSWR	400 MHz–600 MHz, VSWR < 3
	600 MHz–18 GHz, VSWR < 2
Impedance	50Ω
Rated power	<4 W CW
Insulation	400 MHz–18 GHz, >20 dB

Since there are many experimental data, we will not list them all here, but only give the relationship diagram after processing the data according to the principle of unique variable. Obviously, the greater the field strength, the greater the received signal strength. Here, 600 MHz is listed, and the trend diagram of the communication antenna position at D is shown in Fig. 11.4.

The relationship between received signal strength and frequency is shown in Fig. 11.5 when the field strength is 30 V/m, the communication antenna is 1 m to the right of the anti-interference antenna and the field strength is 100 V/m, and the communication antenna is 5 m behind the anti-interference antenna. It can be seen from the trend graph that the received signal strength decreases with the increase of frequency.

The field strength of 30 V/m, frequency point of 600 MHz, field strength of 30 V/m, frequency point of 1.2 GHz, field strength of 50 V/m, and frequency point of 600 MHz are processed, respectively. The relationship between received signal

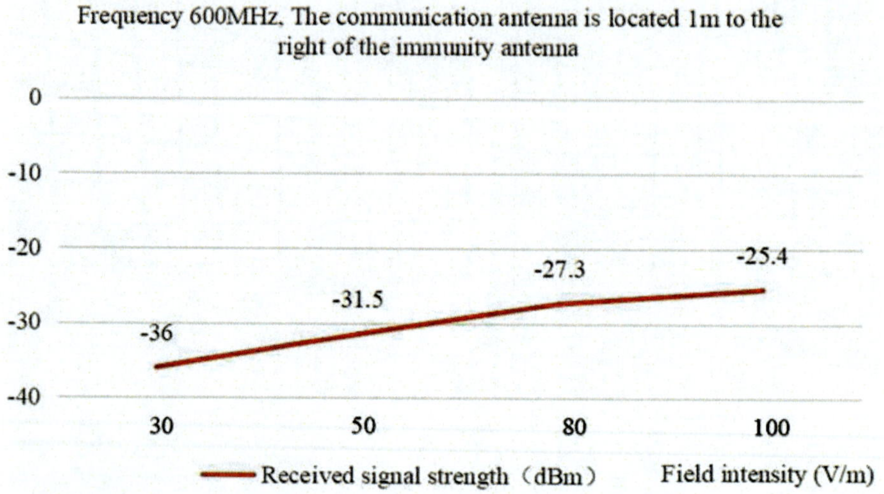

Fig. 11.4 The variation of received signal strength with field strength

strength and frequency is shown in Fig. 11.6. As far as the change trend is concerned, the last 5 m is the position where the received signal strength is the lowest, and the received signal strength decreases with the increase of distance.

To sum up, as far as the test results are concerned, the interference rejection antenna is 1 m to the right, with a field strength of 100 V/m, and the received signal strength is the highest, which is −25.4 dBm, and if no attenuator is added, it is −5.4 dBm, with a margin of 5 dB. The tolerable power of the networked test equipment is generally about 0 dBm, so it will not cause damage to the equipment in theory. However, due to the differences in equipment and line losses in different laboratories, and the increasing frequency of test field strength, it is recommended to place the communication antenna 5 m behind the anti-interference antenna for safety reasons, and if it must be placed 1 m away, it is recommended to add an attenuator in the link.

11.5 Conclusion

Testing and certification of intelligent networked cars is an important link to ensure the safety and comfort of consumers when they using such cars. Furthermore, they can enjoy the convenience brought by intelligent networked cars. However, at present, the detection of automobile intelligence and network connection is still imperfect and nonstandard. In this paper, from the perspective of protecting the test equipment of automobile network connection, some suggestions on the layout of wireless communication test scenarios are given, which can provide reference for perfecting the test in this aspect.

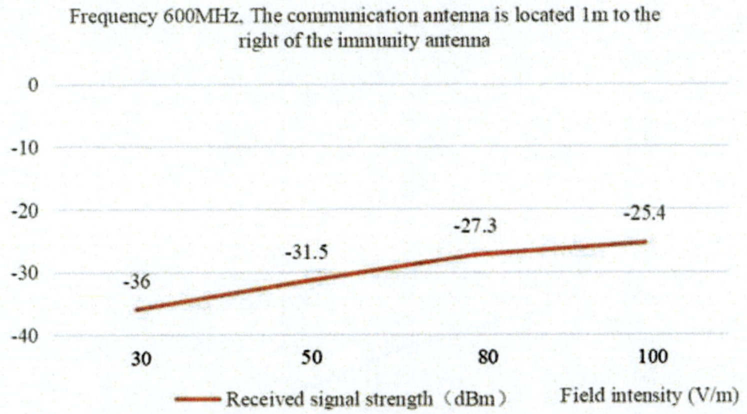

(a) the field strength is 30V/m, and the communication antenna is located 1m to the right of the anti-interference antenna

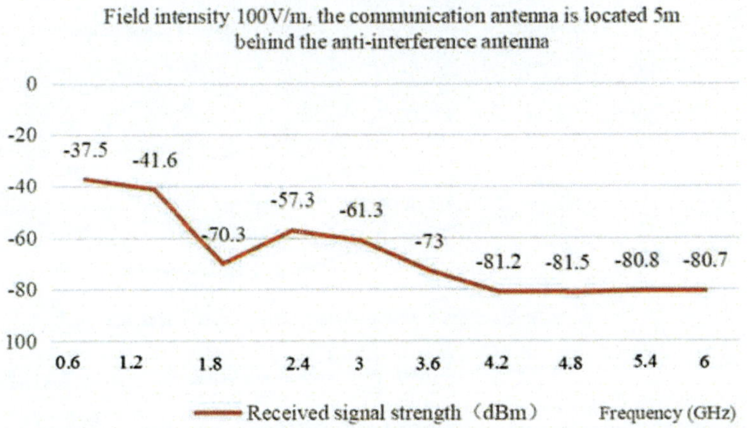

(b) The field strength is 100V/m, and the communication antenna is located 5m behind the anti-interference antenna

Fig. 11.5 Changes of received signal strength with frequency. **a** The field strength is 30 V/m, and the communication antenna is located 1 m to the right of the anti-interferencc antenna. **b** The field strength is 100 V/m, and the communication antenna is located 5 m behind the anti-interference antenna

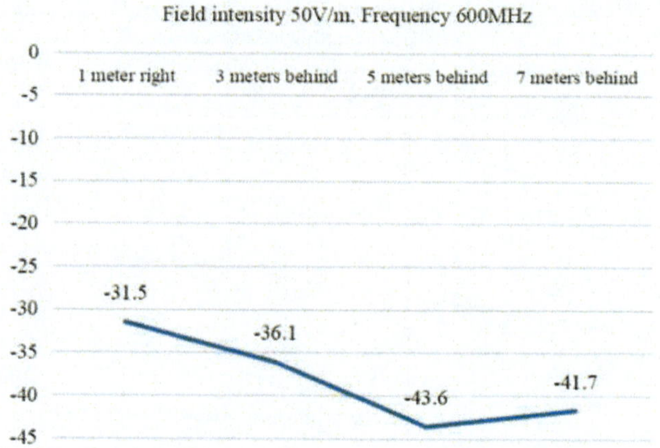

(a) the field strength is 50V/m and the frequency point is 600MHz

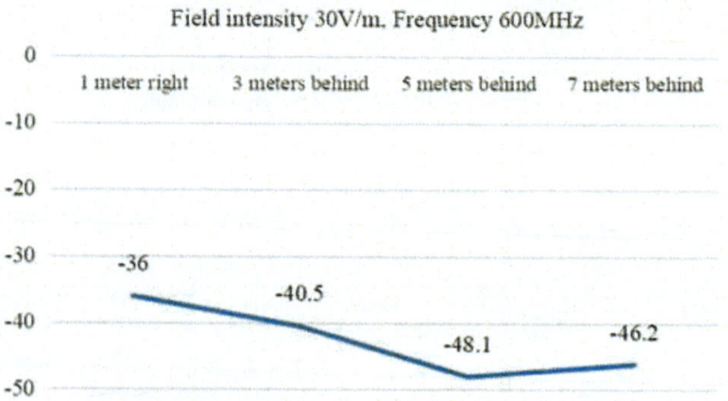

(b) The field strength is 30V/m and the frequency point is 600MHz

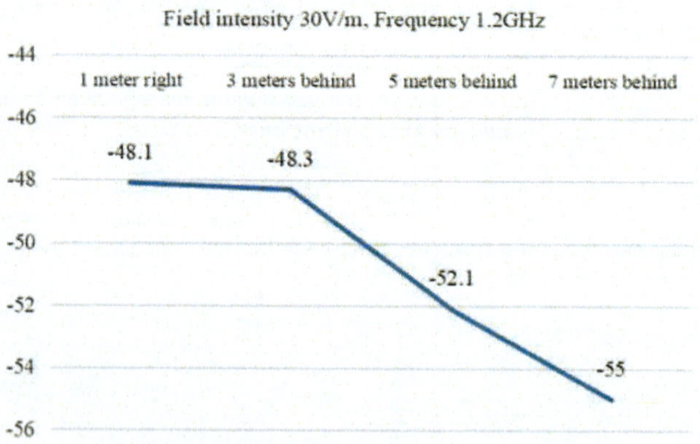

(c) the field strength is 30V/m and the frequency point is 1.2GHz

Fig. 11.6 Changes of received signal strength with distance. **a** the field strength is 50 V/m and the frequency point is 600 MHz. **b** The field strength is 30 V/m and the frequency point is 600 MHz. **c** the field strength is 30 V/m and the frequency point is 1.2 GHz

References

1. Y. Zhong, B. Yang, H. Yang et al., Summary of intelligent networked automobile safety. Inf. Secur. Res. **7**(6), 8
2. J. Wang, X. Ji, X. Zhang et al., Research on networking performance of 5G cellular car networking. Telecommun. Sci. **2020**(1), 49–57 (2020)
3. J. Hao, G. Han, Z. Guo et al., challenge and development of intelligent networked automobile information security, in *14th Annual Meeting of China Intelligent Transportation.0*
4. R. Chen, J. Lei, H. Zhang et al., A test system and method for electromagnetic immunity performance of intelligent networked vehicles. CN110907739A (2020)
5. Y. Zhang, L. Chen, Y. Han, L. Xu, Research on electromagnetic scene construction method for intelligent networked vehicles, in *2020 Annual Meeting and Exhibition of China Automotive Engineering Society.0*
6. S. Mizushina, A. Adachi, Buried radio device. US (2007)

Part II
Intelligent Precision Instrument and Robot Technology

Chapter 12
Geometric Error Modeling of a Special NC Process Device for Precision Two-Dimensional Optical Drum

Xuebing Han, Weidong Feng, and Likun Zhao

Abstract In order to improve the precision of the special NC process device for machining two-dimensional optical drum and ensure its quality, an error model of the special NC process device is established. According to the mechanical topology of the NC process device, to obtain the influence degree of each error component on the total error, the geometric error model is derived by using error analysis technology. The model covers the main error sources of the process unit such as the two axes errors, the parts error and encode error. It can analyze, synthesize, and distribute these errors. An error prediction has been done by the model. An experiment has been conducted on a new NC process device to validate the method. An electronic theodolite and a plane mirror are used to measure the repeat positioning accuracy of the two shafts. The predicted repeated positioning accuracy is compared with the measured results. It shows that integrated geometric error modeling method is effective and applicable in Special NC Process Device.

12.1 Introduction

In the machining of ultra-precision two-dimensional optical drum, a two-dimensional indexing table is a necessary device to make the dividing of the drum around the central axis and the inclination between the machining surface and the central axis [1]. The traditional indexing device has a manual two-dimensional turntable to adjust the angle of the drum during processing [2]. The angle of the turntable must be readjusted after each surface is processed. This device has low positioning accuracy and slow processing efficiency. In order to improve the accuracy and the efficiency, a new NC process device is designed [3]. The device uses a computer-controlled motor to drive the turntable to rotate. It can automatically divide and greatly improve the

Present Address:
X. Han · W. Feng (✉)
Jilin Business and Technology College, Changchun 130507, China

L. Zhao
Changchun Medical College, Changchun 130031, China

G. Liu and F. Cen (eds.), *Advances in Precision Instruments and Optical Engineering*, Springer Proceedings in Physics 270, https://doi.org/10.1007/978-981-16-7258-3_12

processing efficiency of the drum. Compared with the manual turntable, the new NC process device not only increases the motor and encoder, but also has great differences in parts and structure. In order to improve the positioning accuracy of NC process device and ensure the machining quality of drum, it is necessary to establish the accuracy model as the design basis of the device. At present, multi-body dynamics method is a common method to establish error model [4, 5]. This method has high precision but complex calculation. Especially when there are many errors, it needs to be calculated by computer programming [6]. In this paper, the geometric method is used to establish the error model, which has the characteristics of simple principle and convenient calculation.

12.2 Overall Structure of the Device

The whole process device is composed of two parts: two-dimensional turntable and turntable controller. The two-dimensional turntable provides high-precision absolute position information, shafting accuracy, and interface with the machine tool and the processing workpiece. The turntable controller has turntable power supplying, motor driving, position information collection feedback, and shafting locking functions. The overall structure of the two-dimensional turntable (as shown in Fig. 12.1) can be divided into two parts: horizontal axis system and vertical axis system. The horizontal shaft system consists of rolling bearing, rolling motor, rolling encoder, rolling brake, work platform, and some connecting parts. The vertical shaft system is composed of turning bearing, turning motor, turning encoder, turning brake, base, and some connectors.

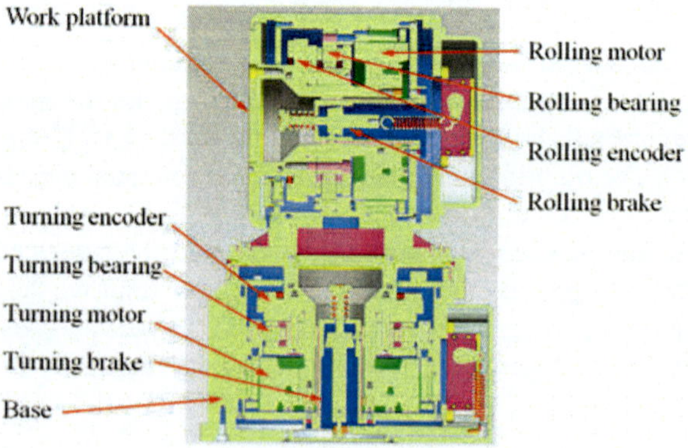

Fig. 12.1 Overall structure of the process device

12.3 Error Modeling of the Device

During the process of the drum, the relative position accuracy of the working face is mainly guaranteed by the shafting accuracy of the process device. There are many factors that affect the accuracy of shafting. The factors mainly include machining error of single part, shape error of part, position error of part, fit clearance, temperature change, influence of lubricant, friction, wear, and elastic deformation, etc. All of those have inevitable consequences for the accuracy of shafting. According to the structure and working principle of the process device, the factors that affect the repeated positioning of the shafting mainly include the swaying error of the two shafting, the system accuracy of the encoder, the machining dimension error, and shape error of the two shaft supporting parts, etc.

12.3.1 Swaying Errors Analysis of the Horizontal Axis

12.3.1.1 Principle of the Errors

The swaying errors are mainly caused by the circular runout of the bearing's inner ring and the motor's rotor. The horizontal axis coordinate system is established as shown in Fig. 12.2.

T point is the projection of the center point of the drum shaft on the processing surface of the drum. According to the processing principle of the two-dimensional optical drum, all points on the same process plane have the same angular relationship in the horizontal axis coordinate system, without considering the machining error of the machine tool. That is all of the points have the same horizontal axis angle. Therefore, the horizontal axis angle of *T* point is the horizontal axis angle of the process plane. The angle error of the machined refraction surface can be calculated by the angle error of horizontal axis at *T* point.

In Fig. 12.2, the x-axis is turning center of the horizontal axis. The y-axis is in a same plane with the x-axis and *T* point. The y-axis is perpendicular to the x-axis. *O* point is the left support of horizontal shaft. ρ is a distance between the center point of the drum and *O* point. *H* is the rotation angle of the horizontal shaft. The space

Fig. 12.2 The horizontal axis coordinate system

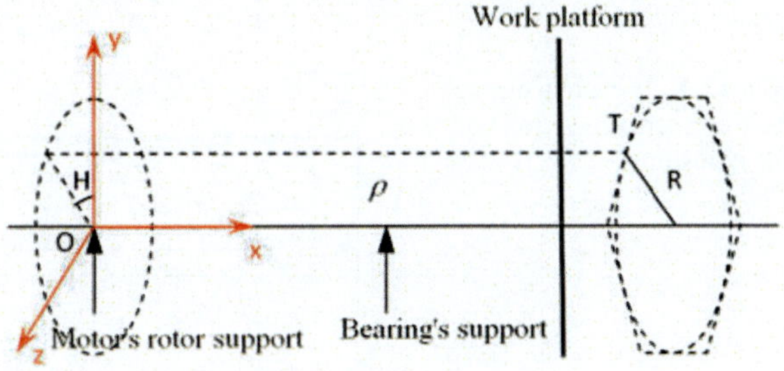

position of T point can be calculated by the following formula.

$$\begin{bmatrix} x \\ y \\ z \\ 1 \end{bmatrix} = \begin{bmatrix} \rho \\ R\cos H \\ R\sin H \\ 1 \end{bmatrix} \tag{12.1}$$

The circular runout of the bearing's inner ring and the motor's rotor can make the horizontal axis coordinate system rotate and translate along y and z axes.

12.3.1.2 Machine Error of the Drum from Rotating Errors

The rotating of the horizontal axis coordinate system around y and z axes can bc simplified as shown in Fig. 12.3. L is the distance of the bearing's inner ring and the motor's rotor. δ is the deflection distance produced by the superposition of two errors of the circular runout.

The angle error of the horizontal shaft caused by the bearing's inner ring and the motor's rotor can be calculated by the following formula.

$$\varepsilon = \mathrm{arctg}(\delta/L) \tag{12.2}$$

θ is horizontal axis rotate angle error by ε. The actual space position of T point can be given by homogeneous coordinate equation. Actual position of T point with error rotating around the y-axis.

$$T' = \begin{bmatrix} x' \\ y' \\ z' \\ 1 \end{bmatrix} = \begin{bmatrix} \cos\theta_y & 0 & -\sin\theta_y & 0 \\ 0 & 1 & 0 & 0 \\ \sin\theta_y & 0 & \cos\theta_y & 0 \\ 0 & 0 & 0 & 1 \end{bmatrix} \begin{bmatrix} x \\ y \\ z \\ 1 \end{bmatrix} = \begin{bmatrix} \rho\cos\theta_y - R\sin H\sin\theta_y \\ R\cos H \\ \rho\sin\theta_y + R\sin H\cos\theta_y \\ 1 \end{bmatrix} \tag{12.3}$$

where θ_y is the rotation error of the horizontal axis around the y-axis, caused by the radial runout of the bearing inner ring and the motor's rotor. The formula (12.4) is deduced from (12.2) and (12.3). H' is the actual angle of the horizontal axis.

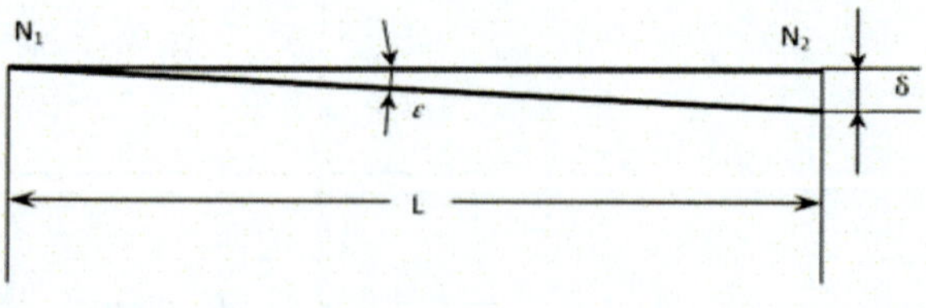

Fig. 12.3 The angle error between actual position and ideal position

$$H = \arctan\left(\frac{z'}{y'}\right) = \arctan\left(\frac{\rho \sin\theta_y + R \sin H \cos\theta_y}{R \cos H}\right) \approx \tan H \cos\theta_y \quad (12.4)$$

Because θ_y is a very small angle and $\cos\theta_y$ is very close to 1, which can be considered approximately θ_y has no effect on the angle of the processing surface of the drum.

The actual space position of T point rotating around the z-axis is

$$T' = \begin{bmatrix} x' \\ y' \\ z' \\ 1 \end{bmatrix} = \begin{bmatrix} \cos\theta_z & \sin\theta_z & 0 & 0 \\ -\sin\theta_z & \cos\theta_z & 0 & 0 \\ 0 & 0 & 1 & 0 \\ 0 & 0 & 0 & 1 \end{bmatrix} \begin{bmatrix} x \\ y \\ z \\ 1 \end{bmatrix} = \begin{bmatrix} \rho\cos\theta_z + R\sin\theta_z \cos H \\ -\rho\sin\theta_z + R\cos\theta_z\cos H \\ R\sin H \\ 1 \end{bmatrix} \quad (12.5)$$

The formula (12.6) can be obtained from (12.1) and (12.5).

$$H' = \arctan\left(\frac{R\sin H}{-\rho\sin\theta_z + R\cos\theta_z\cos H}\right) \quad (12.6)$$

The formula (12.6) is similar to formula (12.4), and it can be considered that the error of the roll around z-axis is approximately considered θ_Z has no effect on the angle of the drum surface.

12.3.1.3 Machine Error of the Drum from Translating Errors

λ_y and λ_z is the translation error of the horizontal axis along the y-axis and the z-axis, caused by the radial runout of the bearing inner ring and the motor's rotor. In order to simplify the calculation, only when the angle error is the largest, that is, when H is 0.

$$\theta_{H-t-z} = H' - H = \arctan(\lambda_z / R) \quad (12.7)$$

At the limit position, H is 90°, the error from the translation error of the horizontal axis along the y-axis is

$$\theta_{H-t-y} = \arctan(R/\lambda_y) - 90° \quad (12.8)$$

12.3.2 *Swaying Errors Analysis of the Vertical Axis*

The swaying errors of vertical axis are mainly caused by bearing end face runout and radial runout. The end face runout error produces the angle error of *H* angle, which has a great impact on the machining surface of the drum. It can be calculated by the following formula:

$$\theta_{H-V-e} = \arcsin(\alpha/r) \tag{12.9}$$

In the formula, α is the value of end face runout, and r is the bear's radius.

The radial runout error also produces the angle error of *H* angle. The formula (12.10) can be obtained using the similar method of the horizontal axis. In the formula, β is the value of radial runout.

$$\theta_{H-V-r} = \arctan(\beta/R) \tag{12.10}$$

The calculation method of angle error caused by radial runout of vertical axis is similar to that of horizontal axis. The radial runout can produce the translation of the vertical axis coordinate system along y and z axes. The translation along y-axis has no effect on the angle of the drum surface. The error from the translation along z-axis is formula (12.11).

$$\theta_{A-t-z} = \arctan(\alpha_z/R_1) \tag{12.11}$$

12.3.3 *Other Errors Analysis of the Two Shafts*

In addition to the error of bearing support, the machining dimension errors and shape errors of parts also affect the accuracy of shaft system. The main part processing errors affecting the shafting accuracy include: part size processing error, roundness error, cylindricity error, coaxiality error, perpendicularity error, etc.

The encoder is an important accessory to ensure the accuracy of the shafting. The resolution, accuracy, and repeatability of the encoder have a great immediate impact on the performance of the device.

12.4 Error Prediction of the Device

The bearings in horizontal and vertical shafts are preliminarily selected. The bearing's end face runout and radial runout are all less than 2 μM in the horizontal shaft. The thrust ball bearing in the vertical shaft, those are all less than 4 μM. Therefore, λ_y and

Table 12.1 The errors value

Error source	Formula	Calculated value
Horizontal axis	$\theta_{H-t-y} = \arctan(R/\lambda_y) - 90°$ $\theta_{H-t-z} = \arctan(\lambda_z/R)$	0.02″ 0.6″
Vertical axis	$\theta_{H-V-e} = \arcsin(\alpha/r)$ $\theta_{H-V-r} = \arctan(\beta/R)$ $\theta_{A-t-z} = \arctan(\alpha_z/R_1)$	0.2″ 4″ 4.6″
Parts	$\theta_{H-P} = \sqrt{\theta^2_{H-t-y} + \theta^2_{H-t-z} + \theta^2_{H-V-e} + \theta^2_{H-V-r}}$ $\theta_{A-P} = \arctan(\alpha_z/R_1)$	1″ 1″
Encode	$\theta_{H-E} = 1''$ $\theta_{A-E} == 1''$	1″ 1″

λ_z are all 2 μM. α and β are all 2 μM. Design dimension of r, R, and R_1 is 180, 200, and 180 mm. Part size processing error, roundness error, cylindricity error, coaxiality error, and perpendicularity error are approximate to 0.3 um ($\lambda_y = \lambda_z = \alpha = \beta =$ 0.3 um) based on the parts processing capacity. The accuracy of the selected encoders is $\pm 1''$. The calculated values of the errors are in Table 12.1.

The overall error of horizontal shaft and vertical shaft can be combined with errors in Table 12.1.

$$\delta_H = \sqrt{0.02^2 + 0.6^2 + 0.2^2 + 4^2 + 1^2 + 1^2} = 4.3'' < 5''$$

$$\delta_A = \sqrt{4.6^2 + 1^2 + 1^2} = 4.8'' < 5''$$

12.5 Test of the Model

The developed and manufactured NC process device based on the model was tested. The repeated positioning accuracy of NC process device was tested by an electronic theodolite and a plane mirror. The model of the electronic theodolite is TM5100A, and its measuring precision is 0.04″. The results are the repeated positioning accuracy of horizontal shaft is 2.88″, that of vertical shaft is 3.12″. The accuracy of the measured results is higher than the design accuracy. The main reason is that in order to simplify the calculation, the error values are the maximum value at the limit position. If the standard deviation of the errors is used in the model, the prediction result will be more accurate.

12.6 Conclusions

The accuracy model of the NC process device for ultra-precision machining of two-dimensional optical drum is established. The error sources of the NC process device are analyzed in detail. The errors of the horizontal and vertical shaft system are analyzed and calculated, which have a great impact on the repeated positioning accuracy. The important parts and key control components are designed and selected, and the error synthesis prediction is carried out. Compared with the measured results of the developed device, the model has good calculation accuracy and can be used in the design of NC process device.

References

1. G. Zhang, Y. Dai, Z. Lai, A novel force-based two-dimensional tool centre error identification method in single-point diamond turning. Precis. Eng. **70**, 92–109 (2021)
2. W.T. Zhang, J.Q. Wang, G.Q. Shi, Positioning accuracy analysis of precision turntable for diamond fly-cutting two-dimensional drum machining. New Technol. New Technol. **11**, 40–42 (2011)
3. G. Shi, W. W. Feng, G. Shi et al., Development of a servo dividing device for fly cutting the two-dimensional drum. Manuf. Automat. **40**(03), 27–29 (2018)
4. D.Y. Wang, *Analysis and Research on Positioning Error of Precision Turntable for 2D Rotary Drum Machining* (Changchun University of Science and Technology, 2015)
5. S. Guo, X. Mei, G. Jiang, Geometric accuracy enhancement of five-axis machine tool based on error analysis. Int. J. Adv. Manuf. Technol. **105**, 137–153 (2019)
6. Y. Liu, M. Wan, W.-J. Xing, Q.-B. Xiao, W.-H. Zhang, Generalized actual inverse kinematic model for compensating geometric errors in five-axis machine tools. Int. J. Mech. Sci. **145**, 299–317 (2018)

Chapter 13
Analysis and Control of Measurement Accuracy of Super High-Rise Building Engineering

Jiabin Yan

Abstract Super high-rise buildings have the characteristics of large investment, long construction period, various functions, complex system, super high structure, and difficult construction. In the early stage of project construction, sufficient analysis and research are required. According to the specific characteristics of the building, the surrounding environmental traffic, geology and hydrology, climate conditions and other conditions, combined with the subjective conditions of construction and construction, a comprehensive balance analysis is required to make various resource allocation plans and implement measures. In terms of dynamic deformation monitoring, through three technical means of measuring robot, high-frequency GPS, and photogrammetry, real-time monitoring and data analysis of buildings are carried out, and the deformation conditions of buildings are predicted in advance, and corrective measures are taken to realize the actual situation once. The research on the construction measurement technology of super high-rise buildings has a positive role in promoting the development of super high-rise building construction technology itself, and has also played a role in promoting and reforming domestic construction methods and construction concepts. Through the application research of this technology, it has reserved scientific and technological innovation talents for engineering projects and enterprises, and provided valuable experience for the construction measurement control of similar large-scale construction projects. It has significant technical, economic, and social benefits.

13.1 Introduction

Super high-rise buildings are large in volume and large in construction area, and the required investment is often over several billion yuan. The capital pressure of the construction unit is very heavy. The financial pressure is reflected in the high cost of the construction period. Once the project is delayed, it will often cause a sharp increase in investment costs and reduce investment returns [1]. The construction

J. Yan (✉)
Engineering College, Yunnan University of Business Management, 296 Haitun Road, Wuhua, Kunming 650106, Yunnan, China

G. Liu and F. Cen (eds.), *Advances in Precision Instruments and Optical Engineering*, Springer Proceedings in Physics 270, https://doi.org/10.1007/978-981-16-7258-3_13

difficulty and technical requirements of super high-rise buildings are higher than those of general construction projects. Therefore, the research on super high-rise construction technology is of great significance not only to the entire society but also to the construction industry. Surveying is the leading work of super high-rise building construction. It not only runs through the whole process of super high-rise building construction, but also an important means to connect the spatial position relationship of each divisional and subdivisional project [2]. However, due to many factors such as high measurement accuracy requirements, difficulty in erecting instruments and equipment at high altitude, limited visibility conditions, and the impact of meteorological factors on the swing of super high-rise buildings, etc., all factors will affect the construction measurement work. Controlling its construction process is the most difficult point in super high-rise surveying work [3]. Super high-rise buildings are buildings that extend vertically upwards. This feature determines that the construction of super high-rise buildings can only be carried out layer by layer. The working space is very small and the construction organization is very difficult. It is necessary to effectively use the working time and space to improve the construction efficiency.

Super high-rise buildings have the characteristics of huge investment, high construction period cost, multiple functions, complex system, super high structure, and difficult construction [4]. Therefore, before construction, we must first analyze the characteristics of the project in depth, clarify the key points of the project's construction technology, and then formulate a targeted construction technology route. The general contracting management improves the utilization efficiency of working time and space, and finally achieves the purpose of shortening the construction period and improving the efficiency of investment [5]. Relevant scholars believe that the selection of suitable templates in construction projects and the comprehensive application of multiple templates in the entire construction industry have become the development trend of formwork projects [6]. In view of the application of wood plywood formwork, he believes that there are mainly the following problems [7]. First, there is a problem with the overall quality of the product. There are many domestic manufacturers of wood plywood template, but most of the manufacturers' products are low in price, so their quality is not satisfactory. This resulted in less turnover of the formwork (about 3–5 times), resulting in a large amount of waste of wood. Second, the phenomenon of "small workshops" is widespread. There are more than 4,500 wood plywood formwork manufacturers, and the number is extremely large. However, most manufacturers have small production scales, backward equipment, and relatively low management level. This is also the reason that the wood plywood template produced by the manufacturer is of poor quality and weaker competitiveness. It is difficult to form a large-scale operation, and it is even more difficult to participate in international competition. At present, many construction and scientific research units at home and abroad are conducting research on the application of super high-rise building construction technology, but most of the units are still immature in the research of super high-rise building construction measurement technology, even if some units have already mastered the construction of super high-rise buildings [8]. The subject research of super high-rise building measurement technology is based on the needs of enterprises and the development trend of the construction industry. The comprehensive research from the basic theory of super high-rise building measurement to specific measurement methods has very important engineering practical and popularization value for the construction of super high-rise buildings.

At present, many large-scale construction enterprises with corresponding construction qualifications and construction experience have gradually formed their own characteristics of super high-rise building construction technology through continuous construction and production practice summary. However, the key technologies, methods and measures to deal with the main points and difficulties in the construction of super high-rise buildings have their own characteristics and cannot be directly applied. Dynamic deformation monitoring of super high-rise buildings obtains the vertical displacement, horizontal displacement, deflection and other deformation data of super high-rise buildings under self-weight load, wind load, sunshine, temperature and other external factors through monitoring. This provides a correction basis for construction surveys, so as to grasp the deformation laws of super high-rise buildings, and provide a basis for assessing the safety of buildings.

13.2 Super High-Rise Building Project Implementation Planning

13.2.1 Super High-Rise Building Construction Organization

Super high-rise building is a huge system project with long construction period and difficult organization. Only by strengthening overall planning can the smooth progress of super high-rise building construction be ensured. The effective means to strengthen the overall planning of super high-rise building construction is construction organization design. Construction organization design is to create the necessary production conditions for the completion of super high-rise building construction tasks, and to formulate advanced and reasonable construction technology. It is a basic technical and economic document that guides the preparation and construction of super high-rise buildings. The fundamental task of super high-rise building construction organization design is to make overall planning from five aspects of manpower, capital, materials, mechanical equipment, and construction methods under specific time and space constraints. The organized, planned, and orderly construction of super high-rise buildings ensures the smooth realization of the construction quality, safety, construction period, and cost targets of the entire project.

Construction organization design is also an important means of scientific management of construction projects and an important basis for construction resource organization. It has the dual functions of strategic deployment and tactical arrangements. The design of construction organization can enhance the system of general contract management. Super high-rise buildings have many functions and complex systems. The construction process is a huge system project. Through construction organization and design, the general contractor can take the overall situation and coordinate

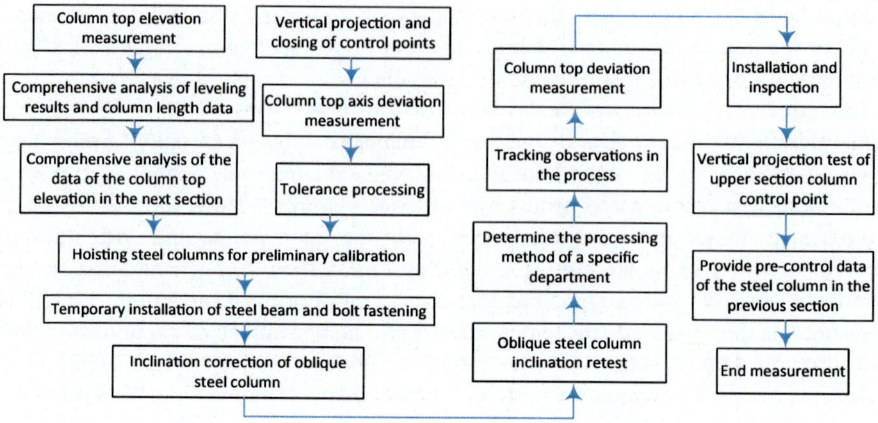

Fig. 13.1 Steel structure measurement process of super high-rise building engineering

all parties, and complex construction activities will have a unified action guide. The design of construction organization can enhance the predictability of general contract management. Super high-rise buildings have high construction technology content and high construction risks. Through the design of construction organization, the general contractor can grasp the various unfavorable conditions that may be encountered in the construction in advance, so as to make various preparations in advance, and make full use of various favorable conditions to eliminate hidden dangers in the construction. Construction organization design can enhance the coordination of general contract management. Due to the large number of units and personnel involved in the construction of super high-rise buildings, the coordination workload is large. Through the construction organization and design, the general contractor can close the relationship between the design and construction, technology and economy, front and rear of the project, and coordinate the various units. In short, through the design of construction organization, the general contractor can significantly improve the organization and management level of the construction of super high-rise buildings. Therefore, the preparation of construction organization design is the core of all work in the preparation stage of super high-rise building construction, and it occupies a very important position in construction organization and management. The steel structure hoisting measurement process of super high-rise building engineering is shown in Fig. 13.1.

13.2.2 Construction Schedule of Super High-Rise Buildings

The construction schedule is an important part of the construction organization design, and it is also an important means for the implementation of the project management plan. The construction schedule plan is the time plan for the construction of a project, which stipulates the start and end time, construction sequence and construction speed of the project construction, and is an effective tool to control the construction period. There are four main types of schedule plans: total schedule plan, unit project schedule plan, sub-project schedule plan, and resource requirement plan.

The overall construction schedule is the manifestation of various construction activities on the construction site in terms of time. The preparation of the overall construction schedule is to make time arrangements for all construction projects on the site according to the construction plan in the construction deployment and the project deployment procedures. Its role is to determine the construction period of each construction project and its main sub-projects, preparations, and full-site projects, as well as the start and completion dates, so as to determine the needs of labor, materials, finished products, semi-finished products, and construction machinery on the construction site. Therefore, the correct preparation of the overall construction schedule is an important condition for ensuring that the construction project is delivered on time and reducing the construction cost of super high-rise buildings.

13.2.3 The Construction Layout and Evaluation Method of Super High-Rise Buildings

The construction layout is the basis for site management and the realization of civilized construction, and is an important content of construction organization design. It has strong technical, economic, and policy characteristics, and requires overall planning and careful treatment. The general construction plan shall be reasonably arranged for the layout of construction machinery and equipment, storage yards of materials and components, on-site processing sites, on-site temporary transportation roads, temporary water supply and power supply lines, and other temporary facilities.

The construction period of super high-rise buildings is long and has obvious phase characteristics. Therefore, the construction layout should be dynamically adjusted in time to meet the requirements of the construction process at each stage. Before compiling the general construction plan, the construction steps should be determined first, and then the construction plan divided by stages should be compiled according to the different stages of the project schedule. Generally, it can be divided into the stages of earth excavation, foundation construction, superstructure construction, and mechanical and electrical installation and decoration. In order to reduce construction investment, attention should be paid to the orderly conversion in the dynamic adjustment of the construction layout, and the adjustment of the main construction temporary facilities (such as main roads, warehouses, offices, and water and electricity lines) should be avoided as much as possible, so as to realize the high sharing of the main construction temporary facilities at all stages.

The analytic hierarchy process provides a simpler evaluation method for decision-making events that cannot be expressed by quantitative analysis. According to the degree of understanding and grasp of the problem in the previous step, the various elements contained in the research problem are classified and grouped according to the principle of whether they have common attributes, and the elements with common properties are regarded as some elements in a new level of analysis. These elements themselves are divided accordingly according to another characteristic to form higher-level factors different from this kind, and so on until finally a goal level of the highest level is formed. This is a general goal level that all the element research and evaluation must achieve.

With the numerical scale, you can start in the lowest level and use the corresponding judgment matrix to express it.

$$B = \begin{pmatrix} b_{11} & \cdots\cdots & b_{1n} \\ b_{21} & \ddots \ \ddots & b_{2n} \\ \vdots & \ddots \ \ddots & \vdots \\ b_{41} & \cdots\cdots & b_{4n} \end{pmatrix} \tag{13.1}$$

The value of b_{ij} in the judgment matrix is determined after comprehensive analysis of the collected data and information, referring to expert opinions and the experience of relevant analysts. Under normal circumstances, in the analytic hierarchy process, people refer to the consistency test as *CI*, and its expression is

$$CI = (\lambda_{\max} - n)/(n - 1) \tag{13.2}$$

For quantitative indicators, fuzzy subsets can be determined:

$$r_{nt}^{t} = \left[r_1 \ r_2 \ r_3 \ r_4 \right] \tag{13.3}$$

Calculate the comprehensive evaluation about the V_t of each evaluation criterion layer:

$$B^t = r_{nt}^t \cdot r_1 \cdot r_4 \tag{13.4}$$

Take the rank matrix C to obtain the total evaluation value:

$$F = C^T \cdot B \cdot r_{nt}^t \tag{13.5}$$

13.3 Dynamic Deformation Monitoring of Super High-Rise Building Engineering Structure

13.3.1 Fully Automatic Robot Measurement

We establish a reference station at a location far away from the building with good visibility, and set up a rear-view station at a location with a wide field of view and unobstructed background, and use forced centering to ensure point accuracy. We set up instruments at the base station every day, and set up a prism at the rear-view station, and perform automatic observation after the positioning is completed. To ensure the comparability of monitoring data, the monitoring time is set from 6 a.m. to 8 p.m., and the monitoring interval is 15 min. The daily data is processed and analyzed by professional software, and the change curve of observation points within a day is obtained, and the change trend of the building body is grasped.

On the whole, the change of the observation point on the core tube is smaller than that of the outer steel structure, indicating that the stability of the core tube structure is stronger than that of the outer steel structure. The displacement change of the observation point changes with time, showing a fluctuating trend, and the whole shows an irregular curve change. The displacement change measurement accuracy of the four monitoring points is shown in Fig. 13.2.

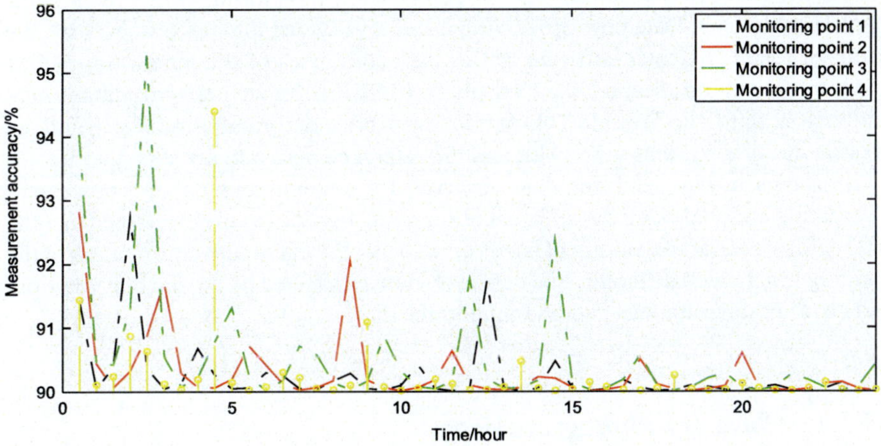

Fig. 13.2 Measurement accuracy of displacement changes of each monitoring point

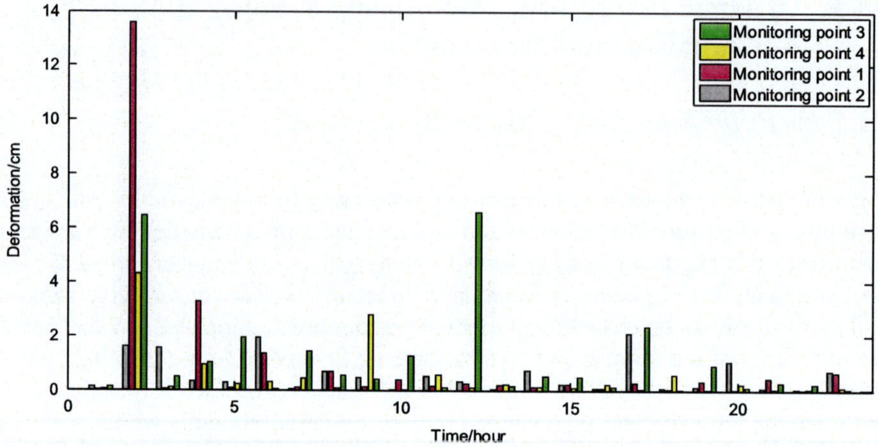

Fig. 13.3 Deformation monitored by GPS

13.3.2 GPS Monitoring Analysis

During the research work, we set up mobile stations on C1, C2, C3, C4 anchor bolts, and collect top deformation data in real time. After the data collection is completed, the data is processed according to the deformation monitoring GPS special software Trimble T4D. Data post-processing is mainly divided into three steps: first, the Trimble Business Center software settles the coordinates of the monitoring points and reference points; second, the Convert To RINEX software performs data format conversion; third, the Trimble T4D software performs deformation analysis, and the coordinates of the monitoring point and the reference point are settled.

As shown in Fig. 13.3, the data observed for a continuous day are processed. Comparing and analyzing the different observation results, it can be obtained that the total deformation at each moment is less than 14 cm; the maximum deformation of the building is not much different. The data is obviously affected by on-site construction, and obvious abnormalities can be found in the data.

13.3.3 Close-Up Photogrammetry

Close-range photogrammetry can accurately record the deformation information of any point of the object in an instant, and then obtain the instant point position relationship; the field time is short, the labor intensity is small, and the operation method is more flexible depending on the purpose of the operation. The image information is rich, the display ability is objective, and it can provide complete and instant three-dimensional spatial information; it can measure the dynamic target under the condition of the synchronization device; it does not need to touch the measured object.

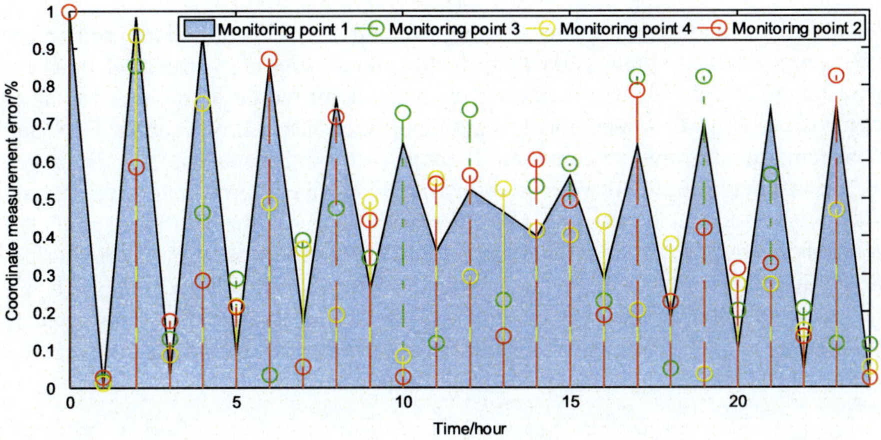

Fig. 13.4 Coordinate measurement error of each monitoring point

The requirements are high, but compared with the traditional geodetic survey, it can greatly reduce the workload of the field; the photo can be stored for a long time, which is conducive to inspection, analysis, and comparison.

Two high-precision camera lenses are set up at the control point on the side of the building every day, and the monitoring points are taken synchronously. During this period, the tripod must not be moved or touched. A special shutter release is used to ensure that the position of the camera does not change. We take photographic observations of the same side building for several consecutive days, and analyze the deflection and deformation of the super high-rise building through monitoring image processing at different times.

In the actual shooting process, the quality of the photos taken directly affects the processing of the office, which is significantly affected by wind and air visibility, and the accuracy of the results is poor when the weather conditions are bad. According to the changes in the observation results over multiple days, the overall deformation of the building is not large, and the top deformation error is within 1%. The coordinate measurement error of each monitoring point is shown in Fig. 13.4.

13.4 Conclusion

This article analyzes and studies the general ideas, construction essentials, and difficulties of super high-rise building construction planning, and summarizes a set of effective construction planning ideas and methods suitable for the construction and production of the enterprise. Preliminary discussion on the key points and implementation solutions of difficulties in the construction of super high-rise buildings has laid the foundation for the implementation of the next stage of the project. In terms of dynamic deformation monitoring, starting from three aspects of building settlement monitoring, elastic compression deformation monitoring, and structural dynamic deformation monitoring, the method of dynamic deformation monitoring for super high-rise buildings is studied; three technologies such as measuring robots, GPS, and close-range photogrammetry are used. Through comparative analysis, the advantages and disadvantages of the three monitoring methods are explored, and effective data with certain guidance is obtained. The deformation status of the building is predicted in advance, and the actual situation is tested and set up at one time, reducing the possibility of rework. Through the application research of this technology, it not only reserves scientific and technological innovation talents for engineering projects and enterprises, but also provides valuable experience for the construction survey control of the project.

References

1. S. Chen, Y.T. Yue, C. Wang, S.G. Yue, X. Pan, Predictive research on maximum electric load of super high—rise buildings. Build. Electr. **38**(05), 34–38 (2019)
2. P. Ripka, M. Mirzaei, A. Chirtsov, J. Vyhnanek, Transformer position sensor for a pneumatic cylinder. Sens. Actuators A Phys. **294**, 91–101 (2019)
3. Q. Guo, G. Shi, C. He, D. Wang, Compound compensation control for the electro-hydraulic position control system supplied with oil by variable-pressure accumulators. Chin. Hydraul. Pneumat. **9**, 1–8 (2016)
4. Y. Su, H. Zhou, H. Shen, Y. Wang, High-sensitivity and real-time displacement sensor based on polarization properties in fiber. Opt. Fiber Technol. **46**, 24–29 (2018)
5. X. Wang, S. Huang, G. Li, C. Li, W. Zhang, Theoretical analysis of the influence of plane control network on lateral breakthrough error of long immersed tunnel. J. Appl. Geodesy **14**(3), 241–251 (2020)
6. O.U. Lashmanov, A.S. Vasilev, A.V. Vasileva, A.G. Anisimov, V.V. Korotaev, High-precision absolute linear encoder based on a standard calibrated scale. Measurement **123**, 226–234 (2018)
7. J. Ma, Z. Yang, Z. Shi, C. Liu, H. Yin, X. Zhang, Adjustment options for a survey network with magnetic levitation gyro data in an immersed under-sea tunnel. Surv. Rev. **51**(367), 373–386 (2019)
8. H. Liu et al., An accelerometer with integrative intensity-modulated optical encoder and patterned leaf spring for low-frequency vibration monitoring. Sens. Actuators A Phys. **251**, 75–83 (2016)

Chapter 14
Multimedia Monitoring System for Gas Pressure Regulating Station

Renxuan Fu, Liang Xue, and Qinghua Wang

Abstract The current gas pressure regulating station of SCADA system still has some problems that it only obtains the operation data of the regulating station and cannot represent the failure of regulating stations or the on-site situation in alarm status. In order to solve these problems, the 5G network is used to design the remote viewing function, which can realize the video monitoring of all kinds of equipment and every corner of the voltage regulating station, find and deal with the accident in time, and improve the security and reliability of the monitoring system.

14.1 Introduction

Natural gas is a new energy of high quality, high efficiency, and clean in this century. It has attracted the attention of countries all over the world for its advantages of energy saving, environmental protection, economy, and convenience. Accelerating the development and utilization of natural gas is of great strategic importance to improving the energy structure, protecting the ecological environment, improving people's quality of life, and saving energy. In 2019, the natural gas consumption of China has reached over 310 billion cubic meters, an increase of about 10% year-on-year, and is forecast to reach 366 billion cubic meters by 2030, growing year-on-year.

Gas regulating station is an important facility in the city gas transmission and distribution system, which can automatically adjust and stabilize the pressure in the pipe network. According to the pressure of Inlet and outlet pipe, they can be divided into high and medium pressure, high and low pressure, medium and low pressure regulating stations, etc. Moreover, according to the object of service, they can be

R. Fu (✉) · Q. Wang
School of Electrical and Mechanical Engineering, Guangdong Polytechnic of Industry and Commerce, Guangzhou, China
e-mail: rxfu@163.com

L. Xue
Guangdong University of Education, Guangzhou, China

Institute of Data Science, City University of Macau, Macau, China

G. Liu and F. Cen (eds.), *Advances in Precision Instruments and Optical Engineering*, Springer Proceedings in Physics 270, https://doi.org/10.1007/978-981-16-7258-3_14

divided into regional regulating stations supplying a certain area and customer regulating stations serving individual buildings or industrial enterprises. Gas regulating stations involve gas regulators, filters to remove solids in suspension, instruments to measure gas temperature and pressure, and control devices and safety devices. The regulating station is a safety protection device that regulates the gas outlet pressure according to the given conditions to ensure that the outlet pressure does not exceed the specified value, thus ensuring the stability of the gas pressure source for downstream users. Failure in operation directly affects the stability of the downstream gas supply pressure, which can lead to abnormal gas supply pressure leading to the scrapping of industrial products and the failure of civil gas equipment, or to gas stoppages, gas leaks or explosive accidents, causing environmental pollution and panic in the surrounding area.

The SCADA systems with different functions have been built in gas pipeline networks, pressure regulating stations, gate stations, storage and distribution stations to achieve different degrees of "the four remote (telemetry, telematics, remote control, remote regulating)" functions. The SCADA system can collect real-time data for the gas dispatch and control center, and realize the monitoring and control of remote equipment. However, the SCADA system can only obtain data from the Remote Monitoring Station (RMS) and cannot represent other status of the RMS. The other status involves indoor flooding, theft conditions, entry of people in unattended RMS, and even whether the position of the actuator has changed during remote operation. Therefore, it is necessary to add a "remote viewing" function on the basis of the traditional "four remotes", which is completed by the remote video monitoring system. The data reported by the SCADA system cannot directly represent the operating status of the equipment on site, and the remote video monitoring system can make up for this deficiency. The implementation of the "remote view" function in the SCADA system improves the safety level of unmanned or unattended RTU. The video monitoring system enables the operators of the higher-level dispatch or monitoring center operators to directly "inspect" various equipment and corners of each unattended RTU, and monitor and record the safety of the unattended RTU and the operation of the equipment. That helps discover and handle the accident in time, improve the safety and reliability of the monitoring system, and provide relevant video information for analyzing the accident afterwards. It also enables the supervisor to strengthen the supervision and management of safety production, to keep abreast of the live situation, to implement more scientific production scheduling and decision-making, to greatly improve the management level of gas production scheduling, and to deal with emergencies at the regulating station.

At present, gas companies generally adopt the independent operation mode of SCADA system and video monitoring system. That operation mode cannot meet the need of the increasing level of automation in gas system dispatching. There is an urgent need to combine the functions of the two systems organically and form a multimedia SCADA monitoring system by which video monitoring can be linked using the data provided. The data information of the SCADA can also be represented visually through the video monitoring.

14.2 System Architecture Design

14.2.1 System Components

The architecture of the multimedia SCADA monitoring system is shown in Fig. 14.1.

The system is composed of three parts: monitoring terminal of the regulating station, wireless network, and control center.

The monitoring terminal of the regulating station consists of data acquisition terminal, power supply, 5G router, webcam, etc.

The 5G wireless communication network means the transmission channel for data or video.

The control center is comprised of a firewall router, data acquisition server, video server, communication server, database server, operator station, engineer station,

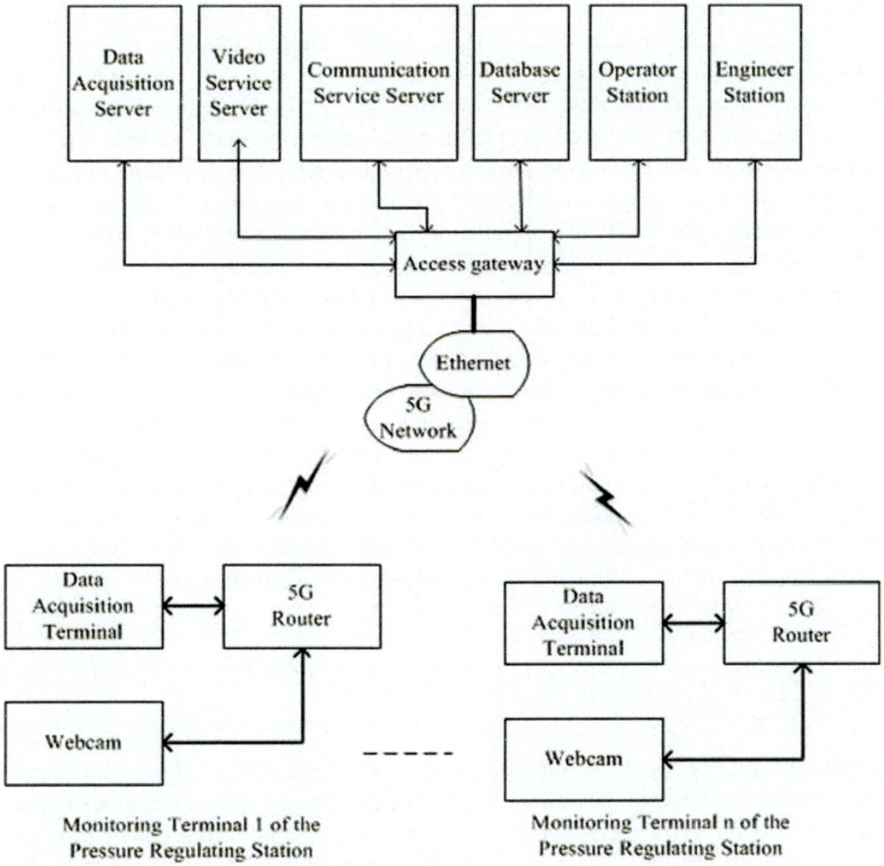

Fig. 14.1 System architecture diagram

etc. The communication server of the control center must have a static IP address assigned by the network administrator, and the administrator is able to map a static public IP to the IP of the communication server by changing the setting of the router. The port number can be set as 5002 or other, and through the setting of the router for the bandwidth of the IP segment where the central communication server is located should be at least 10 M.

14.2.2 Design of the Communication Network

The main transmission methods of video monitoring systems are two types: wireless and wired. Viewing from reliability and stability, the wired transmission method has advantages, but the wiring is difficult and the investment cost is high. To choose a wireless network is a good choice for users who do not need 24-h uninterrupted monitoring.

5G wireless communication networks can meet the requirements of both data and video image transmission [1–3]. 5G is different from 3G/4G networks in that there is only one 5G communication standard in the world, designed with three major application scenarios to support different functional application needs. URLLC is highly reliable, low latency and extremely high availability, including industrial applications, traffic safety, control, remote manufacturing, remote training. eMBB is able to meet high bandwidth application scenarios such as 4K/8K UHD video and VR/AR high traffic applications. mMTC is able to meet the requirement of data connection and transmission of a large number of low-power terminals.

In 4G communication network, monitoring video has the problems of the blurring of image quality, low transmission rate, poor night shooting, etc. However, 5G network has good bearing capacity for 4K/8K ultra-HD video and its transmission peak can reach 10 GBit/s high-speed transmission rate, improve the blurring of image quality, low transmission rate, and other problems. Also, it can provide faster and more high-definition monitoring image and data, which can complete the video capture, content encoding and network transmission of the video monitoring system. It has a broader application space in the fields of telemedicine, remote live broadcast, security monitoring, real-time display, and streetscape acquisition.

14.2.3 How the System Works

In order to save the video transmission traffic, when there is an alarm, the center automatically connects to the video server of the regulating station raised an alarm. The real-time image of the scene is transmitted to the monitoring center through the 5G router. When there is no alarm or the center does not monitor the video of the regulating station, the 5G router does not transmit the video image and does not generate video traffic.

14.3 System Function Design

The multimedia monitoring system of gas regulating station integrates the SCADA system and the video monitoring system, which has the functions of SCADA system and video monitoring, and also has the function of integration and innovation, realizing the data integration and linkage of SCADA and video monitoring [4, 5]. When the SCADA system carries out remote control operation, the multimedia monitoring system will switch to the video of the monitoring system, so that the operator can understand whether the operated object is abnormal.

When the SCADA system detects an important alarm, the video of the remote monitoring system will be switched to confirm the alarm. The system functions are described as the following:

14.3.1 Data Acquisition

The system collects data such as inlet pressure, outlet pressure, and temperature and flow rate of the regulating station. The system can flexibly set the cycle of data acquisition. A shorter cycle of acquisition is set for frequent changes and important regulating station. A longer cycle of acquisition is set for slow changes and less important regulating station. That not only satisfies the system's requirements for data acquisition, but also saves wireless transmission data traffic.

14.3.2 Data Storage

It means saving data, forming data reports, viewing real-time data, querying historical data, and performing data analysis.

14.3.3 Linkage of Electronic Map

The map view allows quickly locating each regulating station, obtaining key operational data such as pressure and flow, and understanding the operational profile of each regulating station to achieve a clear overview and control the overall situations.

The network video monitoring system uses the electronic map to arm the surveillance front end, which can conveniently browse the front-end lens and indicate the alarm information. The front-end monitoring point can be deployed and controlled on the electronic map through the creation of an electronic map, and the corresponding monitoring point can be easily found through the geographic information on the map.

The created electronic map can also be used as a unit element for layout management and inspection group management.

14.3.4 System Alarm

The system has a powerful alarm function that generates an audible and visual alarm when the pressure of the regulating station and other parameters are abnormal, notifying the operator in time and triggering video monitoring for fault.

The system supports the display, storage and querying of real-time and historical alarms, and can achieve functions such as audible alarms, WeChat alarms and email alarms. The system does not simply record alarm information, but also records all data and conditions related to the alarm into a database for the accident tracing and the operation analysis of equipment.

Alarm displays are available in the form of real-time alarm table, historical alarm table, and alarm enquiry table.

Alarms can be prioritized. The different priority alarms can be displayed in the window with different colors and the alarm messages can be sorted by priority.

14.3.5 Trend Curve

The system provides a wealth of curve functions, so that users can conveniently view the real-time and historical curves of pressure, flow, etc.

Trend curve supports the online switching of the real-time or historical curve. Moreover, it supports flexible display and analysis of changes in real-time monitoring data/historical data, trend comparison of similar data in the same period, and trend comparison of the same data in different periods.

14.3.6 Smart Report

The system provides an embedded report system, which allows engineers to set up any report format and realize various operations, data conversion, statistical analysis and printing through the report function provided by the system. The system allows for both real-time and historical data reporting, as well as online report format modification and EXCEL export support.

14.3.7 Hierarchical Management

The system provides an advanced user-based security management. Each user must be allowed or prohibited from monitoring in the monitoring system, according to the pre-defined access rights and scope of use. Also, the system does not allow users to exceed the limits of data access and control remote devices at will for the safety of system operation.

The system divides the operator rights into three levels: system senior administrator, administrator, and operator. The system administrator can not only perform all operations on all monitoring outlets, but also its another important duty is to assign all kinds of operation rights to each operator, i.e., all operators' rights are assigned by the system administrator and in the control signal can be occupied according to the user's rights than its lower level user control rights. Specific operational functions that can be set include switching on and off real-time video recording procedures, system parameter configuration, real-time monitoring, playback viewing, log viewing, etc.

If you need to change users while the system is running, you do not have to shut down the system, just use the re-login function. At this point, the system remains operational and any current actions in progress (e.g., video recording, etc.) are not affected in any way.

The system administrator can do management at any designated (authorized) terminal on the network.

14.3.8 System Log

The system automatically records the user's login to the system and the actions in the monitoring system (e.g., automatic recording of an operator's remote operations), allowing for easy traceability of incidents and regulating the operator's behavior.

14.3.9 Video Surveillance

Users can transmit full-color images, sound, and full-duplex data in real time over a standard Ethernet network, either locally or remotely. Any computer on the network can become the monitoring host when authorized. In addition, multiple authorized network users can monitor in real time or play back one or more monitoring sites.

The system adopts a highly parallelized architecture which can simultaneously and independently process each channel of image and sound efficiently without any influence on each other, ensuring that the megapixel HD camera has the exclusive high-performance indicator of 25 frames per second. At the same time, the display of the image is completely separated from the video processing, so the user can open and close the monitoring screen at any time. In addition, the system can perform tasks such as alarm and video recording when the screen is not displayed. It will not happen that the recording function that was originally opened is closed due to the closing of the screen.

14.3.10 Video Display

When viewing the live screen, users can choose different modes of viewing, i.e., 1/4/6/7/8/9/10/12/13/16 modes. Users can choose the viewing mode according to the specific outlets and double-click a screen to zoom in until it is viewed in full screen.

In the case of a large number of monitoring points and the user wants to play on a single screen, the user can choose the group patrol mode to monitor conveniently. The multi-channel screen can be grouped at will according to the requirement, and the on-screen time of each group can be set. The group patrol function can be started or stopped at any time according to user needs.

Users can also set a certain screen layout to save according to their own usage habit and requirement, which will be automatically open next time. The user can set the virtual lens as needed to achieve "one wall with a panoramic view".

14.3.11 Video Recording Function

The default recording rate is 25 frames per second (PAL), but the user can adjust the recording rate to reduce the load on the hard disk and the system. Each video data recording parameter can be set independently of each other without any interference.

System recording functions include real-time recording, timed recording, motion detection pre-trigger recording, and alarm pre-trigger recording.

Live video mode

The system provides a real-time recording function that can easily and quickly activate the recording function for a specified monitoring channel when required until the user disengages the recording.

Timed recording mode

The system provides a timer recording function, which can set any monitoring channel recording start and end period by date and time. Users can assign the recording actions that should be performed at various times of the day according to their own industry characteristics and requirements, so that the system can automatically operate according to a preset schedule. That function allows the system to be left unattended in many cases to reduce the human resources consumption of the user.

14.3.12 Motion Detection Pre-trigger Recording Mode

The megapixel HD camera has a built-in image motion sensor that automatically detects significant movement of objects within the monitored frame and triggers an alarm if the range of movement exceeds the limit. The parameters of the image motion sensor can be set in the menu to meet the requirements of a particular application and the alarm can be triggered locally in one or several key areas within the camera's visible monitoring range, and the range of movement of the moving object in the frame can be set as required.

The image motion detection triggers the linkage of the alarm and the video recording functions. After the user activates the motion detection pre-triggered recording function, the system links the recording action with the motion detection triggering alarm action. Once the user-specified monitoring channel detects the movement of an object, the system starts to record in the monitoring channel according to the video parameters set by the user. It will stop recording until the alarm is lifted. Motion detection can also be linked with display output, client information prompt, client voice prompt, going to the PTZ preset position, client image output, and alarm output channel switch output.

14.3.13 Intelligent Retrieval and Playback

The system provides intelligent and fast retrieval and playback of video data. When playing back video data, you can search by time, location, camera lens, alarm event and other factors, which greatly reduces the search time and complexity. With that, users can quickly find the video they need which has a variety of playback methods, according to the user requirements for normal speed, fast, slow, frame-by-frame playback, or others. The playback screen can be manually resized by dragging and dropping, or quickly scaled the screen to original size, doubled and full screen with the click of a button. The flexible operation mode enhances the viewing effect when the user plays back the video and facilitates the identification and forensics.

The system also supports a snapshot retrieval function, which allows you to break down the retrieved video time within a certain period, so that the target video to be queried can be found in the shortest possible time.

14.3.14 PTZ Camera Control

To achieve the control function of the high-speed ball, in the graphical interface of the management software of the network video centralized monitoring system, there are command buttons for rotation, draw near/far, focus, aperture, etc. When the user clicks the button of mouse, the system sends a control command to the Pan-Tilt or lens through the Pan-Tilt decoder to realize the control function of the Pan-Tilt and lens.

14.3.15 Linkage Control

When the SCADA system is operated remotely, in order to ensure the safety of the operation, the video in the monitoring system will be linked to switch, so that the operator can understand whether the operated object is abnormal.

If the SCADA system detects the alarm of important equipment, it will link to the switch of the remote viewing video.

When the SCADA information is synthesized and anomalies are detected in the important equipment, the remote viewing video should be switched.

If the SCADA system conducts a normal inspection, whether to switch the remote viewing video by setting can be considered.

If the remote viewing detects an environmental alarm, or the processing result of the video monitoring information generates an environmental alarm, not only the remote viewing video must be switched, but also the alarm processing of the SCADA system must be linked.

14.4 Conclusion

The multimedia monitoring system of the gas pressure regulating station not only realizes the data monitoring and control function of the SCADA system, but also realizes the video monitoring function of the pressure regulating station. In addition, it realizes the linkage function of the SCADA system and the video monitoring system. Moreover, it improves the management efficiency of the pressure regulating station, enhances the safety and reliability of the monitoring system, reduces operating costs, produces better social and economic benefits, and has good application prospects.

References

1. Y. Sun, Analysis and research of 5G-based intelligent video surveillance system. Electr. World **16**, 173–174 (2020)
2. L. Hu, L. Zhang, Design of intelligent video surveillance system based on 5G. Telecommun. Eng. Technol. Standardizat. **12**, 55–58 (2018)
3. G. Hong, Analysis and research of 5G-based intelligent video surveillance system. Sci. Edu. J. (Electronic Edition) **27**, 263 (2019)
4. B. Zhong, Design and application of image monitoring system to SCADA system linkage. China New Technol. New Prod. **21**, 6–7 (2011)
5. R. Li, Interconnection of image monitoring system and SCADA system. Rural Electr. **2**, 40–42 (2009)

Chapter 15
Two-Position Initial Alignment Method Under Large Misalignment Angle Based on Adaptive Cubature Kalman Filter

Xiaofei He, Xiaorui Zheng, Liwei Qiu, and Lili Shi

Abstract The precision of the initial alignment is one of the most critical factors determining navigation accuracy. The linear initial alignment model results in a sizeable linear truncation error when external environmental factors, such as carrier swing, vibration, and gust disturbance. The sizeable linear truncation error reduces the accuracy of the traditional linear two-position Kalman filter initial alignment method. This paper uses a nonlinear initial alignment model with a large misalignment angle to solve this problem. We propose an adaptive cubature Kalman filter method based on an iterative noise covariance matrix. This algorithm improves the estimation accuracy by adaptive updating of the noise covariance matrix. It tunes the filter gain and assists with the process of time and measurement updating. In addition, the noise covariance estimate is only related to the innovation of the last moment. These factors make this algorithm easy to calculate. A two-position test shows that the accuracy of the heading angle is improved by 79%. At the same time, the convergence speed is shortened by 4 min when compared with the initial alignment method based on the adaptive cubature Kalman filter. The estimations of the gyro bias and accelerometer bias are closer to the actual value to verify our proposed algorithm further.

X. He (✉)
National University of Defense Technology, Changsha, China
e-mail: afnuaa@163.com

X. He · X. Zheng · L. Qiu · L. Shi
Beijing Aerospace Times Optical-Electronic Co., Ltd, Beijing, China
e-mail: zxr1327260499@163.com

L. Qiu
e-mail: 13581694158@139.com

L. Shi
e-mail: shilili0802@sina.com

G. Liu and F. Cen (eds.), *Advances in Precision Instruments and Optical Engineering*, Springer Proceedings in Physics 270, https://doi.org/10.1007/978-981-16-7258-3_15

15.1 Introduction

External environmental factors, such as carrier swing, vibration and gust disturbance, carrier mobility, and the low inertial device level, produce a significant error in the coarse alignment. For this situation, both of the simple linear error model for the inertial navigation system and the traditional linear two-position Kalman Filter are inapplicable [1]. Carrying out a two-position initial alignment method under a large misalignment angle can not only enhance the robustness and adaptability of the model but also improve the initial alignment accuracy. It achieves this accuracy by using a multi-position alignment to boost the system observability [2].

A nonlinear Kalman filter is used during the initial alignment at a large misalignment angle to estimate the attitude angle and the level of the inertial device. It can use techniques such as an Extended Kalman Filter (EKF), Unscented Kalman Filter (UKF) and Central Difference Kalman Filter (CDKF), Gauss–Hermite Kalman Filter (GHKF), Cubature Kalman Filter (CKF), or a Particle Filter (PF). The CKF has a relatively higher numerical stability and less computation when it is compared with the Sigma sampling algorithm. Theoretically, it is the closest approximation algorithm to a Bayesian filter when there is a strict theoretical derivation [3].

Although CKF has those advantages, they are not helpful when there is an adverse environment. A high noise level of the inertial navigation measurement system produces significant uncertainties and modifications. To solve this problem when the GNSS signal is occluded in a SINS/GNSS integrated navigation system, a CKF algorithm has been proposed, which adapts to the concept of Mahalanobis distance standard to measure the noise covariance matrix and gain [4]. However, a Mahalanobis distance standard relies on empiricism and lacks a proper theoretical derivation. An enhanced RCKF was proposed for the SINS/GNSS integrated navigation system [5]. This method uses a Gaussian process orthogonal matrix to transfer Sigma sampling points. However, this algorithm needs extra time for training. In addition, an anti-interference CKF algorithm was proposed to solve the uncertainty of the system model and the divergence produced by the non-Gaussian measurement system [6]. This algorithm adjusts the prior covariance matrix and the measurement noise covariance matrix by using several factors. However, it lacks a strict formula derivation process.

In this paper, an adaptive CKF (ACKF) algorithm is proposed to compensate for the system and measurement noise in the initial alignment process when there is a significant misalignment angle. The filtering algorithm can track these noise sources by including the system and measurement noise in the covariance matrix iteration. In addition, it is not necessary to store this information as it is only used in the next initial stage. Therefore, this algorithm possesses lower computational complexity and improved real-time performance.

15.2 The Nonlinear Initial Alignment Error Model

In an adverse environment, the initial alignment error model is inapplicable for a linear model as it has a nonlinear large misalignment angle. This model considers the real attitude conversion matrix. The error model formula is given below.

15.2.1 Attitude Error Propagation Equation

$$\dot{\boldsymbol{\varphi}}^n = (\boldsymbol{I} - \boldsymbol{C}_n^p)\omega_{in}^n + \delta\omega_{in}^n - \boldsymbol{C}_b^n\varepsilon^b \tag{15.1}$$

In (15.1), $\dot{\boldsymbol{\varphi}}^n = [\dot{\varphi}_E\dot{\varphi}_N\dot{\varphi}_U]^T$ represents the attitude error differential equation of the northeast celestial direction. The variable, $\boldsymbol{I}$, represents the identity matrix, and the quantities, $\boldsymbol{n}$, $\boldsymbol{P}$, $\boldsymbol{i}$, and $\boldsymbol{b}$ represent the navigation coordinate frame, platform coordinate frame, inertial coordinate frame, and Earth coordinate frame, respectively. $\boldsymbol{C}_n^p$ is the direction cosine matrices from the nth frame to pth frame and $\boldsymbol{C}_b^n$ represents the matrix from the bth frame to nth frame. ω_{in}^n is the angular velocity and $\delta\omega_{in}^n$ is its error. $\varepsilon^b = [\varepsilon_x\varepsilon_y\varepsilon_z]^T$ represents the projection of the random gyro drift for the bth frame.

15.2.2 Velocity Error Propagation Equation

$$\begin{aligned}\delta\dot{\boldsymbol{V}}^n &= (\boldsymbol{I} - \boldsymbol{C}_p^n)\boldsymbol{C}_b^n\boldsymbol{f}^b - (2\delta\omega_{ie}^n + \delta\omega_{en}^n) \times \boldsymbol{V}^n \\ &\quad - (2\omega_{ie}^n + \omega_{en}^n) \times \delta\boldsymbol{V}^n + \boldsymbol{C}_b^n\nabla^b\end{aligned} \tag{15.2}$$

In (15.2), $\delta\dot{\boldsymbol{V}}^n = [\delta\dot{V}_E\delta\dot{V}_N]^T$ represents the differential equation of the velocity error in the northeast celestial direction. $\boldsymbol{f}^b$ means specific force information that is measured by acceleration. $\delta\omega_{ie}^n$ represents the angular velocity error of the projection of eth frame for the nth frame relative to the ith. ω_{en}^n represents the angular velocity error of the projection of n frame under n frame relative to eth frame. $\delta\omega_{en}^n$ is the error. $\boldsymbol{V}^n = [V_E V_N]^T$ represents the northeast celestial velocity and $\delta\boldsymbol{V}^n$ is its error. $\nabla^b = [\nabla_x\nabla_y]^T$ represents the accelerometer's constant bias.

15.2.3 Inertial Device Error Propagation Equation

$$\begin{cases}\dot{\varepsilon}_x = \dot{\varepsilon}_y = \dot{\varepsilon}_z = 0 \\ \dot{\nabla}_x = \dot{\nabla}_y = 0\end{cases} \tag{15.3}$$

Based on the inertial navigation error equations (15.1) and (15.3), the state equation and measurement equation are as follows:

$$\dot{x} = f(x, t) + Gw \tag{15.4}$$

$$z = Hx + v \tag{15.5}$$

In (15.4), the state vector is, $x = [\varphi_E \varphi_N \varphi_U \delta V_E \delta V_N \varepsilon_x \varepsilon_y \varepsilon_z \nabla_x \nabla_y]^T$, $f(x, t)$ is the nonlinear state equation, G is the system noise-driven array, and w is the system white noise. The variance satisfies the relationships: $E(ww^T) = Q$. In (15.6), $z = [0; 0]$ and $H = [0001000000; 0000100000]$. The quantity v represents the measurement noise and its variance satisfies the relationship: $E(vv^T) = R$.

15.3 The Introduction of CKF

The nonlinear initial alignment error model under a large misalignment angle is given in Sect. 15.2. By using the above ten-order error model (15.1–15.5) and including the Kalman filter, the initial attitude angle, the bias of gyro and accelerometer can be estimated.

A basic CKF is introduced in the following. The core of the CKF algorithm is to use the spherical radial cubature rule to solve a Gaussian multidimensional nonlinear weighted integral. That is an approximate rule that can be used with third-order, fifth-order, or even higher terms according to the requirements for polynomial approximation [7]. A third-order rule is the most widely used method as its use requires a reasonable amount of computation. This rule has been adopted in this paper.

Consider the following discrete nonlinear mathematical model:

$$\begin{cases} x_k = f(x_{k-1}) + Gw_{k-1} \\ z_k = H_k x_k + v_k \end{cases} \tag{15.6}$$

In these equations, x_k is the state vector and the measurement vector is z_k. Both terms are for the k moment. $f(x_{k-1})$ is a nonlinear state equation. G is the system noise-driven array. w_{k-1} is the system white noise. H_k is the measurement matrix. v_k is the measurement noise.

There are nine steps of CKF in evaluating these equations:

(a) Initialization

$$\begin{cases} \overline{\boldsymbol{x}}_0 = E[\boldsymbol{x}_0] \\ \boldsymbol{P}_0 = E[(\boldsymbol{x}_0 - \overline{\boldsymbol{x}}_0)(\boldsymbol{x}_0 - \overline{\boldsymbol{x}}_0)]^T] \end{cases} \tag{15.7}$$

$k = 1, 2, \ldots.$

(b) Generating the cubature point

$$\begin{cases} \boldsymbol{\chi}_{k-1,i} = \overline{\boldsymbol{x}}_{k-1} + \sqrt{n}\sqrt{\boldsymbol{P}_{k-1}}^{(i)}, i = 1, 2, \ldots, n \\ \boldsymbol{\chi}_{k-1,i} = \overline{\boldsymbol{x}}_{k-1} - \sqrt{n}\sqrt{\boldsymbol{P}_{k-1}}^{(i)}, i = n+1, n+2, \ldots, 2n \end{cases} \tag{15.8}$$

where the superscript (i) represents ith column of the matrix $\boldsymbol{P}$.

(c) Propagating the cubature point using the state equation

$$\boldsymbol{\chi}_{k/k-1,i} = \boldsymbol{f}(\boldsymbol{\chi}_{k-1,i}), i = 1, 2, \ldots 2n \tag{15.9}$$

(d) Calculating the prior estimation $\overline{\boldsymbol{x}}_{k/k-1}$ and covariance $\boldsymbol{P}_{k/k-1}$

$$\begin{cases} \overline{\boldsymbol{x}}_{k/k-1} = \frac{1}{2n}\sum\limits_{i=1}^{2n} \boldsymbol{\chi}_{k/k-1,i} \\ \boldsymbol{P}_{k/k-1} = \frac{1}{2n}\sum\limits_{i=1}^{2n} \boldsymbol{\chi}_{k/k-1,i}\boldsymbol{\chi}_{k/k-1,i}^T - \overline{\boldsymbol{x}}_{k/k-1}\overline{\boldsymbol{x}}_{k/k-1}^T + \boldsymbol{G}\boldsymbol{Q}\boldsymbol{G}^T \end{cases} \tag{15.10}$$

(e) Regenerating the cubature point

$$\begin{cases} \boldsymbol{\chi}_{k/k-1,i} = \overline{\boldsymbol{x}}_{k/k-1} + \sqrt{n}\sqrt{\boldsymbol{P}_{k/k-1}}^{(i)}, i = 1, 2, \ldots, n \\ \boldsymbol{\chi}_{k/k-1,i} = \overline{\boldsymbol{x}}_{k/k-1} - \sqrt{n}\sqrt{\boldsymbol{P}_{k/k-1}}^{(i)}, i = n+1, n+2, \ldots, 2n \end{cases} \tag{15.11}$$

(f) Propagating the cubature point using the measurement equation

$$\xi_{k/k-1,i} = \boldsymbol{h}(\boldsymbol{\chi}_{k/k-1,i}), i = 1, 2, \ldots 2n \tag{15.12}$$

In (15.12), $\boldsymbol{h}$ is the measurement equation.

(g) Calculating the prior predicted measurement vector

$$\overline{z}_{k,k-1} = \frac{1}{2n}\sum_{i=1}^{2n}\xi_{k/k-1,i} \tag{15.13}$$

(h) Calculating the variance $\boldsymbol{P}_{zz}$ and the covariance $\boldsymbol{P}_{xz}$

$$\begin{cases} \boldsymbol{P}_{zz} = \frac{1}{2n}\sum_{i=1}^{2n}\xi_{k/k-1,i}\xi_{k/k-1,i}^{T} - \overline{z}_{k/k-1}\overline{z}_{k/k-1}^{T} + R_k \\ \boldsymbol{P}_{xz} = \frac{1}{2n}\sum_{i=1}^{2n}\boldsymbol{\chi}_{k/k-1,i}\xi_{k/k-1,i}^{T} - \overline{\boldsymbol{x}}_{k/k-1}\overline{z}_{k/k-1}^{T} \end{cases} \tag{15.14}$$

(i) Calculating the gain $\boldsymbol{K}_k$, state estimation $\boldsymbol{x}_k$, and variance $\boldsymbol{P}_k$

$$\begin{cases} \boldsymbol{K}_k = \boldsymbol{P}_{xz}\boldsymbol{P}_{zz}^{-1} \\ \overline{\boldsymbol{x}}_k = \overline{\boldsymbol{x}}_{k/k-1} + \boldsymbol{K}_k(z_k - \overline{z}_{k/k-1}) \\ \boldsymbol{P}_k = \boldsymbol{P}_{k-1} - \boldsymbol{K}_k\boldsymbol{P}_{zz}\boldsymbol{K}_k^{-1} \end{cases} \tag{15.15}$$

15.4 The Derivation of ACKF

An introduction to the algorithm flow of CKF is discussed in the previous section. Equations (15.10) and (15.14) indicate that the system noise covariance matrix and the measurement noise covariance matrix in CKF are fixed. However, in the actual environment, the device noise is random. The fixed system noise and the measurement noise covariance matrix affect the filter estimation through the gain matrix. Therefore, we propose an adaptive method (ACKF) for measuring the noise covariance matrix to solve this problem.

The innovation, which is the difference between the latest measurement data and the measurement matrix multiplied by the initial predicted value, is an essential factor affecting the gain and the estimated state [8]. The innovation contains a component of the measurement noise, so many scholars use the innovation to evaluate the measurement noise.

In this section, ACKF is derived using the covariance matching principle and the covariance of the innovation in the window [9]. The estimated residual of the state system is

$$\delta_k = x_k - x_{k/k-1} \tag{15.16}$$

By calculating the mean value and the covariance of the M state residuals before k moment, we obtain the (15.17) and (15.18) [9].

$$\bar{\boldsymbol{\delta}}k = \frac{1}{M}\sum_{i=k-M+1}^{k}\boldsymbol{\delta}_i \tag{15.17}$$

$$\boldsymbol{P}_{\delta_k} = \frac{1}{M-1}\sum_{i=k-M+1}^{k}\left(\boldsymbol{\delta}_i - \bar{\boldsymbol{\delta}}_k\right)\left(\boldsymbol{\delta}_i - \bar{\boldsymbol{\delta}}_k\right)^T \tag{15.18}$$

In (15.17), M represents the sliding window factor, which means the number of residuals.

By taking the expected value of (15.18), we can obtain [10]

$$\boldsymbol{Q}_{\delta_k} = \frac{1}{M}\sum_{i=k-M+1}^{k}\left(\boldsymbol{F}_{i-1}\boldsymbol{P}_{i-1}\boldsymbol{F}_{i-1}^T - \boldsymbol{P}_i\right) + \boldsymbol{Q}_k \tag{15.19}$$

$$\begin{aligned}\boldsymbol{Q}_k = {} & \frac{1}{M-1}\sum_{i=k-M+1}^{k}\left(\boldsymbol{\delta}_i - \bar{\boldsymbol{\delta}}_k\right)\left(\boldsymbol{\delta}_i - \bar{\boldsymbol{\delta}}_k\right)^T \\ & - \frac{1}{M}\sum_{i=k-M+1}^{k}\left(\boldsymbol{F}_{i-1}\boldsymbol{P}_{i-1}\boldsymbol{F}_{i-1}^T - \boldsymbol{P}_i\right)\end{aligned} \tag{15.20}$$

From (15.20), we can derive

$$\begin{aligned}\boldsymbol{Q}_{k-1} = {} & \frac{1}{M-2}\sum_{i=k-M+1}^{k-1}\left(\boldsymbol{\delta}_i - \bar{\boldsymbol{\delta}}_k\right)\left(\boldsymbol{\delta}_i - \bar{\boldsymbol{\delta}}_k\right)^T \\ & - \frac{1}{M-1}\sum_{i=k-M+1}^{k-1}\left(\boldsymbol{F}_{i-1}\boldsymbol{P}_{i-1}\boldsymbol{F}_{i-1}^T - \boldsymbol{P}_i\right)\end{aligned} \tag{15.21}$$

By multiplying the left and right sides of (15.21) with $\frac{M-2}{M-1}$ and subtracting (15.20), we obtain

$$\begin{aligned}\boldsymbol{Q}_k = {} & \frac{M-2}{M-1}\boldsymbol{Q}_{k-1} + \left(\boldsymbol{\delta}_i - \bar{\boldsymbol{\delta}}_k\right)\left(\boldsymbol{\delta}_i - \bar{\boldsymbol{\delta}}_k\right)^T \\ & - \frac{1}{M(M-1)^2}\sum_{i=k-M+1}^{k-1}\left(\boldsymbol{F}_{i-1}\boldsymbol{P}_{i-1}\boldsymbol{F}_{i-1}^T - \boldsymbol{P}_i\right) \\ & - \frac{1}{M}\left(\boldsymbol{F}_{i-1}\boldsymbol{P}_{i-1}\boldsymbol{F}_{i-1}^T - \boldsymbol{P}_i\right)\end{aligned} \tag{15.22}$$

When M is large enough, the denominator $\frac{1}{M(M-1)^2}$ tends to zero. Therefore, the third monomial in (15.22) can be omitted

$$\boldsymbol{Q}_k = \frac{M-2}{M-1}\boldsymbol{Q}_{k-1} + \left(\boldsymbol{\delta}_i - \bar{\boldsymbol{\delta}}_k\right)\left(\boldsymbol{\delta}_i - \bar{\boldsymbol{\delta}}_k\right)^T$$

$$-\frac{1}{M}\left(\boldsymbol{F}_{i-1}\boldsymbol{P}_{i-1}\boldsymbol{F}_{i-1}^{T}-\boldsymbol{P}_{i}\right) \tag{15.23}$$

The estimation process of the system mean residual $\overline{\boldsymbol{\delta}}_k$ can be derived by (15.17) to obtain.

$$\overline{\boldsymbol{\delta}}_{k-1}=\frac{1}{M-1}\sum_{i=k-M+1}^{k-1}\boldsymbol{\delta}_i \tag{15.24}$$

We then multiply $\frac{M-1}{M}$ on both sides of (15.24) and then subtract (15.17) and obtain

$$\overline{\boldsymbol{\delta}}_k=\frac{M-1}{M}\overline{\boldsymbol{\delta}}_{k-1}+\frac{1}{M}\boldsymbol{\delta}_k \tag{15.25}$$

Equations (15.23) and (15.25) are recursive estimations of the system noise covariance matrix. These equations show that the estimation of the system noise covariance does not need to store the residuals and the mean of the residuals from all M moments. We only need the noise covariance and the mean of residuals at the last moment to achieve a recursive estimation of the noise covariance matrix. This procedure can improve the computation speed and the real-time capability of this algorithm. The derivation process of the adaptive measurement noise covariance is the same as the system noise covariance matrix, which is directly given below:

$$\begin{aligned}\boldsymbol{R}_k\approx{}&\frac{N-2}{N-1}\boldsymbol{R}_{k-1}+\left(\boldsymbol{\eta}_k-\overline{\boldsymbol{\eta}}_k\right)\left(\boldsymbol{\eta}_k-\overline{\boldsymbol{\eta}}_k\right)^T\\&-\frac{1}{N}\boldsymbol{H}_k\boldsymbol{P}_{k/k-1}\boldsymbol{H}_k^T\end{aligned} \tag{15.26}$$

$$\overline{\boldsymbol{\eta}}_k=\frac{N-1}{N}\overline{\boldsymbol{\eta}}_{k-1}+\frac{1}{N}\boldsymbol{\eta}_k \tag{15.27}$$

In (15.26) and (15.27), N represents the sliding window factor and $\overline{\boldsymbol{\eta}}_k$ represents the mean value of the innovation.

In combination with Sect. 15.3, the ACKF flow chart is shown in Fig. 15.1.

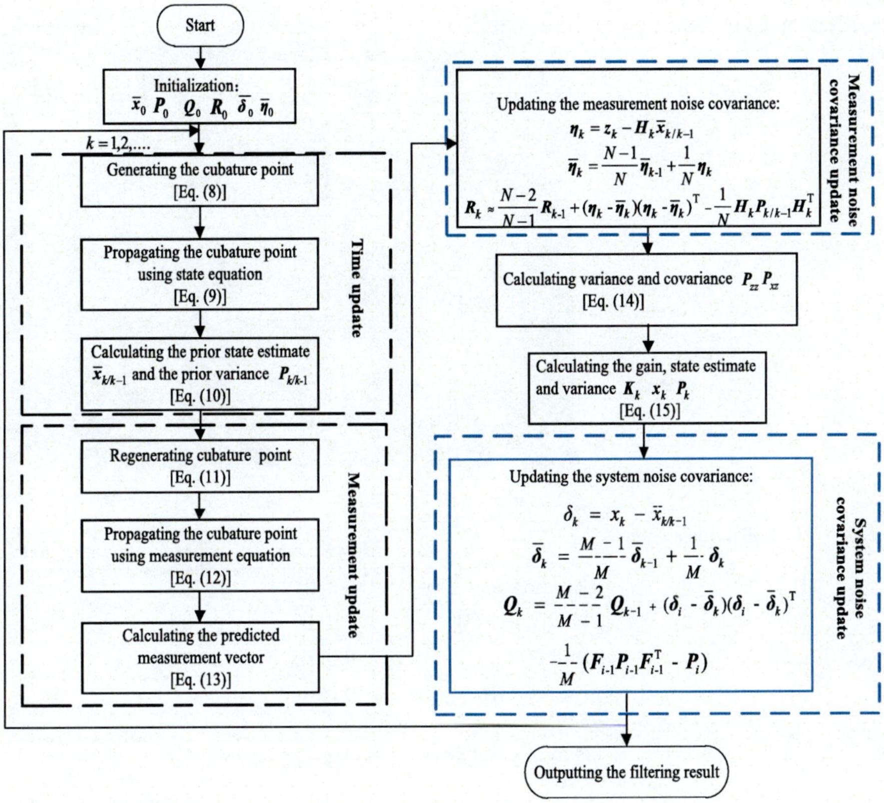

Fig. 15.1 ACKF flow chart

15.5 Results and Analysis

To verify the effectiveness of the proposed method, an inertial measurement unit (IMU) was fixed on a biaxial turntable. Figure 15.2 shows the experimental arrangement. At first, we did a static test for 2 min. The next step was to rotate the turntable by 180° clockwise around the celestial axis of the IMU. Finally, we did a static test for 20 min.

To evaluate the accuracy of this two methods (CKF and ACKF), the bias of gyro and acceleration was calibrated by the 19-position. The attitude angle of IMU was measured by the sighting instrument, which was used as the benchmark to evaluate the accuracy. To simulate the large misalignment angle in the actual environment, the misalignment angle was superimposed on the rough alignment attitude angle, as showed in Table 15.1.

The initial alignment results of the two methods are shown in Figs. 15.3, 15.4, 15.5.

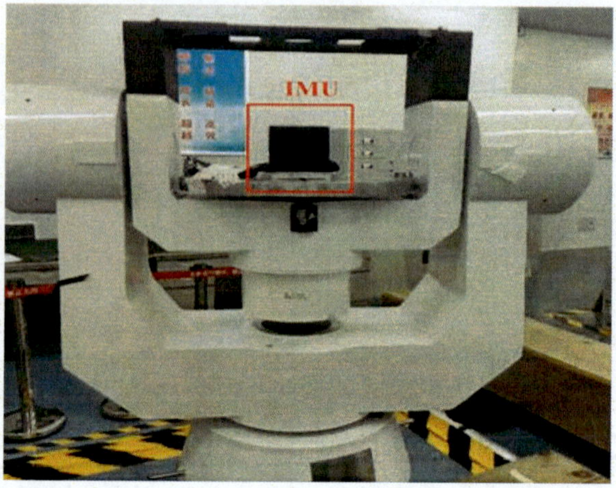

Fig. 15.2 Two-position test of IMU

Table 15.1 Benchmark data

	Benchmark values	Misalignment angle
Heading angle (°)	180.0101	0.5
Pitch angle (°)	0.0710	0.1
Roll (°)	0.0186	0.1
Bias of gyro X-axis (°/h)	−0.0841	/
Bias of gyro Y-axis (°/h)	−0.0342	/
Bias of accelerometer X-axis (mg)	0.3170	/
Bias of accelerometer Y-axis (mg)	−0.4442	/

The estimation result of ACKF for the heading angle and pitch angle is closer to the benchmark. The convergence speed is faster than CKF, because the bias estimation of the gyro X-axis (East) and the accelerometer Y-axis (North) is closer to the actual value by ACKF. The bias of gyro X-axis converged in about 150s (the second position). For the roll angle, the estimated results from these two methods are similar.

Previously, we made a qualitative comparison of the two methods. Table 15.2 shows is a quantitative analysis, which includes the RMSE results for each parameter error.

Table 15.2 shows that the two-position initial alignment method based on ACKF improves the heading angle to 0.08° compared with the CKF and also enhances the horizontal attitude angle of 3.96″ and 0.36″. In a constant drift estimation of gyro devices, ACKF is 0.47 ′/h and 0.02 ′/h superior to CKF. In this estimate of the

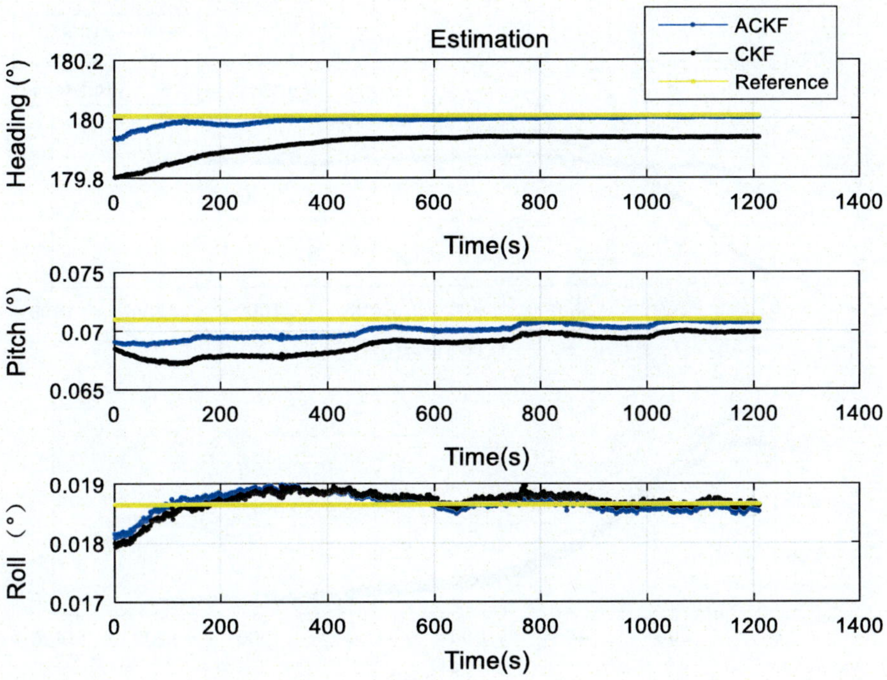

Fig. 15.3 Estimation of attitude angle

constant bias of horizontal accelerometer, ACKF is 0.2 μg and 12.9 μg better than CKF, respectively.

In summary, the ACKF method, which is based on a two-position initial alignment method, is superior to the traditional CKF from both the qualitative and quantitative analyses in terms of convergence speed and accuracy. This improvement demonstrates the effectiveness of ACKF.

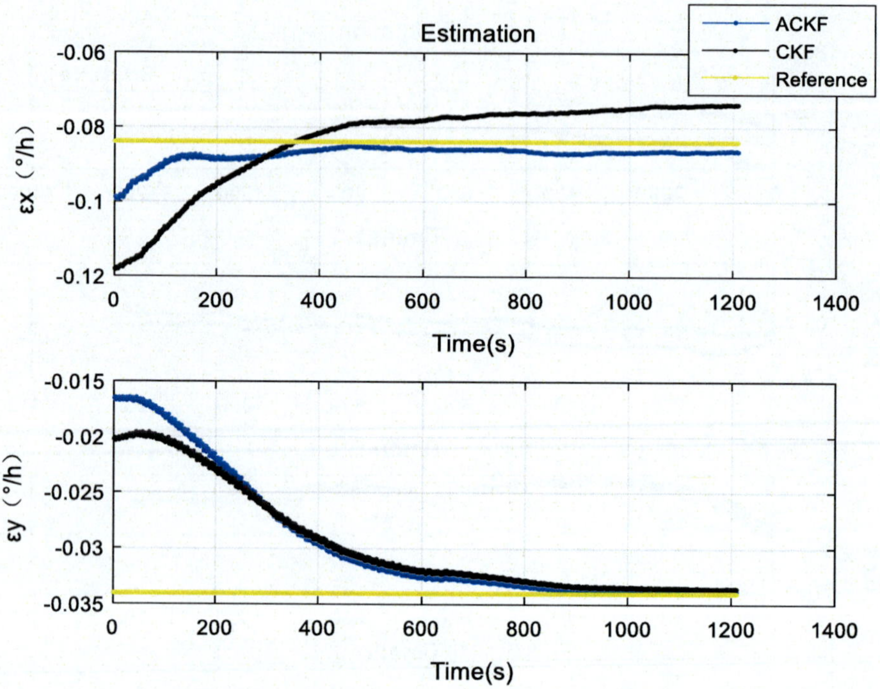

Fig. 15.4 Estimation of the bias of Gyros

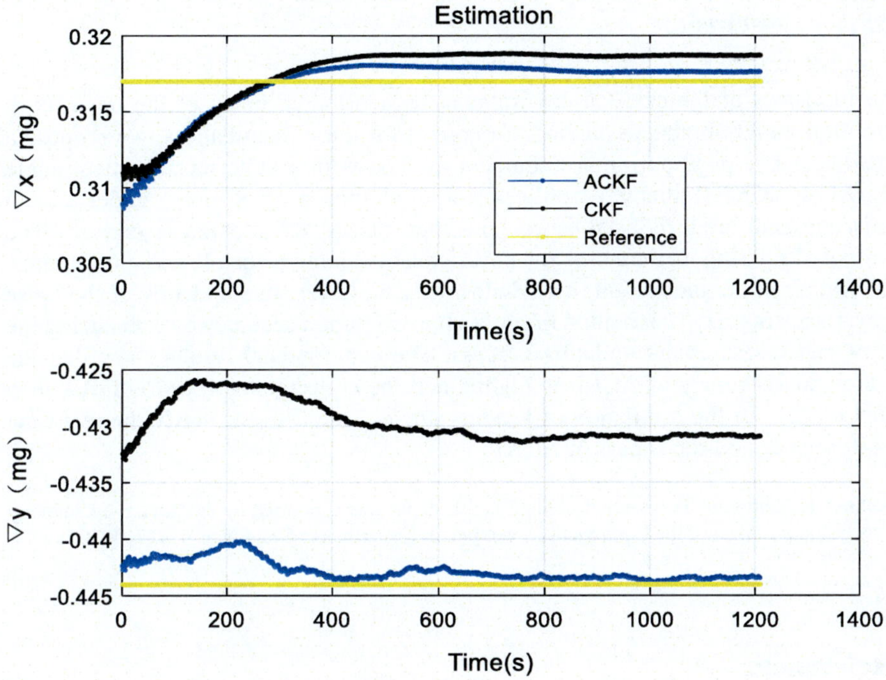

Fig. 15.5 Estimation of the bias of accelerometers

Table 15.2 Statistical results of the two alignment methods

Estimation	CKF (RMSE)	ACKF (RMSE)
Heading error (°)	0.1009	0.0211
Pitch error (″)	8.28	4.32
Roll error (″)	67.32	66.96
X-Gyro bias error (°/h)	0.0120	0.0041
Y-Gyro bias error (°/h)	0.0065	0.0062
X-Acc bias error (mg)	0.0023	0.0021
Y-Acc bias error (mg)	0.0144	0.0015

15.6 Conclusion

Disturbances of the external environment reduce the accuracy of the linear two-position initial alignment method. To correct for these disturbances, we propose a nonlinear two-position initial alignment method with a large misalignment angle based on ACKF. Initially, a nonlinear initial alignment error model under a large misalignment angle is established, and then the ACKF method is derived. The proposed filtering method uses the covariance matching principle and a covariance derivation of the innovation in a window to achieve iterative updating of the noise covariance matrix. The method improves the estimation accuracy by influencing the time and measurement updating process. The experimental results show that the estimation accuracy of ACKF is higher, and the convergence speed is faster than that of CKF in the estimation of heading angle. These results verify the real-time performance and robustness of ACKF.

Acknowledgements This work described in this paper was supported by the Inertial Technology Professional Project of the Equipment Development Department (Grant Nos. 41417040101).

References

1. G.J. Zhang, B. Hu et al., Nonlinear initial alignment based on quaternion error model, in *2018 IEEE CSAA Guidance, Navigation and Control Conference (GNCC)* (IEEE, 2018)
2. A. Zhang, Y. Liu, C. Wang, Research for multi-position alignment on stationary base of SINS, in *International Conference on Computers, Communications, Control and Automation (CCCA 2011)*, Hong Kong (2011)
3. M.V. Kulikova, G.Y. Kulikov, Square-rooting approaches to accurate mixed-type continuous-discrete extended and fifth-degree cubature Kalman filters. IET Radar Sonar Navigat. **14**(11) (2020)
4. B. Gao, G. Hu, X. Zhu et al., A robust cubature Kalman filter with abnormal observations identification using the Mahalanobis distance criterion for vehicular INS/GNSS integration **19**(23) (2019)
5. B. Cui, X. Wei, X. Chen et al., Performance enhancement of robust cubature Kalman filter for GNSS/INS based on Gaussian process quadrature. IEEE Access **99**, 1 (2020)
6. S. Guo, L. Chang, Y. Li et al., Robust fading cubature Kalman filter and its application in initial alignment of SINS. OPTIK **202** (2020)
7. W. Bo, P. Liu, A new cubature Kalman filter improved by backward iterative algorithm, in *International Conference on Intelligent Computation Technology & Automation* (IEEE, 2016)
8. S. Zou, J. Li et al., A nonlinear transfer alignment of distributed POS based on adaptive second-order divided difference filter. IEEE Sens. J. (2018)
9. I. Hashlamon, K. Erbatur, An improved real-time adaptive Kalman filter with recursive noise covariance updating rules. Turk. J. Electr. Eng. Comput. Sci. **24**, 524–540 (2016)
10. N.J. Hao, L.C. Yuan, Adaptive ensemble Kalman filter with recursive noise estimation. Control Decis. (2018)

Chapter 16
Rapid Detection and Measurement Method of Pixel-Level Crack Size Based on Convolutional Neural Network

Wentong Guo

Abstract Cracks on highways are important factors threatening the safe operation of highways. Timely and accurate detection of cracks can effectively avoid traffic accidents. In this paper, a fast detection and measurement method of pixel-level crack size based on convolutional neural network is proposed. The two-stage decision method is used to connect YOLOv3 and Mask R-CNN image recognition algorithms to locate and segment the cracks existing in expressway pavement. Finally, the topological characteristics of the target are extracted to complete the crack pixel size extraction.

16.1 Introduction

In the normal use of traffic infrastructure, it is very important to maintain a good running state for driving safety. Crack is one of the important factors that threaten the normal and safe operation of infrastructure [1]. Timely and accurate detection of crack development and propagation can effectively avoid the occurrence of major disaster accidents. Therefore, crack detection has important value in the field of transportation facilities.

In recent years, with the in-depth study of image processing technology, the use of computer equipment to identify and detect cracks has made significant progress. Peng et al. [2] proposed an improved Otsu threshold segmentation algorithm to remove the mark in the road image, and then used the adaptive iterative method to segment the image after removing the mark to obtain the crack image. Xu et al. [3] improved the selection of filtering parameters in Canny iteration method, which effectively improved the detection accuracy of bridge cracks. Fernandez et al. [4] and others used the decision tree heuristic algorithm to classify the crack image, but this work is not integrated into the real system in the simulation environment. Shi et al. [5] proposed the CrackForest model, which uses the random forest method to deal with, so as to reduce the influence of noise on the accuracy of crack detection.

W. Guo (✉)
School of Water Conservancy Engineering, Zhengzhou University, Zhengzhou 450001, China
e-mail: guowentong@stu.zzu.edu.cn

G. Liu and F. Cen (eds.), *Advances in Precision Instruments and Optical Engineering*, Springer Proceedings in Physics 270, https://doi.org/10.1007/978-981-16-7258-3_16

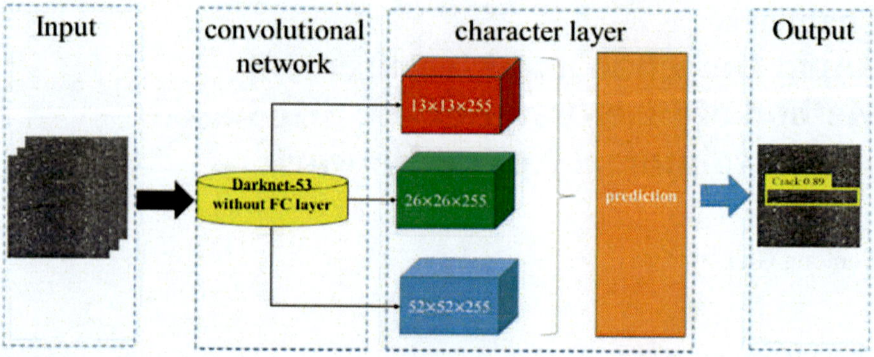

Fig. 16.1 YOLOv3 crack identification process

In this paper, a fast pixel-level crack size detection method based on convolutional neural network is proposed. Firstly, the crack is quickly located and identified by YOLOv3 recognition algorithm. On the basis of the target image, the Mask R-CNN segmentation algorithm is used to secondary process the crack and extract its topological characteristics. Finally, the pixel size information of the crack is obtained.

16.2 Principle of YOLOv3 Algorithm

The image recognition steps based on YOLOv3 algorithm [6] are mainly divided into three parts: image input, convolution network processing, and output of prediction feature layer. Firstly, the 3-channel image with the size of 416×416 is input. By using the network structure of Darknet-53, convolution operation is carried out in the first 52 layers of Darknet-53 network and a large number of residual layers are used in the network. Finally, the detection feature layer with the size of 13×13, 26×26, 52×52 was obtained by up-sampling and feature fusion for image recognition. Figure 16.1 is the implementation process of the Yolov3 algorithm.

16.3 Principle of Mask R-CNN Algorithm

Figure 16.2 is a crack detection network structure framework based on Mask RCNN segmentation algorithm [7], which mainly consists of CNN convolution network area, candidate region network, RoI Align layer, and Mask branch. Firstly, the image is input into the feature extraction network to obtain the feature map. A fixed ROI is set for each pixel position in the feature map and the ROI region is transmitted to the RPN network for binary classification to obtain the candidate box of the target object region. Finally, the target candidate region is classified by multiple categories.

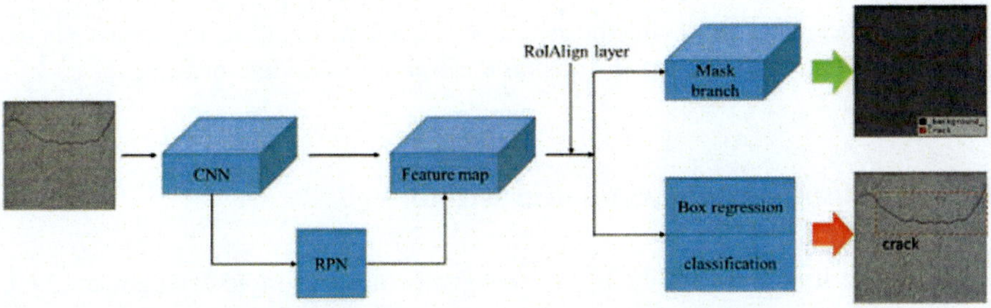

Fig. 16.2 Mask R-CNN crack segmentation process

The candidate box regression and the introduction of FCN generate Mask, and the segmentation task is finally completed.

16.4 Two-Stage Decision Method

In order to solve the problem of low detection accuracy of YOLOv3 algorithm and slow segmentation speed of Mask R-CNN algorithm, the author adopts two-stage decision method [8]. Firstly, the YOLOv3 algorithm with high detection speed is used for image recognition of the collected sample video to quickly locate the crack frame. Then, the Mask R-CNN algorithm is used for image processing of the crack frame after initial recognition to obtain the specific topology information of the crack in the target frame. Finally, the pixel size of the crack is extracted. The Mask R-CNN algorithm has high recognition accuracy and can be used for secondary processing

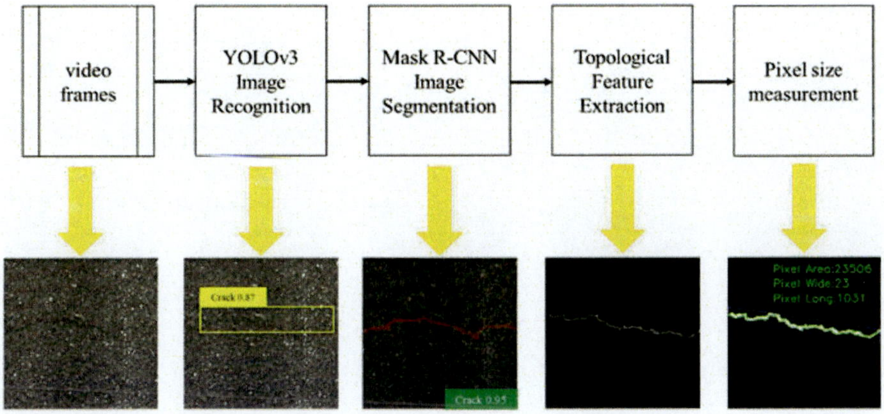

Fig. 16.3 Two-stage process of crack identification

of crack frames, thus improving the accuracy of crack recognition. Figure 16.3 is the application process of two-stage decision method in crack identification.

16.5 Methods Application and Results

In order to verify the feasibility of the pixel-level crack size rapid detection method, the cracks existing on some highways were tested. The experimental results are shown in Table 16.1. It can be seen that the average recognition speed of YOLOv3 reaches 45 FPS, which can meet the requirements of real-time detection. At the same time, the method can realize the work from crack recognition to crack pixel size information.

16.6 Conclusions

In this paper, a fast detection and measurement method of pixel-level crack size based on convolutional neural network is proposed. The experimental results show that the recognition speed of this method is 45 FPS, which can meet the requirements of real-time crack recognition. At the same time, through the Mask R-CNN algorithm for image segmentation, the topological characteristics of the crack can be extracted and the crack pixel information can be obtained. In the future work, the author will study how to obtain the real crack area by camera calibration ratio and develop the corresponding road crack detection equipment.

Table 16.1 Crack test results

Partial test diagram	YOLOv3 image recognition	Mask R-CNN image segmentation	Pixel size measurement
	 	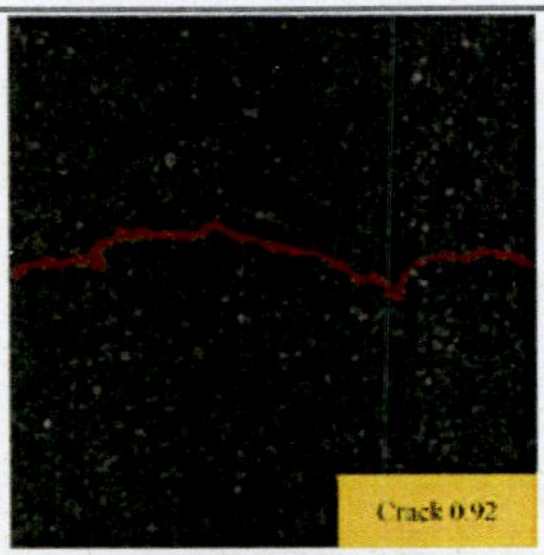 	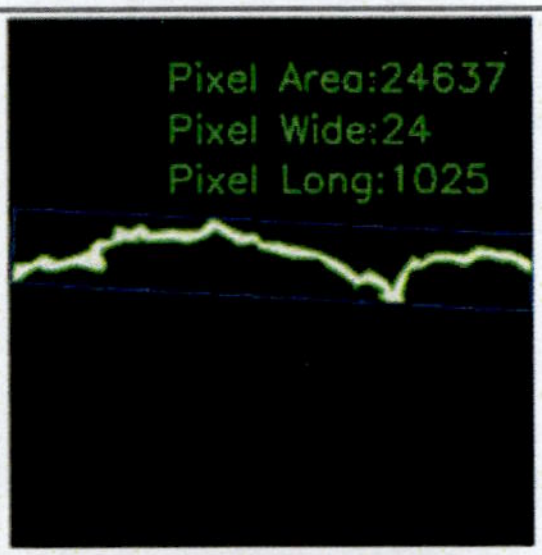
	 	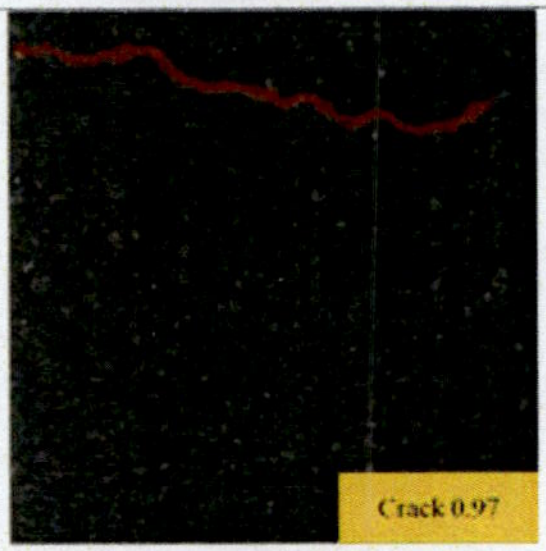 	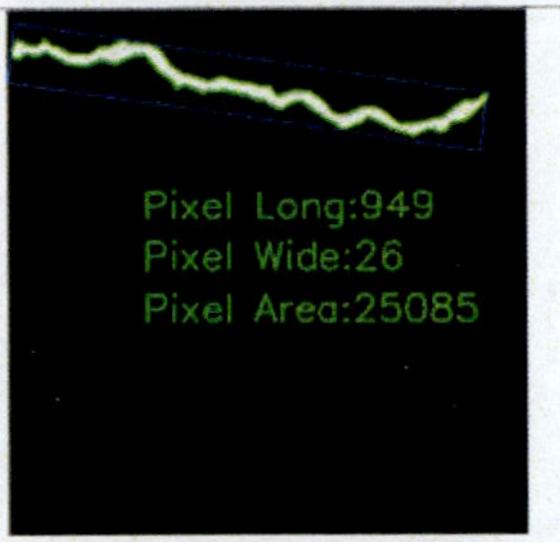
	 	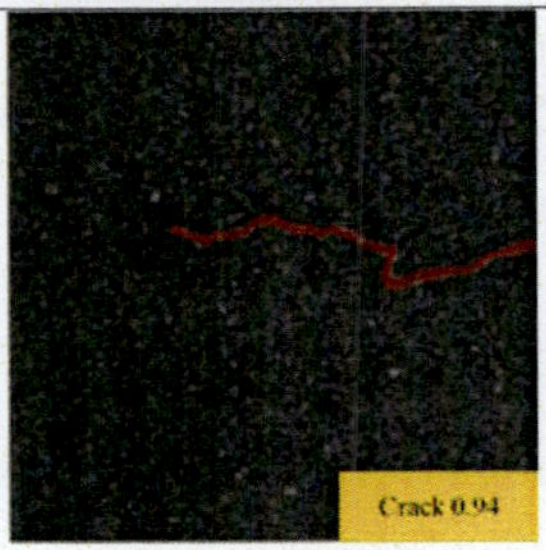 	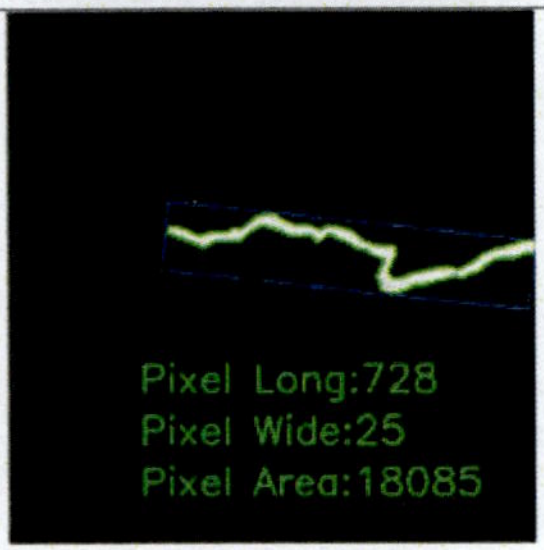

References

1. R. Ran, X.H. Xu, S.H. Qiu, X.P. Cui, Y.B. Ou, A survey of crack detection methods based on deep convolutional neural network. Comput. Eng. Appl. **2021**(3), 1–16 (2021)
2. L. Peng, W. Chao, L. Shuangmiao et al., Research on crack detection method of airport runway based on twice-threshold segmentation, in *2015 Fifth International Conference on Instrumentation and Measurement, Computer, Communication and Control (IMCCC)* (IEEE, Qinhuangdao, China, 2015), pp. 1716–1720
3. H. Xu, Y. Tian, S. Lin et al., Research of image segmentation algorithm applied to concrete bridge cracks, in *2013 IEEE Third International Conference on Information Science and Technology (ICIST)* (IEEE. Jiangsu, China, 2013), pp. 1637–1640
4. A. Cubero-Fernandez, F.J. Rodriguez-Lozano, R. Villatoro et al., Efficient pavement crack detection and classification. EURASIP J. Image Video Process. **1**, 1–11 (2017)
5. Y. Shi, L. Cui, Z. Qi et al., Automatic road crack detection using random structured forests. IEEE Trans. Intell. Transp. Syst. **17**(12), 3434–3445 (2016)
6. J.Y. Xie, R. Liu, Research progress of target detection algorithm based on deep learning. J. Shaanxi Normal Univ. Natural Science Edition. **47**(05), 1–9 (2019)
7. D.L. Kou, Y.C. Qian, Z.W. Zhang, Research progress of target detection framework based on deep learning. Comput. Eng. Appl. **55**(11), 25–34 (2019)
8. Y.Q. Zhao, Y. Rao, S.P. Dong, J.Y. Zhang, Overview of deep learning target detection methods. J. Image Graph. **25**(04), 629–654 (2020)

Chapter 17
Uncertainty Analysis of Efficiency Test of Drive Motor System for Electric Vehicle

Chenchen Dong, Weiwei Zhu, Feng Chen, Zhongjie Zhang, Guowei Lu, and Ting Liu

Abstract The drive motor system for electric vehicles is the power source of electric vehicles. System efficiency, as an important index, can reflect the comprehensive performance of the drive motor system. The efficiency of the drive motor system is generally tested on a single motor test bed. Based on the evaluation and expression method of measurement uncertainty, this paper analyzes the error sources affecting the system efficiency of electric vehicle drive motor, establishes the mathematical model of system efficiency measurement uncertainty, and obtains the calculation formula of system efficiency measurement uncertainty, the standard uncertainty and the synthetic uncertainty of each component. The current and torque components are sensitive to the uncertain results. Through the analysis of the test data, it is found that the current and torque components are sensitive to the measurement uncertainty. At the same speed, the uncertainty of measurement decreases with the increase of torque.

C. Dong · Z. Zhang · T. Liu
Key Laboratory of New Energy Automotive Drive Systems for Zhejiang Market Regulation, Zhejiang Fangyuan Testing Group Co., Ltd, Hangzhou, China

W. Zhu
Hangzhou Xiaoshan Technician College, Hangzhou, China

F. Chen (✉)
Key Laboratory of New Energy Automotive Drive Systems for Zhejiang Market Regulation, College of Mechanical Engineering, Zhejiang University of TechnologyZhejiang Fangyuan Testing Group Co., Ltd, Hangzhou, China
e-mail: chenf929@126.com

G. Lu
Hangzhou Weiheng Technology Co., Ltd, Hangzhou, China

G. Liu and F. Cen (eds.), *Advances in Precision Instruments and Optical Engineering*, Springer Proceedings in Physics 270, https://doi.org/10.1007/978-981-16-7258-3_17

17.1 Introduction

With the shortage of oil energy and the worsening of environmental pollution, the transmission of fuel vehicles has been unable to adapt to the development of the automobile industry. The development of new energy vehicles, especially electric vehicles is imminent. Driven by various national policies, the research of electric vehicles has received more and more attention [1].

The drive motor system is the core component of the drive system of electric vehicles. The drive motor system is usually composed of the drive motor and the drive controller. The input power of the drive controller is DC. The efficiency of the drive motor system will affect the range of the vehicle. The system efficiency test is of great significance to evaluate the comprehensive performance of the drive motor system [2, 3]. The system efficiency of the drive motor used in electric vehicles is usually carried out on a single motor test bench. At present, the research on the system efficiency of the drive motor used in electric vehicles is mostly about the theoretical research on the influencing factors [4]. There are few studies on the accuracy and uncertainty analysis of the measurement efficiency of the system efficiency test bench of the drive motor for electric vehicles.

As the test process will be affected by various conditions, all test data results will have a certain deviation from the true value. Therefore, in order to increase the relative effectiveness of measurement results, uncertainty analysis should be added to the evaluation of measurement results [5, 6], which makes the measured results more close to the true value. JJF 1059.1-2012 "Evaluation and expression of measurement uncertainty," the specification for analysis and evaluation of uncertainty of measurement results, is an indispensable technical document for the analysis of laboratory test data [7–9].

To test the electric car with a drive motor system, according to the JJF 1059.1-2012 uncertainty analysis, explore the electric car with the efficiency of the drive motor system dynamic measurement uncertainty evaluation theory and method, analyzing influencing factors of uncertainty, the drive motor for electric vehicle system efficiency is the accuracy of the measured results provide a reference.

17.2 Drive Motor System Efficiency Test Introduction

17.2.1 Introduction to the Test Bench

According to the structural characteristics of the drive motor system for electric vehicles, the test bench adopts the structure controlled by a single motor, as shown in Fig. 17.1. Wherein the dynamometer (M_o) analog drive motor (M_i) load, speed measurement sensor, and torque sensor are used to collect speed (n_o) and torque signal (T_o), the power analyzer (*PA*) measure the input voltage (U_i) of the drive controller (*C*) and the current (I_i). The control mode of the test bench adopts the dynamometer

Fig. 17.1 Drive motor test bench for electric vehicles

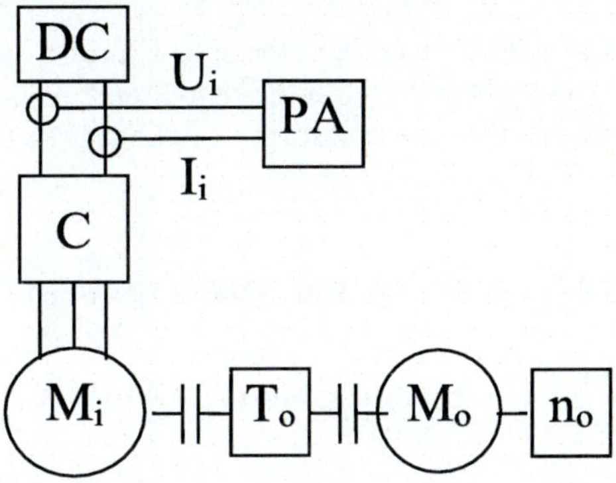

Fig. 17.2 Schematic diagram of test bench structure

motor as the speed mode and the driving motor under test as the torque mode. The structure of the test bench is shown in Fig. 17.2, and the parameters of the test bench are shown in Table 17.1.

17.2.2 Test Conditions and Requirements

Test according to GB/T 18,488.1-2015 "Drive motor system for electric vehicles part 1: technical requirements" [10], GB/T 18,488.2-2015 "Drive motor systems for electric vehicles—part 2: test methods [11], the test direction is forward, and the test sample has been debugged and calibrated before testing. Three typical test conditions, 3000 r/min speed, 24, 54, and 88 N m torque, were selected as the test conditions for uncertainty analysis.

Table 17.1 Test bench parameters of automobile drive motor system

Parameter	The numerical
Maximum speed of drive motor (r/min)	15,000
Rated power of drive motor (kW)	250
Drive torque sensor	T40B/500 N m
Accuracy of torque measurement	±0.05% FS
Accuracy of speed measurement	±1 r/min
Power analyzer	WT1804E
Accuracy of voltage measurement	±0.2% FS
Accuracy of current measurement	±0.05% FS
High- and low-temperature environment warehouse	−40 °C to +150 °C

17.2.3 *Recording of Test Data*

The test bench collects the speed, torque, and voltage and current signals measured by the power meter at the output end, respectively, and the software automatically records the data as test results within 0.5 s of the test bench.

17.3 Sources and Mathematical Models of Uncertainty

17.3.1 *Source Factors of System Efficiency Uncertainty*

Drive motor system efficiency test will be influenced by various factors, which cause drive motor system efficiency measurements appear error source of main influencing factors are: the output speed torque measurement repeatability error introduced, speed accuracy error introduced by the torque sensor, speed torque error, introduced resolution measurement repeatability error introduced by the voltage of the input current, voltage, and current accuracy of the introduction of error, error of the voltage current resolution introduced, the introduction of test temperature error, error of the installation precision is introduced into, and so on. The temperature during the test conforms to the requirements of GB/T 18,488.2-2015 standard, and the installation of samples meets the design requirements. The test temperature and installation accuracy can be ignored for the test results.

17.3.2 Mathematical Model of System Efficiency

According to the characteristics of the structure of the drive motor system for electric vehicles composed of drive motor and drive controller and the input power is DC, the mathematical model for calculating the system efficiency is shown in formula (17.1).

$$\eta = \frac{P_{\text{out}}}{P_{\text{in}}} = \frac{1000}{9550} \times \frac{T_o n_o}{U_i I_i} \times 100\% \tag{17.1}$$

Type:

η System efficiency of the measured driving motor system, %;
P_{out} The output power of the driving motor system under test, kW;
P_{in} The input power of the driving motor system under test, kW;
T_{o} Output torque, N m;
n_{o} Output speed, r/min;
U_{i} Input voltage, V;
I_{i} Input current: A.

17.4 Calculation of Synthetic Standard Uncertainty

According to JJF 1059.1-2012, each factor affecting the measurement will be transferred to the final result in its own way. For this transfer, a corresponding mathematical model must be established to find out the transfer factor so that the total uncertainty of the final measurement result can be synthesized [12, 13].

System efficiency measurement model and uncertainty source factor analysis, system efficiency measurement uncertainty input factors are U_i, I_i, T_o, and n_o. The uncertainty of each component is mainly derived from the measurement repeatability (Class A uncertainty), the accuracy of the sensor and the uncertainty introduced by the resolution (Class B uncertainty). The standard uncertainty of each component is calculated as shown in formula (17.2).

$$U_{\text{di}} = \sqrt{U_{\text{d1}}^2 + U_{\text{d2}}^2 + U_{\text{d3}}^2} \tag{17.2}$$

Type:

$U_{\text{d}i}$ U_i, I_i, T_o, n_o standard uncertainty of each component;
U_{d1} Class A uncertainty introduced by measurement repeatability;
U_{d2} Class B uncertainty introduced by accuracy;
U_{d3} Class B uncertainty introduced by resolution.

The components of each influencing factor are measured separately by their respective sensors, so the synthetic standard uncertainty of system efficiency is calculated as shown in formula (17.3).

$$U_c = \sqrt{c_1^2 U_{Ui}^2 + c_2^2 U_{Ii}^2 + c_3^2 U_{To}^2 + c_4^2 U_{no}^2} \tag{17.3}$$

Type:

c_1–c_4 Sensitivity coefficient of each component;
U_{Ui} Uncertainty introduced by component U_i;
U_{Ii} Uncertainty introduced by component I_i;
U_{To} Uncertainty introduced by component T_o;
U_{no} Uncertainty introduced by component n_o.

17.5 Calculation of Standard Uncertainty of Each Component

17.5.1 Test Data

The test data of the system efficiency at 3000 r/min speed and 24, 54 and 88 N m torque are shown in Tables 17.2, 17.3 and 17.4.

17.5.2 Class A Uncertainty

The class A uncertainty caused by the measurement repeatability is calculated according to the "Bessel formula" (17.4).

Table 17.2 Test data of low torque 24 N m

The serial number	T_o (N, m)	n_o (r/min)	U_i (V)	I_i (A)	η (%)
1	24.888	3000.1	349.56	24.83	90.09
2	24.761	2999.8	349.48	24.77	89.86
3	24.761	2999.8	349.48	24.77	89.86
4	24.873	3000.4	349.46	24.87	89.91
5	24.873	3000.4	349.46	24.87	89.91
6	24.767	2999.8	349.52	24.77	89.85
7	24.767	2999.8	349.52	24.77	89.85
8	24.699	2999.8	349.48	24.73	89.75
9	24.699	2999.8	349.48	24.73	89.75
10	24.267	2999.6	349.53	24.3	89.73
On average	24.736	2999.93	349.50	24.74	89.86

Table 17.3 Test data of middle torque 54 N m

The serial number	T_o (N, m)	n_o (r/min)	U_i (V)	I_i (A)	η (%)
1	55.202	2999.5	349.32	54.07	91.80
2	55.736	3000.5	349.44	54.55	91.87
3	55.163	2999.8	349.25	53.99	91.89
4	55.163	2999.8	349.25	53.99	91.89
5	55.110	3000.1	349.56	54.03	91.67
6	55.110	3000.1	349.56	54.03	91.67
7	55.479	3000.0	349.42	54.37	91.74
8	55.479	3000.0	349.42	54.37	91.74
9	55.652	2999.9	349.28	54.58	91.70
10	55.652	2999.9	349.28	54.58	91.70
On average	55.375	2999.96	349.38	54.26	91.77

Table 17.4 Test data of high torque 88 N m

The serial number	T_o (N, m)	n_o (r/min)	U_i (V)	I_i (A)	η (%)
1	88.924	3000.0	349.21	86.24	92.76
2	88.710	2999.8	349.27	86.15	92.61
3	88.710	2999.8	349.27	86.15	92.61
4	88.978	3000.1	349.30	86.24	92.79
5	88.978	3000.1	349.30	86.24	92.79
6	88.891	3000.1	349.27	86.21	92.74
7	88.891	3000.1	349.27	86.21	92.74
8	88.839	3000.1	349.27	86.16	92.74
9	88.839	3000.1	349.27	86.16	92.74
10	88.747	2999.9	349.29	86.07	92.73
On average	88.851	3000.01	349.27	86.18	92.72

$$s(x) = \sqrt{\frac{1}{n-1}\sum_{i=1}^{n}(x_i - \overline{x})^2} \tag{17.4}$$

The actual test value is continuously measured for 10 times. Therefore, the calculation of class A standard uncertainty of each influencing factor in each test group is shown in formula (17.5), and the class A standard uncertainty of the three groups of test results is shown in Table 17.5.

$$U_{\mathrm{d1}}(x) = \frac{s(x)}{\sqrt{10}} \tag{17.5}$$

Table 17.5 Class A uncertainty U_{d1}

Torque (N m)	T_o (N, m)	n_o (r/min)	U_i (V)	I_i (A)
24	0.0564	0.0870	0.0105	0.0516
54	0.0793	0.0819	0.0379	0.0817
88	0.0319	0.0407	0.0080	0.0174

17.5.3 *Influence of Class B Uncertainty*

Uncertainty caused by sensor accuracy

According to Table 17.1 of test bench equipment parameters, the accuracy of torque sensor is ±0.05%FS, the accuracy of speed sensor is ±1 r/min, the accuracy of voltage measurement is ±0.2%FS, and the accuracy of current measurement is ±0.05%FS. The uncertainty U_{d2} of class B can be obtained according to the half width of uncertainty and uniform distribution. The calculation is shown in formula (17.6), (17.7), (17.8), and (17.9), and the B standard of the three groups of test results is uncertain U_{d2} see Table 17.6.

$$U_{d2}(T_{oi}) = \frac{\overline{T_{oi}} \times 0.05\%}{\sqrt{3}} \tag{17.6}$$

$$U_{d2}(n_{oi}) = \frac{\overline{n_{oi}} \times 1}{\sqrt{3}} \tag{17.7}$$

$$U_{d2}(U_{ii}) = \frac{\overline{U_{ii}} \times 0.2\%}{\sqrt{3}} \tag{17.8}$$

$$U_{d2}(I_{ii}) = \frac{\overline{I_{ii}} \times 0.2\%}{\sqrt{3}} \tag{17.9}$$

Uncertainty caused by resolution

After the torque and speed signals are collected by the sensor and processed by the test bench test software, the display resolution of the torque signal is 0.001 N m, the speed signal is 0.1 r/min, the voltage display resolution is 0.01 V, and the current display resolution is 0.01A. According to JJF 1059.1-2012, the uncertainty caused

Table 17.6 Class B uncertainty U_{d2}

Torque (N m)	T_o (N, m)	n_o (r/min)	U_i (V)	I_i (A)
24	0.0071	0.5774	0.4036	0.0071
54	0.0160	0.5774	0.4034	0.0157
88	0.0256	0.5774	0.4033	0.0249

Table 17.7 Class B uncertainty U_{d3}

Torque (N m)	T_o (N, m)	n_o (r/min)	U_i (V)	I_i (A)
24	0.00029	0.0290	0.0029	0.0029
54	0.00029	0.0290	0.0029	0.0029
88	0.00029	0.0290	0.0029	0.0029

by instrument resolution is 0.29δ (δ is instrument resolution). The class B uncertainty U_{d3}, the calculation is shown in formula (17.10), (17.11), (17.12), and (17.13), and Table 17.7.

$$U_{d3}(T_{oi}) = 0.29 \times 0.001 = 0.00029 \tag{17.10}$$

$$U_{d3}(n_{oi}) = 0.29 \times 0.1 = 0.029 \tag{17.11}$$

$$U_{d3}(U_{ii}) = 0.29 \times 0.01 = 0.0029 \tag{17.12}$$

$$U_{d3}(I_{ii}) = 0.29 \times 0.01 = 0.0029 \tag{17.13}$$

17.5.4 Synthesize the Standard Uncertainty Component

According to formula (17.2), the standard uncertainty calculation results of each component of the test results of the three groups are shown in Table 17.8.

Table 17.8 Standard uncertainty of each component U_d

Torque (N m)	T_o (N, m)	n_o (r/min)	U_i (V)	I_i (A)
24	0.0569	0.5846	0.4037	0.0521
54	0.0809	0.5839	0.4052	0.0832
88	0.0409	0.5795	0.4034	0.0305

Table 17.9 Sensitivity coefficient of each component

Torque (N m)	$c\ (U_{To})$	$c\ (U_{no})$	$c\ (U_{Ui})$	$c\ (U_{Ii})$
24	0.03633	0.00030	0.00257	0.03632
54	0.01657	0.00031	0.00263	0.01691
88	0.01044	0.00031	0.00265	0.01076

17.6 Synthesis Uncertainty Analysis and Extended Uncertainty

17.6.1 Sensitivity Factor

The sensitivity coefficient is the function η with respect to T_o, n_o, U_i, and I_i, the coefficient of the partial derivative. Which is T_o, n_o, U_i, and I_i, the degree of influence on synthesis uncertainty U_c. The sensitivity coefficient of each input is calculated as shown in formula (17.14)–(17.17), and the sensitivity coefficient of each component of the three groups of test results is calculated as shown in Table 17.9.

$$c(U_{To}) = \frac{\partial(\eta)}{\partial U_{To}} = \frac{1000}{9550} \times \frac{n_o}{U_i I_i} \tag{17.14}$$

$$c(U_{no}) = \frac{\partial(\eta)}{\partial U_{no}} = \frac{1000}{9550} \times \frac{T_o}{U_i I_i} \tag{17.15}$$

$$c(U_{Ui}) = \frac{\partial(\eta)}{\partial U_{Ui}} = -\frac{1000}{9550} \times \frac{T_o n_o}{U_i^2 I_i} \tag{17.16}$$

$$c(U_{Ii}) = \frac{\partial(\eta)}{\partial U_{Ii}} = -\frac{1000}{9550} \times \frac{T_o n_o}{U_i I_i^2} \tag{17.17}$$

17.6.2 Composite Uncertainty

The combined standard uncertainty is calculated in formula (17.3), and the combined uncertainty U_c of the three groups of test results is shown in Table 17.10.

Table 17.10 Synthesis uncertainty U_c

Torque (N m)	U_c (%)
24	0.29938
54	0.22232
88	0.12121

Table 17.11 Extended uncertainty U and test results

Torque (N m)	U (%)	System efficiency test results (%)
24	0.60	89.86 ± 0.60
54	0.44	91.77 ± 0.44
88	0.24	92.72 ± 0.24

17.6.3 Extended Uncertainty

The extended uncertainty U is derived from the resultant uncertainty U_c times the inclusion factor k. Take the inclusion probability $P = 95\%$ and inclusion factor $k = 2$, then the extended uncertainty is shown in formula (17.16). The extended uncertainty U of the three groups of system efficiency test results and the system efficiency test results are shown in Table 17.11, and the calculated results are revised to about 0.01%.

$$U = k \times U_c = 2U_c \tag{17.18}$$

17.6.4 Analysis of Factors Affecting Uncertainty

According to the calculation formula and calculation results of the sensitivity coefficient, it can be seen that, under the three torques, I_i and T_o, the influence on the uncertain results is most sensitive, followed by U_i, n_o the sensitivity to uncertain results is the lowest. Therefore, the accuracy of current measurement and torque measurement should be improved as far as possible to reduce the impact of uncertainty on the measurement results.

It can be seen from the test results of the extended uncertainty under the three torques that the measurement uncertainty decreases gradually with the increase of the torque at the same test speed (3000 r/min).

17.7 Summary

According to the evaluation and expression method of measurement uncertainty, the influence sources of system efficiency error of electric vehicle drive motor are analyzed, and the mathematical model of system efficiency measurement uncertainty is established. According to the mathematical model, the calculation formula of the measurement uncertainty of system efficiency and the standard uncertainty and synthetic uncertainty of each component are derived. The main factors affecting the uncertain input of system efficiency measurement are T_o, n_o, U_i, and I_i. The uncertainty of each component is mainly derived from sensor repeatability, sensor accuracy, and resolution. The calculation results of the sensitivity coefficient show that the current and torque components are highly sensitive to the uncertain results. According to the final calculation results of uncertainty, it is found that the measurement uncertainty of the system efficiency test of the driving motor system becomes smaller with the increase of torque at the same speed.

Acknowledgements This research was funded by Contract (20190316) from Science and technology planning project of Zhejiang market supervision and Administration Bureau.

References

1. J. Zhu, Research and development of pure electric vehicle drive system test bench. Shanghai University of Engineering Technology
2. Q. Yu, Research on efficiency improvement and test of drive motor system of pure electric vehicle. J. Hubei Inst. Automot. Technol. **1**, 18–21 (2016)
3. Q. Song, Z.F. Wang, C.N. Zhang, Research on efficiency test method of electric vehicle motor drive system. Micro Spec. Mot. **2006**(06), 4–6 (2006)
4. C.P. Zhang, C.N. Zhang, Q. Song. Research on drive work area efficiency test of electric vehicle motor drive system. Micromotor **2007**(01), 84–86 (2007)
5. C.G. Zhang, Some problems in evaluation of measurement uncertainty. China Test Meas. **33**(5), 24–28 (2007)
6. F. Li, X.B. Yang, Y.X. Zhang, et al., Application of uncertainty evaluation method in measuring motor efficiency. Electr. Meas. Instrum. **54**(8), 109–113 (2017)
7. Evaluation and expression of measurement uncertainty: JJF 1059.1-2012. China Quality Inspection Press, Beijing (2012)
8. W.W. Jin, C.J. Wang, G. Chen et al., Research and design of efficiency test system with low uncertainty for high efficiency and super efficient motor. Mot. Control. Appl. **37**(4), 1–5 (2010)
9. H.G. Hu, W.H. Ji, Measurement equation, observation equation and uncertainty evaluation. China Test Meas. **46**(9), 7–12 (2020)
10. Drive motor systems for electric vehicles—part 1: technical requirements: GB/T 18488.1-2015. Standards Press of China, Beijing (2015)
11. Drive motor systems for electric vehicles—part 2: test method: GB/T 18488.2-2015. Standards Press of China, Beijing (2015)
12. S.A. Li, Attention to JJF 1059.12-2012 new specification for evaluation and expression of measurement uncertainty. Ind. Metrol. **23**(3), 49–50 (2013)
13. General metrology terms and definitions: JJF 1001-2011. China Quality Inspection Press, Beijing (2011)

Chapter 18
Fluorescence Characteristics of DAST in Methanol Solutions

Yuanting Feng, Xiangdong Xu, Jinrong Liu, Minghui Xu, Junjie Hu, Ke Xiong, and Xiaomeng Cheng

Abstract As an important optical material, 4-*N*,*N*-Dimethylamino-4′-*N*′-methyl-stilbazolium tosylate (DAST) is applied widely, but the effects of its concentration in a solution on the fluorescence remain unclear. To investigate this, DAST methanol solutions with various DAST concentrations were prepared and measured. Results reveal that blue-shifted fluorescence peaks and enhanced fluorescence intensity were observed when the DAST concentration was increased from 3.12 to 12.5 μM. But interestingly, the fluorescence peak red-shifts and the fluorescence intensity decreases when the DAST concentration was further increased from 12.5 to 250 μM, due to the aggregation of DAST molecules. Beyond 250 μM, both peak position and intensity almost remained unchanged.

18.1 Introduction

In the past decades, fluorescence has been applied widely in biomedical imaging [1], bioengineering [2], food and environmental detection [3], etc. So, the fluorescent materials have attracted extensive attention [1–4]. Compared with inorganic materials, organic materials show the advantages of higher biocompatibility, adjustability of properties by organic functional groups, and broader spectral responses. Of all organic nonlinear optical (NLO) materials, 4-*N*,*N*-dimethylamino-4″-N″-methyl-stilbazolium tosylate (DAST) occupies an important position, owing to its high electro-optic coefficient, large nonlinear coefficients, and fast optical nonlinear response [5–7]. The chemical structure of DAST is shown in Fig. 18.1. There is yet another optical characteristic of DAST, i.e., its fluorescence, requires further exploration. The facts that the absorption spectrum of DAST covers the entire range of visible light and its emission band mainly locates at the wavelength of 550–750 nm [8] suggest great potential applications of DAST in environmental, medical, and

Y. Feng · X. Xu (✉) · J. Liu · M. Xu · J. Hu · K. Xiong · X. Cheng
State Key Laboratory of Electronic Thin Films and Integrated Devices, School of Optoelectronic Science and Engineering, University of Electronic Science and Technology of China (UESTC), Chengdu 610054, People's Republic of China
e-mail: xdxu@uestc.edu.cn

G. Liu and F. Cen (eds.), *Advances in Precision Instruments and Optical Engineering*, Springer Proceedings in Physics 270, https://doi.org/10.1007/978-981-16-7258-3_18

Fig. 18.1 The chemical structure of DAST

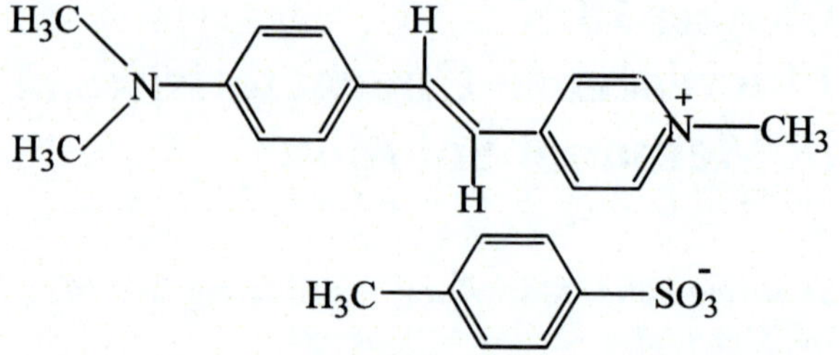

biological applications. However, to the best of our knowledge, the fluorescence properties of DAST at different concentrations are still unclear.

In this work, we present the measurement results of the fluorescence properties of DAST solutions. Particularly, the fluorescence properties of DAST methanol solutions with various DAST concentrations, ranging from a highly diluted state (3.12 μM) to a high concentration state (833 μM), were investigated.

18.2 Experiments

First, DAST powers were synthesized, puried, and characterized according to the processes previously described [9]. Then, the stock solution with the DAST concentration of 1250 μM was prepared by adding appropriate amount of DAST powders to methanol solvent, followed by ultrasonic treatment for 30 min. After that, the stock solution was diluted with methanol to 833 μM. Furthermore, DAST solutions with various DAST molarities of 250, 125, 83.3, 50, 25, 12.5, 6.25, 3.12 μM, were similarly prepared. The optical properties of the as-prepared DAST solutions were measured by a fluorescence spectrometer (Hitachi FL-4600) and an UV–vis spectrophotometer (UV-2600).

18.3 Results and Discussion

Figure 18.2 displays the fluorescence emission spectra of various DAST methanol solutions with the excitation wavelength of 480 nm. When the DAST concentration was 3.12 μM, the solution emitted a peak at about 605 nm. When the DAST concentration was increased, only one fluorescence peak was similarly measured, but notably, both the position and intensity of the fluorescence peak changed. To better reflect such changes, the fluorescence peak wavelength and the intensity as a function of the DAST concentration were plotted in Fig. 18.3a and b, respectively. One can see that below 12.5 μM, the fluorescence peak blue-shifted and the intensity increased as the DAST concentration increased (Fig. 18.3a). Rather differently, weakened intensity and red-shift of the fluorescence peak were observed when the

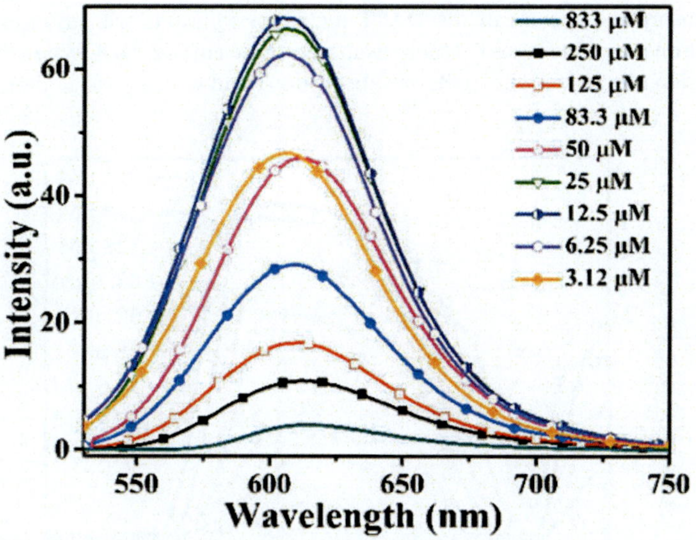

Fig. 18.2 Fluorescence emission spectra of DAST solutions with different DAST concentrations

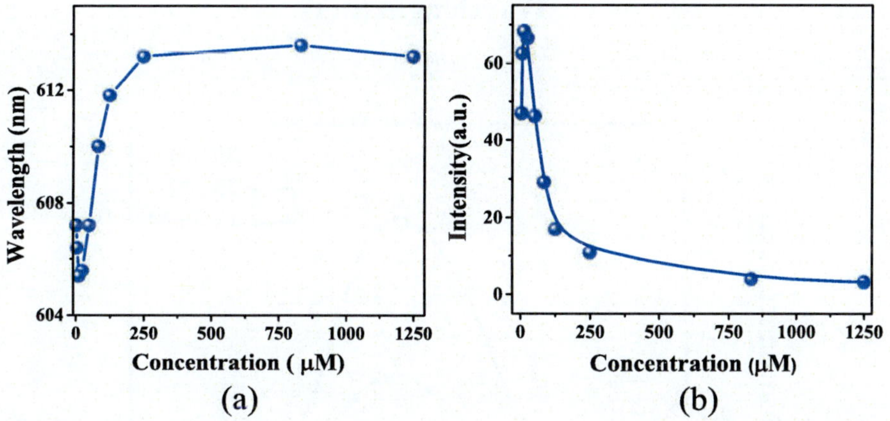

Fig. 18.3 **a** Fluorescence peak wavelength, and **b** fluorescence intensity as a function of the DAST concentration

DAST molarity was further increases from 12.5 to 250 μM. Beyond 250 μM, both the peak position and intensity almost remained unchanged, indicating weak effects of the DAST concentration on the fluorescence properties in this case. We expect that this spectral feature might be attributed to the molecule aggregation at high DAST concentration (>12.5 μM), as observed by Zehentbauer et al. in Rhodamine 6G [10].

To verify our expectation that both the changes of the wavelength and intensity of the fluorescence peak are significantly affected by the molecule aggregation,

UV–vis absorption spectra of the DAST methanol solutions with different DAST concentrations were measured, which results are shown in Fig. 18.4. When the DAST concentration was below 12.5 μM, one absorption band with a peak at about 476 nm

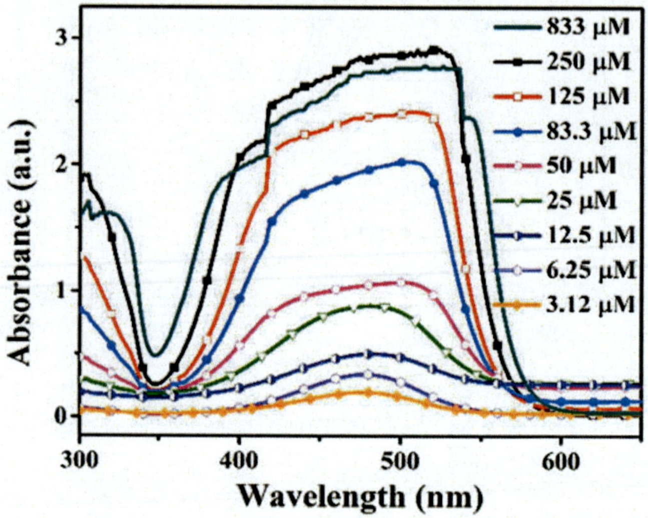

(a)

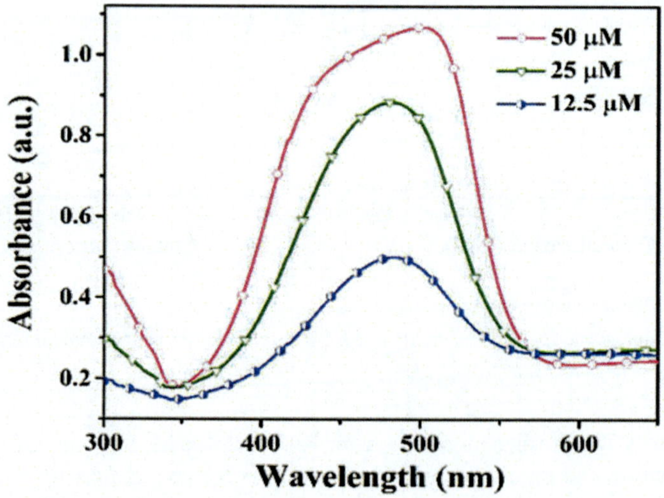

(b)

Fig. 18.4 UV–vis absorption spectra of DAST solutions **a** with different DAST concentrations, **b** magnified spectra of the solutions with 12.5, 25, and 50 μM DAST concentration was measured. In this case, the Lambert–Beer law [11] works, indicating that the DAST molecules in such solution are non-aggregated. With the increase of the DAST concentration, the absorption increased and the full width at half-maximum (FWHM) broadened, as displayed in Fig. 18.4a. When the DAST concentration was increased to 50 μM, two shoulders centered at 418 and 501 nm were measured (Fig. 18.4a), in which one shoulder showing a blue-shifted absorption compared to the monomer absorption band is H aggregates, while the other showing a red-shifted absorption is attributed to J aggregates [12]. Both of such two aggregations will lead to weakened or quenched fluorescence [12]. In addition, the absorption remains strong in the range of 530–575 nm due to the spectral broadening, and particularly, this absorption peak is overlapped with the emission peak (530–730 nm) of DAST solution (Fig. 18.2). Such partial overlapping of these two spectra leads to reabsorption [13] in this case. This suggests that the reabsorption effect is another origin for inducing the decrease of the intensity and red-shift of the fluorescence peak. As comparison, Fig. 18.4b shows the absorption spectra of DAST solutions at the DAST concentrations of 12.5, 25, and 50 μM. When the DAST concentration was higher than 25 μM, aggregation of the DAST molecules occurs, thus leading to the decrease of the fluorescence intensity (Fig. 18.3b), as indicated by Fig. 18.3b. This phenomenon confirms that the change of the fluorescence peak is attributed to the molecule aggregation, as we expected. This work not only points out the relationship between the fluorescence properties of DAST solutions at various DAST concentration but also discloses the mechanism for the changes of both the wavelength and intensity of the fluoresence peak.

18.4 Conclusions

In summary, the effects of DAST concentration on both the fluorescence and UV–vis absorption spectra were investigated. At low DAST concentrations, ranging from 3.12 to 12.5 μM, most DAST molecules are isolated from each other; hence, the fluorescence intensity enhances with the increase of the DAST molality, and only one absorption peak at around 476 nm appears in this case. With the increase of the DAST concentration, we observed both red-shift and blue-shift of the absorption peak, due to J aggregates and H aggregates, respectively. The DAST molecule aggregation causes the reduced fluorescence intensity and reabsorption. Therefore, stronger fluorescence signals can be measured from the methanol solutions with low DAST concentrations (<12.5 μM). The results presented in this work are helpful for optimizing the DAST concentration for enhancing or weakening the fluorescence signals for academic research or practical applications.

Acknowledgements This work was financially supported by the National Natural Science Foundation of China (NSFC 61377063) and Sichuan Science and Technology Program (2018TZDZX0008).

References

1. H. Cho, Y. Guo, D.E. Sosnovik, L. Josephson, Imaging DNA with fluorochrome bearing metals. Inorg. Chem. **52**, 12216–12222 (2013)
2. D. Kankanamge, K. Ratnayake, K. Senarath, M. Tennakoon, E. Harmon, A. Karunarathne, Optical approaches for single-cell and subcellular analysis of GPCR-G protein signaling. Anal. Bioanal. Chem **441**, 4481–4508 (2019)
3. D. Huang, C. Niu, G. Zeng, X. Wang, X. Lv, A highly sensitive protocol for the determination of Hg^{2+} in environmental water using time-gated mode. Talanta **132**, 606–612 (2015)
4. D.D. Nolting, J.C. Gore, W. Pham, Near-infrared dyes: probe development and applications in optical molecular imaging. Curr. Org. Synth. **8**, 521–534 (2011)
5. M. Thakur, J. Xu, A. Bhowmik, L. Zhou, Single-pass thin-film electro-optic modulator based on an organic molecular salt. Appl. Phys. Lett **74**, 635–637 (1999)
6. K. Fan, X.D. Xu, Y. Gu, Z.L. Dai, J. Xu, Organic DAST single crystal meta-cavity resonances at terahertz frequencies. ACS Photon. **6**, 1674–1680 (2019)
7. M.L. Zheng, F. Katsumasa, W.Q. Chen, Two-photon excited fluorescence and second-harmonic generation of the DAST organic nanocrystals. J. Phys. Chem. C **115**, 8988–8993 (2011)
8. K. Kumar, R.N. Rai, S.B. Rai, One- and two-photon-pumped luminescence studies on DAST and UDAST organic dye molecules. Appl. Phys. B **96**, 85–94 (2009)
9. X.D. Xu, Z.Q. Sun, K. Fan, Y.D. Jiang, R. Huang, Y.J. Wen, Conversion of 4-*N,N*-dimethylamino-4′-*N*′-methyl-stilbazolium tosylate (DAST) from a simple optical material to a versatile optoelectronic material. Sci. Rep. **5**, 12269s (2015)
10. F.M. Zehentbauer, C. Moretto, R. Stephen, T. Thevar, J.R. Gilchrist, D. Pokrajac, Fluorescence spectroscopy of Rhodamine 6G: concentration and solvent effects. Spectrochim. Acta A **121**, 147–151 (2014)
11. C.W. Tan, P.P. Zhang, X.X. Zhou, Z.X. Wang, Z.Q. Xu, W. Mao, Quantitative monitoring of leaf area index in wheat of different plant types by integrating NDVI and beer-lambert law. Sci. Rep. **10**, 929 (2020)

12. J. Qi, H. Xiongwei, X. Dong, Y. Lu, W. Wu, Towards more accurate bioimaging of drug nanocarriers: turning aggregation-caused quenching into a useful tool. Adv. Drug. Deliver. Rev **143**, 206–225 (2019)
13. E.P.J. Merkx, T.G. Lensvelt, E. van der Kolk, Modelling and optimization of UV absorbing photovoltaic windows using a thin film AlN:Eu^{3+} luminescence library. Sol. Energy Mater. Sol Cells **200**, 110032 (2019)

Chapter 19
Research on Deflection Quality Assessment of Assembly Car Body Based on Deflection Measurement and Improved Principal Component Analysis Method

Jiafu Cai, Guixiong Liu, Baitian Ouyang, and Fushen Cai

Abstract In order to solve the problem of low manufacturing efficiency of the assembly car body, the importance of deflection measurement and quality assessment for the assembly car body was emphasized. Based on the laser displacement sensor, the deflection of each control point was measured with high precision. The deflection quality of the assembly car body was assessed with the deflection parameters of each control point as the assessment index. The deflection quality assessment method of the assembly car body based on the Improved Principal Component Analysis (IPCA) method was proposed. The IPCA method was used to assess ten sets of car body deflection data measured with high precision. The results showed that this method can effectively assess the deflection quality of assembly car body and will have a guiding role in the production of assembly car body.

19.1 Introduction

The deflection of the assembly car body is an important structural performance index in the manufacturing process of the assembly car body of rail transit vehicles [1]. To ensure the normal operation of the assembly car body, it is necessary that the finished assembly car body has a certain upper deflection [2]. The deflection quality of the assembly car body would include the information of multiple deflection control points on the assembly car body, and the information of multiple control points will cause problems such as complicated calculation and difficult analysis. A reasonable assessment of the deflection quality for the assembly car body would reflect the advantages or disadvantages of the assembly car body manufacturing. At

J. Cai · G. Liu (✉)
South China University of Technology Guangzhou, Guangdong, China
e-mail: megxliu@scut.edu.cn

B. Ouyang · F. Cai
Guangdong Jiangmen Supervision Testing Institute of Quality & Metrology, Jiangmen, China

G. Liu and F. Cen (eds.), *Advances in Precision Instruments and Optical Engineering*, Springer Proceedings in Physics 270, https://doi.org/10.1007/978-981-16-7258-3_19

present, many research scholars at home and abroad have proposed many assessment methods, such as Analytic Hierarchy Process (AHP), Fuzzy Comprehensive Appraisal (FCA) method, Principal Component Analysis (PCA) method, grey decision method, and so on [3–5]. Hu et al. (2018) studied the seismic disaster risk assessment method of hydraulic fracturing area, based on the AHP and FCA, using seismic activity frequency, magnitude frequency coefficient to assess the earthquake risk. The results showed that the assessment method can effectively distinguish the earthquake risk [6]. Shi et al. (2017) built a wind power utilization level assessment index system based on improved analytic hierarchy process (IAHP) and FCA. The wind resource characteristics, wind turbine type, and wind turbine utilization rate were seen as assessment indicators. However, the study only selected one wind farm, so the generality of the model could not be verified [7]. Liu et al. (2020) established a drill bit selection model based on the IPCA. The test showed that thc drill bit selection model adopted the IPCA had the advantages of data dimensionality reduction [8]. Qian et al. (2016) established an FCA model of water quality based on PCA and FCA. The result showed that the FCA model of water quality based on PCA can fully consider each assessment. The interaction between factors can effectively assess the quality of groundwater [9]. Based on the idea of assessment method, this paper studied the deflection quality assessment method of assembly car body based on the IPCA. Through the assessment and analysis of the deflection measurement data and the adjustment force data index of each control point on the assembly car body, the comprehensive assessment results of car body deflection were given. The result had a guiding role in the production and manufacturing of the assembly car body.

19.2 Design for Quality Assessment of Assembly Car Body

In the manufacturing process of the assembly car body, the points at the end, middle, and support points on the edge beams of the car body are seen as the deflection control points. When the deflection of control points is within the qualified range, the overall deflection quality of the assembly car body is qualified. It is necessary to measure and adjust the deflection of control points on assembly car body for many times, resulting in low efficiency of the assembly car body manufacturing. In order to promote the development of assembly car body manufacturing, it is necessary to assess the deflection quality of the assembly car body. Through the quality assessment of the assembly car body, the direction of the assembly car body manufacturing would be guided. In this paper, the deflection measurement data and the adjustment force data of each control point on the assembly car body were seen as quality assessment index. The deflection measurement data of each control point was obtained by the laser displacement sensor, and the adjustment force data of each control point was calculated by finite element simulation analysis method.

19.3 Assessment Method of Deflection Quality for Car Body

19.3.1 Obtainment of the Deflection Parameters of Each Control Point

The deflection data was obtained by deflection measurement at the positions of deflection control points on the edge beams of the assembly car body. Figure 19.1 is a schematic diagram of the deflection control points of the assembly car body. The deflection data was measured by a high-precision laser displacement sensor. The measurement mechanism of the laser displacement sensor was shown in Fig. 19.2. The laser beam emitted from the laser spot was projected onto the surface of the measured object. The laser beam was received by the receiving lens at an angle after diffuse reflection and would form a light spot on the surface of the photosensitive element. The light spot position would change with the distance of the measured object changing, so the distance between the measured object and laser displacement sensor could be measured. The reference distance and measurement range of the laser dis-placement sensor could be determined according to the position range of the light spot. The laser displacement sensor had high accuracy in distance measurement, which could be used to measure the deflection of each control point of the assembly car body. In this paper, the HG-C 1400 laser displacement sensor was selected for deflection measurement and the measurement accuracy of deflection control point deflection could reach up to 0.3 mm. The deflection measurement data was transmitted to the computer through analog-to-digital conversion technology and wireless sensing technology. Suppose the deflection measurement data of each control point of the assembly car body was $\omega_a(a \in [1, 10])$. The adjustment force data of each control point was calculated by the finite element simulation analysis method and the adjustment force data of each control point of the assembly car body was set as F_a.

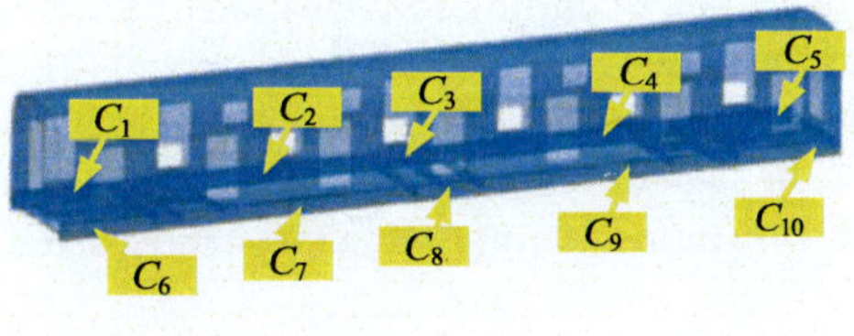

Fig. 19.1 A schematic diagram of the deflection control points of the assembly car body

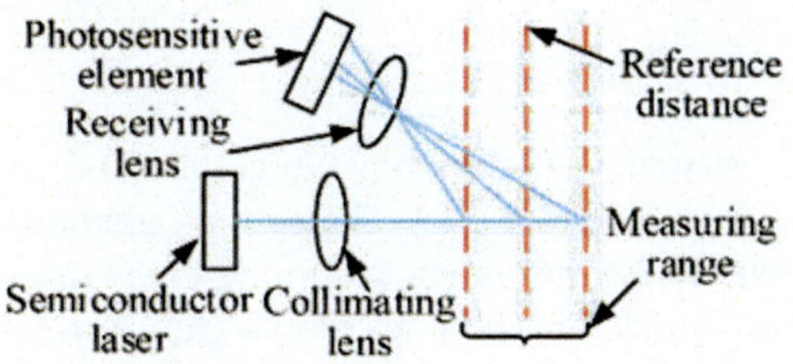

Fig. 19.2 Measurement mechanism of the laser displacement sensor

19.3.2 Improved Principal Component Analysis Method

This paper used the deflection measurement data and the quality of the adjustment force of each control point as the assessment index, to assess the overall deflection quality of the assembly car body. PCA could eliminate the influence of the correlation between the assessment indicators and could clarify the relationship between the variables while minimizing the loss of the original main information [10, 11].

The steps of the traditional principal component analysis method were as follows:

① Performing the same trend transformation on the assessment index data matrix X(X is a matrix of $m \times n$). The larger the assessment index data value, the worse the deflection quality of the car body.

② Standardizing the same-trend matrix to eliminate the influence of the dimensions between different indicators and obtaining the standardized processing matrix as $Z = (z_{ij})_{m \times n}$.

③ Calculating the correlation coefficient matrix $R = (r_{ij})_{n \times n}$ of matrix Z.

④ Solving out the correlation coefficient matrix, achieving n eigenvalues $\lambda_1 \geq \lambda_2 \geq \cdots \geq \lambda_n \geq 0$, and the corresponding orthogonalized unit eigenvector was $t_1, t_2, \cdots, t_n$, where $t_k = (t_{1k}, t_{2k}, \cdots, t_{nk})^{\mathrm{T}}$.

⑤ Calculating the expression of the variance contribution rate α_k and cumulative variance contribution rate γ_k of the principal component Y_k.

According to $\gamma_K \geq 0.90$, the first K index variables was taken as the principal components and the variance contribution rate of each principal component was taken as the weight, then the comprehensive score of the sample $score$ was

$$score = \alpha_1 Y_1 + \alpha_2 Y_2 + \cdots + \alpha_K Y_K \tag{19.1}$$

The traditional PCA converted more correlation indicators into fewer irrelevant indicators. However, the traditional PCA would lose the difference information of the variation degree of each assessment index when the data was standardized and the variance contribution rate used to determine the weight relationship of each index to the sample, was highly subjective [12]. Therefore, this paper adopted IPCA by improving the data standardization process on PCA.

In step ②, choosing a dimensionless method that can better retain the indicator information and improving the standardized processing formula as shown in (19.2):

$$z_{ij} = x_{ij} \Big/ \sqrt{\sum_{i=1}^{p} x_{ij}^2} \tag{19.2}$$

In step ⑤, the entropy weight method was used to determine the scoring weight of each indicator [13]. The entropy method was an objective weighting method that determined the weight according to the information passed to the decision-maker by each indicator. The steps for weighting principal component in PCA were as follows:

① Forming a new matrix by the scores of each sample under principal component and the new matrix was standardized by translating the data, so the scores in the matrix were all greater than 0. The new matrix was $Y' = (y'_{ik})_{m\times K}$, where y'_{ik} represented the score of the i-th sample under the k-th principal component index.

② Calculating the proportion p_{ik} of the score of the i-th sample under the k-th principal component index:

$$p_{ik} = y'_{ik} / \sum_{i=1}^{m} y'_{ik} \tag{19.3}$$

③ Calculating the entropy e_k of the k-th principal component index:

$$e_k = -\frac{1}{\ln m}\sum_{i=1}^{m} p_{ik} \ln p_{ik} \tag{19.4}$$

④ Calculating the weight α'_k of the principal component index of item k:

$$\alpha'_k = (1 - e_k) / \sum_{k=1}^{n} (1 - e_k) \tag{19.5}$$

⑤ The score of the i-th sample under the entropy method was $score'_i$:

$$score'_i = \sum_{k=1}^{n} \alpha'_k p_{ik} \tag{19.6}$$

19.4 Example of Deflection Quality Assessment of Assembly Vehicle Body

In order to verify the feasibility of the method, the deflection quality of the assembly car body was assessed, using the deflection parameters in the actual manufacturing process of the car body. Table 19.1 was the deflection measurement data of each control point of the assembly car body and Table 19.2 was the adjustment force data of each control point calculated by the finite element simulation analysis. The qualified data of deflection was set as (−3.0, 0.0, 13.0, 0.0, −3.0, −3.0, 0.0, 13.0, 0.0, −3.0) and the adjustment error of each point was set as ±0.1.

Combining with the data in Tables 19.1 and 19.2, the IPCA was used for assessment. Table 19.3 listed the eigenvalues, variance contribution rate, and cumulative variance contribution rate corresponding of each principal component.

Table 19.1 Table of deflection measurement data of each control point

Car body number	Deflection measurement data of each deflection control point ω_a (mm)									
	C_1	C_2	C_3	C_4	C_5	C_6	C_7	C_8	C_9	C_{10}
1	−6.2	−2.3	8.5	−1.8	−7.3	−6.6	−3.0	8.8	−2.8	−6.7
2	−4.9	−1.2	13.6	−1.5	−5.2	−4.4	−1.5	12.1	−1.5	−5.0
3	−2.1	2.2	12.8	−0.5	−4.5	−3.5	1.2	11.6	−0.3	−2.9
4	−2.8	−0.3	12.5	−0.5	−3.4	−2.9	−0.3	13.4	−0.2	−3.2
5	−5.3	1.2	12.0	2.2	−4.3	−4.9	1.8	9.5	−0.3	−6.3
6	−5.4	−1.1	7.6	−2.4	−7.2	−5.2	0.0	14.6	2.3	−3.1
7	−2.9	0.2	12.4	−0.1	−3.0	−2.8	0.0	12.9	−0.2	−3.0
8	−5.2	0.0	14.3	−1.2	−6.0	−3.6	1.1	13.8	1.3	−4.7
9	−6.5	−3.0	8.4	−2.6	−7.2	−4.6	−0.7	7.5	0.4	−6.6
10	−3.2	2.0	9.2	−0.8	−4.8	−2.8	0.6	10.0	−0.9	−4.4

According to the principle of $\gamma_K \geq 0.90$, the first k index variables were taken as the principal components and the principal components Y_1, Y_2, Y_3, Y_4, and Y_5 were taken as the principal component indexes to construct the matrix Y'.

The entropy method was used to weight the index and the table of comparison of comprehensive assessment result was shown in Table 19.4. The higher the comprehensive score, the worse the deflection quality of the car body.

According to the table of comparison of the comprehensive assessment results, it could be seen that the deflection quality of the fourth and seventh groups of assembly car body was better. It reflected that the deflection of the two groups of assembly car body was closer to the qualified deflection data in actual manufacturing and it was easier to adjust to the qualified range of deflection of the assembly car body. The deflection quality of the fifth and ninth group assembly car body was worse. Through the assessment of multiple sets of assembly car body deflection data, the deflection quality of the assembly car body was better when $score' \leq 0.060$.

19.5 Conclusion

(1) In view of the low manufacturing efficiency of the assembly car body, the importance of the quality assessment of the deflection of the assembly car body was emphasized. The deflection quality assessment method of the assembly car body was proposed. The laser displacement sensor was used to measure the deflection of each control point of the assembly car body with high precision and the adjustment force data of each control point of the assembly car body was calculated.

(2) In this paper, the assessment process of the PCA was analyzed, the shortcomings of the principal component analysis method were improved. By using the

Table 19.2 Table of adjustment force data of each control point

	Adjustment force data of each deflection control point F_a (MPa)									
Body number	C_1	C_2	C_3	C_4	C_5	C_6	C_7	C_8	C_9	C_{10}
1	129.4	−535.1	1145.7	−863.6	433.0	188.9	−228.1	575.7	−332.4	64.9
2	159.4	349.3	−869.1	390.6	170.7	−0.5	146.6	−176.2	96.6	21.6
3	4.1	−768.4	487.0	0.0	284.9	404.1	−819.0	825.3	−130.4	−312.2
4	−39.9	0.0	41.8	0.0	14.0	0.0	162.5	−220.1	72.1	3.6
5	582.9	−946.1	1329.4	−1153.7	205.1	283.5	−1701.1	2038.3	−1061.9	542.2
6	−25.2	−1174.4	1759.0	−861.4	529.6	492.6	93.1	−115.7	−448.7	0.0
7	0.0	−149.3	225.0	-58.4	−21.1	0.0	0.0	0.0	8.0	0.0
8	409.8	0.0	−818.6	333.8	289.1	74.0	−156.2	149.8	−354.9	217.4
9	303.9	−359.2	847.5	−645.5	178.8	23.7	−1146.0	2530.6	−1822.5	545.2
10	390.9	−1747.0	2125.0	−781.9	90.5	80.3	−921.1	1333.8	−490.7	0.0

Table 19.3 Table of eigenvalues, variance contribution rate, and cumulative variance contribution rate of each principal component

Principal component Y_k	Y_1	Y_2	Y_3	Y_4	Y_5	Y_6	Y_7	Y_8	Y_9
Eigenvalues λ_k	9.21	4.25	2.35	1.37	1.27	1.04	0.34	0.13	0.02
Variance contribution rate α_k/%	46.06	21.25	11.77	6.85	6.37	5.22	1.71	0.67	0.10
Cumulative variance contribution rate γ_k/%	46.06	67.31	79.08	85.93	92.30	97.52	99.23	99.90	100.00

Table 19.4 Table of comparison of comprehensive assessment results

Car body number	1	2	3	4	5	6	7	8	9	10
Comprehensive score	0.131	0.089	0.084	0.056	0.148	0.092	0.050	0.090	0.156	0.103
Rank	8	4	3	2	9	6	1	5	10	7

entropy method to realize the weighting of the assessment index weight, the objective weighting of the index weight was realized and the method was used in the assessment of the deflection quality for the assembly car body.

(3) The method of deflection quality assessment for the assembly car body was verified. The deflection quality of the assembly car body was assessed with the deflection parameters at each deflection control point in the actual manufacturing process. The deflection parameters were seen as the assessment index. The results showed that the deflection quality assessment method of assembly car body based on the IPCA could better realize the deflection quality assessment of the assembly car body, which had a guiding role in the production of the assembly car body.

Acknowledgements This paper is one of the stage results of the Guangdong Administration for Market Regulation 2020 Science and Technology Project *Deflection Quality Evaluation and Auxiliary Adjustment Technology of Assembly Car Body Based on Remote Internet Laser Displacement Sensor* (2020ZJ04).

References

1. L. Feng, J. Yang, Y. Liu, et al., Research of the deflection of articulated vehicle body. Locomot. Roll. Stock Technol. **2018**(04), 1–3 (2008)
2. F. Wang, G. Yang, C. Wang, et al., Study of deflection calculation of train car-body. Mach. Build. Autom. **48**(02), 15–22 (2019)

3. A. Darko, A.P.C. Chan, E.E. Ameyaw et al., Review of application of analytic hierarchy process (AHP) in construction. Int. J. Constr. Manag. **19**(5), 436–452 (2019)
4. H.S. Loh, Q. Zhou, V.V. Thai et al., Fuzzy comprehensive evaluation of port-centric supply chain disruption threats. Ocean Coast. Manag. **148**, 53–62 (2017)
5. J. Camacho, A. Pérez-Villegas, P. García-Teodoro et al., PCA-based multivariate statistical network monitoring for anomaly detection. Comput. Secur. **59**, 118–137 (2016)
6. J. Hu, J. Chen, Z. Chen et al., Risk assessment of seismic hazards in hydraulic fracturing areas based on fuzzy comprehensive evaluation and AHP method (FAHP): a case analysis of Shangluo area in Yibin City, Sichuan Province, China. J. Petrol. Sci. Eng. **170**, 797–812 (2018)
7. R. Shi, X. Fan, Y. He, Comprehensive evaluation index system for wind power utilization levels in wind farms in China. Renew. Sustain. Energy Rev. **69**, 461–471 (2017)
8. B. Liu, R. Xu, J. Wang, et al., Bit selection and evaluation model based on improved principal component analysis. China Pet. Mach. **48**(09), 8–14 (2020)
9. C. Qian, W. Mu, K. Wang, et al., Fuzzy comprehensive evaluation of groundwater quality based on principal component analysis. Water Resour. Power **34**(11), 31–35 (2016)
10. I.T. Jolliffe, J. Cadima, Principal component analysis: a review and recent developments. Philos. Trans. R Soc. A Math. Phys. Eng. Sci. **374**(2065), 20150202 (2016)
11. R. Baklouti, M. Mansouri, M. Nounou et al., Iterated robust kernel fuzzy principal component analysis and application to fault detection. J. Comput. Sci. **15**, 34–49 (2016)
12. L. Li, S. Liu, Y. Peng et al., Overview of principal component analysis algorithm. Optik **127**(9), 3935–3944 (2016)
13. M. Zair, C. Rahmoune, D. Benazzouz, Multi-fault diagnosis of rolling bearing using fuzzy entropy of empirical mode decomposition, principal component analysis, and SOM neural network. Proc. Inst. Mech. Eng. C J. Mech. Eng. Sci. **233**(9), 3317–3328 (2019)

Chapter 20
Random-Optimal Differential Evolution Neural Network Model for Inverse Calculation of Demolition Robot

Jianzhong Huang, Yuwan Cen, Yimo Zong, and Jiahan Bao

Abstract For the inverse calculation of laser-guided autonomous positioning of the demolition robot, a direct mapping model of laser measurement to the driving space of the joint hydraulic cylinder is established by using artificial neural network (ANN) so as to avoid the inverse calculation accuracy depending on the parameters and the accuracy of calibration. In order to improve the convergence rate of the differential evolution (DE) optimizing ANN, a new random-optimal differential evolution (RODE) is proposed to improve the balance of the evolutionary population exploring and exploiting process. Simulation results show that the RODE can significantly improve the convergence speed and maintain good optimization stability and the output precision of optimized inverse calculation ANN can meet the global positioning control requirements of the demolition manipulator.

20.1 Introduction

Demolition robot is a kind of special engineering machinery equipped with hydraulic serial manipulator and hydraulic crushing hammer, which can complete the breaking task in a harsh environment. At present, the demolition robot has no autonomy and programmed movement ability [1]. The manual operation and positioning efficiency

J. Huang (✉)
School of Mechanical Engineering, Hefei University of Technology, Hefei, China
e-mail: 13955586254@139.com

Y. Cen · J. Bao
Engineering Technology Research Center of Hydraulic Vibration Technology, Anhui University of Technology, Maanshan, Anhui Province, China
e-mail: cywan@ahut.edu.cn

J. Bao
e-mail: baojiahan@126.com

Y. Zong
School of Mechanical Engineering, Anhui University of Technology, Maanshan, China
e-mail: zongyimo@163.com

G. Liu and F. Cen (eds.), *Advances in Precision Instruments and Optical Engineering*, Springer Proceedings in Physics 270, https://doi.org/10.1007/978-981-16-7258-3_20

of the demolition manipulator is low. Therefore, it is necessary to make the manipulator autonomous positioning in order to improve the positioning speed and accuracy. In response to this research goal, we have established a set of laser-guided autonomous positioning system of the demolition manipulator.

The joints of demolition manipulator are driven by the hydraulic cylinder. Therefore, in order to realize the laser-guided autonomous positioning control of the manipulator, it is first necessary to establish the inverse calculation model from the three-dimensional laser measurement space to the hydraulic cylinder driving space of the manipulator joint.

The accuracy of inverse calculation simultaneously depends on the accuracy of the redundant inverse kinematic solution of the manipulator and the calibration of the conversion relationship between the measurement system and the robot system. It contains nonlinear characteristics, large geometric parameter errors, and complex calculation process, so it is difficult to obtain effective inverse calculation result.

The artificial neural network (ANN) has global nonlinear mapping characteristics and does not need to establish physical model [2]. The ANN can be adopted to describe the direct mapping relationship between the three-dimensional laser measurement space and the driving space of the joint hydraulic cylinder. The ANN model converts the inverse calculation into the training process of neural network weights, which avoid the calibration of spatial transformation relationship, equation derivation, and numerical pathological problems.

Only by training ANN weight set effectively we can obtain a mapping model that meets engineering requirements. In robot engineering, the BP neural network is more used to solve computational problems [3–5]. The weight optimization of ANN belongs to a high-dimensional real parameter optimization problem. The BP neural network has the disadvantages of low learning efficiency, slow convergence speed, and easy to fall into local minima. It is an effective method to employ intelligent algorithm to optimize ANN weight set [6]. For instance, Köker et al. [7] combined the simulated annealing (SA) algorithm with ANN to solve the inverse kinematic problem of PUMA560; Köker et al. [8] also combined genetic algorithm (GA) and ANN to form a genetic evolutionary neural network to solve the inverse kinematics problem of 6DOF Stanford robot; Kinoshita et al. [9] combined particle swarm optimization (PSO) with ANN for estimation of inverse model. It is also found that SA belongs to a greedy algorithm with increased randomness, so the convergence speed is slow. As a classical evolutionary algorithm, GA has the disadvantages of premature population and easy to fall into local optimization [10]. PSO algorithm is a successful random optimization algorithm, but it is also easy to fall into local extremum [11].

As a new evolutionary algorithm, differential evolution (DE) has the advantages of adaptive global search, simple parameter setting, easy execution, and so on. It has certain advantages in the optimization of ANN weight set [12, 13]. However, the convergence speed of DE algorithm is slow [14].

To solve this problem, this paper proposes a random-optimal differential evolution (RODE) algorithm, that is, the random variation basis vector and the optimal variation basis vector are alternately used in the optimization process. This RODE algorithm can speed up DE convergence while ensuring optimization stability. The RODE is adopted to optimize the inverse calculation ANN model of demolition robot, and verify the effectiveness of the optimized inverse calculation ANN model in the positioning control of the manipulator.

20.2 Inverse Calculation Neural Network Modeling of Laser-Guided Positioning of Demolition Manipulator

This section first introduces the laser-guided autonomous positioning control system of the breaking manipulator. Based on the analysis of the inverse calculation process of manipulator autonomous positioning, an ANN global mapping model for the inverse calculation of demolition robot is established.

20.2.1 Laser-Guided Autonomous Positioning Control System of Demolition Manipulator

The laser-guided positioning control system of the demolition robot is shown in the Fig. 20.1.

As shown in the figure above, the demolition robot is equipped with a 4DOF (Degree of Freedom) hydraulic serial manipulator. We installed computer system on the demolition robot platform. Magnetostrictive sensor is built-in each joint hydraulic cylinder to measure displacement, and the displacement resolution can reach 0.1 mm. Moreover, electro-hydraulic proportional valve controlled cylinder system of the joint adopts closed-loop precise control mode.

A set of three-dimensional laser ranging system is designed and installed on the robot rotary platform. The system is constructed by two-dimensional digital pan–tilt platform and laser rangefinder. The laser ranging system is installed in the manner that the coordinate axis of the measuring coordinate system *O-XYZ* is parallel to the coordinate axis of the manipulator base coordinate system $O\text{-}X'Y'Z'$ (Fig. 20.2).

20.2.2 Basic Principle of Inverse Calculation of Demolition Robot

The mapping relationship between the three-dimensional laser measurement points and the joint driving value of the manipulator is deduced, and the complexity of the inverse calculation process of the laser-guided positioning of the manipulator is analyzed.

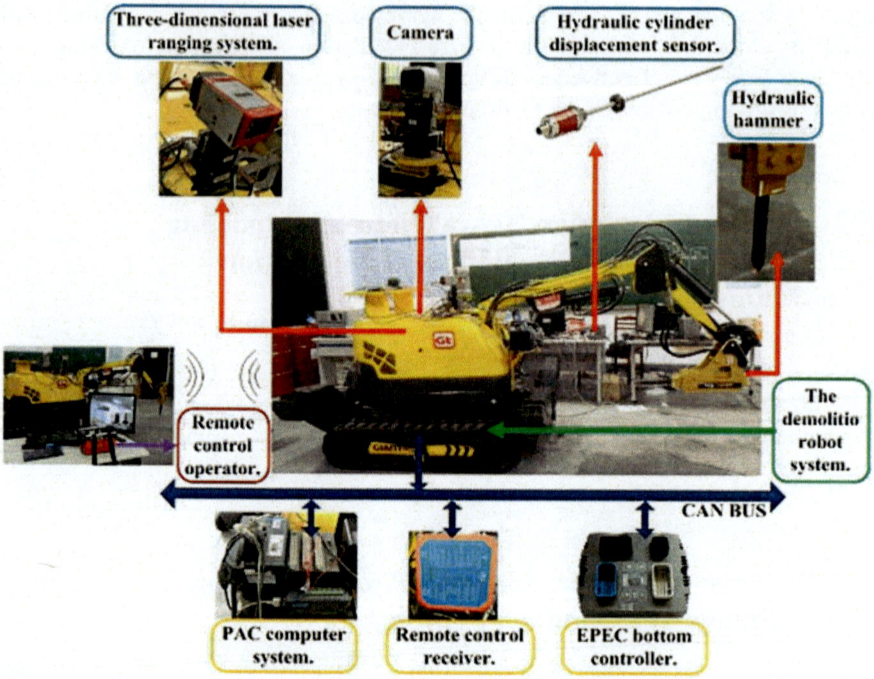

Fig. 20.1 Laser-guided positioning control system for demolition manipulator

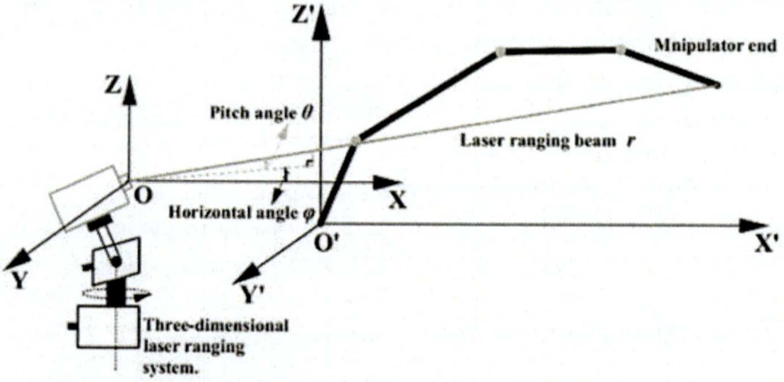

Fig. 20.2 Position relationship between three-dimensional laser ranging system and manipulator

Forward kinematics modeling of demolition manipulator

The structural diagram of 4DOF hydraulic series manipulator assembled by the demolition robot is as follows (Fig. 20.3).

The working space of the demolition manipulator is a plane space $X'O'Z'$. $P \in X'O'Z'$ represents the position vector from the origin of the base coordinate system to the end endpoint of the manipulator. The forward kinematics equation of the end

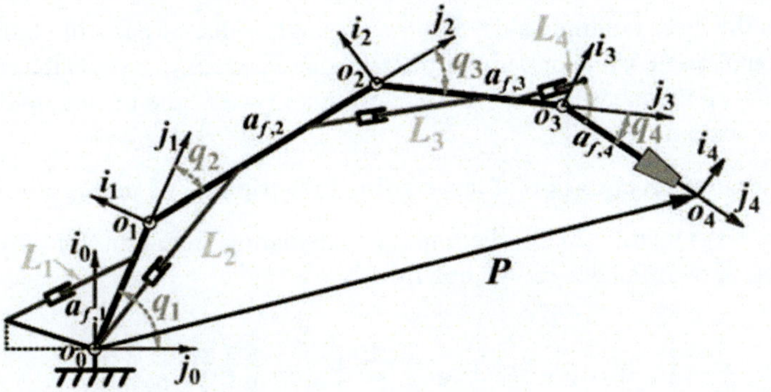

Fig. 20.3 Structure diagram of manipulator of demolition robot

position vector $P = [x' \; z']^T$ can be obtained as follows:

$$\begin{cases} x' = \sum_{i=1}^{4}\left(a_i \cdot \cos\left(\sum_{j=1}^{i} q_j\right)\right) \\ z' = \sum_{i=1}^{4}\left(a_i \cdot \sin\left(\sum_{j=1}^{i} q_j\right)\right) \end{cases} \tag{20.1}$$

It can be seen from the above equation that the joint angle value q constitutes the four-dimensional joint angle space C^4 of the manipulator, and the end position $P \in W^2$ of the manipulator can be uniquely determined by each joint angle $q \in C^4$ [15].

Conversion calculation between target point and laser measured value

The laser measurement system outputs a three-dimensional vector value $(r, \theta, \varphi)^T \in M^3$. It can be seen from Fig. 20.2 that, by operating the pan–tilt platform of the laser ranging system to rotate and pitch, the laser beam can be adjusted to measure the target point in the working space $X'O'Z'$ of the manipulator. The conversion relationship between laser measurement value $(r, \theta, \varphi)^T$ and target point coordinate value $(x', z')^T$ is shown in the following equation:

$$\begin{cases} r = \sqrt{(x' + \Delta x)^2 + \Delta y^2 + (z' + \Delta z)^2} \\ \theta = \arcsin \dfrac{\sqrt{(x' + \Delta x)^2 + \Delta y^2}}{\sqrt{(x' + \Delta x)^2 + \Delta y^2 + (z' + \Delta z)^2}} \\ \varphi = \arctan(\dfrac{\Delta y}{x' + \Delta x}) \end{cases} \tag{20.2}$$

where r is the laser ranging value, θ is the tilt angle value of pan–tilt platform, φ is the horizontal angle value of pan–tilt platform, and $[\Delta x, \Delta y, \Delta z]^T$ is the translation vector between the measurement space coordinate system and the manipulator base coordinate system.

Inverse calculation equation of laser-guided positioning of manipulator

Based on (20.1) and (20.2), the inverse calculation equation for laser-guided positioning of manipulator can be obtained.

$$\begin{cases} r\sin\theta\cos\varphi - \Delta x = \sum_{i=1}^{4}\left(a_i\cdot\cos\left(\sum_{j=1}^{i}q_j\right)\right) \\ r\cos\theta - \Delta z = \sum_{i=1}^{4}\left(a_i\cdot\sin\left(\sum_{j=1}^{i}q_j\right)\right) \end{cases} \tag{20.3}$$

According to the above equation, after the laser beam selects the target point to obtain the measurement vector $(r, \theta, \varphi)^T$, first convert the laser measurement value into the joint angle value vector q, that is, realize the mapping from the measurement space M^3 to the joint angle space C^4, as follows:

$$M^3 \to C^4 \tag{20.4}$$

It can be seen from Fig. 20.3 that the angle q_i of each joint of the hydraulic serial manipulator is actually determined by the stroke value L_i of the cylinder, that is, the joint angle q needs to be converted into the position control parameter L of the actual movement of the hydraulic cylinder, as shown in the following equation:

$$L_i = F_i(q_i)\ i = 1, 2, 3, 4 \tag{20.5}$$

Equation (20.5) describes the nonlinear mapping relationship from joint angle space C^4 to hydraulic cylinder drive space L^4, as follows:

$$C^4 \to L^4 \tag{20.6}$$

Analysis of inverse calculation of laser-guided positioning of manipulator

Referring to (20.3) and (20.5), the following difficulties exist in the inverse calculation of the autonomous positioning of the manipulator:

Equation group (20.3) is a nonlinear underdetermined equation group with infinite solutions. The related solutions often have problems such as no closed solution, complex calculation process and slow solution speed.

The translation vector $[\Delta x, \Delta y, \Delta z]^T$ between O-XYZ and O-X′Y′Z′ must be calibrated.

The demolition robot has errors in the connecting rod parameters a_1, a_2, a_3, and a_4 of (20.3), which brings calculation errors for the inverse calculation.

It can be seen from (20.4) that the geometric parameter error of the mechanism will also affect the calculation error of C4 → L4.

It is essential to get accurate translation vector $[\Delta x, \Delta y, \Delta z]^T$ and connecting rod parameter values to obtain accurate inverse calculation solutions. This must adopt professional equipment for measurement, modeling and calculation, and the calibration process is cumbersome [16].

In summary, it is difficult to obtain the closed solution for the inverse calculation of the laser-guided positioning of the demolition manipulator. There are many factors that produce calculation errors and rely on the calibration of external measuring equipment.

20.2.3 ANN Modeling for Inverse Calculation of Demolition Robot

Based on the mapping relations (20.4) and (20.6), the inverse calculation process is simplified into a direct mapping relationship, as shown in the following equation:

$$f(p) : M^3 \to L^4 \tag{20.7}$$

In robot engineering, multilayer feedforward neural network is an effective method to establish a global mapping model covering the robot workspace 2. Research shows that as long as the number of hidden layer neurons is enough, the single hidden layer feedforward neural network can ensure the ability to describe complex continuous mapping relationships [17]. A hidden layer multilayer feedforward neural network is employed to describe the nonlinear mapping relationship represented by (20.7) to simplify the inverse calculation process of the demolition robot. The inverse calculation ANN modeling is shown in the Fig. 20.4.

The input layer of the ANN model is three-dimensional laser measurement vector m, and the output layer is four-dimensional vector L of the joint cylinder stroke.

20.3 Differential Evolution Neural Network

On the basis of introducing the basic principles of differential evolution (DE), aiming at the weight optimization of the inverse calculation ANN, the coding design of the differential evolution neural network and the derivation of the fitness value equation are completed. Finally, the convergence problem of DE optimizing ANN is analyzed in detail.

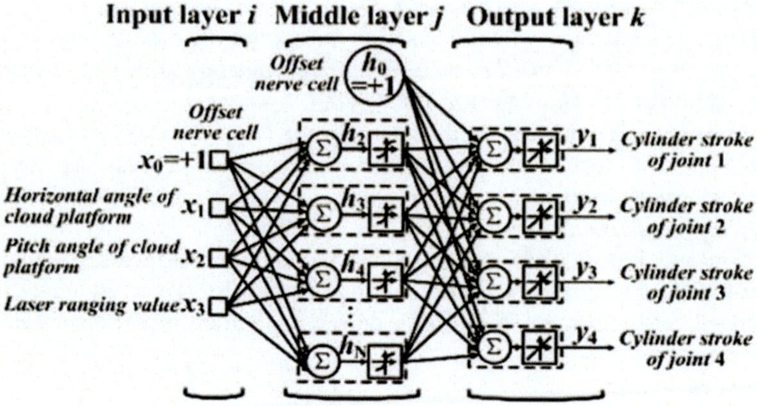

Fig. 20.4 ANN mapping model for inverse calculation of demolition robot

20.3.1 Differential Evolution

The basic principle of DE is as follows [18, 19]:

Differential evolution population

The operation object of DE algorithm is continuous space vector individuals. The t-generation evolutionary population $P_x(t)$ of DE is composed of N_p target vectors $x_i(t)$, and $x_i(t)$ is composed of d-dimensional real number coding, as shown in the following equation:

$$x_i(t) = x_{j,i}(t), \; j = 1, \ldots, D_o \tag{20.8}$$

Evolutionary population initialization

Generation 0 population is randomly generated in the weight decision space, and its initialization result is as follows:

$$x_{j,i}(0) = rand_j(0, 1) \cdot (b_U - b_L) + b_L \tag{20.9}$$

where b_L and b_U are lower and upper bounds of the optimization space, respectively.

Variation operator

The most important operator of DE is the differential mutation operator. It adds a scalable and randomly selected vector difference component to a basis vector $x_{basic,r0}(t)$ to generate the variation vector $v_i(t)$.

$$v_i(t) = x_{basic,r0}(t) + F \cdot (x_{r1}(t) - x_{r2}(t)) \tag{20.10}$$

The value of the vector index should satisfy $i \neq r0 \neq r1 \neq r2$ in each variation operation.

Crossover operator

The test vector $u_i(t)$ is generated in the form of exponent (bin), as shown in the following equation:

$$u_{j,i}(t) = \begin{cases} v_{j,i}(t), rand_j(0,1) \leq Crorj = j_{rand}, \\ x_{j,i}(t), else. \end{cases} \tag{20.11}$$

where the crossover rate *Cr* is set to a real number between 0.9 and 1.

Repairing operator

The operator is used for vector individual constraints.

$$v_{j,i}(t) = \begin{cases} \min\{U_j, 2L_j - v_{j,i}(t)\}, if\ v_{i,j}(t) < L_j \\ \max\{L_j, 2U_j - v_{j,i}(t)\}, if\ v_{i,j}(t) > U_j \end{cases} \tag{20.12}$$

Selection operator

Based on a one-to-one greedy selection mechanism,

$$x_i(t+1) = \begin{cases} u_i(t), f(u_i(t)) \leq f(x_i(t)) \\ x_i(t), else \end{cases} \tag{20.13}$$

where $f(u_i(t)$ is the fitness value of the test vector and $f(x_i(t)$ is the fitness value of the target vector.

20.3.2 Principles of Differential Evolution Neural Network

DE is adopted to optimize the weight of the laser-guided positioning ANN model of the demolition manipulator to construct a differential evolution neural network and carry out the design of evolutionary individual coding and the derivation of fitness function.

Evolutionary individual coding

In the research of evolutionary neural network, evolutionary coding refers to how to describe the connection weight of the neural network for the evolutionary algorithm to search in the corresponding solution space. The operation object of DE is a continuous space real number vector. The neural network weight is directly combined into a real value vector to express chromosomes, see the Fig. 20.5.

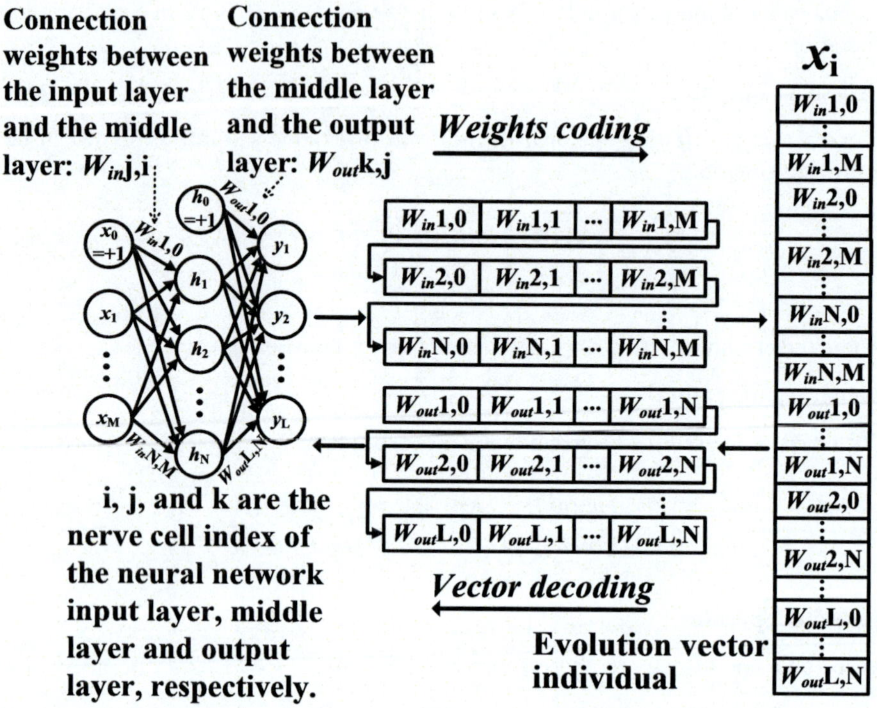

Fig. 20.5 Coding and decoding of differential evolution neural network

Dimension calculation of ANN weight space

Based on Fig. 20.5, the dimensional calculation of the ANN weight optimization space is shown in the following equation:

$$D = (M + 1) \times N + (N + 1) \times L \tag{20.14}$$

Fitness function of differential evolution individual

In the calculation of the evolutionary neural network, the mean square error of the neural network target output and the actual output is used as the fitness of the evolutionary individual and the network parameters are adjusted by the algorithm to minimize the mean square error as follows:

$$fit_i = \frac{1}{Q}\sum_{q=1}^{Q}\sum_{k=1}^{L}\left(t_k - f_2\{w_{ou}k,0 + \sum_{j=1}^{N} w_{ou}k,j \cdot f_1[w_{in}j,0 + \sum_{i=1}^{M}(w_{in}j,i \cdot r_i)]\}\right)^2 \tag{20.15}$$

where Q is the total number of training samples.

20.3.3 Analysis of Convergence Process of Differential Evolution

The effect of selection of the differential variation basis vector on the convergence process of evolutionary population is analyzed. Aiming at the problem of slow convergence of DE, a random-optimal differential evolution (RBDE) is proposed for optimizing inverse calculation neural network model.

Differential variation strategy

The performance of the DE is mainly determined by its evolution strategy and control parameters. Among them, the evolution strategy is mainly determined by the selection method of the variation basis vector. In the DE, the variation basis vector randomly selects the vector individual $x_{rand}(t)$ of the population, see the following equation:

$$v_i(t) = x_{rand}(t) + F \cdot (x_{r1}(t) - x_{r2}(t)) \tag{20.16}$$

In order to improve the convergence speed of differential evolution, the researchers also proposed that the variation basis vector adopts the optimal vector individual $x_{best}(t)$ in the population as follows:

$$v_i(t) = x_{best}(t) + F \cdot (x_{r1}(t) - x_{r2}(t)) \tag{20.17}$$

In the following, the DE that adopt random variation basis vector is referred to as DE_RAND, corresponding to which the DE using optimal basis vector is referred to as DE_BEST. Relevant studies show that the two evolutionary strategies have a significant impact on population convergence, among which DE_RAND is helpful to maintain population diversity, but the convergence speed is slow. On the contrary, DE_BEST usually speeds up the convergence speed of the population and reduces the possibility of stagnation, but it is easy to converge locally and reduce the success rate of the algorithm [19].

Population average distance

The population average distance (PAD) can reflect the convergence state of the evolutionary population, that is, the diversity of the population. The average distance between the individual and the population center of gravity is calculated in (20.18) and (20.19).

$$\overline{x}(t) = \frac{1}{N_p}\sum_{i=1}^{N_p} x_i(t) \tag{20.18}$$

$$d_{ave}(t) = \frac{1}{N_p}\sum_{i=1}^{N_p} \|x_i(t) - \overline{x}(t)\| \tag{20.19}$$

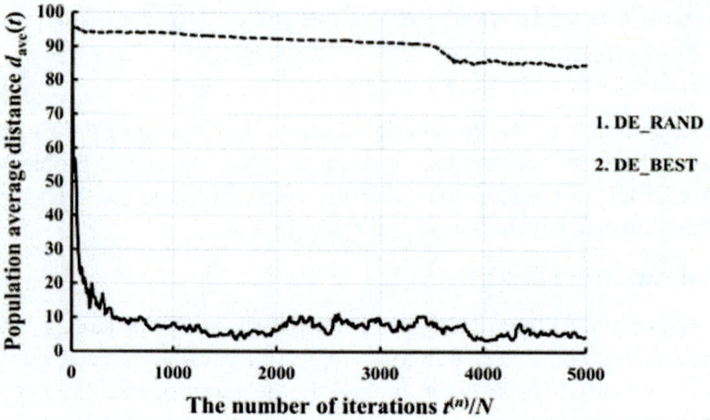

Fig. 20.6 Convergence process of differential evolution strategy

The population of evolutionary algorithm is distributed in a large space at the initial stage. With the iteration of the algorithm, the difference between individuals of the population gradually decreases, and its average distance gradually decreases to a small value, indicating that the population converges to an extreme point.

Comparative study on convergence process of differential evolution strategies

DE_RAND and DE_BEST are used to optimize the inverse calculation of the ANN model. The comparison of the convergence process is shown in the Fig. 20.6:

The following conclusions can be drawn from the analysis of the above figure:

The PAD of DE_RAND decreases very slowly, and the effective convergence of the population cannot be achieved in the high-dimensional real number space, and the search efficiency is low.

In contrast, DE_BEST is significantly better than the DE_RAND algorithm in convergence speed, but the population diversity is rapidly reduced in the early stage of optimization, which affects the global search.

The variation basis vector selection strategy of differential variation operation plays a leading role in the convergence speed of the differential evolution population.

20.3.4 Random-Optimal Differential Evolution Neural Networks

This section presents random-optimal differential evolution (RODE), which balances the exploration and exploitation process of DE population, and adopts RODE algorithm to optimize the inverse calculation of the ANN model of the demolition robot.

Random-optimal difference variation operator

Aiming at the problems of DE, in order to speed up the convergence speed of DE, while maintaining the global search ability of DE, combine the advantages of the DE_RAND and DE_BEST variation strategies in the convergence process, and propose a new random-optimal differential mutation operator.

Based on (20.16) and (20.17), RODE alternately uses random basis vector and optimal basis vector for variation operation, see the following equation:

$$v_i(t) = \begin{cases} x_{rand}(t) + F \cdot (x_{r1}(t) - x_{r2}(t)), t\% \, 2 == 0 \\ x_{best}(t) + F \cdot (x_{r1}(t) - x_{r2}(t)), t\% \, 2 == 1 \end{cases} \tag{20.20}$$

It can be seen from the above equation that the random-optimal difference variation operator does not increase the complexity of DE variation operation.

Algorithm flow of RODE optimizing inverse calculation ANN

The specific steps of using RODE algorithm to optimize the inverse calculation of the ANN model are as follows:

Step 1: Set the number N of neurons in the middle layer of ANN and the maximum number of iterations $G_{\max}$, and calculate the weight space dimension D by the (20.14).

Step 2: Use (20.9) to initialize the population to form the target vector group;

Step 3: (20.20) and (20.11) are used to perform vector variation and crossover operations, and (20.12) is used to constrain the boundary of the variation vector to generate test vector group.

Step 4: Input the training sample set of the demolition robot, and employ (20.15) to calculate the fitness value $f(x_i(t))$ of the target vector and the fitness value $f(u_i(t))$ of the test vector.

Step 5: Based on the individual fitness value, use the selection (20.13) to complete the selection operation and produce the individual $x_i(t+1)$ of the new target vector group.

Step 6: Judge whether the maximum number of iterations $G_{\max}$ or the optimized fitness value accuracy requirement is reached, if the evolution requirement is met, the algorithm flow is ended. Otherwise, the calculations from Step 3 to Step 6 are performed again.

Analysis of convergence process of random-optimal differential evolution

The RODE algorithm is used to optimize the inverse calculation ANN model, and the convergence process of evolutionary population is shown in the Fig. 20.7.

Analysis of the above figure shows that:

Compared with DE_RAND, RODE significantly accelerates the convergence speed of DE.

Compared with the DE_BEST strategy, RODE slows down the convergence speed in the early stage of optimization, but the convergence speed exceeds DE_BEST in the later stage of optimization. At the end of optimization, RODE has converged to the extreme point.

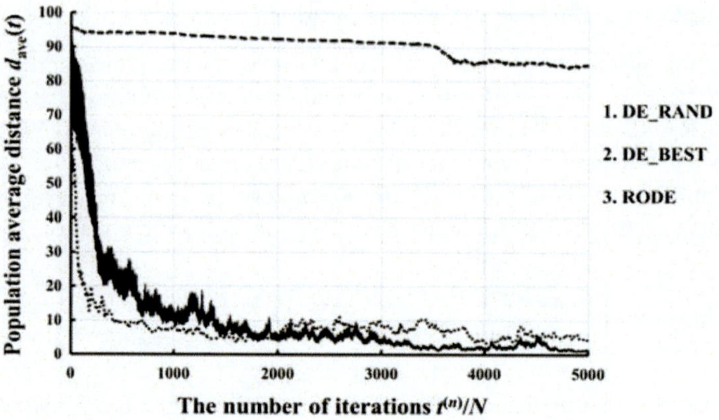

Fig. 20.7 Convergence process of random-optimal differential evolution

Analyzing the convergence process of RODE shows that it achieves the goal of slowing population convergence first and then fast population convergence. Comparing two evolutionary strategies of DE_RAND and DE_BEST, it better balances the exploration and exploitation process of the evolutionary population.

20.4 Algorithm Simulation Experiment of Random-Optimal Differential Evolution Inverse Computational Neural Network

This section first analyses the performance of RODE in the optimization of neural network weight based on algorithm simulation experiments, and then verifies the effectiveness of the inverse calculation neural network model of the demolition robot optimized by RODE.

20.4.1 The Training Samples of Inverse Calculation ANN

The training data example of the ANN mapping model for the inverse calculation of the laser-guided positioning control of the demolition robot is shown in the following Table 20.1.

In the above table, the voltage value is the output value of the joint cylinder displacement sensor, which is proportional to the cylinder stroke.

Table 20.1 The training sample example of inverse calculation ANN

Serial Number	Laser measurement value			Measured cylinder stroke			
	Laser ranging value/mm	Horizontal angle of pan–tilt°	Pitch angle of pan–tilt°	Stroke of cylinder 1/V	Stroke of cylinder 2/V	Stroke of cylinder 3/V	Stroke of cylinder 4/V
1	3060	−3.08	17.25	0.964	0.925	2.493	0.933
2	3094	−2.99	12.98	1.011	0.938	3.423	1.741
3	995	18.68	12.77	1.006	1.797	0.962	4.370
4	2018	−8.19	36.51	1.001	1.748	1.787	3.115
5	1226	14.40	16.07	1.003	1.736	1.787	4.377
---	---	---	---	---	---	---	---

20.4.2 *Comparative Analysis of the Results of Optimizing Inverse Calculation ANN*

In the inverse calculation ANN optimization simulation experiment, the DE population size N_p is set to 200, the weight range is [−15, 15], the number of hidden layer neurons is set to 15, and 500 sets of data are employed as training samples.

In order to verify the algorithm stability and optimization accuracy of the RODE, the fitness value results output by RODE, DE_RAND and DE_BEST optimizing inverse calculation ANN are compared and analysed. In addition, genetic algorithm (GA) and particle swarm optimization (PSO) are adopted to optimize the inverse calculation ANN to further prove the calculation performance of RODE.

Each optimization method is set to run 20 times, and the maximum iteration number is 5000 generations. The optimization results of various algorithms are shown in Table 20.2.

Analysis of the above table shows that:

The standard deviation of DE_RAND is large, and the average value of the optimization results is the lowest. This is because its convergence speed is too slow, and it does not converge and output the optimized solution within a limited number of iterations.

The average value of RODE output is better than DE_BEST, and the standard deviation is the lowest, indicating that the new RODE has good optimization stability.

GA is slightly better than DE_RAND in the inverse calculation of the ANN model optimization, and the optimization performance is lower than DE_BEST, PSO, and RODE.

In conclusion, the RODE not only significantly accelerates the convergence speed of the DE, but also maintains the stability of the algorithm, which embodies the stable optimization performance in the optimization of the neural network.

Table 20.2 Comparison of fitness values of optimizing inverse calculation ANN

Serial number	Algorithm name	Optimal fitness value	Worst fitness value	Average value of fitness value	Standard deviation of fitness value
1	DE_RAND	1.72×10^{-3}	1.54×10^{-2}	7.51×10^{-3}	3.30×10^{-3}
2	DE_BEST	8.76×10^{-4}	1.06×10^{-3}	9.93×10^{-4}	5.29×10^{-5}
3	RODE	8.24×10^{-4}	8.81×10^{-4}	8.50×10^{-4}	1.36×10^{-5}
4	GA	1.00×10^{-3}	2.67×10^{-3}	1.28×10^{-3}	2.18×10^{-3}
5	PSO	9.88×10^{-4}	1.50×10^{-3}	1.02×10^{-3}	7.02×10^{-4}

20.4.3 Verification of Calculation Results of Inverse Calculation ANN

In order to verify the effectiveness of the RODE inverse calculation ANN algorithm, 150 sets of verification samples in the workspace of the manipulator are collected. Input these data into the optimized ANN model, and calculate the stroke of the joint cylinder corresponding to the target points of laser measuring.

Analysis of calculation results of joint cylinder stroke

The measured value of cylinder stroke corresponding to the target point is compared with the calculated value, as shown in the Table 20.3.

For 150 verification samples, it can be seen from Table 20.4 that the average values of the difference between the measured and calculated strokes of 4 joint cylinders are 5.01 mm, 4.97 mm, 3.86 mm and 4.76 mm, respectively.

End error analysis of stroke calculation value of joint cylinder

Based on the forward kinematics mapping (20.1), the measured value and calculated value of the joint cylinder stroke are converted into the end point positions $(X_k, Y_k)^M$ and $(X_k, Y_k)^C$ of the manipulator respectively, and the error value of both is End_{diff}, see below (Fig. 20.8).

As can be observed in Fig. 20.9, the average value of the calculated difference of the manipulator end position is 2.35 cm. As a kind of construction machinery, the drill rod diameter of the end hydraulic hammer of the demolition robot is 4 cm. Therefore, for the inverse calculation ANN mapping model is obtained by RODE offline optimization, its calculation accuracy can meet the calculation requirements for the precise positioning of large manipulators in practical engineering applications.

20.5 Conclusion

In order to realize the autonomous positioning of manipulator, a laser-guided autonomous positioning control system of demolition robot is constructed. Aiming at the inverse calculation of autonomous positioning, an ANN global mapping model is established to simplify the inverse calculation process of laser-guided autonomous positioning of demolition manipulator.

Aiming at the hyperparameter optimization problem of the inverse calculation of the ANN model weight set, DE is used to optimize the ANN weights. In order to speed up the convergence of DE, a new random-optimal differential evolution (RODE) is proposed. Convergence research shows that RODE can not only accelerate the convergence speed of De but also better balance the exploration and exploitation process of DE optimization.

Table 20.3 Example of inverse calculation results

Serial number	Cylinder 1 Stroke		Cylinder 2 Stroke		Cylinder 3 Stroke		Cylinder 4 Stroke	
	Calculation/mm	Measure/mm	Calculation/mm	Measure/mm	Calculation/mm	Measure/mm	Calculation/mm	Measure/mm
1	22.89	26.19	163.69	167.08	68.42	64.69	81.83	80.82
2	23.56	27.65	166.57	170.28	70.18	67.41	177.53	176.28
3	28.04	25.56	159.24	155.22	133.36	134.88	249.55	249.64
4	27.82	25.04	157.67	153.31	132.18	134.32	345.99	345.93
5	88.13	83.97	247.74	243.94	270.48	273.13	84.76	85.84
⋮	⋮	⋮	⋮	⋮	⋮	⋮	⋮	⋮

Table 20.4 The average of the calculated difference of the joint cylinder stroke of the verified sample

Parameter name	Calculated difference of cylinder			
	Cylinder 1	Cylinder 2	Cylinder 3	Cylinder 3
Minimum value	1.83	1.89	1.05	0.52
Maximum value	6.95	7.91	5.98	5.58
Average value	5.01	4.97	3.86	4.76

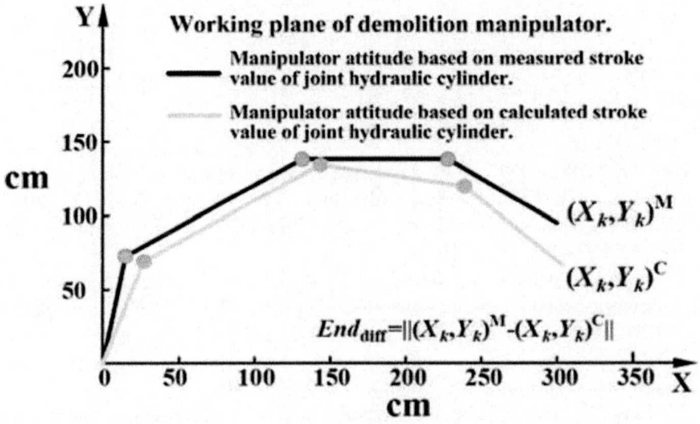

Fig. 20.8 Schematic diagram of calculation difference of manipulator end position

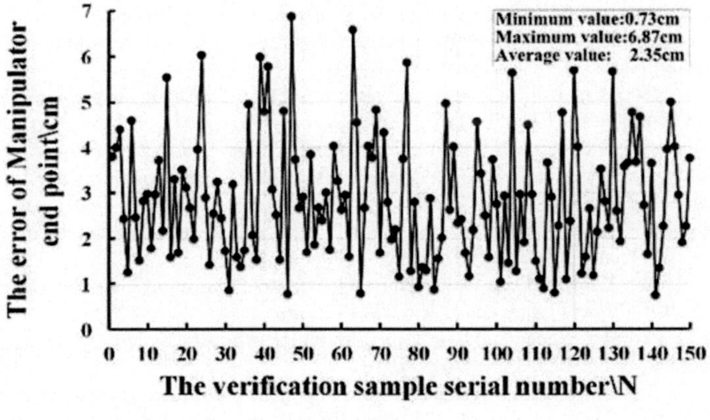

Fig. 20.9 Calculation error of manipulator end position

The algorithm simulation experiment proves that the output accuracy of the optimized inverse calculation ANN meets the positioning requirements of the demolition manipulator. The inverse calculation method of the hydraulic manipulator based on the ANN global nonlinear mapping model can provide a reference scheme for autonomous positioning control of other large engineering manipulators.

Acknowledgements This research was supported by the Major Special Project of Science and Technology in Anhui Province (201903a05020029).

References

1. D. Derlukiewicz, M. Ptak, S. Kozioek, Proactive failure prevention by human-machine interface in remote-controlled demolition robots. Adv. Intell. Syst. Comput. **445**, 711–720 (2016)
2. Z. Mao, T.C. Hsia, Obstacle avoidance inverse kinematics solution of redundant robots by neural networks. Robotica **15**, 3–10 (1997)
3. A.T. Hasan, N. Ismail, A.M.S. Hamouda, I. Aris, M.H. Marhaban, H.M.A.A. Al-Assadi, Artificial neural network-based kinematics Jacobian solution for serial manipulator passing through singular configurations. Adv. Eng. Softw. **41,** 359–367 (2010)
4. B. Daya, S. Khawandi, M. Akoum, Applying neural network architecture for inverse kinematics problem in robotics. J. Softw. Eng. Appl. **3**, 230–239 (2010)
5. Y. Cai, Z. Qiang, X. Xi, A. Rahmani, Inverse kinematics identification of a spherical robot based on BP neural networks, in *2011 6th IEEE Conference on Industrial Electronics and Applications*, Beijing (2011), pp. 2114–2119
6. T.K. Gupta, K. Raza, Optimization of ANN architecture: a review on nature-inspired techniques. *Machine Learning in Bio-Signal Analysis and Diagnostic Imaging* (2019), pp. 159–182
7. R. Köker, A neuro-simulated annealing approach to the inverse kinematics solution of redundant robotic manipulators. Eng. Comput. **29**, 507–515 (2013)
8. R. Köker, A genetic algorithm approach to a neural-network-based inverse kinematics solution of robotic manipulators based on error minimization. Inf. Sci. **222**, 528–543 (2013)
9. K. Kinoshita, K. Watanabe, M. Isshiki, Estimation of inverse model based on ANN and PSO with adaptively varying acceleration coefficients, in *2014 Proceedings of the SICE Annual Conference*, Sapporo (2014), pp. 281–286
10. K.F. Man, K.S. Tang, S. Kwong, Genetic algorithms: concepts and applications [in engineering design]. IEEE Trans. Ind. Electron. **43**, 519–534 (1996)
11. B. Liu, L. Wang, Y.H. Jin, D.X. Huang, Improved particle swarm optimization combined with chaos. Chaos Solitons Fractals **25**, 1261–1271 (2005)
12. F. Neri, V. Tirronen, Recent advances in differential evolution: a survey and experimental analysis. Artif. Intell. Rev. **33**, 61–106 (2010)
13. A.P. Piotrowsk, Differential evolution algorithms applied to neural network training suffer from stagnation. Appl. Soft Comput. **21**, 382–406 (2014)
14. L. Tang, Y. Dong, J. Liu, Differential evolution with an individual-dependent mechanism. IEEE Trans. Evol. Comput. **19**, 560–574 (2015)
15. B. Siciliano, L. Sciavicco, L. Villani, G. Oriolo, *Robotics: Modelling, Planning and Control* (Springer Publishing Company, Incorporated, London, 2009)
16. A. Neves, A.J. Pinho, D.A. Martins, B. Cunha, An efficient omnidirectional vision system for soccer robots: from calibration to object detection. Mechatronics **21**, 399–410 (2011)
17. G. Cybenko, Approximation by superpositions of a sigmoidal function. Math. Control Sig. Syst. **2**, 303–314 (1989)

18. M. Ali, M. Pant, V.P. Singh, An improved differential evolution algorithm for real parameter optimization problems. Int. J. Recent. Trends Eng. **1**, 63–65 (2009)
19. K. Price, R.M. Storn, J.A. Lampinen, *Differential Evolution: A Practical Approach to Global Optimization* (Springer, Berlin, Heidelberg, 2005)

Chapter 21
Analysis and Performance Evaluation on Mechanical Property of Nuclear Pump Liquid Annular Seals

Li Song, XiaoHui Luo, and Can Zhao

Abstract The stability performance of nuclear pump liquid annular seal was impacted by the parameters including the operating factors and other mechanical factors. The liquid annular which used in nuclear power station was taken as one analysis part, the technical method and model of six degrees of freedom were introduced to explain the dynamic influence on seal during the gravity impact load station. The tested data including liquid seal force and pressure parameter with the change of time was analyzed. The relationship of the factors such as sealing clearance, liquid viscosity, and pumps speed was investigated as well. From the research, it can be found that maximum pressure and seal force was repaid rising and descend quickly subsequently. If pumps sealing clearance was increased, the parameter of sealing force and pressure will be increased as well. If the pump liquid viscosity was increased, the parameter of displacement in axis direction and sealing pressure will be decreased as well. In addition, it can be see that pumps speed factor has no relationship with the parameter of sealing force in gravity direction and sealing pressure.

21.1 Introduction

The mechanical seals were played a major role in the pumps, especially in the nuclear power station. The annular seal was regarding as the core component, it can eliminate and restrict the leakage. Convention components structure of pumps annual seal were installed in the middle between pumps annular seal parts and inter-stage seal

---

L. Song (✉) · X. Luo (✉) · C. Zhao
Equipment Procurement & Supply Division, China Nuclear Power Engineering Co. Ltd., Shenzhen, Guangdong, China
e-mail: songli2009@cgnpc.com.cn

X. Luo
e-mail: luoxiaohui1985@126.com

C. Zhao
e-mail: zhaocan@cgnpc.com.cn

G. Liu and F. Cen (eds.), *Advances in Precision Instruments and Optical Engineering*, Springer Proceedings in Physics 270, https://doi.org/10.1007/978-981-16-7258-3_21

parts [1]. Much studies and tests had been carried and completed in the field of the annular seal parts as deep as hydraulic dynamic analysis. The pumps rotor and seal parts shall suffer huge impact force. Many scholars including Ma Jin Kui, Rao, TICHY, Li Zhen, etc., had studied related test and research about the performance characteristics and condition of seal fluids once suffering transient impact force [2, 3]. For instance, the scholars named Ma Jin Kui had get the relationship and formula on the parameter of thinnest oil film thickness, maximum pressure of film, and axial direction pressure of the pumps under different pulse. The scholars named Rao had get the parameter and curve of rotor in the condition of critical speed by analyzing, and the characteristic of dynamic response in shock condition was summarized as well. The scholars named TICHY has studied deeply on the detail standard how the impact load and load in horizon direction will influence axial trajectory of pumps. The resonance performance of pump annular bearing in various load was studied by Li Zhen. Yan Et team had summarized and build the conclusion and Jeffcott rotor model on fluid analysis of pumps seal component with computer software [4].

To get the further performance and characters of nuclear pumps annual seal [5], this paper will introduce the dynamic response characteristic of pump annular. Factors including seal force, structure clearance, rotor speed, and film pressure will be investigated and discussed.

21.2 Analysis Element

The common structure of pump liquid annular can see Fig. 21.1, the simplified analysis structure and model can see Fig. 21.2.

The detail dynamic formula of annular seal is as follows.

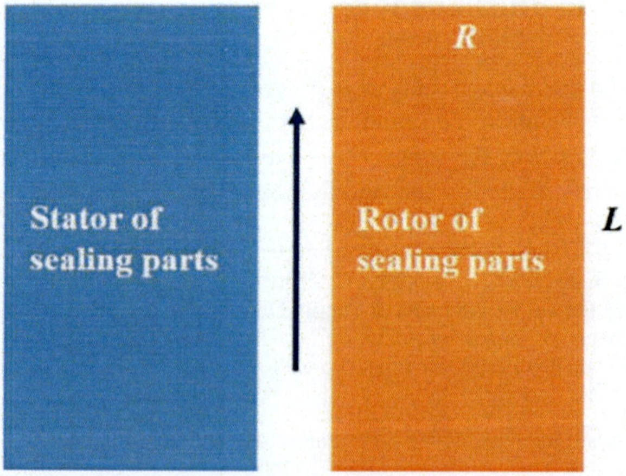

Fig. 21.1 The common structure of pump liquid annular [6]

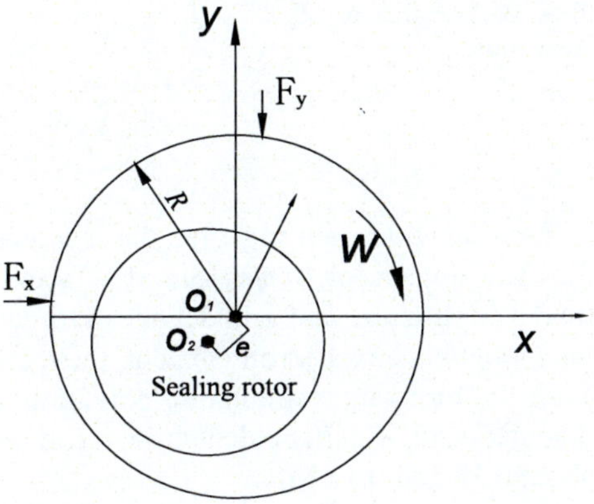

Fig. 21.2 The simplified mechanical diagram

$$M\ddot{x} = F_x + Q_x$$

$$M\ddot{y} = F_y + Q_y + Mg$$

The parameter of horizontal acceleration at pump rotor center is expressed as $\ddot{x}$; the parameter of vertical acceleration at pump rotor center is expressed as $\ddot{y}$; The parameter of sealing force in vertical direction is expressed as F_y; the parameter of impact load in horizontal direction is expressed as Q_x; the parameter of impact load in vertical direction is expressed as Q_y; the parameter of pump rotor mass and gravitational acceleration are expressed as *M and g, respectively.* In order to get the further solution for above equation, individual parameter of axial position need to be breakdown as follows:

$$\dot{x}(\tau + \Delta\tau) = \dot{x}(\tau) + \ddot{x}(\tau)\Delta\tau$$

$$\dot{y}(\tau + \Delta\tau) = \dot{y}(\tau) + \ddot{y}(\tau)\Delta\tau$$

$$x(\tau + \Delta\tau) = x(\tau) + \dot{x}(\tau + \Delta\tau)\Delta\tau$$

$$y(\tau + \Delta\tau) = y(\tau) + \dot{y}(\tau + \Delta\tau)\Delta\tau$$

In above equation, the parameter of time step is expressed as $\Delta\tau$; the parameter of total time is expressed as τ. The equation will be periodic and stop till the time is terminated.

21.3 Calculation and Simulation

From the simulation model build by software, the parameters were set as below (Table 21.1).

Table 21.1 Parameter of simulation

Parameter	Value (mm)	Parameter	Value
Sealing radius gap	0.25	Rotor mass	25 kg
Sealing length	50	Rotor speed	3000 r/min

From previous investigation, the seal pressure difference is not distinct caused by transient impact load, therefore, this factor was ignored in this paper. The element model of structure and grid detail can be seen in Figs. 21.3 and 21.4. After meshing the model via hexahedron element, there are around 1.7 million elements in totally. With the turbulence simulation condition, setting the wall surface to be adiabatic boundaries [7, 8], after calculation, it can be seen that the inlet and outlet pressure is about 0.18 and 0.12 MPa.

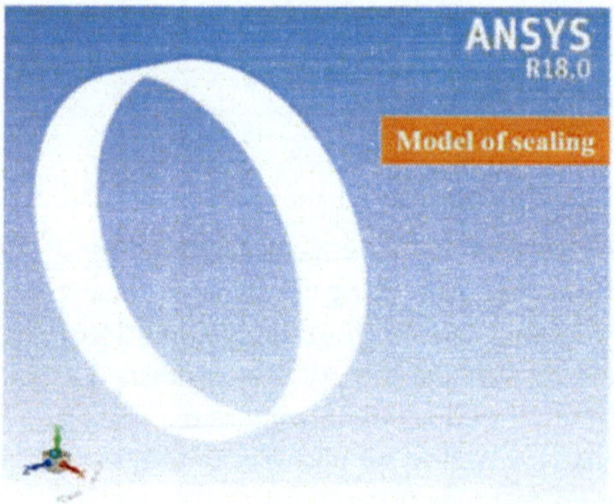

Fig. 21.3 The simulation model of sealing

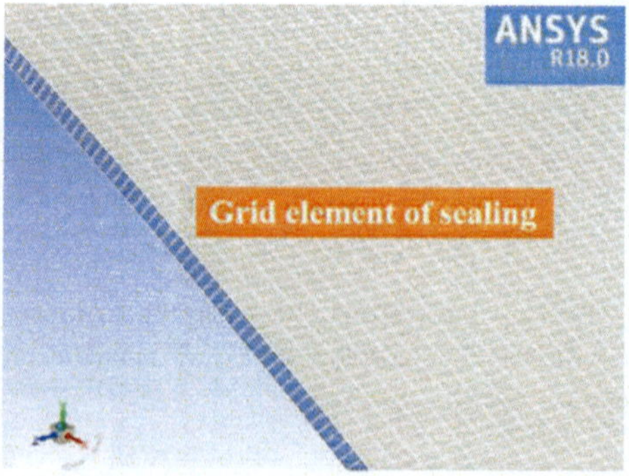

Fig. 21.4 Grid element of annular sealing

21.3.1 Influence of the Annular Sealing Radius

To simulate the different radius parameter of sealing including 0.10, 0.20, and 0.30 mm, the analyzed curve and result which can be seen in Figs. 21.5, 21.6, 21.7 and 21.8. From the result, it can be summarized that when radius parameter is larger, the sealing trajectory in axis direction, seal pressure in horizontal direction and gravity direction are rising and maximum sealing pressure in horizontal and gravity direction will be rising as well. In addition, once the radius parameter is larger, the transient impact will be rising simultaneously.

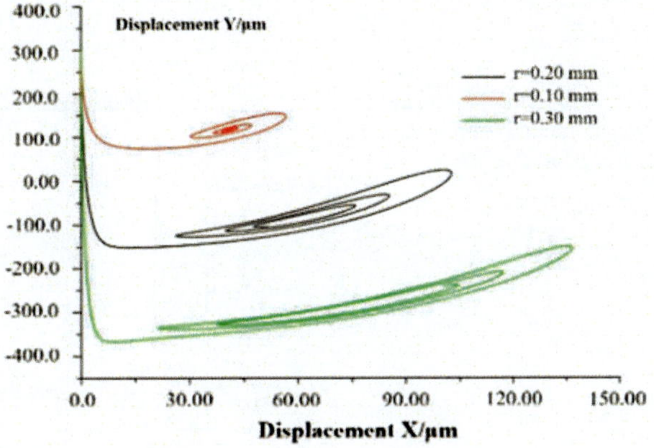

Fig. 21.5 Simulation on different sealing radius

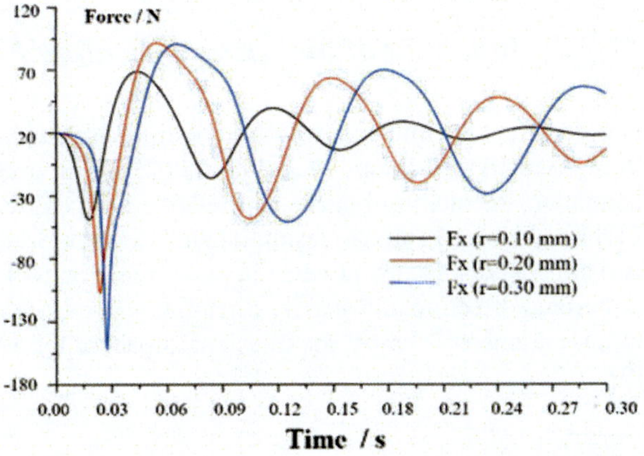

Fig. 21.6 Simulation result of horizontal pressure

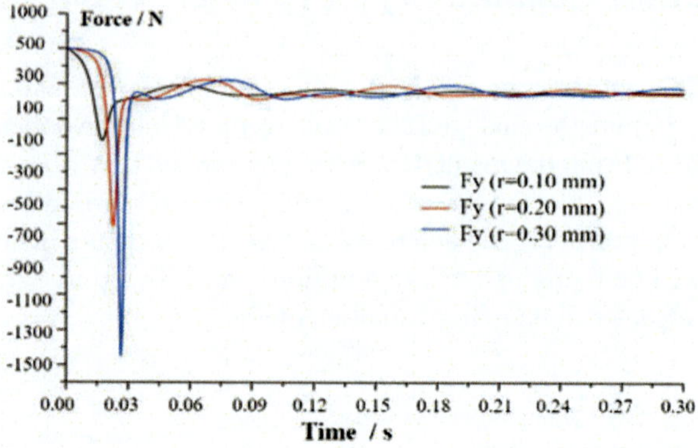

Fig. 21.7 Simulation result of sealing pressure in gravity direction

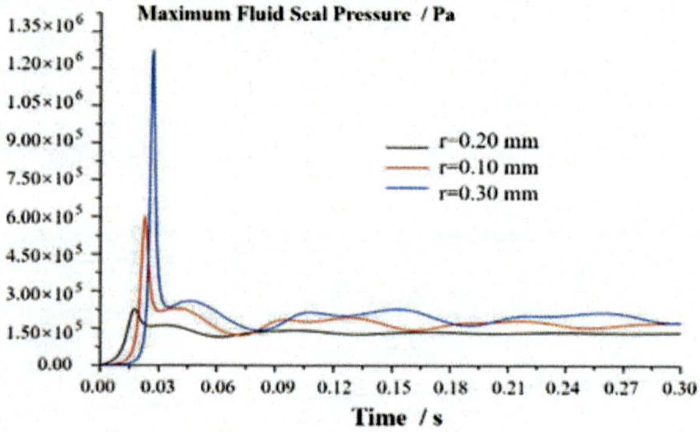

Fig. 21.8 Simulation result on parameter of maximum sealing force

21.3.2 Influence of the Annular Sealing Fluid Viscosity

In order to get the influence of different fluid viscosities for annular sealing, different fluid viscosity conditions including 0.002, 0.020, and 0.080 Pa s are analyzed by computer; the analyzed curve and result which can Fbe seen in Figs. 21.9, 21.10, 21.11 and 21.12. From the result, it can be summarized that when the fluid viscosity is rising, the displacement value in axial direction will reduce, and the sealing force in horizon direction and gravity direction will reduce as well. In addition, once the radius parameter is larger, the transient impact of force will be rising simultaneously [9].

21.3.3 Influence of the Annular Sealing Rotor Speed

In order to get the influence of different sealing rotor speeds for annular sealing, different speed conditions including 1500, 3500, and 5500 r/min are analyzed by computer; the analyzed curve and result which can be seen in Figs. 21.13, 21.14,

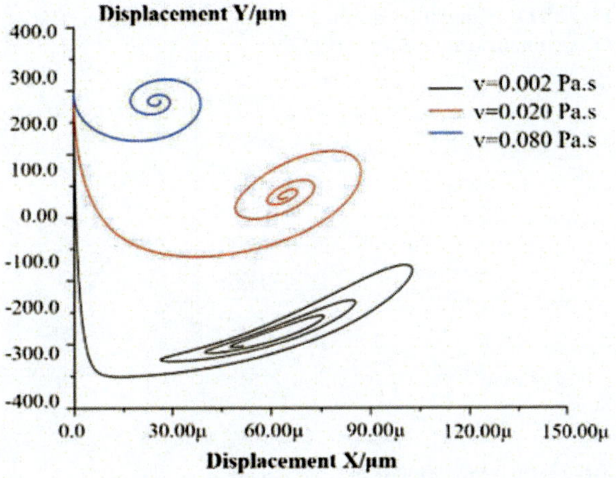

Fig. 21.9 Simulation result of axial displacement

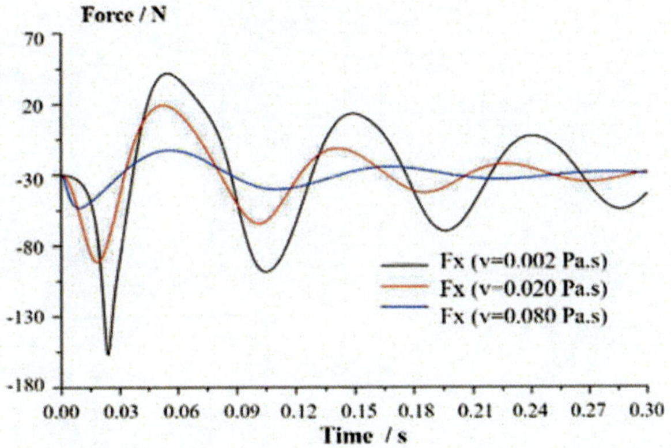

Fig. 21.10 Simulation result of sealing pressure in horizontal direction

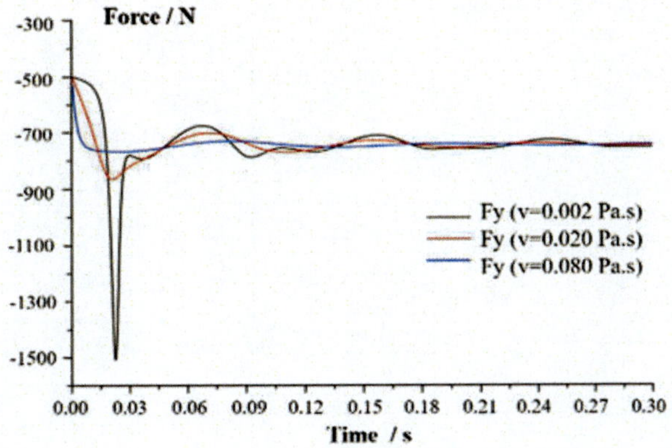

Fig. 21.11 Simulation result of sealing pressure in gravity direction

Fig. 21.12 Simulation result of maximum sealing force

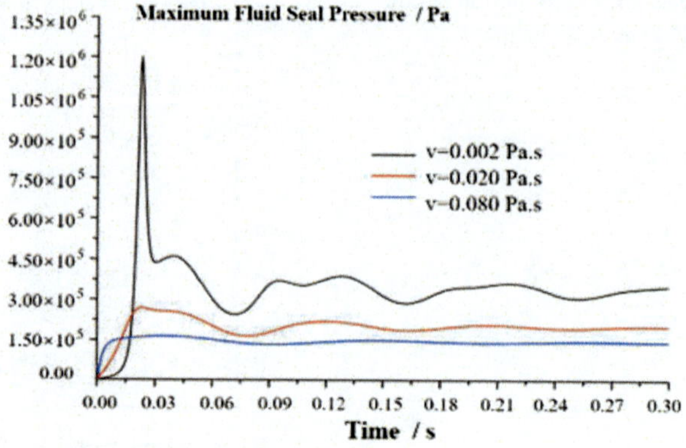

Fig. 21.13 Simulation result of axial displacement

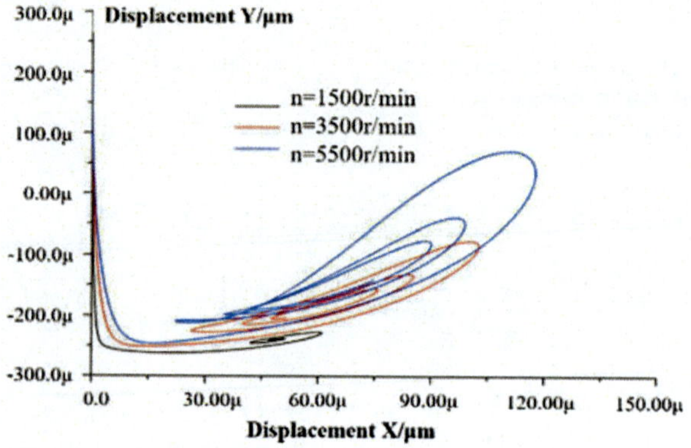

Fig. 21.14 Simulation result of sealing pressure in horizontal direction

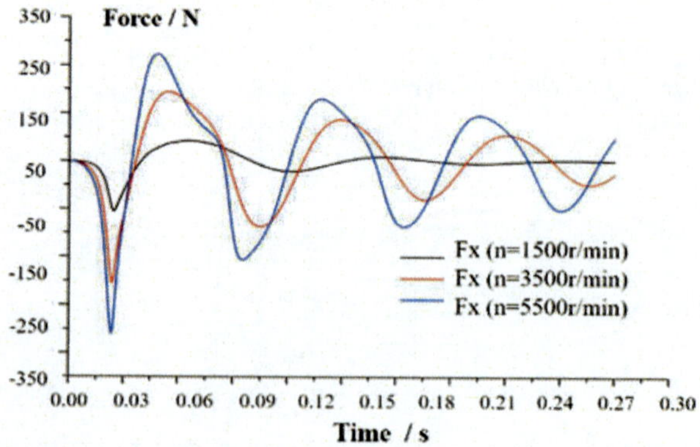

21.15 and 21.16. From the result, it can be summarized that when the sealing speed is rising, the displacement value in axial direction will be reduced, while the force in gravity direction is not line with the change of speed, it will almost be stable.

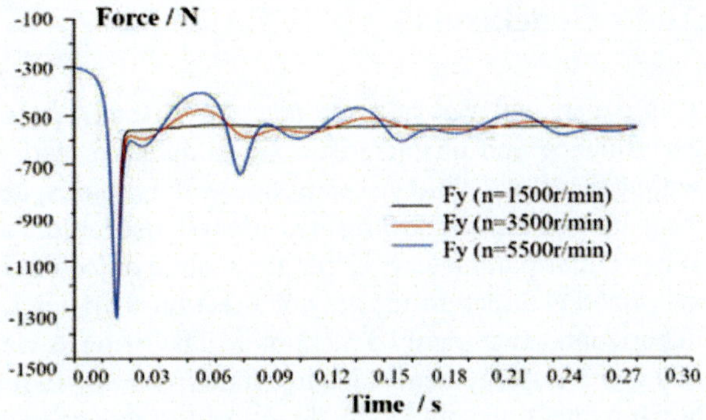

Fig. 21.15 Simulation result of sealing pressure in gravity direction

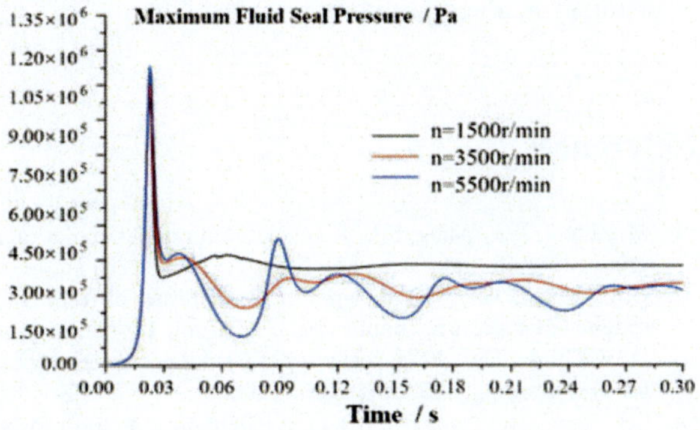

Fig. 21.16 Simulation result of maximum sealing force

Meanwhile, the impact value of annular sealing under different speeds seems to have not changed although time is last [10].

21.4 Conclusion

This paper analyzed and summarized different parameters' influence on sealing performance and characteristic via simulation, and the parameters are including sealing clearance, fluid viscosity, and sealing rotor speed. From the simulation result, it can be seen that the sealing pressure will rise rapidly and then decline subsequently. When radius parameter is larger, the sealing trajectory in axis direction, seal pressure in horizontal direction and gravity direction are rising and maximum sealing pressure in horizontal and gravity directions will be rising as well. When the fluid viscosity is rising, the displacement value in axial direction will reduce, and the sealing force in horizon direction and gravity direction will reduce as well. Regarding the parameter of rotor speed, the simulation result has proven that there will be no obvious influence on sealing force and pressure.

References

1. M. Zhang, X.F. Wang, S.L. Xu et al., Numerical simulation of the flow field in circumferential grooved liquid seals. Adv. Mech. Eng. **5**, 1–10 (2013)
2. G. Chochua, T.A. Soulas, Numerical modeling of rotor dynamic coefficients for deliberately roughened stator gas annular seals. J. Tribol. **129**(2), 335–341 (2006)
3. D.W. Childs, The SSME seal test program: leakage tests for smooth, hole-pattern and helically-grooved stators. NASA, Alabama (1986)
4. L.T. Tam, An iterim report on the calculation method for a multi-dimensional whirling seal. Chamber of North America, Alabama (1986)
5. Guangdong Nuclear Power Training Centre. 900MW PWR nuclear power plant system and equipment (Atomic Energy Press, Beijing, 2004)
6. J.F. Dietzen, Calculating rotor dynamic coefficients of seals by finite-difference techniques. J. Tribol. **109**(3), 388–394 (1987)
7. J.F. Gulich, *Centrifugal Pumps* (Springer, Berlin, 2010)
8. D.W. Childs, *Turbomachinery Rotor Dynamics: Phenomena, Modeling, and Analysis* (Springer Press, Texas, 1993)
9. H.K. Versteeg, W. Malalasekera, *An Introduction to Computational Fluid Dynamics* (Longman Group Ltd, British, 1995)
10. C.E. Brennen, A.J. Acosta, Fluid-induced rotor dynamic forces and instabilities. Struct. Control Health Monit. **13**(1), 10–26 (2006)

Chapter 22
Research on Eddy Current Inspection Test for Defects of Aluminum Alloy ∏-Type Components in High-Speed Train Car Body

YanDe Li, XueZhi Zhang, and Kai Song

Abstract Taking the aluminum alloy ∏-type complex components as the test object, a special multi-channel eddy current detection probe was developed to realize the detection of multiple parts of a single probe. It was carried out that engineering detection experiments and the effect of lift-off height on detection signals under the same detection parameters. The results show that the attenuation of the detection signal amplitude and the deflection of the lift-off signal phase are linearly related to the increase of the lift-off height. It greatly shortens the time required for inspection, saves the cost of removing surface coating and re-covering the coating.

22.1 Introduction

During the operation of rail vehicles, the aluminum alloy ∏-type complex components are subjected to varying degrees of alternating loads due to the suspension of other large equipment. Long-term fatigue stress is very easy to initiate cracks. Once the cracks grow rapidly, it is very easy to cause unpredictable damage [1]. In recent years, various inspection methods have been used at home and abroad to conduct a lot of research on the inspection of aluminum alloy components [2, 3]. But they have not achieved large-area in situ inspection of components, which is inconvenient to apply in the engineering maintenance stage.

Multi-channel eddy current testing has the advantages of fast response speed, large effective scanning area, and high detection sensitivity, etc. [4, 5]. It can realize

Y. Li (✉) · X. Zhang
Zhuzhou CRRC Times Electric Co., Ltd., Zhuzhou, Hunan, China
e-mail: aliyd1@csrzic.com

X. Zhang
e-mail: zhangxz3@csrzic.com

K. Song
Key Laboratory of Nondestructive Testing, Ministry of Education, Nanchang Hangkong University, Nanchang, Jiangxi, China
e-mail: kevin.song@foxmail.com

G. Liu and F. Cen (eds.), *Advances in Precision Instruments and Optical Engineering*, Springer Proceedings in Physics 270, https://doi.org/10.1007/978-981-16-7258-3_22

efficient detection of key parts of aluminum alloy and make effective judgments on cracks or corrosion pits due to fatigue damage [6, 7]. In this paper, a special multi-channel eddy current detection probe is developed for ∏-type complex components, artificial grooves are made to simulate cracks in key parts of the components, engineering inspections are carried out, the suspected defects are re-inspected through penetration testing, and the lift-off height is studied to test the results. Finally, the linear relationship between the lift-off height and the amplitude and phase of the detection signal is obtained.

22.2 Principle of Multi-channel Eddy Current Testing

Eddy current testing uses the principle of electromagnetic induction to induce an eddy current field in a conductive material in an alternating magnetic field. The eddy current penetration depth formula is given as

$$\frac{J_\delta}{J} = e^{-x\sqrt{\pi f \mu \sigma}} \tag{22.1}$$

where J represents the eddy current density on the surface of the workpiece; f represents the frequency of the test coil (Hz); δ represents the electrical conductivity (S/m); J_δ represents the eddy current density in the workpiece at a depth of δ from the surface; μ represents the permeability of the vacuum; if the inspected workpiece is a ferromagnetic material, μ represents the relative permeability; δ represents the penetration depth (in m). It can be seen from the above formula that the eddy current intensity is different at different depths. Therefore, the lift-off height will affect the eddy current detection, which leads to a change in the impedance of the detection coil, causing a defect signal characteristic in the detection signal.

Multi-channel eddy current testing adopts time-division multiplexing technology. Through transmission at different times, the signals of each channel do not overlap each other on the time axis, so that a single communication terminal can transmit multiple signals, which enables different signals to be transmitted at different times. A transmission line realizes the transmission of multiple data channels, as shown in Fig. 22.1.

22.3 Experiment Method

22.3.1 Test Materials

The aluminum alloy ∏-type artificial groove comparison sample was designed and manufactured, and the specimen is shown in Fig. 22.2. The parameters of manual

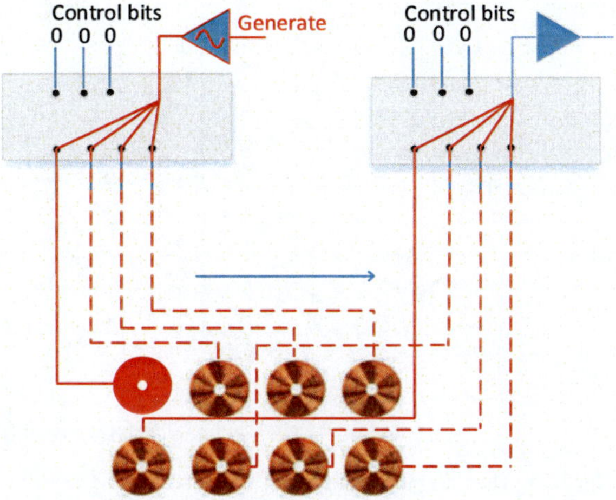

Fig. 22.1 Time division multiplexing

Fig. 22.2 Physical drawing of test block

notch are 5 mm × 0.2 mm × 0.2 mm (length × width × depth), which are located in the R angle area and the plane area, respectively. The five defects simulate the possible flaws in different key parts of the actual component in service, as shown in Fig. 22.3.

22.3.2 Design and Manufacture of Probe

For the aluminum alloy components, a special multi-channel eddy current detection probe was developed. The probe shell is a rectangular plastic steel with rounded corners. The rounded corners can fit well with the R corner of the specimen. The probe contact end has four eddy current coil positioning holes, the eddy current coil is placed in the positioning hole. There is an opening on the back of the rectangular plastic steel, and the time-sharing chip circuit is placed in the opening at the front

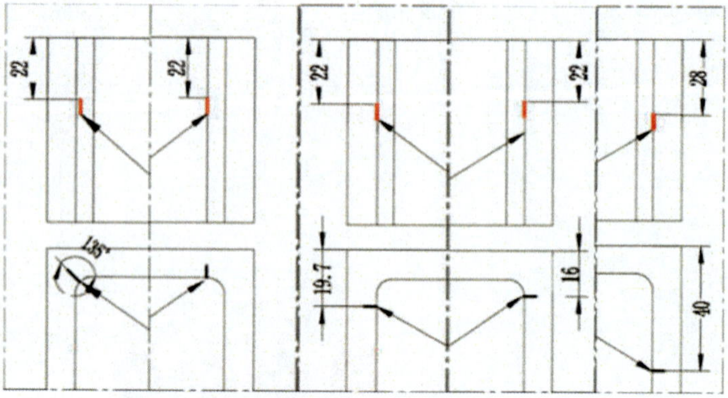

(a) Defects 1 and 2 (a) Defects 3and 4 (c) Defects5

Fig. 22.3 The diagram of defect

end of the probe shell. The terminal of the eddy current coil is connected to the time-sharing circuit, and the probe is shown in Fig. 22.4.

22.3.3 *Test Verification*

The test frequency is set to a single variable, the test frequency is adjusted to 50–400 kHz, and the channel 4 of the four-channel eddy current test probe is used to scan the defect 5 of the comparison sample. The actual fitting effect of the probe is shown in Fig. 22.5. We have ensured uniform speed to avoid amplitude variation caused by different scanning speed. Adjust the lift-off signal to the horizontal direction and record the change in the amplitude of the detection signal. The detection result is shown in Fig. 22.6.

It can be seen from Fig. 22.6: when the frequency is 50–200 kHz, the vertical component of the defect impedance graph increases with the increase in the excitation frequency; when the frequency is 200–400 kHz, the vertical component of the defect signal decreases with the increase in the excitation frequency, and its distribution is normally distributed. It can be seen that the detection sensitivity is highest when the frequency is 200 kHz, so the four-channel eddy current detection probe selects the best frequency 200 kHz for scanning. The results of scanning using this frequency are shown in Fig. 22.7. The presence of cracks causes the conductivity around the area to change. The conductivity changes the eddy current changes in the test block, directly changing the amplitude and phase of the detection signal, forming a non-horizontal direction sudden changes in the detection signal. On the one hand, the change in electrical conductivity causes the voltage of the detection coil in the detection probe to change; on the other hand, the penetration depth of eddy current detection also changes with the change in electrical conductivity.

Fig. 22.4 The diagram of detection probe

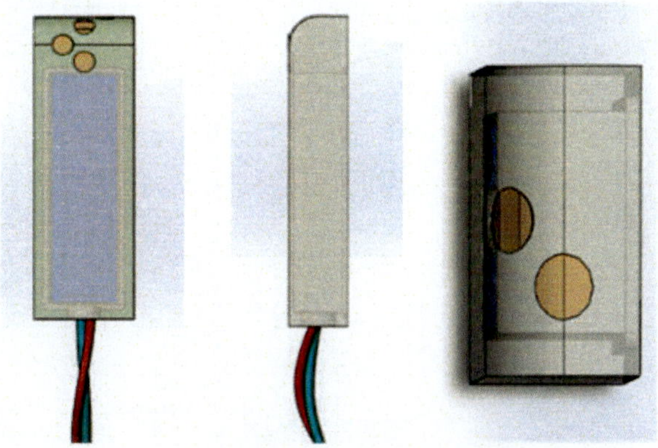

(a) front view (b) Left view (c) Top view

(d) Fitting diagram

Fig. 22.5 The actual fitting effect of the probe

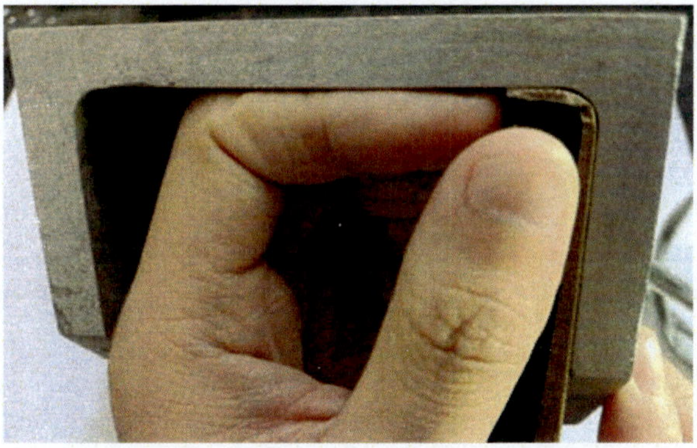

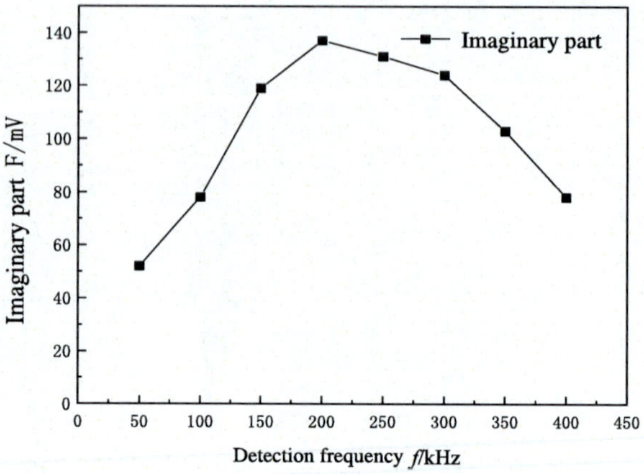

Fig. 22.6 Defect detection effect at different frequencies

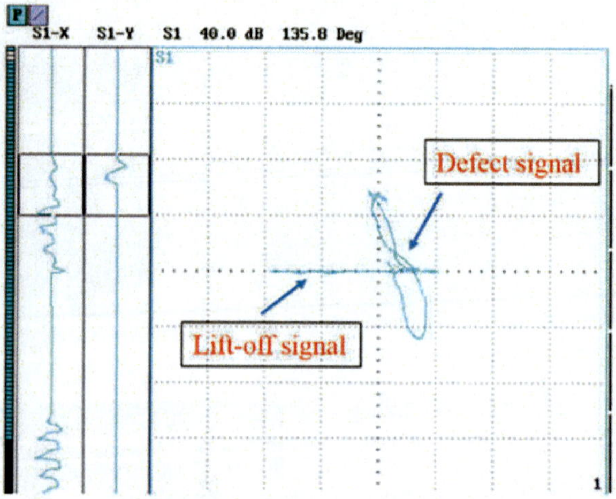

Fig. 22.7 The detection signal of reference block

22.4 Testing and Research

22.4.1 The Effect of Lift-Off Height on Detection Signal

In actual engineering inspection research, there are often coatings on the surface of the workpiece to protect the workpiece. Performing all paint stripping operations on the inspected workpiece will greatly increase the workload. The lift-off heights of the detection probes at different inspection positions are different, which will directly affects the phase of the lift-off signal.

In order to study the effect of lift-off height on the detection signal, the same defect was scanned at different lift-off heights. Use the channel 4 at the plane of the detection probe to scan the manual groove 5 with the parameter of 5 mm × 0.2 mm

× 0.2 mm (length × width × depth) on the R-type aluminum alloy test block at different lift-off heights. The surface is covered with different layers of Teflon tape to simulate different heights of lift-off. The single layer thickness of Teflon tape is 0.12 mm. When the Teflon tape is not attached to the detection surface of the probe, set the lift-off signal to the horizontal direction, and gradually increase the number of Teflon layers, keeping the detection parameters constant and the scanning speed constant, analyze the amplitude and phase change characteristics of the detection signal, and the detection result is shown in Fig. 22.8.

It can be seen from Fig. 22.8 that, as the lift-off height increases, the phase of the lift-off signal deviates, and the amplitude of the defect signal decreases accordingly. Continue to increase the lift-off height, and use the probe close to the sample surface to detect the signal amplitude and phase as the reference base point. Continue to increase the lift-off height until the lift-off height is 1.2 mm.

It can be seen that with the increase of the lift-off distance, the phase change of the detection signal is much smaller than the change of the lift-off signal phase.

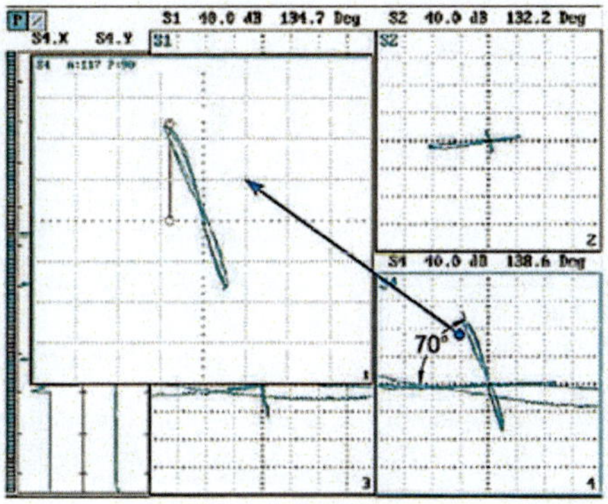

(a) Lifting height is 0 mm

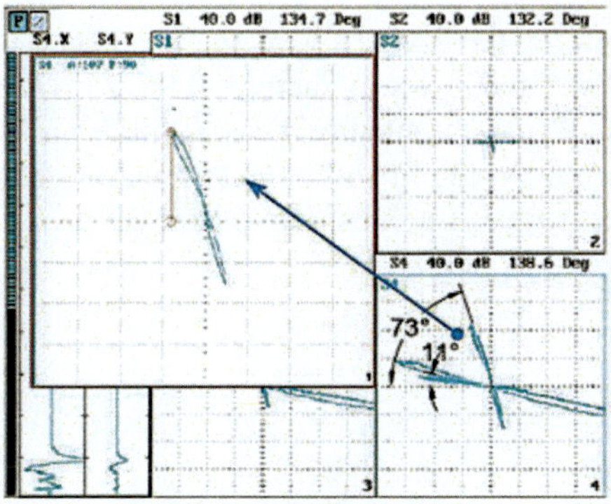

(b) Lifting height is 0.12 mm

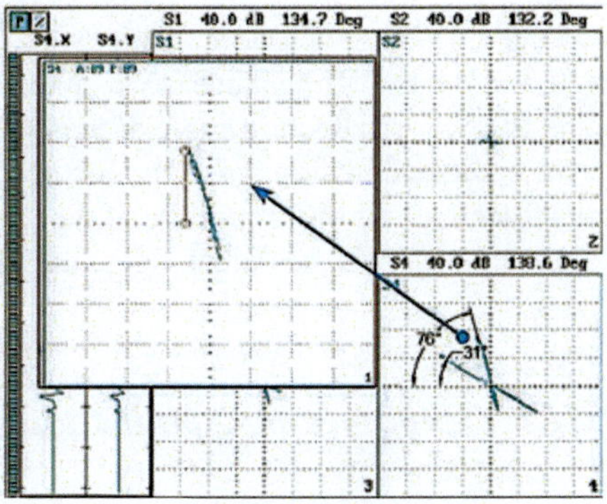

(c) Lifting height is 0.48 mm

Fig. 22.8 Detection signal at different lift-off heights

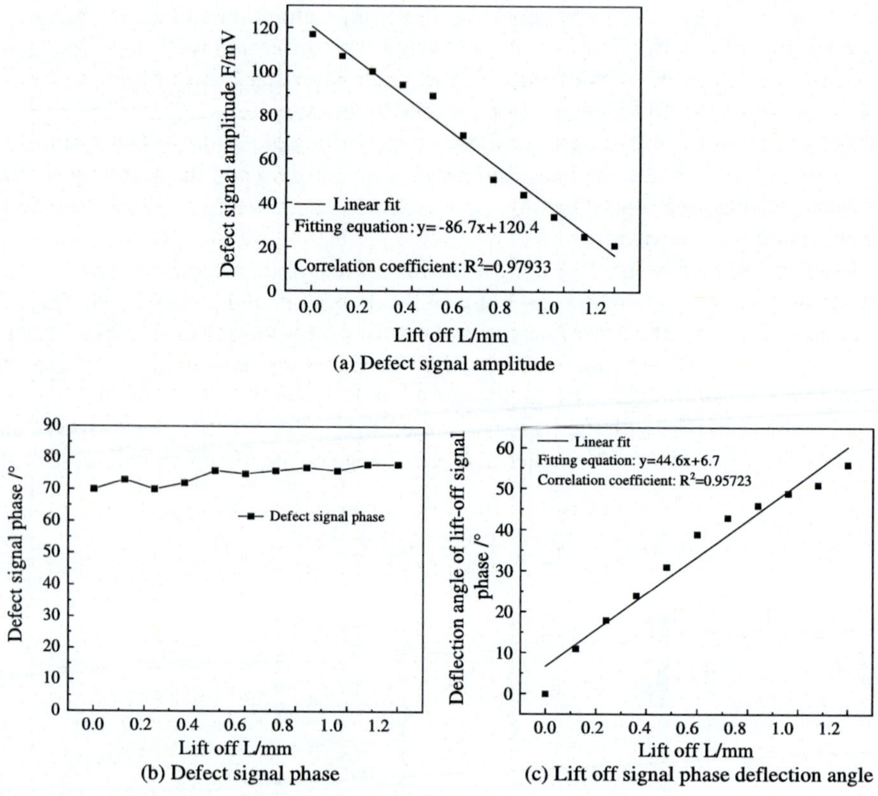

Fig. 22.9 Detecting signal amplitude and phase at different lift-off heights

From Fig. 22.9, for the change of lift-off height, the linear relationship between defect signal amplitude and lift-off height is $y = -86.7x + 120.4$ and its fitting factor is 0.97933. The linear relationship between the lift-off signal phase deflection angle and the lift-off distance is $y = 44.6x + 6.7$ and its fitting factor is 0.95723. The larger the phase deflection angle of the lift-off signal, the smaller the phase difference between the defect signal and the lift-off signal. As the lift-off height increases, the rate of decrease in the amplitude of the defect signal is greater than that of the phase difference between the defect signal and the lift-off signal.

22.4.2 Engineering Application

Using the special eddy current detection probe developed for aluminum alloy ∏-type components to carry out engineering application test research. In the engineering inspection, the upper surface of the aluminum alloy ∏-type component has a protective coating on the surface of the workpiece, with a thickness of 200–250 μm, as shown in Fig. 22.10,

When inspecting the inspection workpiece, a suspected defect signal was found at the left corner R of the component as shown in Fig. 22.11. There are suspected defect signals in all four channels. In order to verify the detection results, penetrant inspection is carried out on the left corner R. First, we polished the paint with steel brush and sandpaper, then cleaned the polished area with cleaning agent, wiped the

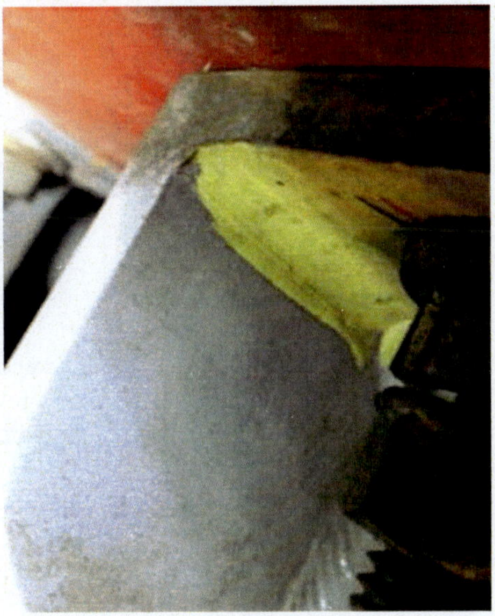

Fig. 22.10 The upper surface coating

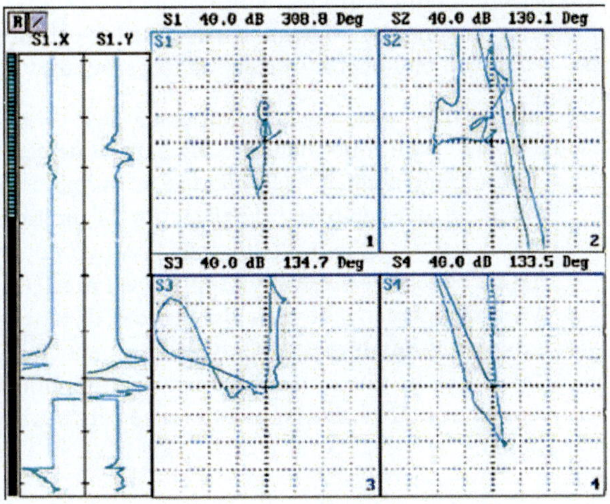

Fig. 22.11 Left R angle eddy current test

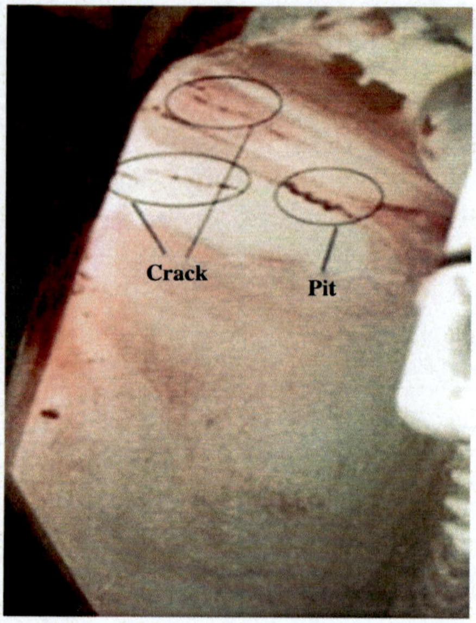

Fig. 22.12 Penetration test result

penetrant in the suspected area in one direction with cotton cloth, and finally sprayed the developer for observation. The test result is shown in Fig. 22.12. The results show that there is a crack in this place, which is consistent with the eddy current test results.

22.5 Conclusion

This paper develops a special multi-channel eddy current detection probe for ∏-type complex components, studies the effect of lift-off distance on the detection signal, and carries out engineering detection experiments. The following conclusions are obtained:

(1) The dedicated four-channel eddy current detection probe has been developed. The arrangement of each channel is suitable for both the R-angle area and the plane area, realizing the detection of multiple parts of a single probe. The optimization of the test parameters carried out enables the defect detection to have a better detection signal-to-noise ratio.

(2) The relationship between the lift-off distance and the detection signal is explored through experiments and it is concluded that as the lift-off distance increases, the rate of decrease in the amplitude of the defect signal is greater than that of the phase angle difference between the defect signal and the lift-off signal.

(3) In the engineering test and research, the four-channel eddy current detection probe has a good detection effect on the fatigue defects of the workpiece and the penetration test re-inspection is consistent with the eddy current test results.

Acknowledgements This study was co-supported by the National Natural Science Foundation of China (NO: 51865033), for which the authors are grateful.

References

1. H. Liu, C. Guo, S. Zang, Analysis of tension property of stainless steel-aluminium alloy clad. China Mech. Eng. **18**(07), 862–865 (2007)
2. Z. Li, Z. Wang, Y. Chen, Defect detection system for warm extrusion molded aluminum alloy components. Sensor World 17(02), 25–27+15
3. W.L. Haworth, A.F. Hieber, R.K. Mueller et al., Fatigue damage detection in 2024 aluminum alloy by optical correlation. Metall. Mater. Trans. A Phys. Metall. Mater. Sci. **8**(10), 1597–1604 (1977)
4. L.E.I. Hua, Analysis of eddy current nondestructive testing method for metal components. Mod. Manuf. Technol. Equip. **08**, 137–138 (2018)
5. M. Seyfpour, S. Ghanei, M. Mazinani, M. Kashefi, C. Davis, Nondestructive examination of recovery stage during annealing of a cold-rolled low-carbon steel using eddy current testing technique. Nondestruct. Test. Eval. **33**(2), 165–174 (2018)
6. Y. Jiang, J. Huang, Time division multiplexing fiber grating vibration sensor array. J. Photon. **37**(01), 73–76
7. Altera Corporation; Patent Issued for Routing-Efficient Time Division Multiplexing (TDM) Data Path Circuitry: United States of America, USPTO 10,250,347[P].2019–04–02

Chapter 23
A Simple Fire Extinguishing Demonstration System Based on Single-Chip Micyoco

Haiyin Qing, Yanbao Wu, Changjun Wu, and Haoyu Song

Abstract The system uses STC89C52 microcontroller as the main control chip, with infrared transmission and infrared reception as the transmission device, in order to achieve the powder and flame of the fire extinguisher sprayed to simulate, reproduce the fire extinguishing step. The system makes full use of the principle of "sound, light, and electricity", greatly restores the real fire drill scene, through the use of fire extinguishers, the real reproduction of the fire extinguishing process, to achieve the purpose of fire extinguishing drills. This not only reduces energy consumption but also improves presentation efficiency. If it can be widely used, it will be an indispensable tool in fire propaganda departments, schools, enterprises and institutions, government agencies, social organizations, the research and development of this project will bring people a lot of conveniences, if the whole society generally uses this product, it can improve the efficiency of personnel and social safety.

23.1 Introduction

Fire Protection Law of the People's Republic of China, as amended by the Fifth Meeting of the Standing Committee of the Eleventh National People's Congress, The society has set off a round of learning the new fire law upsurge, various forms of fire law publicity and implementation activities have been carried out in various places. The construction of legal system of fire control is an important link to carry out fire control work in an all-round way. In order to cooperate in full swing now hereby stable cross activities, the Ministry of Education jointly the Ministry of Public Security issued a 28 ministries file Fire safety management regulations of colleges and universities, May 1, 2009 and clearly requires colleges and universities should carry out fire control safety education and training, to strengthen the fire drill and

H. Qing (✉) · Y. Wu · C. Wu · H. Song
School of Electronics and Materials Engineering, Leshan Normal University, Leshan, China
e-mail: qinghaiyin123@163.com

C. Wu
e-mail: 534772165@qq.com

G. Liu and F. Cen (eds.), *Advances in Precision Instruments and Optical Engineering*, Springer Proceedings in Physics 270, https://doi.org/10.1007/978-981-16-7258-3_23

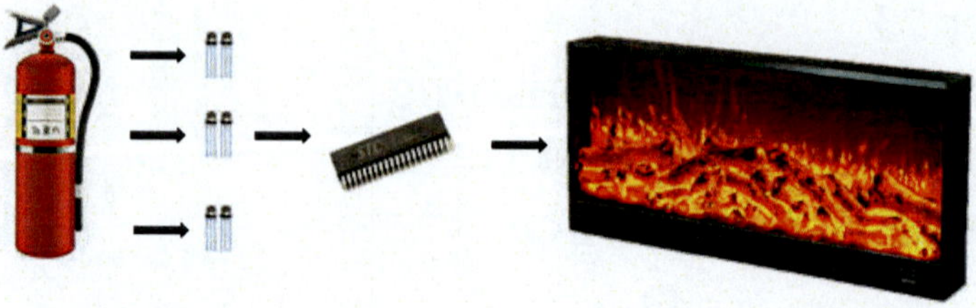

Fig. 23.1 Flow chart of the system

improve the fire safety of staff and students consciousness and save your survival skills. Because of the limitation of funds and foreign exchange, fire training in most colleges and universities is mainly confined to a variety of simple fire extinguishing drills based on fire equipment and familiar with the campus consumption equipment for all kinds of fire extinguishers, namely the basis of commonly used, which include dry powder fire extinguishers, CO_2 fire extinguisher and foam fire extinguisher, fire demonstration mainly is to use some open flame or smoke simulation, using different fire extinguishers to put out the fire. In this process, although the teachers and students can learn some basic fire common sense, to be familiar with a number of fire equipment, but as a result of the training itself is set spraying and fire extinguisher itself out of the dust to the campus and the whole social environment more than heavy pollution of the premise, developed a set of simulated fire demo system idea was born.

The concept of fire extinguishing the demo system is introduced; the system makes full use of the principle of sound and light, greatly reducing the real fire drill scenario, through the use of fire extinguishers, recreating the extinguishing process, to achieve the purpose of fire drill. Such not only reduced the energy consumption and improve the efficiency of the demo system concept, which is shown in Fig. 23.1.

23.2 Product Design

The software part of the system consists of the transmitter and the receiving control. Transmitter is mainly used to continuously send infrared signals and LED white light to the receiver. The receiving control terminal is mainly used to receive infrared and white light signals sent by the transmitter. The signal is analyzed and processed by the single-chip microcomputer to judge the fire extinguisher scanning flame. After the fire extinguishing is completed, the system enters the standby state and broadcasts the voice "the fire extinguishing has been completed".

The hardware part of the system consists of the transmitter and the receiving control. The transmitting end is mainly composed of infrared transmitting tube and strong light LED lamp. The receiving control end is mainly composed of infrared receiving module, photosensitive resistor sensor module, audio module, flame control module, and so on. At the same time, the STC89C52 single-chip microcomputer is used as the main control chip.

23.3 System Composition

At present, SCM is a powerful assistant for developers and an indispensable tool in practical application. It is widely used in various fields [1]. As an important representative of embedded systems, it plays an important role in the development of electronic technology [2].

The maximum operating clock frequency of the STC89C52 single-chip microcomputer is 80 MHz, the sheet contains 4 K Bytes of ROM, the device is compatible with the standard MCS-51 instruction system and 80 C51 pin structure. A universal 8-bit CPU and an ISP Flash memory unit are integrated into the chip, Cooperate with PC end control program can download program into the single-chip computer, So you don't have to buy a universal programmer, And faster than the editor [3]. STC89C52 MCU pin diagram is shown in Fig. 23.2.

Because the system needs to simulate the sound of the flame, we use the MP3 decoding chip to store the audio file and select the corresponding audio file directly from the single-chip computer instruction [4].

MY2480-16P is a small micro-integrated voice module developed by Shenzhen Maiyou Technology Co. Ltd. Using MY2480-24TS MP3 main control chip, support MP3, WAV format double decoding, module maximum support 16 MB FLASH can also be connected to the U disk or USB data line computer to replace FLASH audio files. This module built-in 3 W power amplifier can directly drive 3 W horns, which is more convenient to use.

Simulated flame box: black oak box with 75 * 20 * 45 CM, Internal emptying, A groove of 75 CM * 5 CM on the front. Two threads with a distance difference of 70 CM are drilled from the groove 3 CM, and then the two sides are left with a gap of

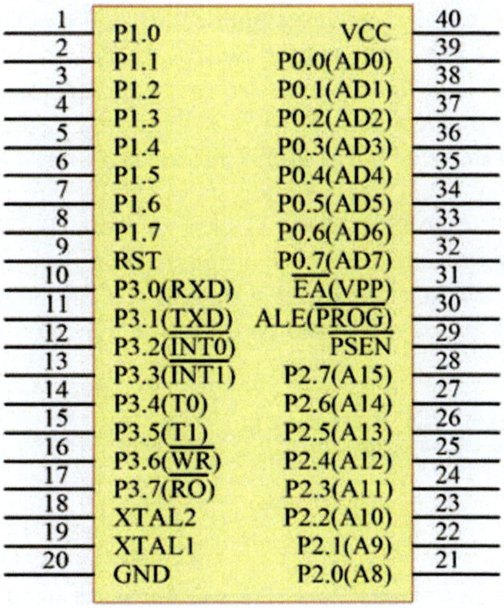

Fig. 23.2 STC89C52 single-chip microcomputer pin diagram

5–5 and a platform with a width of 4 CM is placed at a height of 6 CM above the gap. The specific components of the simulated flame box are: (1) rotating motor, rotating motion when electrified; (2) electronic carbon strips, made up of wooden blocks, and light transmission; (3) fire system, from the silver-plated reflector projection to the reflective screen formed; (4) LED light, XML-T6L2U2; model (5) infrared receiver, type VS1838B bandwidth 2–5 kHz; (6) flat glass, GB4871-3 mm; model; (7) wooden boxes, the model is 75/20/45.

The infrared receiver is one of the most commonly used electronic components in electronic equipment. According to its transmission function, it can be divided into pulse type and level type. The level infrared receiver outputs a continuous low (or high) level when receiving a millisecond infrared pulse signal, which is easy to receive and process, but its transmission distance is limited. The pulse infrared receiver only responds to the arrival of the carrier signal in milliseconds. Therefore, the use of pulse infrared receiving control as a long-distance, output continuous pulse signal module is a lot of electronic system design scene needs [5].

The external receiving circuit is usually composed of infrared receiving diode and amplifier circuit. The amplifier circuit is usually composed of an integrated block and several resistors and capacitors and needs to be encapsulated in a metal shield box, so the circuit is more complex. The volume is small. The receiver head is a special infrared receiving circuit, which integrates the infrared receiver tube with the amplifier circuit. It is small (the size is equivalent to a medium power transistor) and has good sealing, high sensitivity, and low price. The market price is only a few Yuan and it has only three pins, which are the positive pole of the power supply, the negative pole of the power supply and the output end of the signal, and its working voltage is about 5 V. As long as it is connected to the power is a complete infrared receiving amplifier, very convenient to use.

Its main functions include amplification, frequency selection, and demodulation. After its receiving amplification and modulation, the original signal will be output directly at the output. So that the circuit to achieve the most simplified! Sensitivity and anti-interference are very good and can be said to be an ideal device to receive infrared signals. IRM-3638T infrared receiver tube product parameters are shown in Table 23.1. IRM-3638T infrared receiver tube appearance size diagram is shown in Fig. 23.3.

An optical sensor is a sensor that uses Guang Min element to convert the optical signal into an electrical signal. Its sensitive wavelength is a near-visible wavelength, including infrared wavelength and ultraviolet wavelength. Optical sensors are not only limited to the detection of light but also can be used as detection elements

Table 23.1 IRM-3638T IR receiver product parameters

Name of name	Parameters
Supply voltage	0–6 V
Working temperature	−25 to 80 °C
Storage temperature	−40 to 80 °C
Welding temperature	260 °C

Fig. 23.3 IRM-3638 T external dimensions of infrared receiver

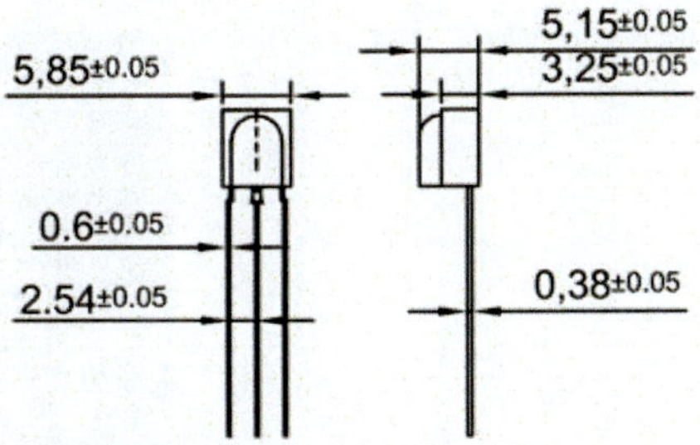

to form other sensors to detect many non-electric quantities, as long as these non-electric quantities can be converted into changes in optical signals. The optical sensor is one of the most widely used sensors, which plays an important role in automatic control and non-electric measurement.

23.4 Physical Debugging

According to the hardware design, use Altium Designer software to complete the schematic drawing, the following is the control end. The physical schematic diagram is shown in Fig. 23.4.

According to the hardware design, the Altium Designer software is used to complete the drawing of the PCB diagram, and the receiving control end can be drawn, which can save the cost. Some of the PCB components are encapsulated by patch, pursuing small volume and portable; as shown in Fig. 23.5.

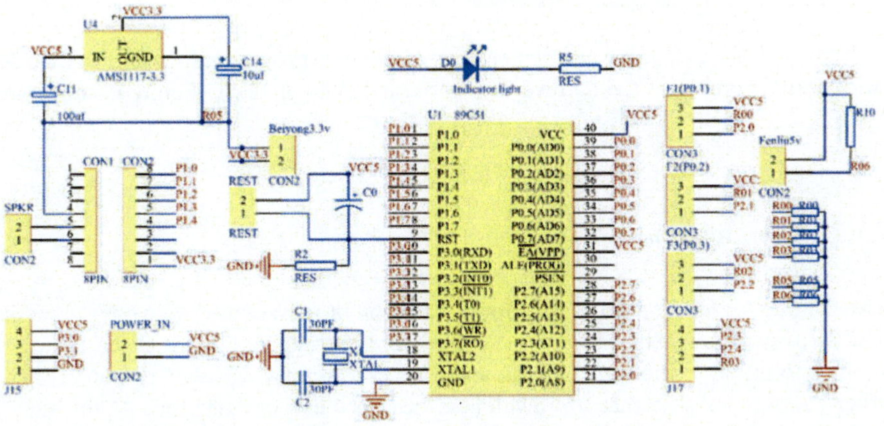

Fig. 23.4 Physical schematic diagram

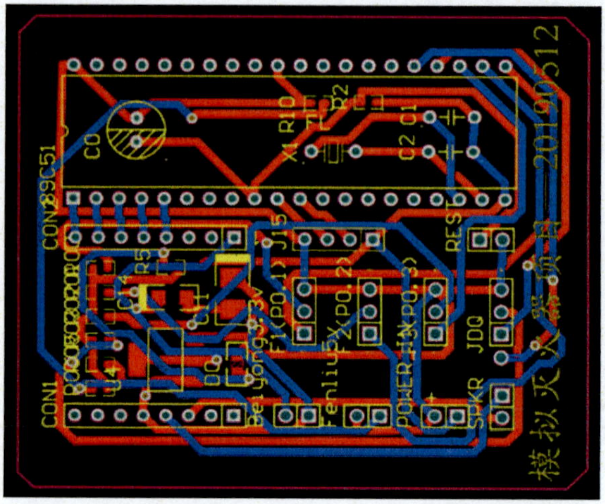

Fig. 23.5 Physical PCB

Fig. 23.6 Physical diagram of the system

The software design of the system completes the programming, and the hardware design completes the hardware construction. The physical picture is shown in Fig. 23.6.

23.5 Conclusion

The existing firefighting demonstration is based on the real combustion environment and the use of fire extinguishers to achieve fire extinguishing, which has caused serious pollution to the environment and time consuming, waste of resources, and other shortcomings so that fire awareness has not been popularized. At the same time, per capita participation is very low.

The system has a simple structure, stable and reliable operation, and strong anti-interference ability. The successful development of the system improves the efficiency of fire demonstration. For the first time, the system combines "LED lamp flame" with "infrared induction lamp fire extinguisher" to design a realistic fire scene and the product is efficient, reliable, and running for a long time, which can reduce energy consumption and achieve the effect of energy saving and emission reduction. To sum up, it has the following advantages:

(1) Environmental protection: no wood-burning, zero air pollution, and reusable;
(2) Safety: without open fire, children and adolescents can use it safely;
(3) High stability: using an infrared sensor, the anti-interference ability is improved significantly and the reliability is high, and it can run for a long time;
(4) Low power consumption: the circuit is composed of basic combinational circuits, which can reduce energy consumption and achieve the effect of energy saving and emission reduction;
(5) Low installation cost: the original circuit is simple, the equipment is exquisite, the installation is convenient, can improve the personnel installation work efficiency.

The system is cheap, easy to operate, can significantly improve working efficiency and reduce costs. If it can be widely used, it will be an indispensable tool in fire control propaganda departments, schools, enterprises and institutions, government agencies, and social organizations. If the whole society uses this product generally, it can improve the work efficiency and social safety of the personnel. Therefore, it has a wide market prospect and great social significance.

Acknowledgements The authors would like to thank the Chincsc Meridian Project for the use of data. This work was supported by the Sichuan provincial University Key Laboratory of Detection and Application of Space Effect in Southwest Sichuan. This work was supported in part by the National Natural Science Foundation of China under Grant 41804148, in part by the Application Foundation of Science and Technology Department of Sichuan Province under Grant 2019YJ0302, and in part by the Natural Science Research Foundation of Leshan Normal University under Grant LZDP013, Grant ZZ201803, and Grant DGZZ202001.

References

1. Z. Tang, Application of SCM in electronic field. Shenghua Electron. Technol. Softw. Eng. **24**, 242 (2018)
2. Z. Chaoping, Q. Yin, S. Xu, A portable infusion monitoring system based on single chip microcomputer. Electron. Test. **14**, 10–13 (2019)
3. N. Xiaoyan, H. Xu, M. Xuchang, L. Xiwen, S. Lei, Application and development of single-chip microcomputer. A Brief Discuss. Hebei Agric. Mach. **12**, 53 (2018)
4. T. Haizhou, Y. Yungang, Y. Wei et al., Intelligent anti-theft tracking system for battery car. Des. Dev. **7**, 22–23 (2019)
5. C. Zhenhua, Design of remote anti-fall infrared receiving module. Fujian Comput. **11**, 133–216 (2018)

Chapter 24
LC–MS/MS Determination of 25-Hydroxyvitamin D in Human Serum Based on Covalent Organic Frameworks Magnetic Solid-Phase Extraction Materials

Xinquan Liu, Yu Ai, and Kang Du

Abstract This paper established a method for the test on 25-hydroxy vitamin D2/D3 in human serum based on a novel covalent organic material-based magnetic solid-phase extraction-liquid chromatography tandem mass spectrometry. At the same time, an automated pre-processing platform was developed that can process samples quickly, efficiently, and with multiple throughputs, reducing manual operations and errors, and providing a fully automatic sample pre-processing system that can realize large-scale sample detection in clinical laboratories. Experiments have investigated the effects of different materials and elution solvents on the extraction efficiency of the target. Under the best conditions, 25-hydroxyvitamin D2/D3 has a good linearity in the linear range, and the sensitivity, precision, and accuracy can all satisfy the testing requirements of clinical biological samples. This method is of great significance for clinical laboratories to improve testing efficiency.

24.1 Introduction

Vitamin D is a fat-soluble vitamin necessary for body metabolism, which not only affects calcium and phosphorus metabolism but also has a wide range of physiological effects. It is an essential substance for maintaining human health, cell growth, and development and is closely related to many diseases [1–3]. 25-hydroxy vitamin D, including 25-hydroxy vitamin D2 (25-OH VD2) and 25-hydroxy vitamin D3 (25-OH VD3), is one of the main metabolic forms of vitamin D in the body [4, 5]. Because

X. Liu (✉) · K. Du (✉)
School of Precision Instrument and Optoelectronics Engineering, Tianjin University, Tianjin 300072, China
e-mail: liuxinquantju@163.com

K. Du
e-mail: dukang@tjbskj.cn

Y. Ai
Tianjin Boomscience Technology Co., Ltd., Tianjin 300401, China

G. Liu and F. Cen (eds.), *Advances in Precision Instruments and Optical Engineering*, Springer Proceedings in Physics 270, https://doi.org/10.1007/978-981-16-7258-3_24

of its long half-life and stable existence, it is considered to be a marker of vitamin D nutritional level [6]. Liquid chromatography-tandem mass spectrometry (LC–MS/MS) for the detection of vitamin D has the characteristics of strong specificity and high accuracy and is considered to be the "gold standard" assay for evaluating its nutritional status [7–10]. However, due to the complex sample pre-processing operation steps and different analyte pre-processing procedures, the detection throughput is low, which limits its application in the detection of large quantities of samples in clinical laboratories [11–13]. To alleviate the pressure of labor, space, and equipment investment caused by the increasing demand for clinical testing, and to reduce the quality risk caused by manual operation, it is urgent to improve the efficiency of LC–MS/MS method for detecting serum 25-hydroxy vitamin D.

Magnetic-solid phase extraction (M-SPE) based on magnetic micro–nanoparticles is a dispersive solid-phase extraction technology that uses magnetic or magnetizable materials as the adsorbent matrix [14–16]. Compared with the traditional solid-phase extraction (SPE) technology, this kind of micro–nanoparticles can be completely exposed to and fully contacted with the system to be tested, so it can absorb and extract the substance to be tested from the system to be tested in a short time, realizing the high-fold enrichment of trace compounds in the sample. In addition, the magnetic particles can be easily separated and collected from the system to be tested by an external magnetic field, which eliminates the complicated pumps, valves, pipelines, and other components of the traditional automatic SPE system and avoids tedious filtration or centrifugation process [17–19].

In this experiment, covalent organic frameworks based on magnetic nanospheres (MCOFs) with different pore diameters were selected. This material is independently developed and synthesized by the research group of Teacher Tang Anna in Nankai University, which can design the pore size of the substance based on the size of the target analyte molecule, so as to achieve maximum efficiency of adsorption and extraction of the substance to be measured. At the same time, the automatic pretreatment platform-the magnetic solid-phase extraction instrument can be designed with a variety of fluxes. It can be extracted by a single tube or 8 $\times$ 96 samples, and its operation is simple and fast. The extraction time of 8 $\times$ 96 samples is less than 20 min, which greatly improves the efficiency of the experiment. The automatic processing of the whole process of extraction of the test substance reduces the quality risk caused by manual operation. Using LC–MS/MS to achieve the quantitative detection of 25-hydroxyvitamin D in human serum greatly improves the detection efficiency of the sample.

24.2 Materials and Method

24.2.1 *Synthesis of Dual-Porous Magnetic Covalent Organic Composite Materials*

In the experiment, Fe_3O_4 nanoparticles were synthesized by the solvothermal reduction method, and then the amino-functionalized magnetic silicon nanospheres ($Fe_3O_4@SiO_2-NH_2$) were synthesized. Finally, 4,4′,4″,4‴-(ethene-1,1,2,2-tetrayl) tetraaniline (ETTA) and [1,1′-biphenyl]-4,4′-dicarbaldehyde (BPDA) monomers were synthesized by solvothermal method to synthesize heterosporous MCOFs materials modified with COFs materials [20].

24.2.2 *Synthesis of Dual-Porous Magnetic Covalent Organic Composite Materials*

Using MCOFs as a carrier, using the π bond, hydrophobic effect, and pore retention provided by the surface material of MCOFs in an aqueous environment to absorb 25-hydroxy vitamin D, and destroy the force in an organic environment, making MCOFs and 25-hydroxy vitamin D to be separated [21–23], and then the entire extraction and purification process of 25-hydroxy vitamin D is realized by moving MCOFs or transferring liquid. The specific experimental process is as follows: adding MCOFs suspension (100 μl MCOFs solution + 300 μl activation solution) to the first row of the 96 deep well plates; adding 100 μl serum sample, 180 μl methanol, 20 μl saturated zinc sulfate solution, and 200 μl sample diluent to the second row; adding 200 μl eluent to the third row and fourth row; add 100 μl eluent to the fifth row. Then the designed program was input on the solid-phase extraction instrument, and the whole experiment was completed at room temperature. After the program was finished, the eluent was injected into LC–MS/MS for testing.

24.2.3 *LC–MS/MS*

A Shimadzu liquid chromatography (Shimadzu LC-20AD) tandem triple-quadrupole mass spectrometer (AB Triple Quad 4000 Mass Spectrometry) was used for sample detection. This experiment adopted electronspray ionization (ESI), multiple reaction monitoring (MRM) for selective ion monitor, and the following mass spectrometry detection conditions were applied: Atomizing gas (Gas1), 70 psi; Ionspray voltage (IS), 5000 V; Curtain gas (CUR), 40 psi; Collision gas (CAD), 3 psi; Ion source temperature (TEM), 500 °C. The transition pairs of each compound and its internal standard were as follows: 25-OH VD3, 401.4 → 365.4, 25-OH VD2, 413.3 → 337.4, d6-25-OH VD3, 407.4 → 371.4, d6-25-OH VD2, 419.3 → 355.4. Chromatographic separation was performed on a C18 column (Phenomenex Company, 50 × 3 mm, 2.6 μm) at 30 °C. Solution A was 0.05% acetic acid in methanol, and solution B was 0.05% acetic acid in ultra-pure water. The gradient elution program was as follows: 80% A at 0 min and then held for 0.7 min; 80% A at 1.1 min; 97% A at 1.1 min and then held for 1.2 min for column equilibration. For LC analysis, 10 μL extract was injected, and the flow rate was maintained at 0.6 mL min^{-1}. A representative spectrum of 25-hydroxy vitamin D is shown in Fig. 24.1.

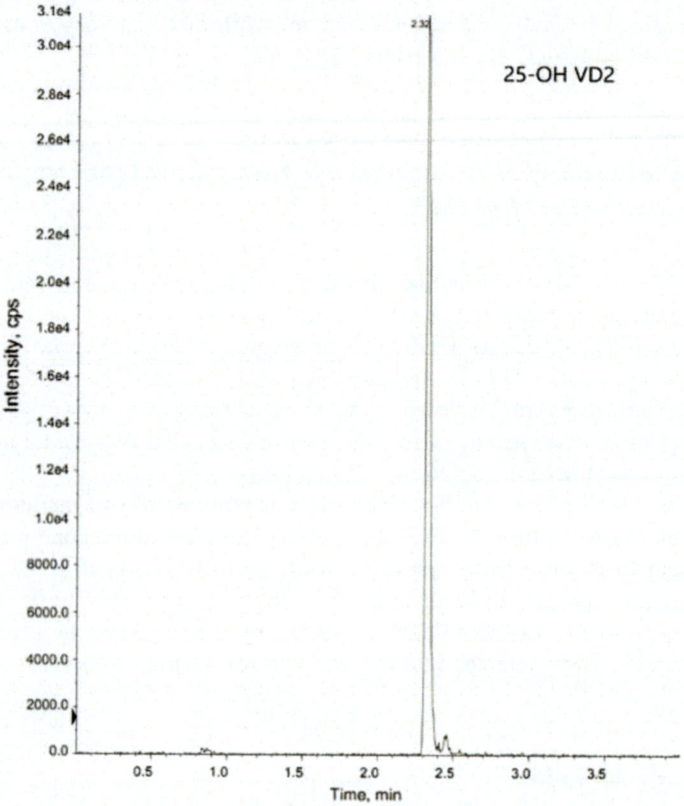

Fig. 24.1 LC–MS/MS representative chromatogram of 25-hydroxy vitamin D

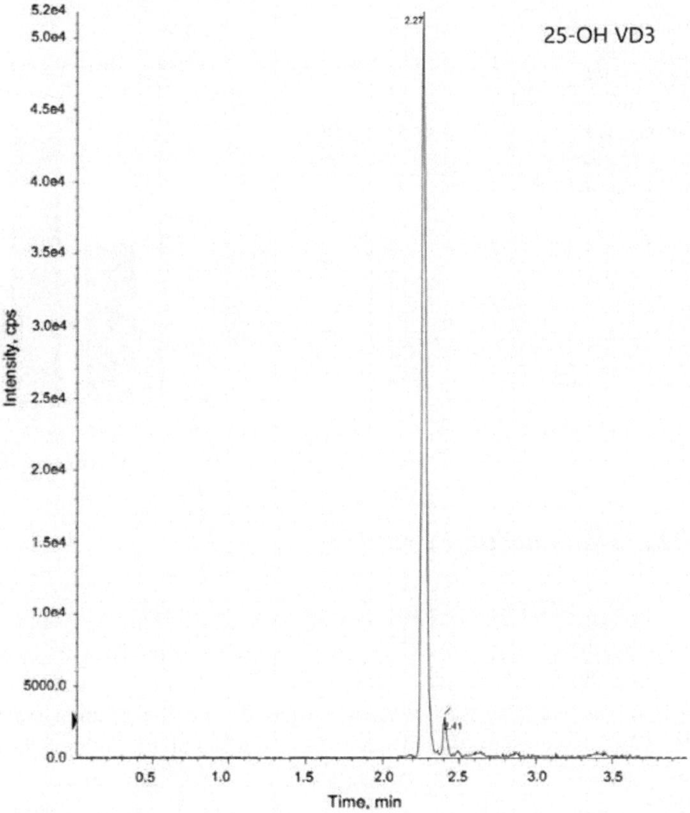

Fig. 24.1 (continued)

24.3 Results and Discussion

24.3.1 Methodological Evaluation

The detection linear ranges of 25-OH VD2 and 25-OH VD3 were 0.5–50.0 ng/mL and 1.0–100.0 ng/mL, respectively, and the linear correlation coefficients (r^2) obtained were all greater than 0.99; both the inter-day and intra-day precisions were less than 15% and the recoveries in the 80–120% range.

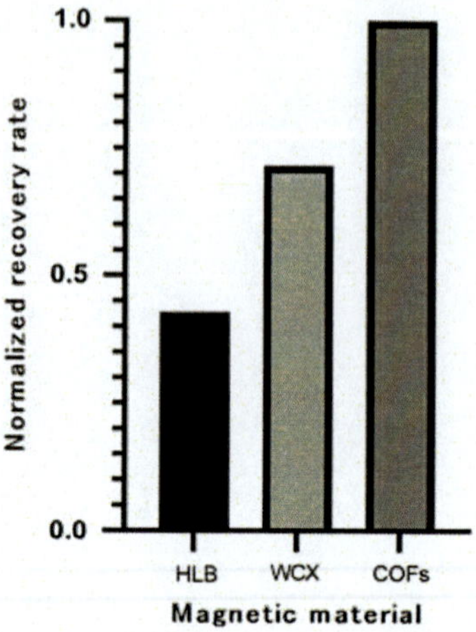

Fig. 24.2 The extraction recovery rate of three magnetic materials: HLB, WCX, and COFs to 25-hydroxyvitamin D in human serum, normalized based on the recovery rate of COFs material

24.3.2 *Material Selection*

This study compared the extraction recovery rate of 25-hydroxyvitamin D from human serum by three magnetic solid-phase extraction materials of hydrophilic-lipophile balance (HLB), weak cation exchange (WCX), and COFs. The extraction performance of various materials was investigated, and the comparison result is shown in Fig. 24.2. ChemBio 3D software was used to calculate the size of 25-OH VD2 molecule as 0.67 × 1.69 nm and the size of 25-OH VD3 molecule as 0.58 × 1.69 nm. Based on the molecular size of 25-hydroxy vitamin D, the heteropore was independently developed and synthesized by the cooperating unit COFs magnetic material, and the size of the magnetic core was about 200 nm. After the COFs monomer was wrapped, the size of the magnetic particles was about 300–400 nm, and the size of the double pores was 3.2 and 1.3 nm, which can effectively intercept 25-hydroxy vitamin D molecules, and remove the protein and other organic macromolecules in the serum. Therefore, the extraction capacity of this COFs magnetic material for 25-hydroxyvitamin D in human serum is stronger than that of the other two commercial extraction materials.

24.3.3 *Optimization of Eluent*

The experiment was compared with methanol, n-hexane, acetone, ethyl acetate, ethyl acetate methanol, and acetone methanol solution as eluents. The experimental results are shown in Fig. 24.3. When ethyl acetate methanol solution is used as the eluent, the recovery rate is the highest. 25-hydroxy vitamin D is weakly polar, n-hexane is a non-polar solvent (polarity is 0.06), ethyl acetate and acetone are weakly polar solvents (polarity are 4.3 and 5.4, respectively), and methanol is a polar solvent (polarity

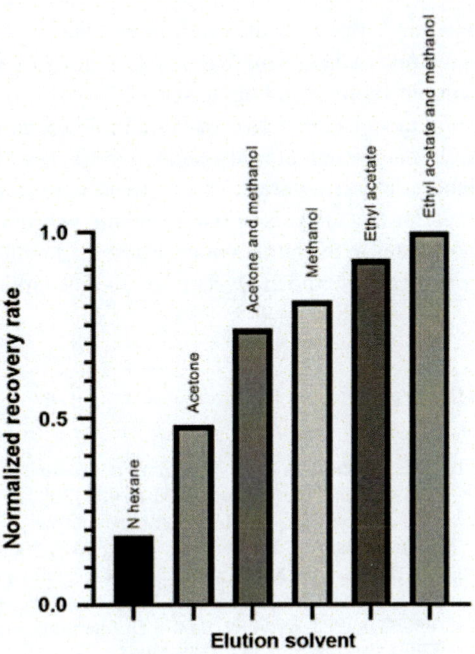

Fig. 24.3 Comparison of the extraction recovery rate of 25-hydroxyvitamin D in human serum with various eluents, normalized based on the recovery rate of ethyl acetate methanol solvent

is 0.06) [24]. According to the principle of similar mutual solubility, the polarity of the mixture of ethyl acetate and methanol is more suitable for the extraction of 25-hydroxy vitamin D, so the elution capacity is the strongest.

24.4 Conclusion

This experiment explored a new type of magnetic covalent organic framework material with different pore diameters for the extraction and recovery of 25-hydroxyvitamin D in serum. Combined with an automated pre-processing platform for magnetic solid-phase extraction, it can avoid the possibility of large differences between different batches and different experimenters due to manual intervention, which can improve throughput, consistency, and versatility. At the same time, the selected COFs magnetic materials can be designed with different pore sizes according to the molecular size of the target analyte to maximize the extraction of the target analyte.

Vitamins in the human body are essential nutrients for the human body and play a vital role in the physiology and pathology of the human body. However, since the current common clinical testing methods can only detect one to two specific vitamins, it is impossible to detect the content of multiple vitamins in the sample at the same time. In addition, the content of vitamins in the serum is widely distributed, such as the content of vitamin K that is below the nanogram level, and the content of vitamin E that is at the microgram level. Therefore, it is very challenging to require a method that can simultaneously efficiently and accurately quantify a variety of fat-soluble vitamins in human serum. To solve this problem, this study also explored the simultaneous extraction of fat-soluble vitamins A, D, E, and K in human serum with COF magnetic materials, and the initial results have been achieved. In the future, the laboratory will work with cooperators to design a variety of COFs magnetic materials for the analysis of other organic substances such as water-soluble vitamins and hormones in the human body, combined with an automated pre-processing platform, to better serve the analysis of clinical samples.

References

1. M.F. Holick, T.C. Chen, Vitamin D deficiency: a worldwide problem with health consequences. Am. J. Clin. Nutr. **87**(4), 1080S-S1086 (2008)
2. A. Prentice, G.R. Goldberg, I. Schoenmakers, Vitamin D across the lifecycle: physiology and biomarkers. Am. J. Clin. Nutr. **88**(2), 500S-506S (2008)
3. N. Binkley, R. Ramamurthy, D. Krueger, Low vitamin D status: definition, prevalence, consequences, and correction. Endocrinol. Metab. Clin. North Am. **39**(2), 287–301 (2010)
4. F.R. Greer, 25-Hydroxyvitamin D: functional outcomes in infants and young children. Am. J. Clin. Nutr. **88**(2), 529S-533S (2008)
5. L. Chi, S. Li, X. Shang, B. Jiang, Correlation between serum 25-hydroxyvitamin D level and diabetic ret-inopathy: a protocol for systematic review and meta-analysis. Medicine (Baltimore) **100**(4), e23697 (2021)
6. M. Satoh, T. Ishige, S. Ogawa, M. Nishimura, K. Matsushita, T. Higashi, F. Nomura, Development and validation of the simultaneous measurement of four vitamin D metabolites in serum by LC-MS/MS for clinical laboratory applications. Anal. Bioanal. Chem. **408**(27), 7617–7627 (2016)
7. M. Vogeser, Quantification of circulating 25-hydroxyvitamin D by liquid chromatography-tandem mass spectrometry. J. Steroid Biochem. Mol. Biol. **121**(3–5), 565–573 (2010)
8. S.K. Panda, I. Aggarwal, H. Kumar et al., Magnetite nanoparticles as sorbents for dye removal: a review. Environ. Chem. Lett. **19**, 2487–2525 (2021)
9. A. Akbarzadeh, M. Samiei, S. Davaran, Magnetic nanoparticles: preparation, physical properties, and applications in biomedicine. Nanoscale Res. Lett. **7**(1), 144 (2012)
10. Y. Huang, A. Keller, Magnetic nanoparticle adsorbents for emerging organic contaminants. ACS Sustain. Chem. Eng. **1**, 731–736 (2013)
11. M. Soylak, O. Ozalp, F. Uzcan, Magnetic nanomaterials for the removal, separation and preconcentration of organic and inorganic pollutants at trace levels and their practical applications: a review. Trends Environ. Anal. Chem. **29**. 2214-1588 (2021)
12. Y. Zhang, X. Jin, X. Ma, Y. Wang, Chiral porous organic frameworks and their application in enantioseparation. Anal. Methods **13**(1), 8–33 (2021)
13. X. Lin, X. Wang, J. Wang, Y. Yuan, S. Di, Z. Wang, H. Xu, H. Zhao, P. Qi, W. Ding, Facile synthesis of a core-shell structured magnetic covalent organic framework for enrichment of organophosphorus pesticides in fruits. Anal. Chim. Acta. **1101**, 65–73 (2020)
14. Y. Gao, G. Liu, M. Gao, X. Huang, D. Xu, Recent advances and applications of magnetic metal-organic frameworks in adsorption and enrichment removal of food and environmental pollutants. Crit. Rev. Anal. Chem. **50**(5), 472–484 (2020)
15. J.J. Jarju, A.M. Lavender, B. Espiña, V. Romero, L.M. Salonen, Covalent organic framework composites: synthesis and analytical applications. Molecules **25**(22), 5404 (2020)

16. X. Zhang, G. Li, D. Wu, B. Zhang, N. Hu, H. Wang, J. Liu, Y. Wu, Recent advances in the construction of functionalized covalent organic frameworks and their applications to sensing. Biosens. Bioelectron. **145**, 0956–5663 (2019)
17. H.L. Qian, C.X. Yang, W.L. Wang, C. Yang, X.P. Yan, Advances in covalent organic frameworks in separation science. J. Chromatogr. A **1542**, 1–18 (2018)
18. X. Wang, N. Ye, Recent advances in metal-organic frameworks and covalent organic frameworks for sample preparation and chromatographic analysis. Electrophoresis **38**(24), 3059–3078 (2017)
19. S. Zhang, Q. Yang, C. Wang, X. Luo, J. Kim, Z. Wang, Y. Yamauchi, Porous organic frameworks: advanced materials in analytical chemistry. Adv. Sci. **5**(12), 2198–3844 (2018)
20. W. Li, H.X. Jiang, Y. Geng, X.H. Wang, R.Z. Gao, A.N. Tang, D.M. Kong, Facile removal of phytochromes and efficient recovery of pesticides using heteropore covalent organic framework-based magnetic nanospheres and electrospun films. ACS Appl. Mater Inter. **12**(18), 20922–20932 (2020)
21. M. He, X. Ou, Y. Wang, L.D. Chen, B. Chen, B. Hu, Porous organic frameworks-based (micro)extraction. J. Chromatogr. A **1609**, 460–477 (2020)
22. N. Manousi, G.A. Zachariadis, Recent advances in the extraction of polycyclic aromatic hydrocarbons from environmental samples. Molecules **25**(9), 2182 (2020)
23. W. Wang, Z. Li, S. Zhang, X. Yang, X. Zang, C. Wang, Z. Wang, Triazine-based porous organic framework as adsorbent for solid-phase microextraction of some organochlorine pesticides. J. Chromatogr. A **1602**, 83–90 (2019)
24. C. Reichardt, T. Welton, *Solvents and Solvent Effects in Organic Chemistry*, 4th edn. (Wiley, Wiley-VC, 2010)

Chapter 25
Research on Car Locking Device to Prevent the Car from Moving Accidentally

Shuangchang Feng, Jie Chen, Yanchun Liang, and Yiwen Ju

Abstract With the rapid development of economy, the number of high-rise buildings is increasing, people use elevators more and more frequently, maintenance work is heavy, but due to elevator design defects, illegal operation and other reasons, maintenance workers have an increasing number of elevator accidents, among which accidents caused by the accidental movement of elevator car are the most serious and frequent. When the car roof is overhauled, the elevator is in a stop operation state. Due to the failure of mechanical components or control system, the elevator car moves unexpectedly and leaves the flat floor area, which will cause damage to maintenance and inspection personnel and cause serious safety accidents. In order to solve the above problems, this paper designs a car lock device, which can slide up and down to prevent the car from moving accidentally, which can prevent the accidental movement of the car caused by the failure of the driving host or control system, and ensure the safety of the maintenance and inspection personnel.

25.1 Introduction

In the context of the vigorous development of urban construction, the number of high-rise buildings is gradually increasing, and the demand for elevators in life is also increasing. The number of elevator accidents is increasing; especially the elevator accident movement accidents are more serious [1]. Relevant reports show that because of the accidental movement of the elevator car caused by the elevator injuries occur from time to time, so in order to reduce the accident of the elevator car accidental movement, the relevant personnel need to pay attention to the design of safety protection device. If the elevator is a non-machine room elevator, the top of the car is generally used as the working platform when it is overhauled in the shaft [2]. Maintenance personnel need to stand on the top of the car, so it is necessary to ensure that the car is absolutely locked to prevent accidental movement of the car.

S. Feng (✉) · J. Chen · Y. Liang · Y. Ju
Elevator Inspection Department, Shanghai Institute of Special Equipment Inspection and Technical Research, Shanghai, China
e-mail: fengsc@ssei.cn

G. Liu and F. Cen (eds.), *Advances in Precision Instruments and Optical Engineering*, Springer Proceedings in Physics 270, https://doi.org/10.1007/978-981-16-7258-3_25

Compared with electrical locking, the safety factor of mechanical locking is higher. Therefore, in the prior art, mechanical locking is usually used to lock the car and guide rail. There is an important defect in the existing mechanical car locking device, that is, the locking device needs to install a certain connecting device on the guide rail, that is, a fixed plate. Therefore, the lift car can only be locked at the fixed plate installation position of the guide rail. In the process of maintenance, if it is necessary to lift the car frequently, the convenience of the locking device will be greatly reduced [3]. In order to improve the above problems, this paper designs a car locking device, which does not need to install fixed parts on the elevator guide rail and can quickly and conveniently stay in any position, which is convenient to use, safe and reliable.

25.2 Common Car Moving Protection Device

The commonly used car accidental movement protection device mainly includes the detection system and the brake stop component [4]. The detection system tests the car's accidental movement distance. When the car accidental movement exceeds the preset value, the brake stop component is triggered to realize the car brake stop. The brake stop component mainly acts on the suspension system or traction wheel, which can brake the car quickly and effectively. The brake and stop parts of the existing car accidental movement protection device usually act on the following positions: car, counterweight, suspension wire rope, traction wheel and traction wheel axle with only two supports. The braking part not only makes the car stop but also makes the car stop all the time. The stop part can be shared with the uplink overspeed protection device and the downlink overspeed protection device; when the brake parts act on the car and counterweight, the car safety tongs and counterweight safety tongs can be used to stop the car. At this time, the protection device to prevent the car from accidental movement is composed of a monitoring system, speed limiter safety tongs; When the brake and stop parts act on the position of the hanging wire rope, the rope clamp can be used to achieve the brake and stop. At this time, the protection device to prevent the car from accidental movement is composed of the monitoring system and the rope clamp; brake and stop components acting on the traction wheel can be realized by using the brake. At this time, the protection device to prevent the car from accidental movement is composed of monitoring system and brake.

25.3 Structural Design

The structure design of a car locking device that can slide up and down to prevent accidental movement of car is shown in Fig. 25.1. The structure includes sliding sleeve, fixing sleeve, connecting plate, car connector, anti-stripping edge, connecting pin, bolt, connecting fork, locking block, lock fork, locking bolt, locking block driving bolt, etc.

Fig. 25.1 Structure model

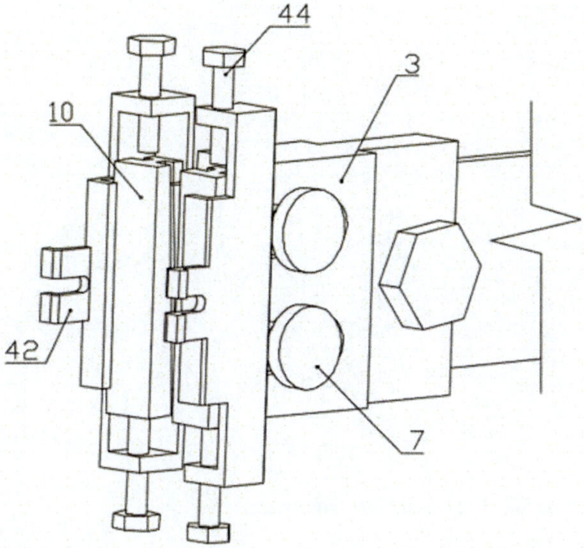

Figure 25.2 is an explosion diagram designed in this paper to prevent the car locking device from moving accidentally. Figure 25.3 is the explosion diagram in unlocked state. Among them, the attached drawings are as follows: 3-sliding sleeve; 4-fixing sleeve; 5-connecting plate; 6-car connection; 7-connecting pin; 8-bolts; 9-connecting fork; 10-lock block; 31-lock the fork; 32-connection slot; 33-locking bolt I; 34-locking hole I; 41-bolt mounting groove; 42-anti stripping edge; 43-edge blocking; 44-locking block drive bolt; 51-locking hole 2; 61-slot; 62-locking hole 3; 63-mounting holes; 621-locking bolt II; 101-sliding groove.

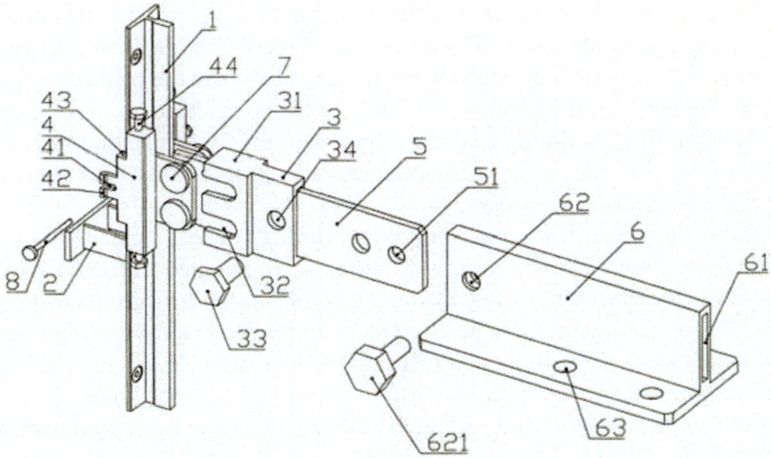

Fig. 25.2 Exploded view (locked)

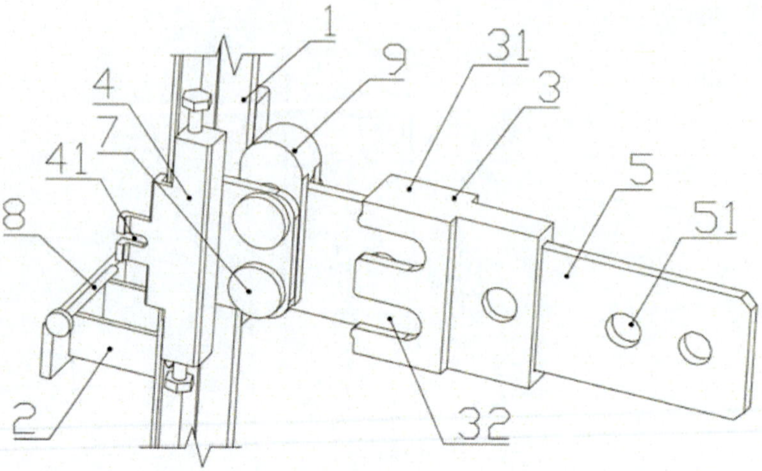

Fig. 25.3 Exploded view (not locked)

25.4 Working Principle

Figure 25.2 is a schematic diagram of the device. The specific connection method is as follows: the elevator guide rail 1 is installed on the inner wall of the elevator shaft through the guide rail bracket 2, and the fixing sleeve 4 and the locking block 10 are both connected with the fixing sleeve 4 through the sliding groove 101, and the sliding groove 101 is inclined. When the locking block 10 slides upward with respect to the fixed sleeve 4, the locking block 10 is inclined to slide upward with respect to the fixed sleeve 4. The locking block 10 moves in the direction of elevator guide rail 1, and the inclination angle of sliding groove 101 is 5°. One end of the fixing sleeve 4 is provided with a retaining edge 43, which is used to hook the mounting surface of elevator guide rail 1. The other end of the fixing sleeve 4 is connected with the connecting plate 5 through the connecting pin 7, and the connecting plate 5 is between the two fixing sleeves 4. Two ends of the connecting pin 7 are provided with a retaining ring. A sliding sleeve 3 is provided on the connecting plate 5, the front end of the slide sleeve 3 is provided with a locking fork 31, and the locking fork 31 has a connection groove 32.

The connecting plate 5 is installed in the car through the car connector 6, the car connector 6 has a slot 61 in the center, and the bottom is provided with the installation hole 63. The car connector 6 is installed in a part of the car through the installation hole 63. The connecting plate 5 can slide freely in slot 61. The connecting plate 5 is realized by inserting the locking bolt 621 into the lock hole No. 62 and the locking hole 251 adjustment of relative position of car connector 6. Lock holes 251 usually have more than one. When using, when the locking device is connected with guide rail 1, the sliding sleeve 3 is on the outside of the fixing sleeve 4, and the connecting pin 7 is in the inner part of the connection groove 32. The other end of the fixing sleeve 4 is connected with two anti-release edges 42 through bolts 8, so that the two fixing sleeves 4 are firmly wrapped in the guide rail 1, and the retaining edge 43 hooks the mounting surface of the guide rail 1. Finally, the locking block drive bolt 44 is operated to make the locking block 10 squeezes the guide rail 1 and realize the connection and locking with the guide rail 1. When the connecting plate 5 is subjected to a downward pull, the downward pull increases the extrusion pressure between the locking block 10 and the guide rail 1, since the sliding groove 101 is set tilt.

When the car is required to adjust vertically, loosen bolts 8 and sliding sleeve 3, open two fixing sleeves 4, and insert connecting fork 9 between fixing sleeve 4 and connecting plate 5. Connecting fork 9 makes two fixing sleeves 4 open to the maximum range. In this state, the locking block 10 is separated from guide rail 1, and the stop 43 is separated from guide rail 1, and the stop 43 is not within the range of guide rail support 2, at this time, the car can be moved up and down.

25.5 Conclusions

In order to ensure the absolute locking of the car and prevent the car from moving accidentally, this paper designs a car locking device, which does not need to install fixed parts on the elevator guide rail and can quickly and conveniently stop the car at any position. It can effectively avoid casualties and is convenient to use, safe and reliable:

- The device does not need the support of the guide rail installed on the guide rail and can realize the connection at any position on the guide rail. When the device is in the unlocked state, the rib will not be touched by the guide rail bracket and will not affect the up and down operation of the car. It can quickly detect the performance of the hydraulic buffer objectively and accurately and eliminate the interference of human factors.
- When the device is in the locked state, the two fixed sleeves will form a safe and reliable closed fixed cavity, one end of which is fixed by sliding sleeve, and the other end is connected to the guide rail by bolts and ribs, so as to ensure the reliable contact between the internal locking block and the guide rail.
- The contact surface between the locking block and the guide rail of the device is not the connection surface between the guide rail and the car, so it can avoid accidental damage to the guide rail.

Acknowledgements This research was supported by The Science and Technology Planning Project of Shanghai Municipal Administration of market supervision (2021–23).

References

1. J. Li, W. Li, Analysis of technical requirements and research status of elevator car accidental movement protection device. China Spec. Equip. Saf. **33**(2), 9–13 (2017)
2. J. Hao, R. Miao, L. Li et al., Analysis on accidental movement of traction driven passenger elevator car and its protection device. Electromech. Inf. **3**, 72–73 (2018)
3. Y. Yang, W. Yan, W. Yue et al., An elevator accidental moving protection device. Chinese Patent, 106672735 (2017)
4. L. Wenjin, A simple measure to prevent elevator car from moving accidentally. China Elev. **23**(19), 59–61 (2012)

Part III
Laser Technology and Infrared Thermal Imaging

Chapter 26
Design of Optical System for Laser Dazzle Simulation Human Eye Test Target

Niu Jin, Xu Xiping, and Duan Zhenhao

Abstract Aiming at the characteristics of laser dazzling scenes, a simulated human eye target optical system that can test dazzling spots is designed. The system is based on Gullstrand I model eye, combined with the characteristics of the laser dazzling scene to establish an optical model of a simulated human eye target. Then, the design indicators of the optical system are proposed, which can meet the requirements of the laser dazzling fundus spot test, and the design of the optical model can be optimized by changing the optical structure, lens material and system aberration of the model. Finally, a comparison experiment was carried out between the simulated human eye target optical model simulation and the real human eye laser dazzling effect. Experiments show that the laser spot effect obtained by the simulated human eye target optical system matches the real human eye laser dazzling effect $\geq$90%. The system realizes the non-biological simulation test of the effect of laser dazzling fundus spots and provides an experimental research basis for further research on laser dazzling effect.

26.1 Introduction

Laser dazzlers can use small energy to make people's eyes dazzle, temporarily paralyze vision and lose their ability to observe. It is widely used in military and public security fields [1, 2] and can interfere with pilots, motor vehicle drivers, and instrument operators. The visual function of special combat groups such as snipers and snipers can make terrorists or rioters temporarily blind without causing crowd riots and panic, effectively reducing the enemy's combat effectiveness [3, 4]. In recent years, with the sustained development of laser technology, the research about the dazzling effect of laser dazzlers has attracted more and more attention from researchers. Non-lethal laser dazzle is a very complicated physiological phenomenon.

N. Jin (✉) · X. Xiping · D. Zhenhao
Departments of Optoelectronic Engineering, University of Changchun Science and Technology, Jilin, China
e-mail: niujin0615@163.com

G. Liu and F. Cen (eds.), *Advances in Precision Instruments and Optical Engineering*, Springer Proceedings in Physics 270, https://doi.org/10.1007/978-981-16-7258-3_26

How to quantify physiological perception into objective theory to guide the application of laser dazzler is a difficult problem. This paper adopts the artificial human eye target simulation technology to design a human eye-simulating optical system that can meet the needs of laser dazzle non-biological quantitative testing and realizes that the simulated light spot that is consistent with the real human eye visual dazzling effect can be obtained without the participation of biological bodies. This laid the experimental foundation for the detection of laser dazzling effect.

The human eye is a sophisticated optical system and each tissue from the cornea to the retina has different structural and optical characteristics. Different bodies and ages have different characteristics in the structural parameters of the human eye. Through the actual measurement of a large number of human eye structure parameters, the statistical average value is taken as the human eye optical constant, combined with optical parameters such as refractive index and surface shape, a simulated optical model of the human eye can be established. The human eye model can simulate the optical structure of the human eye when it is irradiated by laser, evaluate the human eye's retinal imaging quality, and complete the measurement of the laser dazzling effect.

26.2 Selection of Human Eye Model

26.2.1 Authoritative Eye Models

According to the human eye physiological anatomy experiment and optical theory, a dozen scholars such as Listing, Gullstrand, LeGrand, Lotmar, etc. painstakingly studied, obtained models such as paraxial eye [1], astigmatism to the cornea, and non-uniform refractive index of the lens. Eye models are close to the actual conditions of the real human eye, such as the distributed eye model, the introduction of aspheric theory into the eye model, and the introduction of an adjustable eye model with a gradient index (GRIN) lens.

26.2.2 The Gullstrand I Model Eye

The Gullstrand precision eye model is similar to the actual refractive condition of the eye [2, 3]. The model includes six refractive surfaces, namely, the anterior and posterior surfaces of the cornea, the anterior and posterior surfaces of the lens cortex, and the anterior and posterior surfaces of the lens nucleus. Figure 26.1 shows the structure diagram and main optical data of the Gullstrand I model eye in the non-accommodative state. Figure 26.2 shows the schematic diagram of the software structure of Gullstrand I model eye.

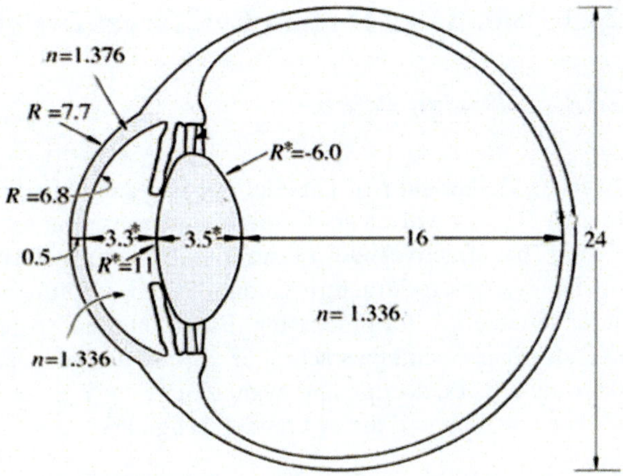

Fig. 26.1 Gullstrand I model eye structure diagram and main optical data

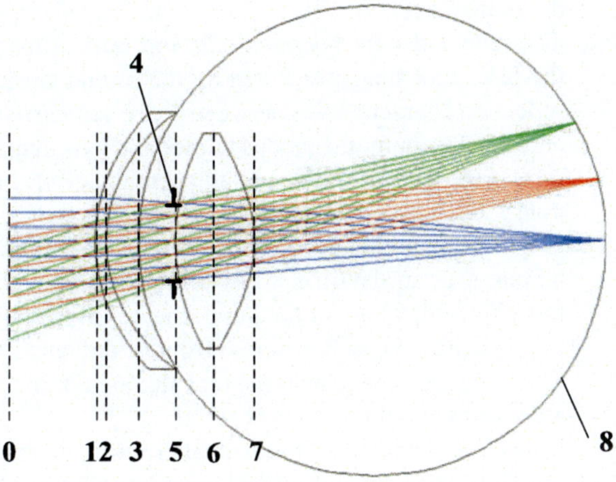

Fig. 26.2 Schematic diagram of the software structure of Gullstrand I model eye

Among them, the 0 surface is marked as the object surface in the optical design software; the 1 surface is a virtual surface, the thickness parameter of this surface is not important, the purpose is to better read the ray-tracing image; the 2 surface is the front surface of the simulated cornea; 3 sides are the interface between the simulated cornea and the anterior chamber; 4 sides are the system aperture diaphragm, which simulates the position of the pupil of the human eye. In order to simulate the decentering effect of the pupil position, the aperture diaphragm offset is set to −0.5 mm; face 5 is the anterior part of the simulated lens, that is, the front surface of the lens; face 6 is the interface between the anterior and posterior parts of the simulated lens; face 7 is the posterior surface of the simulated lens; face 8 is the simulated retina, which is the image plane position of the system.

26.3 Simulated Human Eye Target Design

26.3.1 Design Points

Because Gullstrand No. 1 model eye is a theoretical model eye close to the actual human eye, it involves six refractive surfaces, one retinal surface and six layers of tissue. The refractive index of each tissue layer's material is consistent with the actual human eye, and its structure is complicated, resulting in a very difficult selection of suitable materials for processing. Based on the research of Gullstrand No. 1 model eye, this paper combines actual needs and designs a simplified model eye for laser dazzling effect detection and evaluation.

The key points of the eye model design are:

a. The refractive characteristics are in line with the refractive tissue characteristics of the actual human eye, and the equivalent air focal length should be close to the human eye.
b. The reflected light between adjacent optical materials should avoid affecting the laser spot image received by the retinal simulator, combined with the low reflection characteristics between the refractive tissues of the human eye (cornea and aqueous humor, aqueous humor and lens, lens, and vitreous), set the relative refractive index difference of adjacent materials of the model eye to be no more than 1%.
c. It can simulate the laser dazzling effect of two kinds of non-adjustable state of human eyes in daytime. The variable entrance diaphragm is set according to the adjustable range of the pupil, the adjustment range is 2–8 mm; the entrance pupil position is set 3.3 mm in front of the lens-like lens.
d. The laser spot image near the macular area of the human eye is the key research area, and the field of view is set to.
e. Considering that the inherent aberration of real human eyes affects visual function, it is an important cause of glare, halation, and decreased vision under low-illumination light, moreover, the laser dazzlers are irradiated without precautions, it is too late to adjust when dazzling.
f. Gullstrand No. 1 model eye is an authoritative model eye. In terms of image quality, it is considered that its aberration level is the same as the normal aberration level of the human eye in the non-adjusted state. Therefore, the laser dazzling eye model is set at the incident aperture of 3 mm. (pupil diameter of a normal human eye), the deviation between the paraxial light and off-axis light from the Gullstrand I model eye is no more than 15%; The deviation between the MTF value of the paraxial light and the Gullstrand I model eye is not more than 0.1.

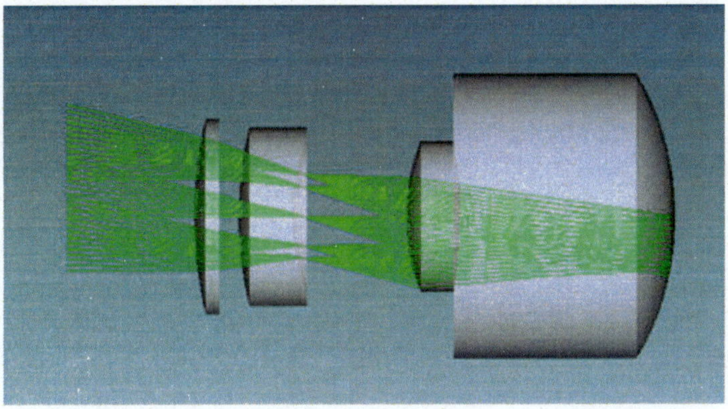

Fig. 26.3 3D shadow map of the laser dazzling eye model

26.3.2 Optimal Design of Laser Dazzling Eye Model

In order to ensure that the laser dazzling eye model is close to the real human eye state, the necessary optimization processing is carried out in the following aspects: By selecting the combination of glass MGF2-E, H-FK95N, H-QK3L, and H-FK95N, the relative refractive index difference between the media is optimized; by adjusting the parameters such as the radius of curvature and the glass spacing, the radius of speckle is optimized at the paraxial zone which is $-5°$–$0°$; the aberration of the simulated human eye optical system is adjusted by the design of software operand, so that the MTF value and the diffraction encircled energy meet the index requirements. In order to ensure the operability of the actual manufacturing and reduce the processing difficulty, a spherical lens is used to replace the aspherical design, and the imaging spot size is close to the aspherical model by changing the distance between the diaphragm and the sixth surface; considering the structural strength of the lens, it can be adjusted by balance the thickness of each piece of lens offsets the part due to the inconsistency of the lens material and the characteristics of the human eye tissue, which causes the human eye aberration error. The optimized laser dazzling eye model is shown in Fig. 26.3.

26.3.3 Design Result Analysis

The optimized laser dazzling eye model, its refractive characteristics conform to the actual human eye, the focal length is 16.30 mm; between the refractive tissue layers of the human eye (cornea and aqueous humor, aqueous humor and lens, lens and vitreous) The relative refractive index difference is optimized to 0.047, 0.049, 0.049, and the maximum deviation does not exceed 1%; when the entrance diaphragm is 3 mm (the pupil diameter of a normal human eye), the laser dazzling eye model's radius of speckle is 9.823, 8.262, 7.110, 6.330, 5.882, 5.736 μm at the paraxial zone, which is −5°–0°. As shown in Fig. 26.4, the maximum deviation between the optimized model spot size and the design index of Gullstrand I eye model is no more than 10%; when the entrance diaphragm is 3 mm, the MTF of the laser dazzling eye model at the field of view −5° is 0.32@60lp/mm, and the MTF curve comparison

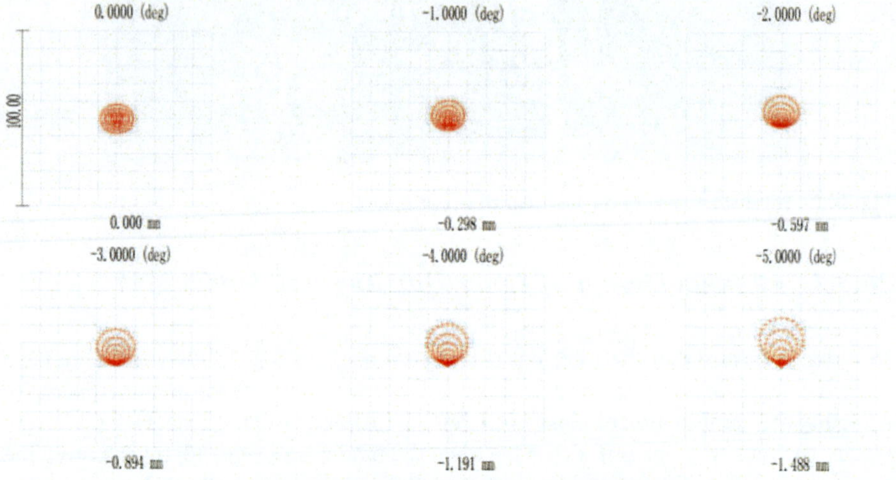

(A) Gullstrand I model eye

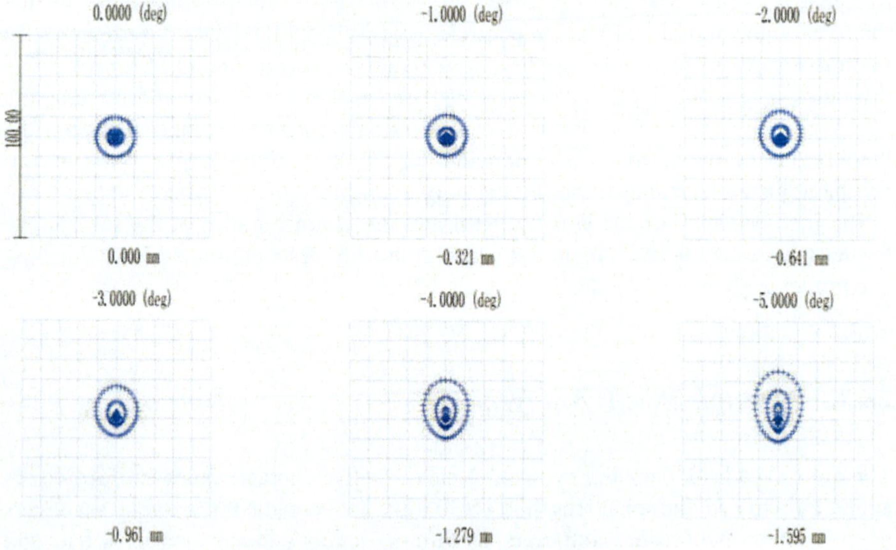

(b) Laser dazzling eye model

Fig. 26.4 Comparison of the diffusion pattern at the retina

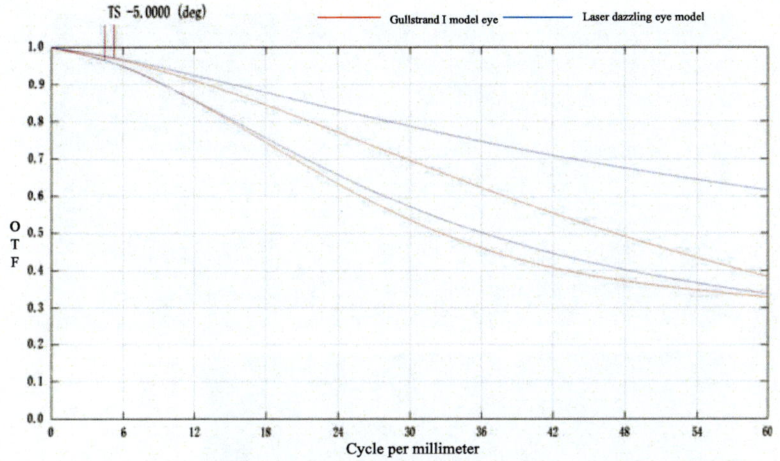

Fig. 26.5 Comparison of MTF curves between the laser dazzling eye model and Gullstrand Model I when the field of view is −5°

chart of Gullstrand Model I is shown in Fig. 26.5. It can be seen that the laser dazzling eye model ensures that the imaging quality of the lens does not affect the imaging detection of the detector. To sum up, the laser dazzling eye model meets the design criteria, is consistent with the Gullstrand precision eye model, and is in line with the real human eye imaging effect.

In the laser dazzling scene, the human eye will be subconsciously made to look into the laser by the sudden high-energy laser radiation and then quickly avoid the light stimulus. Therefore, in order to ensure accurate analysis of the dazzling scene's imaging spot on the retina, it is necessary to select a spot that matches the reality of the retina. The image sensor is used for imaging. In a human-like optical system, the image sensor acts like the retina in the eyeball. It senses the light signal transmitted by the lens and outputs it to the subsequent image processing system. Imitating the human eye optical target, the imaging spot and its energy distribution on the retina simulator after receiving laser irradiation are shown in Fig. 26.6.

26.4 Comparison Test Results

A certain type of high-power police laser dazzler is used as the research object to test and evaluate the dazzle effect. The beam expansion laser's wavelength of the laser dazzler is 532 nm, and the power density at 50 m is 0.068 mW/cm^2. This dose [4] is lower than the maximum exposure allowed by the national safety standards, which can ensure that the tester's retina will not be irreversibly damaged during the

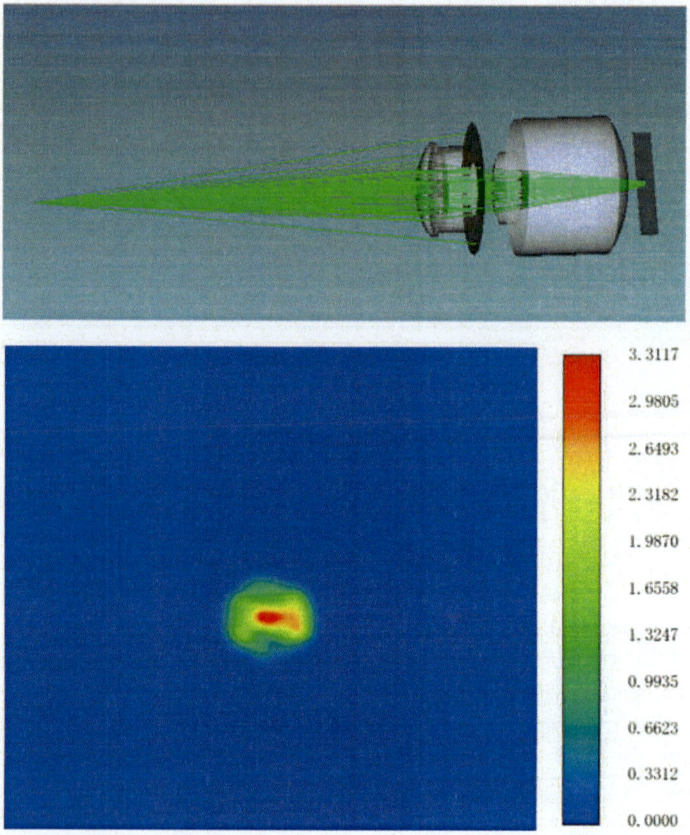

Fig. 26.6 The imaging spot and energy distribution diagram of the human-like optical target

experiment. In order to verify the consistency between the test results of the human-like eye target system and the subjective evaluation of the human eye, a volunteer dazzling effect matching test was also carried out in the experiment.

The test result of the human-like eye target is: the dark environment illuminance is 0.025 lx, the bright environment illuminance is 235 lx, the laser power density at 50 m is 0.068 mW/cm^2, and the bionic glare spot after data processing is shown in Fig. 26.7. Volunteers' subjective perception is that after being irradiated by laser suddenly, a spot shadow with a different shape and color will appear in the center of the field of view, and with time, the spot area will gradually shrink, and the color will become lighter until it disappears completely; look at the resolution plate pattern When the shadow of the light spot covers the target and hinders the recognition of the target, when the shadow of the light spot gradually becomes smaller and lighter, the resolution plate is gradually distinguished until it is completely distinguished, and the human eye returns to the visual state before the irradiation.

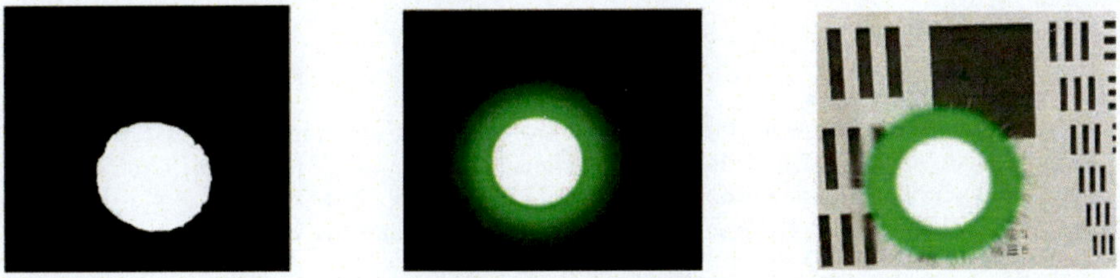

(a) Detector and light spot (b) Optical system display light spot (c) Human-like eye light spot after data processing

Fig. 26.7 Bionic glare points after detection of human-like eye targets and data processing

The human-like eye target system and volunteers were tested and evaluated the laser dazzling effect at the same time, and the matching rate of the dazzling effect reached 90%, which met the requirements for quantitative testing of laser dazzling effect.

26.5 Conclusion

Aiming at the characteristics of laser dazzling scenes, a simulated human eye target optical system that can test dazzling spots is designed. The system is based on Gullstrand I model eye, combined with the characteristics of the laser dazzling scene to establish an optical model of a simulated human eye target. Then, the design index of the optical system that meets the requirements of the laser dazzling fundus spot test is proposed. After the optical structure, lens and system aberration of the model are changed by the material, the design of the optical model can be optimized. The system realizes the human-eye dazzling effect test in the laser dazzling scene with a spectral resolution of 1 nm, a working band of 380–780 nm, an ambient light test range of 1×10^{-4}–1×10^{3} lx. In this way, the objective of performing objective quantitative tests without the aid of biological tests is achieved. The verification and comparison experiment results show that the laser dazzling effect matching rate of the system and real human eyes reaches 90%, which meets the requirements of non-biological testing of laser dazzling effect and provides for the safety and effectiveness research of non-lethal laser dazzlers.

References

1. P. Mouroulis, *Visual instrumentation: Optical design and engineering principles.* (2014)
2. H. Von Helmholtz, J.P. Southall (eds.) *Helmhotz's Treatise on Physiological Optics*, vol. 1. (Optical Society of America, Rochester, New York, 1924),.pp. 334–350
3. H. Guo, Z. Zhaoqi Wang, Q. Zhao, W. Quan, Y. Wang,Individual eye model based on wavefront aberration. Optik **116**, 80–85 (2005)
4. R. Song, J. Hou, S. Chenet al., High power supercontinuum generation in a nonlinear ytterbium-doped fiber amplifier. Opt. Lett. **37**(9), 1529–1531 (2012)

Chapter 27
Effects of Cooperative Target on Laser Rangefinder's Ranging Performance

Yan Jin, Jinfeng Jiang, and Zhijun Chen

Abstract The laser rangefinder's ranging performance theoretical models of diffuse target and cooperative target are established based on the laser ranging equation. The effects of the cooperative target on ranging performance are discussed at different atmospheric visibilities and different beam divergence angles. The results show that cooperative targets can increase the ranging performance, can decrease the laser emission power efficiently, and can ensure personnel safety accordingly.

27.1 Introduction

Laser rangefinder is a kind of equipment [1–4] widely used in laser technology in the military. Laser rangefinders can be used with a variety of tactical weapons to make the first hit rate of more than 80%. It has been widely used in multi-type photoelectric tracking, alerting, and search equipment [5].

The ranging ability, i.e. the maximum range, is an important indicator of a laser rangefinder, and its size depends on the environmental conditions and the characteristics of the rangefinder itself. At present, the research on the influencing factors of ranging ability is mainly focused on the visibility of the atmosphere, the size of the target, and the characteristics of the rangefinder itself [6–8]. Based on the characteristics of laser ranging targets, this paper analyzes the impact of cooperative targets on laser ranging capabilities.

Y. Jin (✉) · J. Jiang
AVIC Manufacturing Technology Institute, Beijing 100024, China
e-mail: tg667788@xzcstudio.com

Z. Chen
Institute of Army Aviation, Beijing 100012, China

G. Liu and F. Cen (eds.), *Advances in Precision Instruments and Optical Engineering*, Springer Proceedings in Physics 270, https://doi.org/10.1007/978-981-16-7258-3_27

27.2 Theoretical Model of Ranging Ability

The laser ranging ability is mainly related to the detection sensitivity of the detector and the laser energy received by the detector. When the equivalent noise power of the laser received by the detector is greater than the equivalent noise power of the detector, the ranging distance can be achieved, otherwise, The equivalent noise power of the laser received by the detector is less than the equivalent noise power of the detector, so the ranging distance cannot be achieved.

27.2.1 *Theoretical Model of Detector Equivalent Noise Power*

For laser rangefinders, the target's laser echo is very weak, in order to improve the receiving sensitivity of the target echo, an avalanche photodiode (APD tube) is usually used as the photodetector. For a typical InGaAs avalanche photodiode, its equivalent noise power is shown in (27.1).

$$\mathrm{NEP} = 330/\sqrt{f} \tag{27.1}$$

Among them, NEP is the equivalent noise power of the detector, and f is the bandwidth of the avalanche photodiode.

27.2.2 *Theoretical Model of Laser Equivalent Noise Power*

Laser equivalent noise power is received by laser rangefinder P as shown in (27.2).

$$P = P_r/\mathrm{SNR} \tag{27.2}$$

among them, P_r The laser power is received by the laser rangefinder.

Assumptions: the propagation of laser light in the atmosphere obeys the laws of geometric optics; the atmosphere is uniform and isotropic; the reflection of the incident light by the target is either diffuse or specular; the energy distribution in the laser beam is approximately uniform or at least axially symmetric; the receiving system and the transmitting system are tightly coupled, and the optical axes are parallel to each other, the laser power intercepted by the target at a distance r P_s for [9]:

$$P_s = \frac{4P_t\tau_t A_s \cos\theta}{\pi^2\theta_t^2 R^2} e^{-\mu R} \tag{27.3}$$

among them, P_t is the laser emission power; τ_r is the transmittance of the transmitting optical system; A_s the target area illuminated by the laser beam; θ is the angle between the average surface normal of the illuminated part and the incident ray; θ_r The beam divergence angle of the optical system; μ the attenuation coefficient of the laser energy through the unit length of the atmosphere. Here, an empirical formula is used to determine the atmospheric attenuation coefficient during the laser transmission. The empirical formula is to calculate the atmospheric attenuation coefficient by using the normalized contrast between the target and the background and the line-of-sight relationship, as shown in (27.4).

$$\mu = \frac{3.91}{V}\left(\frac{\lambda}{0.55}\right)^{-q}$$

$$q = \begin{cases} 1.6, \text{ when } V \text{ is large} \\ 1.3, \text{ moderate} \\ 0.585\,V^{1/3}, \text{ when } V \le 6\,\text{km} \end{cases} \tag{27.4}$$

where v is the visibility of the atmosphere; λ is the laser wavelength.

Diffuse target

When the target surface is rough and irregular, we consider it to be a diffuse reflection target. The rough surface of the diffuse reflection target does not reflect the incident laser light uniformly in all directions. The reflection in the incident direction is the strongest, and the reflection gradually decreases as the angle between the incident light and the incident light increases. $\pi/2$, the reflected light intensity, decreases to almost zero at the time. Because the specific angular distribution of the reflected light intensity varies with the target surface, it is more practical to use a cosine distribution instead of a uniform distribution. In addition, the laser beam energy distribution is also non-uniform, which can be considered the energy distribution in the beam is an axisymmetric Gaussian distribution. After the above analysis, the ranging model for diffuse reflection targets is shown in (27.5).

$$P_r = \frac{4P_t\tau_t\tau_r A_r A_{SM}\rho}{\pi^2\theta_t^2 R^4} e^{-2\mu R} \tag{27.5}$$

where: P_r the laser power received by the laser rangefinder; τ_r is the transmittance of the receiving optical system; A_r effective receiving area for receiving optical system; A_{SM} effective receiving area for diffuse targets; ρ is the diffuse reflection coefficient of the target.

Cooperation goals

For an ideal Lambertian target, the scattering space is hemispherical space, while for an ideal cooperative target, because the beam does not change its properties after reflection, the energy is concentrated after the reflection and returns in the direction of incidence. Therefore, the distance measurement equation for laser ranging on the cooperative target is as follows, (27.6).

$$P_r = \frac{16 P_t \tau_t \tau_r A_r A_{SH} \rho_S}{\pi^2 \theta_t^2 R^4 \theta_S^2} e^{-2\mu R} \tag{27.6}$$

where: A_{SH} effective reception area for cooperation goals; ρ_s the reflection coefficient of the cooperative target; θ_s reflection divergence angle for cooperative targets.

27.2.3 SNR Theoretical Model

The ranging capability index is inseparable from the detection probability and false alarm rate indicators. In order to ensure that the laser rangefinder meets a certain false alarm rate and detection probability, the received laser power should meet a certain signal-to-noise ratio requirement. Under white noise, noise current i_n to fit a zero-mean random variable with a normal distribution, let the root mean square error be I_n, Then the distribution probability is:

$$P(i_n) = \frac{1}{\sqrt{2\pi} I_n} \exp\left(-\frac{i_n^2}{2I_n^2}\right) \tag{27.7}$$

Set the detection threshold to I_t, the laser echo pulse width is τ, the number of false alarms per second caused by the above noise is P_F. It can be expressed by the following formula:

$$\overline{P}_F = \frac{1}{2\sqrt{3}\tau} \exp\left(-\frac{I_t^2}{2I_n^2}\right) \tag{27.8}$$

Let the signal current be I_s, detection probability P_d can be expressed as:

$$P_d = P(I_s + i_n \rangle I_t) = \int_{I_t - I_s}^{\infty} P(i_n) di_n = \frac{1}{\sqrt{2\pi} I_n} \int_{I_t - I_s}^{\infty} \exp\left(-\frac{i_n^2}{2I_n^2}\right) di_n \tag{27.9}$$

The above formula can be expressed as an error function:

$$P_d = \frac{1}{2} + \frac{1}{2} erf\left(\frac{I_s - I_t}{\sqrt{2} I_n}\right) \tag{27.10}$$

According to Formulas (27.8) and (27.10), the signal-to-noise ratio of the laser rangefinder under a certain detection probability and false alarm rate can be obtained.

According to Formulas (27.1), (27.2), (27.4)–(27.6), (27.8), (27.10), the theoretical laser ranging capability index for diffuse reflection targets and cooperative targets can be obtained.

27.3 Impact Analysis of Cooperative Targets on Laser Ranging

27.3.1 *Influence of Cooperative Targets on Ranging Ability Under Different Atmospheric Visibility Conditions*

The relevant parameters of the laser rangefinder for numerical simulation calculation are shown in Table 27.1. Under different visibility conditions, the relationship between the receiver power and the equivalent noise power of the detector at different ranging distances for diffuse reflection targets is shown in Fig. 27.1a. This range can be achieved when the laser power received by the detector is greater than the equivalent noise power of the detector. The relationship between the received power and the equivalent noise power of the detector at different ranging distances for the cooperative target under different visibility conditions is shown in Fig. 27.1b. It can be seen from Fig. 27.1b that when the cooperative target is configured, when the visibility is 5 km, the maximum range of the laser rangefinder can reach more than 15 km. With the increase of the visibility of the atmosphere, the maximum range can reach more than 50 km. Therefore, the laser ranging ability can be significantly improved under the condition of cooperative targets. In addition, the laser rangefinder has

Table 27.1 Parameter of the laser rangefinder

No.	Item	Parameter
1	Laser emission energy E_t	60 mJ
2	Laser pulse width τ	10 ns
3	Transmission optical system transmittance τ_t	88%
4	Laser emission beam divergence angle θ_t	1 mrad
5	Diffuse reflection target effective receiving area A_{SM}	1 m^2
6	Target diffuse reflection coefficient ρ	0.7
7	Cooperation target effective receiving area A_{SH}	9 cm^2
8	Cooperative target reflection coefficient ρ_S	0.92
9	Cooperative target reflection beam divergence angle θ_S	1 mrad
10	Receiving optical system caliber d	80 mm
11	APD bandwidth f	200 MHz
12	Detection probability P_F	95%
13	False alarm rate p_d	1%

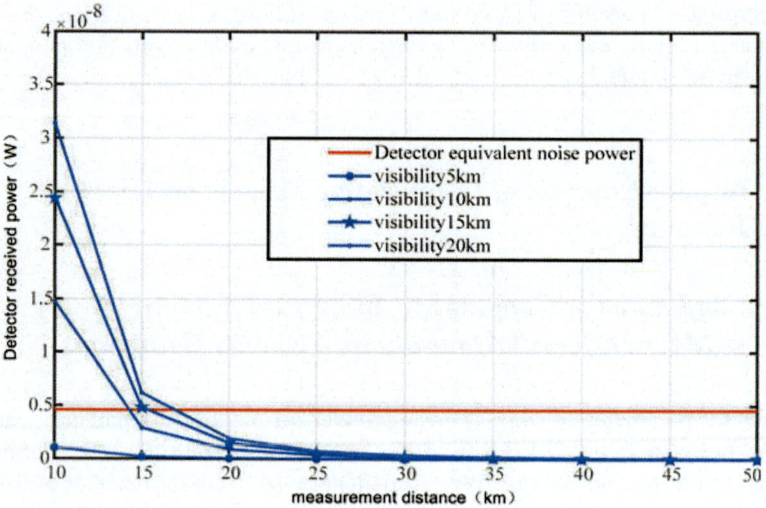

(a) Diffuse target

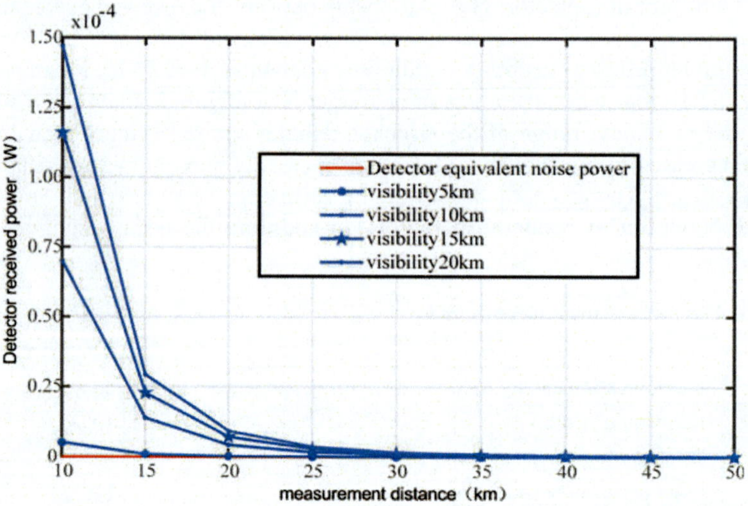

(b) Cooperative target

Fig. 27.1 The received power of laser rangefinder at different visibility

significantly reduced requirements for environmental and atmospheric conditions, enabling the laser rangefinder to work in harsher conditions.

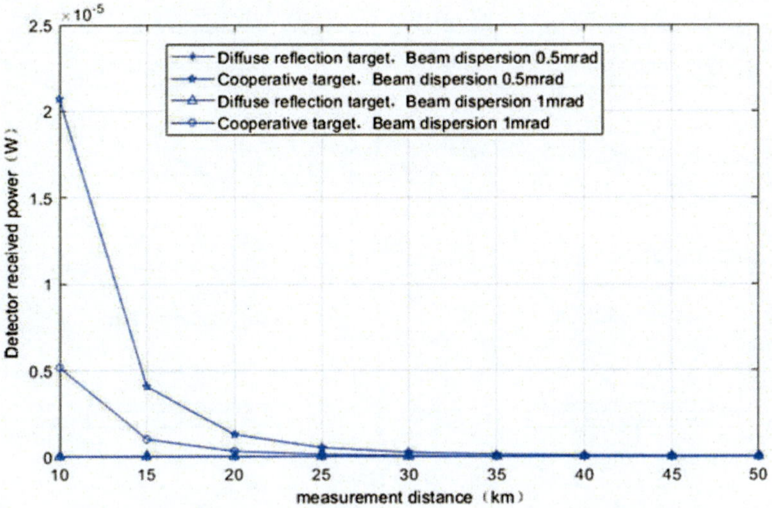

Fig. 27.2 The received power of laser rangefinder at different beam divergence angles

27.3.2 Analysis of the Influence of Cooperative Targets on Ranging Ability Under Different Laser Beam Divergence Angles

Using the parameters in Table 27.1, under the condition of visibility of 5 km and different laser emission beam divergence angles, for diffuse reflection targets and cooperative targets, the receiving power of laser rangefinder detectors with different ranging distances is shown in Fig. 27.2. The smaller the divergence angle of the laser emission beam, the more obvious the difference in laser ranging ability is with and without the cooperative target. When the ranging distance is 10 km and the laser beam divergence angle is 0.5 mrad, configure the cooperative target. The received power of the lower detector is about 10 three times without the cooperation target.

27.3.3 Theoretical Calculation of Laser Emission Power Under the Condition of Configuring Cooperative Targets

According to the analysis in Sect. 3.1, when the same range is required, configuring the cooperation target can reduce the laser transmission power. The reduction of the laser transmission power can reduce the weight and volume of the laser rangefinder on the one hand and can further ensure the use of Personnel safety. According to the parameters of the rangefinder shown in Table 27.1, the Formula (27.6) can be used

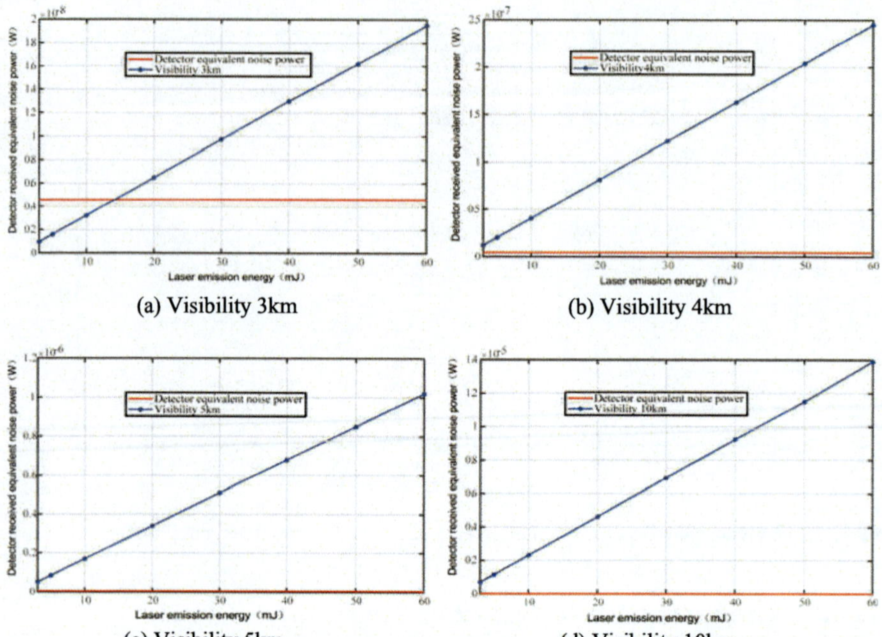

(a) Visibility 3km (b) Visibility 4km

(c) Visibility 5km (d) Visibility 10km

Fig. 27.3 The received noise equivalent power of laser rangefinder at different visibility and different emission power

to obtain the distance of the cooperative target under different laser emission power conditions. The equivalent noise power received by the detector is shown in Fig. 27.3. Figure 27.3 shows that when the atmospheric visibility is greater than 3 km and the laser emission energy is less than 10 mJ, the laser rangefinder can achieve a maximum range of 15 km. According to the laser hazard type determination method [10], when the laser output energy is 10 mJ, the laser can be guaranteed The rangefinder laser is a Class 3A laser, and the laser emitted by the rangefinder will not cause damage to the eyes of the person on the measured target when the measurement distance is greater than 100 m under the condition that the atmospheric attenuation is ignored. Effectively reduce the laser transmission power of the laser rangefinder, thereby effectively reducing the weight and volume of the rangefinder and ensuring human eye safety.

27.4 Conclusion

Based on the characteristics of laser ranging targets, this paper analyzes the influence of cooperative targets on laser ranging capabilities under different atmospheric visibility conditions and different laser emission beam divergence angles. The analysis results show that the configuration of cooperative targets can effectively improve ranging capabilities. The laser rangefinder has significantly reduced the requirements for environmental and atmospheric conditions and can make the laser rangefinder work in harsh conditions; the smaller the laser emission beam divergence angle, the laser ranging ability with and without the cooperation target The more obvious the difference is; in addition, the configuration of cooperation targets can effectively reduce the laser emission power and further provide protection for improving personnel safety.

References

1. R.W. Byren, Laser range finder, in *IR/ED Handbook* (1993), pp. 79–110
2. Li Y, Zukang L, Yu L et al., Numerical simulation and optimal design of the system performance for the airborne laser range finder. Proc. SPIE **4220**, 317–320
3. S. Hao, Industrial application of laser ranging. Prog. Laser Optoelectron. **1**(11), 51–53 (2000)
4. Y. Yeping, Y. Zhaojin, H. Min, Uncertainty analysis of calibration and measurement of main parameters of laser rangefinder. J. Appl. Opt. **26**(4), 56–57 (2005)
5. H. Binxin, Z. Mei, Y. Zutao, Design and implementation of laser range finder range simulation detection system. Comput. Meas. Control **16**(5), 619–623 (2008)
6. W. Haixian, Y. Ai, Study on the influence of atmospheric attenuation coefficient on laser ranging ability. Ship Sci. Technol. **29**(6), 116–119 (2007)
7. L. Guangyu, Z. Tianshu, Influence of resident effect on laser atmospheric transmission. High Power Laser Part. Beams **11**(2), 181–184 (1999)
8. W. Xiusheng, N. Yanxiong, Z. Peng et al., Numerical simulation study of the influence of target shape on the ranging ability of laser rangefinder. Prog. Laser Optoelectron. **42**(11), 28–31 (2005)
9. W. Guanghui, Y. Peigen, *Application of Laser Technology in the Weapon Industry* (Ordnance Industry Press, Beijing, 1995)
10. gjb 470a-1997, Control and protection of military laser hazards

Chapter 28
Analysis and Study of Target Ball Error Accuracy for the Laser Tracker

Chang'an Hu, Song Hu, Junbo Liu, and Haifeng Sun

Abstract Laser tracker is the mainstream equipment in the field of industrial measurement. The analysis and research of its measurement accuracy have a certain scientific research value and use value. In this paper, the measurement results of the metal and glass target balls at $D = 3730$ mm and $D = 2815$ mm were analyzed and analyzed by combining with the measurement system of the laser tracker. The measurement results of the metal and glass target balls at $D = 2766$ mm and $D = 2727$ mm were analyzed by using three-point support base. It is found that the maximum error of the metal target is 14.2 μm, the 1.5-inch glass target is 5.8 μm, and the 0.5-inch glass target is 9.3 μm at $D = 2727$ mm by using the three-point support base. In the case of removing the positioning error and the base error of the target ball, the three target balls all meet the requirements of use. The closer the target ball is to the laser tracker system, the smaller the error value is. Compared with the base supported by plane, the measurement accuracy of the target ball is improved to some extent when the base supported by triangle is used.

28.1 Introduction

In recent years, with the rapid development of modern three-dimensional space technology, laser tracker has become a high-precision large-size measuring instrument with mature technology, high efficiency and stable performance in industrial measuring system. In some major industries, the 3D high precision measurement

C. Hu · S. Hu (✉) · J. Liu · H. Sun
Institute of Optics and Electronics, Chinese Academy of Sciences, Chengdu 610209, China
e-mail: husong@ioe.ac.cn

J. Liu
e-mail: ljbopt@126.com

C. Hu
University of Chinese Academy of Sciences, Beijing 100049, China

National Institute of Measurement and Testing Technology, Chengdu 610021, China

G. Liu and F. Cen (eds.), *Advances in Precision Instruments and Optical Engineering*, Springer Proceedings in Physics 270, https://doi.org/10.1007/978-981-16-7258-3_28

method represented by the laser tracker gradually replaces the traditional measurement method and is widely used in the installation and detection of large precision equipment [1–4].

28.2 Laser Tracker Measuring System

Laser tracker system (LTS) is a high-precision large-size measurement instrument in industrial measurement system. It integrates laser interferometric ranging technology, photoelectric detection technology, precision machinery technology, computer and control technology, modern numerical calculation theory and other advanced technologies. The space moving target is tracked and the three-dimensional coordinates of the target are measured in real time. It has the characteristics of high precision, high efficiency, real-time tracking and measurement, quick installation and easy operation, etc. and is suitable for real-time measurement of geometric quantities in large-scale space.

28.3 Error Factors Affecting Measurement

The main factors affecting the indication error of the laser tracker measurement system are the indication error of the point length, the double-sided indication error, the distance indication error, the spherical reflection target error and the dynamic velocity error.

The point-to-point length indication error of laser tracker refers to the difference between the indication value and the reference value when the laser tracker measures the point-to-point length of the short ruler. The measurement needs to be made in horizontal, vertical, diagonal and any direction specified by the user [5].

Laser tracker double-sided indication error refers to the difference of the indication value of the laser tracker when the fixed point is tested on both sides and the forward-looking/rear-looking measurement at the same point. This measurement needs to be made at three different points.

The distance indication error of the laser tracker refers to the difference between the indication value of the laser tracker and the reference value when the laser tracker measures the long ruler installed along the measuring axis. The measurement needs to be made on at least four different reference lengths.

Laser tracker spherical reflection target error refers to the spherical reflection target installed in the base point, the laser tracker measured the value of the change. The measurement involves rotating the spherical reflector target four positions about the laser incident axis and four positions about the vertical axis of the laser.

Dynamic indication error refers to the maximum speed when the laser tracker can keep tracking ability when the target moves along the circumference.

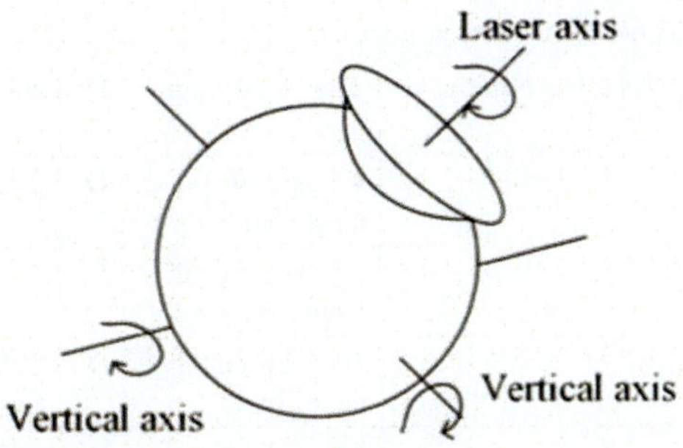

Fig. 28.1 Schematic diagram of spherical reflection target error measurement

28.4 Error Analysis Method of Measuring Target Ball

The reflector of laser tracker generally has cat eye reflector, corner reflector and tool ball reflector and other types. The reflector adopts a spherical structure, embedded in the interior of a hollow cone prism, ranging equivalent reflection center coincides with the center of the ball, in theory, the distance between the prism center and any spherical measuring point is equal to the radius of the ball R. The accuracy of the reflector is very high, each reflector needs to be tested strictly before leaving the factory, and the allowable error is generally (0.01–0.025) mm.

Similar to total station ranging, laser tracker ranging also exists range plus constant and range multiplication constant, which should be corrected. In addition, the error of the center of the ball and the error of the incident angle have a great influence on the measurement of the station construction of multiple laser trackers. This error is not only affected by the error of the prism itself but also affected by the processing error of the base. In order to ensure the measurement accuracy of the laser tracker, it is necessary to test the reflection prism error.

As shown in the figure, the spherical reflection target is installed at the base point (stable and at the same height as the instrument), aligned with the laser incident direction and recorded the value of the laser tracker; rotate four positions around the laser incident to record the laser tracker; swing around two axes perpendicular to the laser axis respectively to record the value of the laser tracker. The measurement results are shown in the Fig. 28.1.

28.5 Example Analysis of Measuring Target Ball Error

In this paper, the 1.5-inch metal target ball, 1.5-inch glass target ball and 0.5-inch glass target ball configured in the laser tracker system are tested at different distances. The measured environmental conditions are: temperature 24.1 °C, air pressure 952.4 hPa, humidity 60.3%RH. The specific test scheme is shown in Table 28.1.

Table 28.1 Test protocol

The serial number	Target ball type	Distance	Type of base support	Number of measurements
1	1.5-inch metal target ball	$D = 3730$ mm	The plane	3
2	1.5-inch metal target ball	$D = 2815$ mm	The plane	3
3	1.5-inch metal target ball	$D = 2766$ mm	Three points	3
4	1.5-inch glass target ball	$D = 3730$ mm	The plane	3
5	1.5-inch glass target ball	$D = 2815$ mm	The plane	3
6	1.5-inch glass target ball	$D = 2766$ mm	Three points	3
7	0.5-inch glass target ball	$D = 2727$ mm	Three points	1

28.5.1 1.5-Inch Metal Target Ball Test

The 1.5-inch metal target ball has the advantage of preventing it from falling and is commonly used in the measurement process. It is necessary to carry out its measurement accuracy. For the 1.5-inch metal target ball, the plane base measurement results are used at the position $D = 3730$ mm and $D = 2815$ mm, respectively, and the three-point support base measurement results are used at the position 2766 mm. It can be seen from the three tables that the maximum measurement error of the metal target ball at $D = 3730$ mm is 15.5 μm, that of the metal target ball at $D = 2815$ mm is 14.4 μm, and that of the metal target ball at $D = 2766$ mm is 14.2 μm. According to the data, the closer the target ball is to the laser tracker system, the smaller the error value is. The datum of the triangular-supported base is a slight improvement over that of the flat-supported base.

The positioning error of the laser tracker measuring system is 15 μm + 6 μm/m, which is about 37, 32 and 31 μm, respectively, when $D = 3730$ mm, $D = 2815$ mm and $D = 2766$ mm. After removing the influence of the positioning error and the base of the target ball, the accuracy error of the target ball in the three places meets the use requirements (Table 28.2).

Figure 28.2 shows the comparison of the three measurement results at $D = 2815$ mm for a 1.5-inch metal target ball. The image clearly shows the measurement error curve. By comparing the figures, it is found that the error curves of the metal target ball in the laser axis, vertical axis 1 and vertical axis 2 are consistent, and the data fluctuate to some extent. The maximum measurement error of the metal target ball at $D = 2815$ mm is 14.4 μm, and the correction point appears in the laser axis of the second measurement (Tables 28.3 and 28.4).

Table 28.2 Measurement results of 1.5-inch metal target ball $D = 3730$ mm (plane support)

The serial number	Rotation axis	Rotation angle/(°)	X/mm	Y/mm	Z/mm	Distance/μm
1	Laser axis	0	1377.6101	3462.2266	−167.1943	15.5
2		180	1377.6245	3462.2245	−67.1890	
3		90	1377.6125	3462.2243	−167.1908	12.4
4		270	1377.6126	3462.2266	−167.1786	
5	Vertical axis 1	40	377.6111	3462.2257	−167.1972	10.5
6		−40	1377.6105	3462.2245	−167.1868	
7	Vertical axis 2	40	1377.6127	3462.2280	−167.1910	6.1
8		−40	1377.6116	3462.2234	−167.1872	

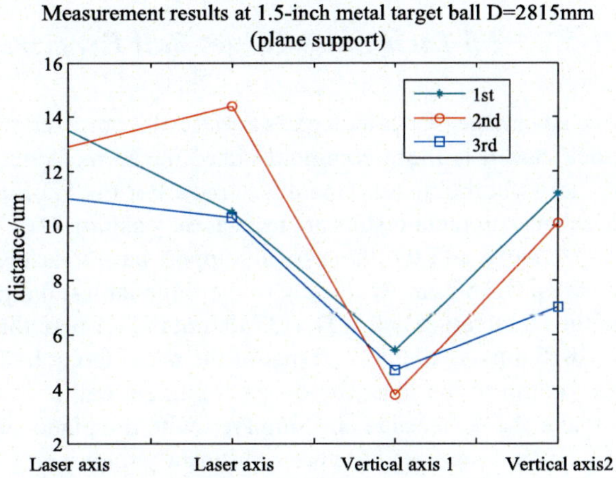

Fig. 28.2 Three measurement results when 1.5 in metal target ball $D =$ 2815 mm (plane support)

Table 28.3 Measurement results of 1.5-inch metal target ball $D = 2815$ mm (plane support)

The serial number	Rotation axis	Rotation angle/(°)	X/mm	Y/mm	Z/mm	Distance/μm
1	Laser axis	0	2070.9713	1898.9193	−173.8509	12.9
2		180	2070.9634	1898.9183	−173.8610	
3		90	2070.9723	1898.9162	−173.8628	14.4
4		270	2070.9634	1898.9207	−173.8525	
5	Vertical axis 1	40	2070.9711	1898.9210	−173.8511	3.8
6		−40	2070.9713	1898.9172	−173.8515	
7	Vertical axis 2	40	2070.9731	1898.9170	−173.8494	10.1
8		−40	2070.9667	1898.9248	−173.8497	

Table 28.4 Measurement results of 1.5-inch metal target ball $D = 2766$ mm (three-point support)

The serial number	Rotation axis	Rotation angle/(°)	X/mm	Y/mm	Z/mm	Distance/μm
1	Laser axis	0	2061.2728	1835.5138	−174.5207	13.3
2		180	2061.2677	1835.5102	−174.5326	
3		90	2061.2770	1835.5066	−174.5307	14.2
4		270	2061.2662	1835.5144	−174.5257	
5	Vertical axis 1	40	2061.2750	1835.5126	−174.5207	3.4
6		−40	2061.2740	1835.5093	−174.5209	
7	Vertical axis 2	40	2061.2735	1835.5107	−174.5211	3.3
8		−40	2061.2758	1835.5127	−174.5224	

28.5.2 1.5-Inch Glass Target Ball Experiment

The advantage of 1.5-inch glass target ball compared with metal target ball is its low price, and it is more commonly used in the measurement process. It is necessary to carry out the measurement accuracy. For the 1.5-inch glass target ball, the plane base measurement results are used at the position $D = 3730$ mm and $D = 2815$ mm, respectively, and the three-point support base measurement results are used at the position 2766 mm. According to the three tables, the maximum measurement error of the metal target ball at $D = 3730$ mm is 11.1 μm, that of the metal target ball at $D = 2815$ mm is 7.0 μm, and that of the metal target ball at $D = 2766$ mm is 5.8 μm. According to the data, the closer the target ball is to the laser tracker system, the smaller the error value is. Compared with the plane-supported base, the data of the triangular-supported base are significantly improved.

The positioning error of the laser tracker measuring system is 15 μm + 6 μm/m, which is about 37 μm, 32 μm and 31 μm, respectively, when $D = 3730$ mm, $D = 2815$ mm and $D = 2766$ mm. After removing the influence of the positioning error and the base of the target ball, the accuracy error of the target ball in the three places also meets the use requirements (Tables 28.5 and 28.6).

28.5.3 0.5" Glass Target Ball Experiment

At a measuring distance of $D = 2727$ mm, the 0.5-inch glass target ball is supported by three points, and the data shown in Table 28.7 show a measurement error of 9.3 μm. A 0.5-inch glass target ball is more difficult to measure than a 1.5-inch target ball due to its relatively small diameter. Through the data, it is found that the measuring accuracy of 0.5-inch glass target ball meets the requirements of use (Table 28.8).

Table 28.5 Measurement results of 1.5-inch glass target ball $D = 3730$ mm (plane support)

The serial number	Rotation axis	Rotation angle/(°)	X/mm	Y/mm	Z/mm	Distance/μm
1	Laser axis	0	1377.6641	3462.2335	−167.2129	3.6
2		180	1377.6664	3462.2321	−167.2153	
3		90	1377.6671	3462.2323	−167.2125	3.3
4		270	1377.6695	3462.2338	−167.2141	
5	Vertical axis 1	40	377.6674	3462.2328	−167.2113	1.8
6		−40	1377.6658	3462.2321	−167.2106	
7	Vertical axis 2	40	1377.6567	3462.2320	−167.2002	11.1
8		−40	1377.6605	3462.2325	−167.2107	

Table 28.6 Measurement results when 1.5-inch glass target ball $D = 2815$ mm (plane support)

The serial number	Rotation axis	Rotation angle/(°)	X/mm	Y/mm	Z/mm	Distance/μm
1	Laser axis	0	2070.9710	1898.9328	−173.8431	1.1
2		180	2070.9713	1898.9321	−173.8439	
3		90	2070.9775	1898.9290	−173.8462	7.0
4		270	2070.9723	1898.9326	−173.8431	
5	Vertical axis 1	40	2070.9709	1898.9318	−173.8462	6.6
6		−40	2070.9714	1898.9322	−173.8397	
7	Vertical axis 2	40	2070.9698	1898.9323	−173.8405	5.0
8		−40	2070.9669	1898.9339	−173.8443	

Table 28.7 Measurement results when 1.5-inch glass target ball $D = 2766$ mm (Three-point support)

The serial number	Rotation axis	Rotation angle/(°)	X/mm	Y/mm	Z/mm	Distance/μm
1	Laser axis	0	2061.2783	1835.5165	−174.5187	5.8
2		180	2061.2752	1835.5164	−174.5237	
3		90	2061.2794	1835.5156	−174.5221	1.6
4		270	2061.2792	1835.5167	−174.5210	
5	Vertical axis 1	40	2061.2785	1835.5159	−174.5188	1.6
6		−40	2061.2772	1835.5160	−174.5179	
7	Vertical axis 2	40	2061.2775	1835.5165	−174.5210	1.3
8		−40	2061.2768	1835.5169	−174.5200	

Table 28.8 Measurement results of 0.5-inch glass target ball $D = 2727$ mm (three-point support)

The serial number	Rotation axis	Rotation angle/(°)	X/mm	Y/mm	Z/mm	Distance/μm
1	Laser axis	0	2048.0392	1792.1104	−187.7215	9.3
2		180	2048.0307	1792.1142	−187.7221	
3		90	2048.0329	1792.1125	−187.7230	6.0
4		270	2048.0302	1792.1120	−187.7176	
5	Vertical axis 1	40	2048.0289	1792.1128	−187.7205	3.5
6		−40	2048.0296	1792.1135	−187.7238	
7	Vertical axis 2	40	2048.0338	1792.1121	−187.7204	7.5
8		−40	2048.0227	1792.1156	−187.7231	

As can be seen from the table, with the change of the attitude of the ball prism, the maximum error of the reflection center of the metal target ball can reach 14.2 μm when the three-point support base is used at the close distance $D = 2766$ mm, and the maximum error of the reflection center of the 1.5-inch glass target ball can reach 5.8 μm when the three-point support base is used at the close distance $D = 2727$ mm. The maximum error caused by the reflection center of the 0.5-inch glass target ball can be 9.3 μm. According to the data, the closer the target ball is to the laser tracker system, the smaller the error value is. Compared with the base supported by plane, the measurement accuracy of the target ball is improved to some extent when the base supported by triangle is used.

28.6 Conclusions

There are many kinds of influence on measurement error of laser tracker system, and the measurement target ball is also a part of the influence on measurement accuracy. The precision of three kinds of commonly used target balls, 1.5-inch metal target ball, 1.5-inch glass target ball and 0.5-inch glass target ball, is analyzed. With the change of the attitude of the ball prism, the maximum error of the reflection center of the 1.5-inch metal target ball is 14.2 μm, and the maximum error of the reflection center of the 1.5-inch glass target ball is 5.8 μm, using the three-point support base at the position of $D = 2766$ mm. The maximum error of the reflection center of the 0.5-inch glass target ball is 9.3 μm when the three-point support base is used at $D = 2727$ mm. In the case of removing the positioning error and the base error of the target ball, the three target balls all meet the requirements of use. The closer the target ball is to the laser tracker system, the smaller the error value is. Compared with the base supported by plane, the measurement accuracy of the target ball is improved to some extent when the base supported by triangle is used. The next step is to develop a tool to further shorten the measurement distance and reduce the impact of the measurement distance on the precision of the target ball.

Acknowledgements The work is supported by the National Natural Science Foundation of China (NSFC) under grants of number 61604154, number 61875201, number 61975211 and number 62005287, and partially supported by the science project of Sichuan province under grant of 2018JY0203 and the Youth Innovation Promotion Association of the Chinese Academy of Sciences (2021380).

References

1. C.A. Hu, S.T. Luo, W.Z. Li et al., Application of laser tracker in industrial measurement field, in *2021 International Conference on Mechanical Engineering, Intelligent Manufacturing and Automation Technology, Guilin*, vol. 1820 (2021), pp. 1–5
2. C.A. Hu, W.B. Du, M.Y. Zhou, Research on process inspection and compensation method for finish machining of large size and high precision conical shaft. Mach. Tool Hydraul. **1**, 144–147 (2019)
3. C.A. Hu, F. Lv, S.T. Luo, Z. Yang, R. Zhang, W.Z. Li, Application of the indoor large-length standard device, in *2020 International Conference on Intelligent Control, Measurement and Signal Processing and Intelligent Oil Field*, vol. 1894 (2021), pp. 012017–012023
4. C.A. Hu, W.Z. Li, Y. Zhou, W.B. Du, Y.H. Peng, J.G. Li, Application of the precision industrial measurement technology in geometric measurement, in *2021 7th International Conference on Manufacturing Technology and Applied Materials*, vol. 1885 (2021), pp. 1–6
5. JJF 1242, *Calibration Specification for Laser Tracker 3-Dimensional Measuring System*. (2010)

Chapter 29
Drowning Person Target Intelligent Recognition Method Based on Fusion of Visible Light and Infrared Thermal Imaging

Jianan Luo, Chunxu Li, and Jie Wen

Abstract Inland waterway, especially mountainous waterway channel, has the characteristics of rapid current, different width of river and large change of water level, which brings great risks to the navigation of ships. Once the persons fall into the water, it is difficult to search and rescue. This study aimed at developing a rapid drowning person recognition method, which establishes a deep learning architecture for infrared and visible image fusion. Compared with the traditional convolution network, the coding network is combined with convolution layer, fusion layer and dense block, in which the outputs of each layer are connected with each other, which can be used to obtain more useful features from the source image in the coding process. The target detection experiment of drowning personnel is carried out in the Lancang River. The results show that the method can accurately identify the target under the conditions of insufficient illumination and fast-moving speed, and the recognition rate is 90%.

29.1 Introduction

Every year, people drown all over the world. Tens of thousands of people die of drowning every year due to the accidental drowning of crew, tourists, capsizing and sinking of ships and so on. The main reason is that the current is turbulent and the water area is large, so it is difficult to find and locate the person falling into the water. With the upgrading of computing hardware and the optimization of artificial intelligence algorithms, image processing and detection have been applied to solve all kinds of problems, but the problem of drowning person detection still needs to be solved urgently.

Image fusion is an enhancement technology. Its purpose is to combine the images obtained by different types of sensors to generate images with stronger robustness or

J. Luo (✉) · C. Li · J. Wen
Intelligent Shipping Center, China, Waterborne Transport Research Institute, Beijing, China
e-mail: marinegis@foxmail.com

G. Liu and F. Cen (eds.), *Advances in Precision Instruments and Optical Engineering*, Springer Proceedings in Physics 270, https://doi.org/10.1007/978-981-16-7258-3_29

richer information, so as to facilitate subsequent processing or help decision-making. Infrared and visible image fusion has advantages in many aspects.

First, their signals come from different forms, which provide different aspects of scene information, that is, the visible image captures the reflected light, while the infrared image captures the thermal radiation [1]. Therefore, this combination is more informative than the single-mode signal. Second, infrared and visible images show the inherent characteristics of almost all objects, which can be obtained by relatively simple equipment [2]. Finally, infrared image and visible image have complementary characteristics, so as to produce robust and informative fusion image. Visible images usually have high spatial resolution and considerable detail and light–dark contrast. Therefore, they are in line with human visual perception. However, these images are easily affected by bad conditions, such as insufficient lighting, fog and other bad weather. Infrared images describing the thermal radiation of objects can resist these interferences but usually have low resolution and poor texture. Due to the universality and complementarity of the images used, visible and infrared image fusion technology has a wider application field than other fusion technologies.

The fusion of visible and infrared images is of great significance for personnel detection, especially for people falling into the water. First of all, if only visible light images are used for detection, people are in the fast flowing and unclear river. In addition, the proportion of people exposed to the water when falling is very small, and the people falling into the river are almost integrated with the river, which is difficult to distinguish between the naked eye and the camera [3]. Even excellent detection algorithms are difficult to detect accurately, and the light conditions are good and fashionable, it cannot be detected at night or in heavy fog [4]. The infrared image can distinguish people from the background well. Because the human body has a higher temperature than the river water, the brightness of the human body reflected in the infrared image will be higher than the river water, so it is more prominent [5]. However, due to the low resolution and lack of texture features, the infrared image can only obtain rough contour information [6].

If there are high-temperature objects similar to the shape of the drowning person in the picture, it is easy to cause misjudgment and missing judgment, as shown in Fig. 29.1. If the visible and infrared images are fused, the image not only highlights the human body but also contains certain texture features, which will greatly improve the detection accuracy and recall.

29.2 Image Acquisition and Registration

29.2.1 Image Acquisition

The study uses a dual light camera that can obtain visible and infrared images at the same time. One side is an optical camera and the other is an infrared thermal imaging camera.

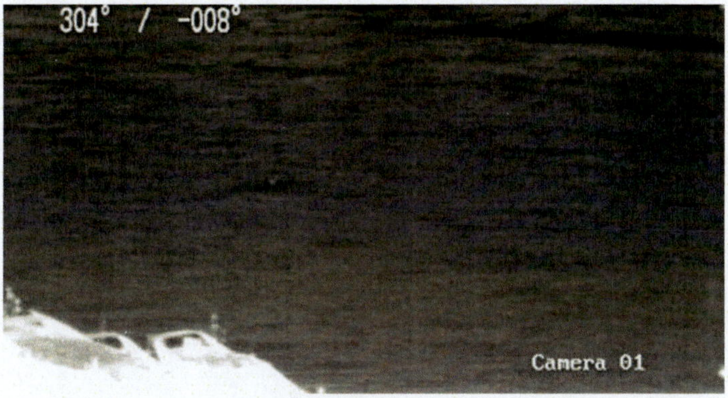

Fig. 29.1 Missing judgment of drowning personnel

29.2.2 Image Registration

Because infrared and visible images are obtained by different sensors, they are usually different in size, perspective and field of view [7]. The above dual light camera will also bring different viewing angles. However, successful image fusion requires strict geometric alignment of the fused image, so it is necessary to register the visible and infrared images before fusion. The registration of infrared image and visible image is a multi-mode registration problem.

For the registration problem here, the feature-based registration method is used. The feature-based method first extracts two groups of salient structures, then determines the correct correspondence between them, and estimates the spatial transformation accordingly, which is then used to align a given image pair.

The first step of feature-based method is to extract robust common features that can represent the original image. Edge information is one of the most commonly used choices in infrared and visible image registration, as shown in Fig. 29.2, because different registration methods can well preserve the size and direction of edge information. Edge mapping can be discretized into point sets. A popular strategy to solve the point matching problem includes two steps: calculating a set of assumed correspondences and then removing outliers through geometric constraints. By calculating feature descriptors at points, the matching between points with too large descriptor

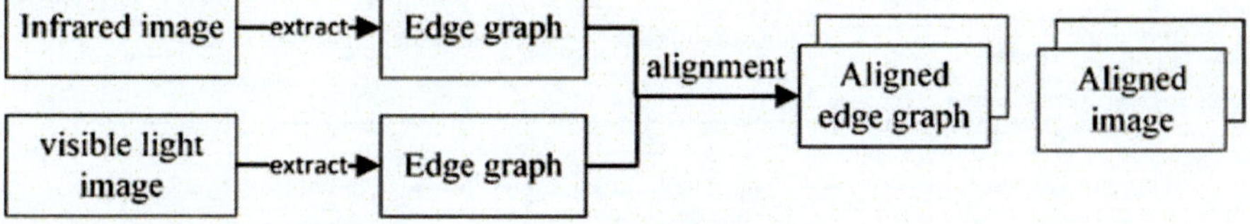

Fig. 29.2 Edge information features and image registration process

difference is eliminated, random sample consistency (RANSAC) is used to remove false matching from the assumed set, and the hypothesis verification method is used to obtain the minimum possible outlier without subset through resampling to estimate the given parameter model.

29.3 Image Fusion

29.3.1 Converged Network

A deep learning architecture for infrared and visible image fusion is adopted. Compared with the traditional convolution network, the coding network is combined with convolution layer, fusion layer and dense block, in which the output of each layer is connected with each other. Using this architecture, we can obtain more useful features from the source image in the coding process, select the appropriate fusion strategy to fuse the features, and finally reconstruct the fused image through the decoder.

As shown in Fig. 29.3, before fusion, the depth features of visible and infrared images are extracted, the first convolution layer extracts rough features, and then three convolution layers (the output of each layer is cascaded into the input of subsequent layers) form dense blocks. Such an architecture has two advantages. First, the size of the filter and the step of convolution operation are 3 respectively $\times 3$ and 1. Using this strategy, the input image can be any size; Second, dense blocks can retain depth features as much as possible in the coding network, and this operation can ensure that all salient features are used in the fusion strategy.

As shown in Fig. 29.4, L1 norm and soft-max operations are applied in the fusion layer. The fusion layer includes a plurality of convolution layers (3×3), the output of the fusion layer will be the input of the convolution layer. This simple and effective architecture is used to reconstruct the final fused image.

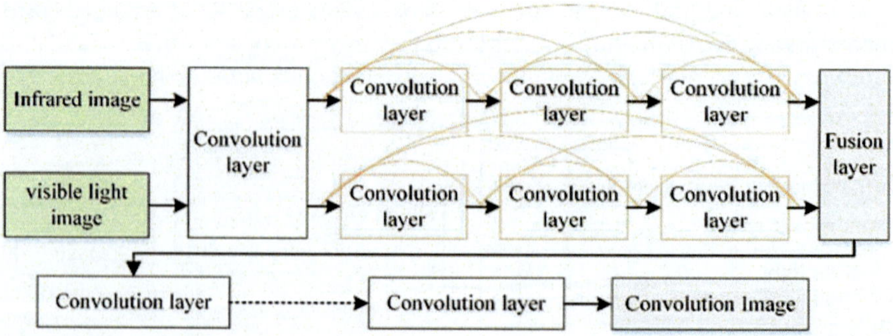

Fig. 29.3 Fusion network structure

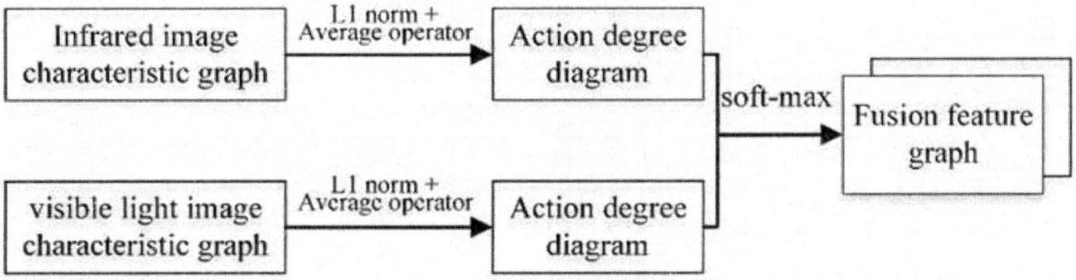

Fig. 29.4 L1 norm and softmax operation are applied in the fusion layer

29.3.2 Loss Function

The loss function of fusion network is composed of pixel loss function L_P and structural similarity loss function L_{ssim} weighting results in:

$$L_p = \|O - I\|^2 \tag{29.1}$$

$$L_{ssim} = 1 - SSIM(O,\ I) \tag{29.2}$$

$$L_{fus} = \lambda L_{ssim} + L_p \tag{29.3}$$

where O and I represent an output image and an input image, respectively. L_p is the Euclidean distance between output O and input I. L_{ssim} represents the structural similarity, which represents the structural similarity of two images. This index is mainly composed of three parts: correlation, brightness loss and contrast distortion. The product of the three components is the evaluation result of the fused image. Since there are three orders of magnitude differences between pixel loss and L_{ssim} loss, in the training phase, the λ set to 1000.

29.4 Detection of Personnel Falling into the Water

29.4.1 Detection Network

The convolutional neural network CNN is used to recognize the target of the drowning person. The central idea of the detection network is to divide the picture into S × S areas. If the center of an object falls on a cell, the cell is responsible for predicting the object. Each cell needs to predict multiple bounding box values, predict a confidence level for each bounding box, and then conduct prediction analysis in units of each cell.

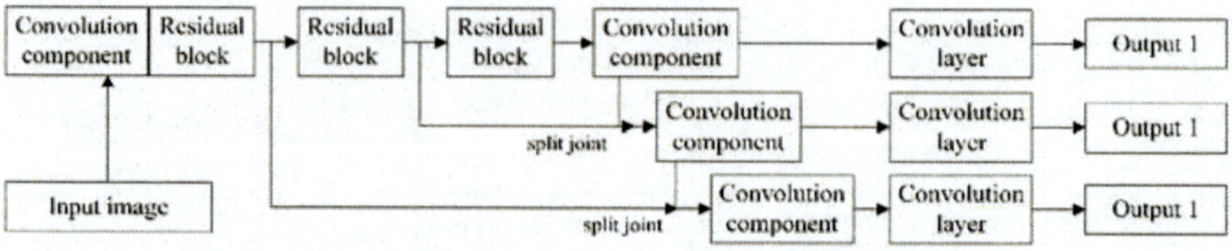

Fig. 29.5 Darknet-53 backbone network

The backbone network adopts the modified darknet-53, as shown in Fig. 29.5. This network has high classification accuracy, fast calculation speed and few network layers. The full connection layer is removed. The network here is a full convolution network, which uses a large number of residual layer hopping connections. In order to reduce the negative gradient effect caused by pooling, the pooling layer is abandoned and the step size of the convolution layer is used to realize downsampling. In this network structure, the convolution with step size of 2 is used for down sampling.

The network outputs three feature maps of different scales, draws lessons from FPN, and uses multi-scale to detect targets of different sizes. The finer the unit, the finer the object can be detected.

Before model training, it is first necessary to make a dataset of fusion images, capture visible and infrared images with a dual light camera, obtain the fusion images through the above registration and fusion process, label the drowning personnel, make a dataset in the format required for training, and select the pretraining model for training, The algorithm model that can identify the drowning person in the visible and infrared fusion image is obtained. Then evaluate the accuracy of the model and optimize it from the aspects of data set and algorithm, so that it can achieve a better recognition effect.

29.4.2 Loss Function

The loss function of the detection model is divided into three parts, L_{box} brought by bounding box, L_{obj} caused by confidence, error L_{cls} brought by category:

$$L_{box} = \lambda_{coord} \sum_{i=0}^{S^2} \sum_{j=0}^{B} \begin{array}{l} 1_{i,j}^{obj} (2 - w_i \times h_i) \Big[(x_i - \hat{x}_i)^2 + (y_i - \hat{y}_i)^2 \\ + (w_i - \hat{w}_i)^2 + \left(h_i - \hat{h}_i\right)^2 \Big] \end{array} \tag{29.4}$$

$$L_{cls} = \lambda_{class} \sum_{i=0}^{S^2} \sum_{j=0}^{B} 1_{i,j}^{obj} \sum_{c \in classes} p_i(c) log(\hat{p}_i(c)) \tag{29.5}$$

$$L_{obj} = \lambda_{nobj} \sum_{i=0}^{S^2} \sum_{j=0}^{B} 1_{i,j}^{nobj} (c_i - \hat{c}_i)^2 + \lambda_{obj} \sum_{i=0}^{S^2} \sum_{j=0}^{B} 1_{i,j}^{obj} (c_i - \hat{c}_i)^2 \quad (29.6)$$

The detection loss function is the sum of the above three errors:

$$L_{dec} = L_{box} + L_{cls} + L_{obj} \quad (29.7)$$

$$L = L_{fus} + L_{dec} \quad (29.8)$$

29.5 Detection Fusion Reverse Guidance

The purpose of common visible infrared image fusion technology is to make the fused image contain as much information of two kinds of images as possible, neither lose the contrast information in the infrared image nor the texture information in the visible image, or make the fused image more in line with the human visual system, Therefore, the loss function of the initial fusion process is defined as the weighted sum of the pixel loss function and the structural similarity loss function.

The focus of this system is to accurately detect the person falling into the water. The result of image fusion is only an intermediate process. Whether it is image fusion or detection process, its optimization should take accurate detection as the ultimate goal. In order to achieve this ultimate goal, the training of image fusion should be modified so that the loss function in the detection process can guide the fusion, and the final detection results will be optimized in the fusion stage.

As shown in Fig. 29.6, first mark the person falling into the water on the registered visible or infrared image. Since the image has been registered and aligned, and the target position after fusion remains unchanged, the mark can be copied to the fusion image as the ground truth. After the fusion image passes through the detection network, the predicted boundary box, classification and confidence are obtained, and the detection error is calculated by comparison with the mark, i.e. L_{dec}, this loss

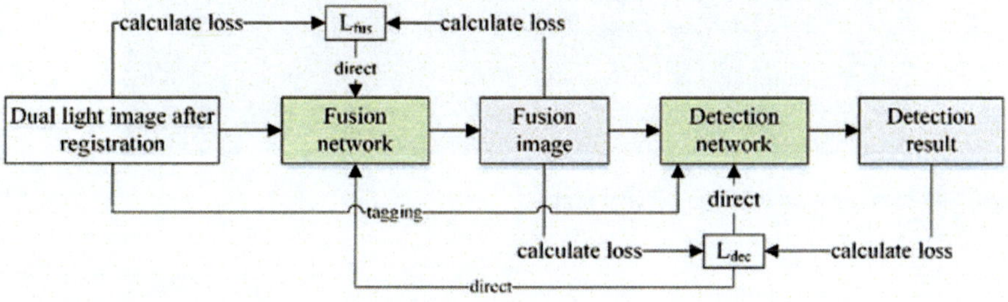

Fig. 29.6 Dual registration image fusion detection

function is used not only to evaluate and optimize the detection network but also to evaluate and optimize the fusion network. It is equivalent to the loss function of the fusion network, and the following corrections are made:

$$L = L_{fus} + L_{dec} \tag{29.9}$$

Furthermore, we developed a waterway operational monitoring system based on this study and demonstrated its application in China's inland waterway, as shown in the figure below. The test shows that the recognition accuracy of the system at night is more than 95% (Figs. 29.7, 29.8, 29.9 and 29.10).

Fig. 29.7 Dynamic identification effect of crew in daytime

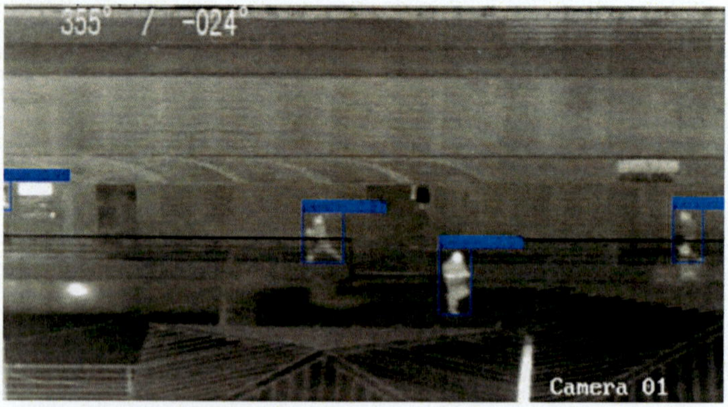

Fig. 29.8 Identification effect of crew at night

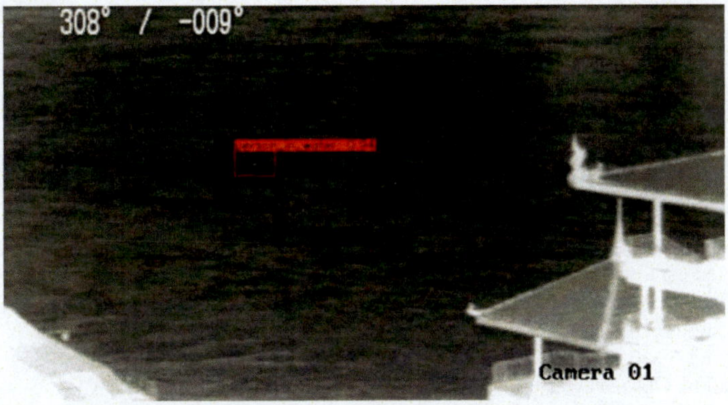

Fig. 29.9 Detection results of people falling into the water under thermal imaging mode (depression angle 9°)

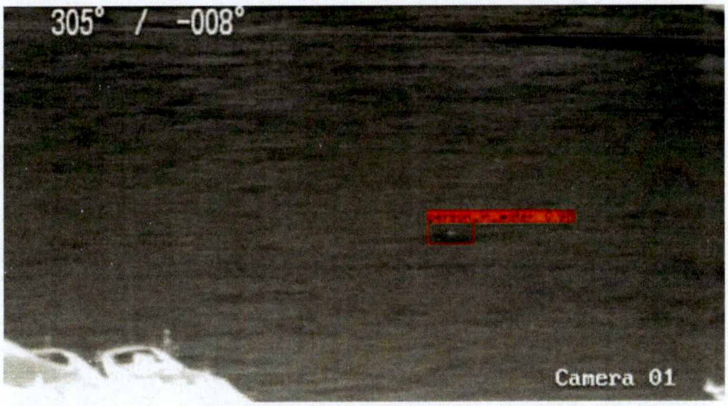

Fig. 29.10 Detection results of people falling into the water under thermal imaging mode (depression angle 8°)

29.6 Conclusion

In this study, a drowning person detection method based on visible light and thermal imaging data fusion is proposed. The method includes image acquisition, image registration, image fusion and target detection. It is a complete, feasible and practical system, which can be used to detect drowning persons for subsequent positioning and rescue.

Acknowledgements This work was supported by China Waterborne Transport Research Institute Fundamental Research Funds under Grant 182102-2021.

References

1. W. Ronggui, W. Jing, Y. Juan, Feature pyramid random fusion network for visible-infrared modality person re-identification. Opto-Electron. Eng. Sichuan **47**, 190669-1–190669-12 (2020)
2. Q. Yonsheng, S. Guobing, W. Yuwu, Research on night environment image enhancement algorithm based on infrared and visible light image fusion. J. Harbin Univ. Commer. (Nat. Sci. Ed.). Haerbin **37**, 422–427 (2021)
3. D. Guipeng, T. Gang, L. Chunying et al., Infrared and visible images fusion based on non-subsampled contourlet transform and guided filter. Acta Armamentarii Beijing (in press)
4. G. Jiamin, L. Aiping, M. Doudou et al., Infrared and visible image fusion combining neighborhood features with IDCSCM. Laser Infrared **50**, 889–896 (2020)
5. Z. Xiaopeng, Feature pyramid random fusion network for visible-infrared modality person re-identification. Opto-Electron. Eng. Sichuan **47**, 190669-1–190669-12 (2020)
6. Y. Yongwu, Application of infrared image and neural network in ship target recognition. Ship Sci. Technol. Beijing **43**, 175–177 (2021)
7. S. Jianhui, Z. Hang, L. Jianju, Infrared and visible light image fusion based on DPN deep learning network. J. Shenyang Ligong Univ. Shenyang **39**, 2023–2027 (2020)

Chapter 30
Integrated Laser In Situ Auxiliary Device

Hongda Li, Guangfeng Shi, Jiyu Gao, Youliang Li, Mingbo Liang, and Siwei Meng

Abstract Hard and brittle materials (germanium, silicon, ceramics, etc.) have the characteristics of good chemical stability, high temperature resistance, corrosion resistance, oxidation resistance, high strength, and hardness. It is widely used in all walks of life. However, due to shortcomings such as high brittleness and low fracture properties, it brings many difficulties to processing. There are mainly brittle fracture zones on the surface of the workpiece and a low material removal rate, which leads to long processing completion time and severe tool wear during processing. In response to the above problems, laser-assisted machining (LAM) is introduced. The surface of the workpiece to be processed by laser radiation is converted from light energy into heat energy to achieve the purpose of softening the material, combined with traditional processing. Laser-assisted machining (LAM) greatly reduces the cutting force, reduces the generation of micro-cracks on the surface of the workpiece, improves the surface finish, and reduces tool wear. In this paper, an integrated laser auxiliary device is designed for laser-assisted technology to realize the integration of "in situ" and "off-site". The device adopts a modular design, which can adjust the laser incident angle and the size of the laser spot radius according to factors such as cutting tools and processing requirements, which not only improves the energy utilization rate of the laser but also facilitates adjustment and installation.

30.1 Introduction

Laser-assisted machining (LAM) is a combination of laser and traditional metal processing technology. The processing area on the workpiece is softened by high-energy laser beam heating to reach the optimal cutting temperature of the material.

H. Li · G. Shi (✉) · J. Gao · Y. Li · S. Meng
Mechanical and Electrical Engineering Department, Changchun University of Science and Technology, Changchun, Jilin, China
e-mail: shiguangfeng@cust.edu.cn

M. Liang
Shenzhen Han's Scanner S&T Co., Ltd., Shenzhen, China
e-mail: bright@hanslaser.com

G. Liu and F. Cen (eds.), *Advances in Precision Instruments and Optical Engineering*, Springer Proceedings in Physics 270, https://doi.org/10.1007/978-981-16-7258-3_30

Cutting at this temperature can make the deformation of the material easier, the cutting force, cutting specific energy, surface roughness, surface damage, and tool wear are reduced, and the processing efficiency and material removal rate are also improved. Tian [1] carried out an experimental study of laser-assisted polishing (Lab) mp35n, annealing, and hardening of AISI 4140. The experimental results show that, compared with the traditional polishing, the workpiece after lab can obtain higher compressive residual stress, improve surface finish, and reduce tool wear. In 2000, Rozzi [2, 3] of Purdue University established the temperature field model of laser-assisted turning Si3N4 ceramic cutting domain for the first time. The finite volume method was used to establish the model, and the three-dimensional transient temperature field of the workpiece was well described. Rozzi successfully predicted the surface temperature of the workpiece, compared it with the actual measured temperature, and verified the effectiveness of the model. In addition, some scholars carried out in-depth research on the composite process test. Dandekar [4] has carried out a laser-assisted cutting test on titanium alloy and tested the surface finish and hardness of the workpiece after the test. The test results show that the hardness of the workpiece after laser-assisted cutting has not changed, and the microstructure of the material is observed under the metallographic microscope, and the grain has no obvious change, but the surface roughness of the workpiece has been improved. Rashid [5] carried out a Lam experiment on ti-10v-2fe-3al alloy material and found that Lam technology can effectively improve the machinability of alloy material. Anderson [6, 7] used Lam technology to process stainless steel P550, Inconel 718, and other materials. From the perspective of cutting specific energy, the effects of cutting parameters on cutting force, tool life, and workpiece surface quality were studied. There is also a research institute that has developed a laser auxiliary device for laser-assisted technology. The laser in situ auxiliary processing device is shown in Fig. 30.1 [8, 9]. The laser is coupled with the tool head of transparent material and acts on the processing area of the processed material.

As shown in Fig. 30.2, the laser off-set-assisted machining device uses the laser to soften the material before the tool processes the workpiece. The existing laser-assisted devices are mono-functional and cannot adjust the laser incidence position flexibly. The laser is susceptible to the effects of cutting fluids and chips during machining, and the machined surface can be affected as a result. Some devices hollow out the inside of the tool holder to ensure beam transmission, which can easily cause instability during machining, and the small welding area between the tool holder

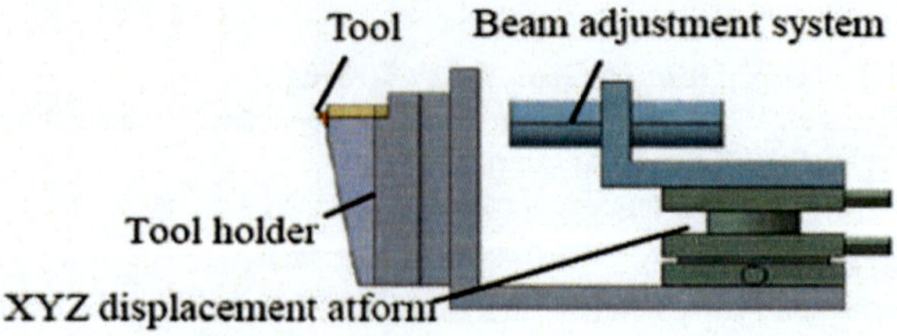

Fig. 30.1 Laser in situ auxiliary processing device

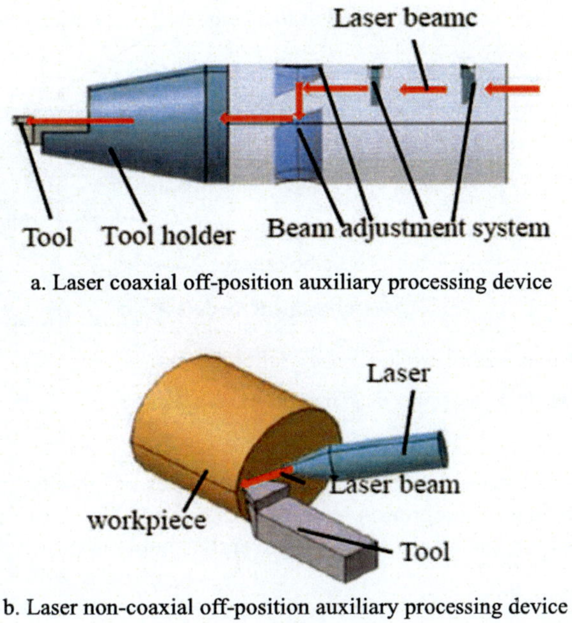

Fig. 30.2 The scheme of laser off-position auxiliary processing device

and tool can lead to insufficient stiffness. From the above, it can be seen that laser-assisted technology plays a vital role in processing difficult-to-machine materials, and a large number of scholars have conducted many in-depth studies. In response to the above phenomena and problems, the research of this paper was carried out. The existing laser-assisted device was optimized and an integrated laser-assisted device was developed.

30.2 The Overall Program

30.2.1 Overall Program Design

In this paper, the integrated laser in situ assist device is designed in three aspects, which are tool holder structure design, laser transmission structure design, and laser optical path design. The tool holder structure adopts a modular design, including three parts: tool holder mounting base plate, tool holder mounting cover plate, and sleeve. The main advantages of this tool holder are flexible installation, easy movement, and the ability to integrate with different types of machine tools. The laser transmission structure consists of three main parts: the convex lens adjustment module, the reflector adjustment module, and the laser adjustment module. The structure of the laser transmission is designed with the advantage of easy adjustment and adjustable laser spot radius size. The laser optical path design includes the optical path design for

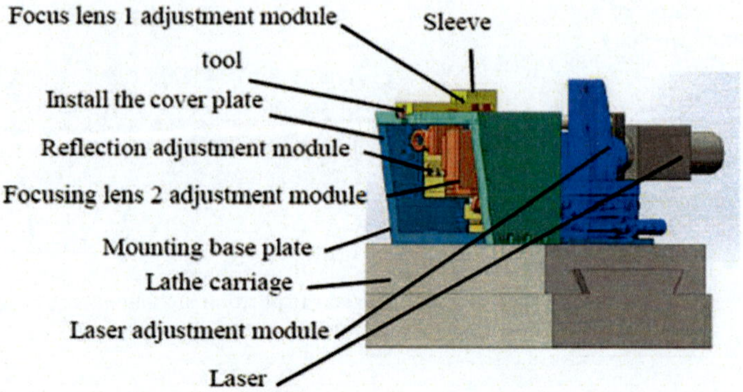

Fig. 30.3 Assembly diagram of laser assist device

two forms of laser incidence. Two types of "in-situ" and one type of "out-of-situ" incidence integration have been achieved. The advantage of laser light path design is high laser utilization. The overall scheme is shown in Fig. 30.3.

30.2.2 Tool Holder Structure Design

The structure design of the base plate for tool holder installation

The tool holder mounting base plate is the foundation of the unit and is connected to the mounting cover to support the overall unit and mount the main components. As shown in Fig. 30.4, the bottom plate of the tool holder is installed with an L-shaped plate, and the cut surface of a certain angle is connected with the closed plate.

Fig. 30.4 Base plate for tool holder mounting

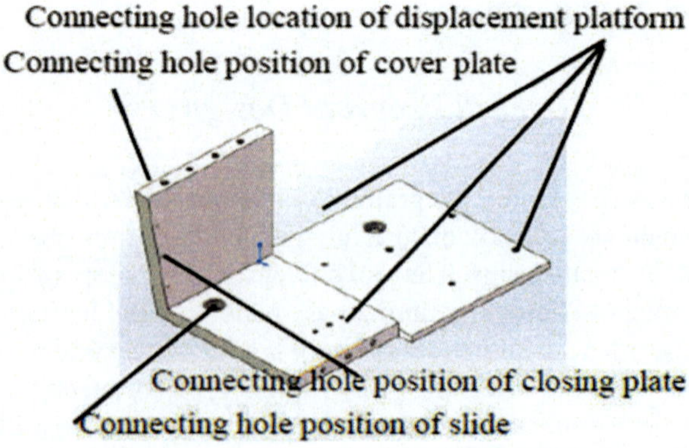

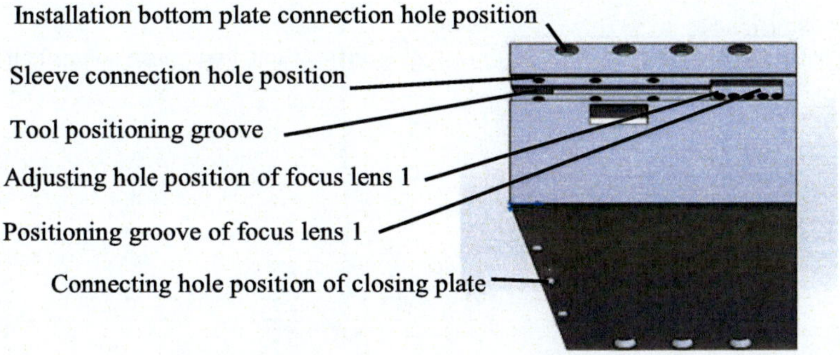

Fig. 30.5 Tool holder mounting cover plate

Tool holder mounting cover plate structure design

The tool holder mounting cover plate should ensure the positioning and clamping of the tool and the positioning and adjustment of the focusing mirror 1. The structure of the tool holder mounting cover is shown in Fig. 30.5, which is machined from an L-shaped plate and cut using the same angle as the tool holder mounting base plate. The mounting holes include the connection holes to the sleeve (M4 bolts), the connection holes to the mounting base plate, the height adjustment holes for the focus mirror 1 and the connection holes for the closure plate. The connection of the sleeve to the mounting cover enables the positioning of the tool for clamping. The distance from the focusing lens to the tool is adjusted by the focusing lens 1 positioning slot.

Sleeve structure design

The function of the sleeve is to position the tool for clamping, to realize the clamping and adjustment of the focusing mirror 1, and to protect the transmission of the laser beam. A diagram of the sleeve structure as shown in Fig. 30.6, with tool positioning slots in the sleeve, connected to the tool holder mounting cover by M4 bolts to enable tool clamping. There is a positioning slot for the focusing mirror 1 inside the sleeve, and the mounting bracket is adjusted in height and horizontal direction by bolts to

Fig. 30.6 Sleeve structure diagram

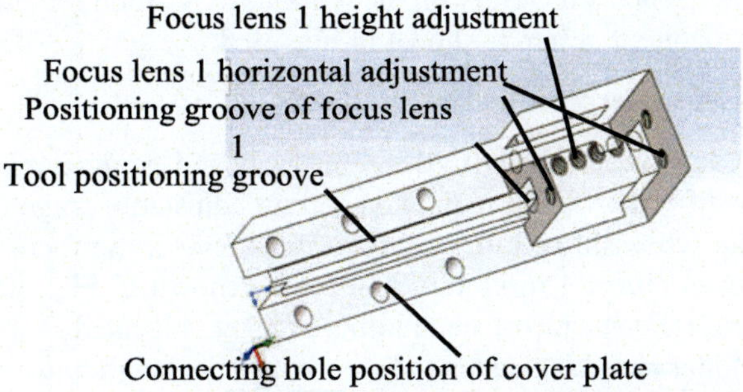

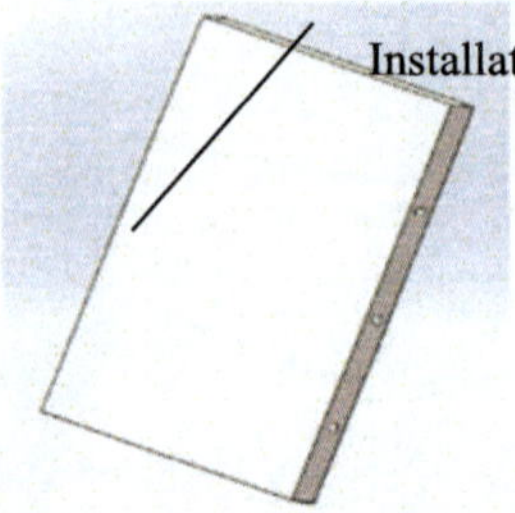

Fig. 30.7 Structural drawing of the closure plate

realize the size and position adjustment of the focusing spot, and the positioning slot of the mounting bracket is marked with a scale for easy adjustment. There is a laser beam channel inside the sleeve, the width of which is smaller than the width of the tool positioning slot. The laser beam can be applied to the workpiece through the tool in situ coaxially, or off situ coaxially.

Closed plate structure design

The closure plate ensures that the cutting fluid and chips do not affect the laser energy during the machining process, achieving high efficiency and quality machining. The closure plate connects the mounting cover and the mounting base plate, as shown in Fig. 30.7.

30.2.3 Laser Transmission Structure Design

Laser adjustment module

Two forms of incidence of the laser beam during laser-assisted processing, depending on the processing requirements, requiring the laser beam to be capable of a wide range of height adjustments. When the laser beam passes through the tool in situ, the change in the position of the incident point affects the laser beam exit position and direction, thus affecting the machining quality. Therefore, the position of the laser beam needs to be precisely adjusted before processing to ensure that the laser energy acts efficiently on the workpiece position required for processing. A specific structure diagram is shown in Fig. 30.8.

Focus mirror adjustment module

When processing with different tools and parameters, the required laser spot size is different, which requires a certain adjustable range for the focusing lens to meet the processing requirements. In this device, the focus mirror 1 is mounted on the focus mirror 1 mounting bracket, as shown in Fig. 30.9, using bolts for fastening (M3). Focus mirror 1 mounting bracket is located in the positioning slot of the tool holder mounting cover and sleeve, and the height position and horizontal position of

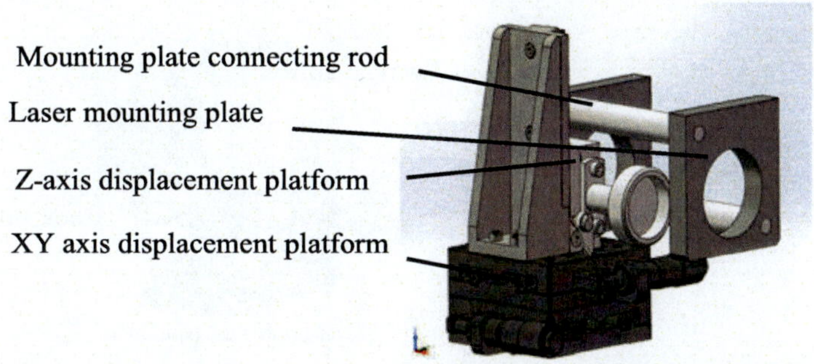

Fig. 30.8 Laser tuning module

Fig. 30.9 Focus lens 1 adjustment module

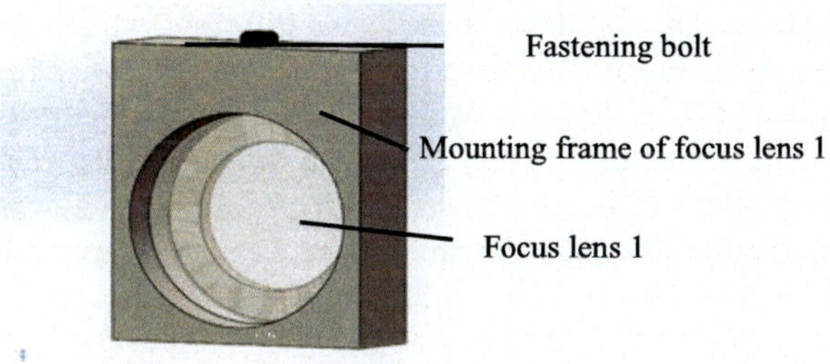

the focusing mirror are adjusted by the adjustment bolts on the mounting cover and sleeve.

The laser beam passing through the focusing mirror 2 is reflected by the reflector and then incident from the bottom of the tool holder at a certain angle. The focusing mirror 2 needs to be angularly adjusted according to the angle of the incident laser beam and also spatially positioned according to the different incident points. The focusing mirror 2 adjustment module is shown in Fig. 30.10.

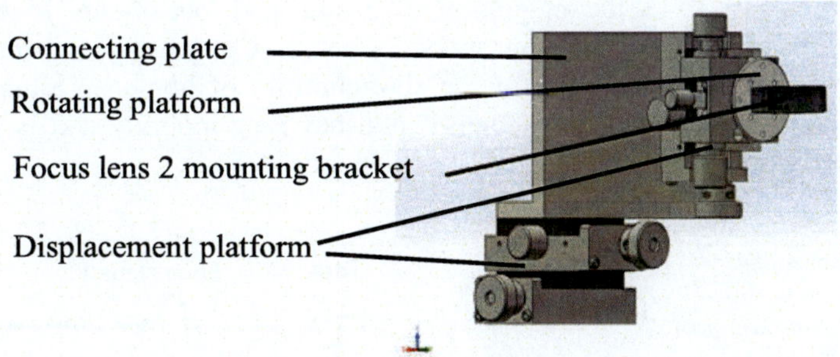

Fig. 30.10 Focusing lens 2 adjustment module

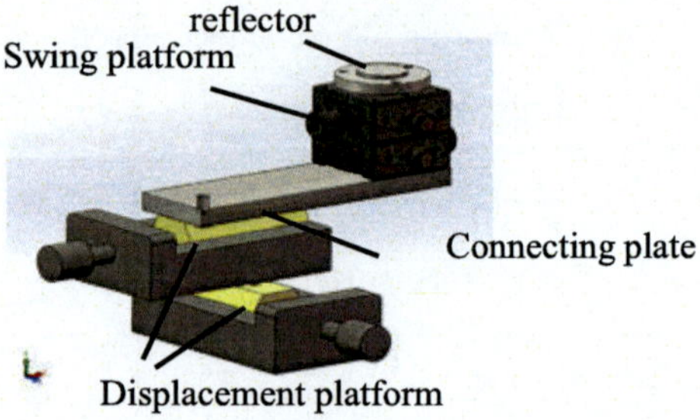

Fig. 30.11 Structure diagram of the reflection adjustment module

Reflector adjustment module

For different parameters tool, different processing requirements, the laser beam incidence angle, incidence position is different, which need to reflect the adjustment module to ensure that the angle and position of the reflector can be easily and flexibly adjusted. The structure of the reflection adjustment module is shown in Fig. 30.11. The reflector has a cooling system to prevent thermal distortion. The device uses air cooling to quickly transfer a large amount of heat brought by the laser beam irradiation to ensure that the reflector can continue to work properly.

30.3 Laser Optical Path Design

Optical path system design, as shown in Fig. 30.12, is a schematic diagram of the beam transmission of the two laser incidence forms of the device optical path system, after the laser beam 1 departs from the laser and is focused by the focusing mirror 1, coaxially and in situ through the tool. After laser beam 2 is emitted by the laser, it is reflected by the reflector, focused by the focusing mirror 2, and then passes through the tool at an angle required by the processing, from the bottom of the tool in situ at an angle.

As shown in Fig. 30.13, the integration of "in situ" and "out-of-situ" is achieved, with two forms of incidence and four modes of action. Compared to existing laser-assisted equipment, it makes up for the shortcomings of homogenization while achieving high utilization of laser energy, and has easy mobility and is suitable for the integration of basically all CNC machine tools.

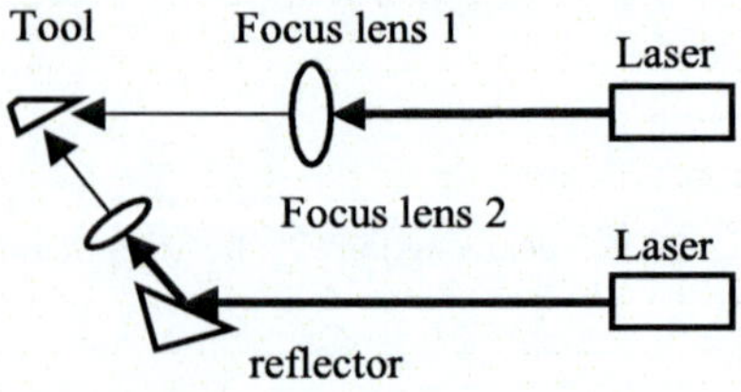

Fig. 30.12 Schematic diagram of laser beam transmission of optical path system

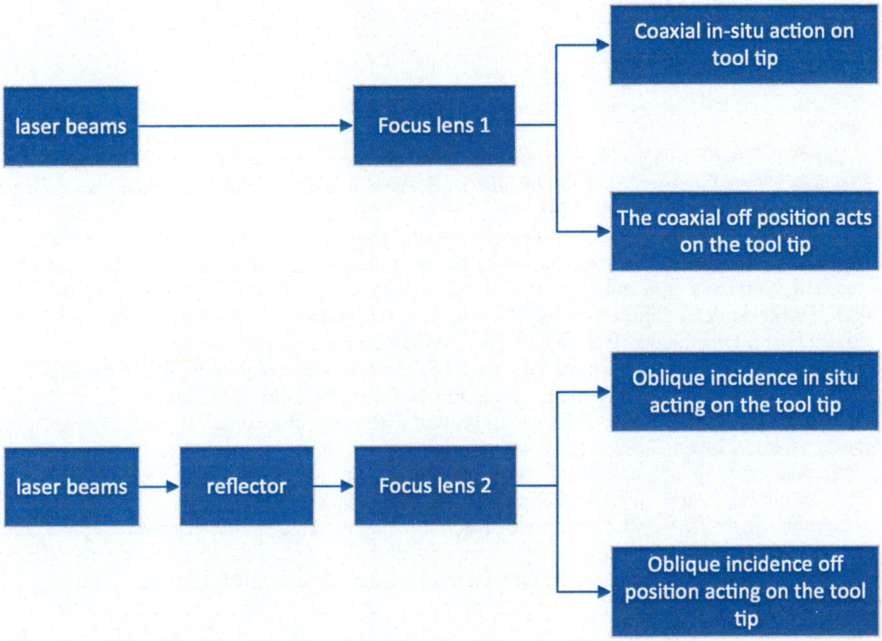

Fig. 30.13 Laser incidence mode

30.4 Conclusion

In this paper, based on the existing laser-assisted device, an integrated laser-assisted turning system is designed and developed to meet the processing requirements of hard and brittle materials. The 3D modeling of the tool holder structure and optical transmission structure in the integrated device is completed by using Soidworks. At the same time, the device has been built and applied to the laboratory test. The device has the following advantages:

(1) The device solves the single defect of the existing laser auxiliary device, and realizes the integration of out of position and in situ auxiliary.
(2) Compared with the laser in situ auxiliary device of a company, the stiffness of the device has been significantly improved.
(3) The laser in situ oblique incidence is realized to avoid the influence of cutting fluid and chips on the optical path.

Acknowledgements This work was supported by the Jilin Province Science Development Fund Project (Approval Number: 20190302123GX, 20180414068GH).

References

1. Y. Tian, C. Shin, Laser-assisted burnishing of metals. Int. J. Mach. Tools Manuf. **47**(1), 14–22 (2007)
2. J.C. Rozzi, F.E. Pfefferkorn, F.P. Incropera, Transient, three-dimensional heat transfer model for the laser assisted machining of silicon nitride: II. Assessment of parametric effects. Int. J. Heat Mass Transf. **43**(8), 1425–1437 (2000)
3. J.C. Rozzi, F.E. Pfefferkorn, F.P. Incropera et al., Transient, three-dimensional heat transfer model for the laser assisted machining of silicon nitride: I. Comparison of predictions with measured surface temperature histories. Int. J. Heat Mass Transf. **43**(8), 1409–1424 (2000)
4. C.R. Dandekar, Y.C. Shin, J. Barnes, Machinability improvement of titanium alloy(Ti-6Al-4V) via and hybrid machining. Int. J. Mach. Tool Manuf. **50**(2), 174–182 (2010)
5. R.A. Rashid, M.J. Bermingham, S. Sun et al., The response of the high strength Ti-10V-2Fe-3Al beta titanium alloy to laser assisted cutting. Precis. Eng. **37**(2), 461–472 (2012)
6. M. Anderson, R. Patwa, Y.C. Shin, Laser-assisted machining of Inconel 718 with an economic analysis. Int. J. Mach. Tools Manuf. **46**(14), 1879–1891 (2006)
7. M.C. Anderson, Y.C. Shin, Laser-assisted machining of an austenitic stainless steel: P550. Proc. Inst. Mech. Eng. Part B J. Eng. Manuf. **220**(12), 2055–2067 (2006)
8. H. Shahinian, D. Zaytsev, J. Navare et al., Micro laser assisted machining (μ-LAM) of precision optics. (2019)
9. D. Kang, J. Navare, S. Yang et al., Observations on ductile laser assisted diamond turning of tungsten carbide. (2019)

Chapter 31
An In-Motion Alignment Method for Laser Doppler Velocimeter-Aided Strapdown Inertial Navigation System

Zhiyi Xiang and Jian Zhou

Abstract With ultra-high velocity measurement accuracy, laser Doppler velocimeter (LDV) is promising to replace the odometer to provide vehicle velocity information in the process of land integrated navigation. This paper investigates the in-motion initial alignment for the LDV-aided strapdown inertial navigation system (SINS). Aiming at the problem that the uncertainty noise in the in-motion alignment process will slow down the alignment speed, a robust square-root unscented quaternion estimator (RSRUSQUE) method is proposed in this paper. The RSRUSQUE method improves the defects of the traditional unscented quaternion estimator (USQUE) method, such as poor noise resistance, slow convergence speed under large misalignment angle and easy to lead to the non-positive definite covariance matrix. This will help to estimate and compensate attitude errors while estimating attitude so as to improve the accuracy of the process model and measurement model, and finally improve the accuracy of attitude estimation. The performance of the proposed scheme is verified by a vehicle field test. The results show that the proposed method has higher alignment accuracy, faster convergence speed and stronger robustness than other compared methods.

31.1 Introduction

Strapdown inertial navigation system (SINS) has been widely used in aerospace, military, industrial and consumer fields because of its self-containment, anti-jamming capability, high sampling rate and good concealment [1]. In recent years, the application of SINS in vehicles has attracted increasing attention. The initial alignment of the SINS is one of the key technologies that affect the accuracy of vehicle navigation, and the accuracy and speed of the initial alignment directly affect the accuracy of the SINS, so the initial alignment of SINS has been a hot research topic.

Z. Xiang · J. Zhou (✉)
College of Advanced Interdisciplinary Studies, National University of Defense Technology, Changsha, China
e-mail: wttzhoujian@163.com

G. Liu and F. Cen (eds.), *Advances in Precision Instruments and Optical Engineering*, Springer Proceedings in Physics 270, https://doi.org/10.1007/978-981-16-7258-3_31

To implement the in-motion initial alignment procedure efficiently, much effort has been devoted to investigating the novel alignment methods. An optimization-based alignment (OBA) method was proposed for GPS-aided high-accuracy SINS [2]. In the OBA method, the attitude matrix is decomposed into two time-varying attitude matrices and a constant attitude matrix. The two time-varying attitude matrices are calculated by the body angular rate and the navigation angular rate, respectively. Finally, the constant attitude matrix is obtained based on the constructed vector observations using Davenport's q-method [3]. However, most of the existing OBA methods do not take the IMU bias into account, which will affect the accuracy of the constructed vector observations. Therefore, the OBA method is not applicable to low-cost SINS. For large initial alignment errors, the unscented-transformation-based unscented Kalman filter (UKF) is used for in-motion initial alignment because of its easy implementation, moderate computational cost and appropriate performance [4]. The UKF in its quaternion application form is proposed to avoid the singularity problem and the norm constraint of the quaternion in practical applications called the unscented quaternion estimator (USQUE) [5]. The USQUE is approved as a method that can replace the OBA method due to its capability of estimating other parameters other than the attitude and handing the noise in the model. USQUE converges slowly with large unknown initial attitude error. As a variant of UKF, it lacks the adaptive ability to system noise and easily leads to a non-positive definite covariance matrix.

Considering that the velocity measurement accuracy of the odometer and DVL is not high enough, our research group has proposed a variety of laser Doppler velocimetry (LDV) structures for vehicle velocity measurement in integrated navigation [6, 7]. As a new type of velocity sensor, LDV has the advantages of high accuracy, rapid dynamic response, non-contact measurement, good directional sensitivity, complete autonomy and good spatial resolution [8]. At present, there are few reports about the application of LDV in the field of navigation. In this paper, a new robust in-motion alignment method, named robust square-root unscented quaternion estimator (RSRUSQUE), is proposed to improve the alignment accuracy and speed.

31.2 In-Motion Alignment for SINS/LDV

In this paper, the local-level navigation frame is denoted as the n frame, the vehicle body frame is denoted as the b frame, the inertial nonrotating frame is denoted as the i frame and the earth frame is denoted as the e frame.

The velocity kinematic equation in the n frame is given by

$$\dot{\boldsymbol{v}}^n = \boldsymbol{C}_b^n \boldsymbol{f}^b - \left(2\boldsymbol{\omega}_{ie}^n + \boldsymbol{\omega}_{en}^n\right) \times \boldsymbol{v}^n + \boldsymbol{g}^n \tag{31.1}$$

where $\boldsymbol{v}^n = \begin{bmatrix} v_E^n & v_N^n & v_U^n \end{bmatrix}^T$ is the ground velocity in n frame, $\boldsymbol{f}^b$ is the specific force in b frame, $\boldsymbol{g}^n$ is the gravity vector in n frame, $(\cdot)\times$ means to solve the antisymmetric matrix, $\boldsymbol{\omega}_{ie}^n$ is the earth rotation rate with respect to the i frame and $\boldsymbol{\omega}_{en}^n$ is the navigational rotating rate in n frame relative to e frame.

According to the coordinate transformation method, it has

$$\dot{\boldsymbol{v}}^n = \left(\boldsymbol{C}_b^n \boldsymbol{v}^b\right)' = \boldsymbol{C}_b^n\left(\dot{\boldsymbol{v}}^b + \boldsymbol{\omega}_{nb}^b \times \boldsymbol{v}^b\right) \tag{31.2}$$

where $\boldsymbol{\omega}_{nb}^b$ is body angular rate with respect to the n frame.

Substituting (31.4) into (31.1) yields

$$\boldsymbol{C}_b^n\left(\dot{\boldsymbol{v}}^b + \left(\boldsymbol{\omega}_{ib}^b + \boldsymbol{\omega}_{ie}^b\right) \times \boldsymbol{v}^b - \boldsymbol{f}^b\right) = \boldsymbol{g}^n \tag{31.3}$$

31.2.1 Process Model

Considering the velocity provided by LDV is accurate, it is necessary to incorporate it into the process model and measurement model. In order to achieve this, denote the initial b frame as an inertial frame. The attitude matrix $\boldsymbol{C}_b^n(t)$ and attitude update equations can be written as

$$\boldsymbol{C}_b^n(t) = \boldsymbol{C}_i^{n(t)} \boldsymbol{C}_{b(t)}^i \tag{31.4}$$

$$\dot{\boldsymbol{C}}_{b(t)}^i = \boldsymbol{C}_{b(t)}^i\left(\boldsymbol{\omega}_{ib}^b \times\right) \tag{31.5}$$

$$\dot{\boldsymbol{C}}_{n(t)}^i = \boldsymbol{C}_{n(t)}^i\left(\boldsymbol{\omega}_{in}^n \times\right) \tag{31.6}$$

where

$$\boldsymbol{\omega}_{in}^n = \boldsymbol{\omega}_{ie}^n + \boldsymbol{\omega}_{en}^n \tag{31.7}$$

Equation (31.6) is the process model of the proposed method.

31.2.2 Measurement Model

Substituting (31.4) into (31.3) and multiplying $\boldsymbol{C}_{n(t)}^i$ on both sides, we get

$$\boldsymbol{C}_{b(t)}^i\left(\dot{\boldsymbol{v}}^b + \left(\boldsymbol{\omega}_{ie}^b + \boldsymbol{\omega}_{ib}^b\right) \times \boldsymbol{v}^b - \boldsymbol{f}^b\right) = \boldsymbol{C}_{n(t)}^i \boldsymbol{g}^n \tag{31.8}$$

Integrating by the time interval of interest on both sides of (31.8), the following can be obtained:

$$\begin{aligned} & \boldsymbol{C}_{b(t)}^{i}\boldsymbol{v}^{b}(t)-\boldsymbol{v}^{b}(0)+\int_{0}^{t}\boldsymbol{C}_{b(\tau)}^{i}(\boldsymbol{\omega}_{ie}^{b}\times\boldsymbol{v}^{b})d\tau-\int_{0}^{t}\boldsymbol{C}_{b(\tau)}^{i}\boldsymbol{f}^{b}d\tau \\ & =\boldsymbol{C}_{n(t)}^{i}\boldsymbol{C}_{n(0)}^{n(t)}\int_{0}^{t}\boldsymbol{C}_{n(\tau)}^{n(0)}\boldsymbol{g}^{n}d\tau \end{aligned} \tag{31.9}$$

Define two vectors as

$$\begin{cases} \boldsymbol{\alpha}(t)=\boldsymbol{C}_{b(t)}^{i}\boldsymbol{v}^{b}(t)-\boldsymbol{v}^{b}(0)+\int_{0}^{t}\boldsymbol{C}_{b(\tau)}^{i}(\boldsymbol{\omega}_{ie}^{b}\times\boldsymbol{v}^{b})d\tau-\int_{0}^{t}\boldsymbol{C}_{b(\tau)}^{i}\boldsymbol{f}^{b}d\tau \\ \boldsymbol{\beta}(t)=\boldsymbol{C}_{n(0)}^{n(t)}\int_{0}^{t}\boldsymbol{C}_{n(\tau)}^{n(0)}\boldsymbol{g}^{n}d\tau \end{cases} \tag{31.10}$$

The measurement model can be given by

$$\boldsymbol{\alpha}(t)=\boldsymbol{C}_{n(t)}^{i}\boldsymbol{\beta}(t) \tag{31.11}$$

31.3 Proposed Robust Square-Root Unscented Quaternion Estimator Algorithm

In this section, an RSRUSQUE algorithm based on the square-root form of an unscented quaternion estimator with adaptive measurement noise covariance matrix is proposed for vehicle in-motion initial alignment.

The direct implementation of a UKF with a quaternions-based state is not suitable because the quaternion estimate is determined using the weighted quaternions averaging operation. Thus, no guarantees can be made that the quaternion will have a unit norm. To represent an attitude error quaternion preserving the constraint of quaternion propagation, the quaternion is used for attitude propagation, and the unconstrained three-component vectors of generalized Rodrigues parameters are used for filtering and local attitude error representation.

Denote the error of the attitude error quaternion by $\delta\boldsymbol{q}=\left[\delta q_0,\delta\boldsymbol{q}_{1:3}^T\right]^T=\left[\delta q_0,\delta\boldsymbol{\rho}^T\right]^T$, and the corresponding GRP representation $\delta\mathfrak{R}$ is given by

$$\delta\mathfrak{R}=f\frac{\delta\boldsymbol{\rho}}{a+\delta q_0} \tag{31.12}$$

where a is a parameter from 0 to 1 and f is a scale factor. The GRP is used to place the singularity of the attitude representation in a certain angle range, and different combinations of a and f have different physical meanings. For example, when $a=0$

and $f = 1$, (31.12) gives the Gibbs vector, and when $a = f = 1$, (31.12) gives the standard vector of modified Rodrigues parameters.

The inverse transformation from $\delta\mathfrak{R}$ to $\delta\boldsymbol{q}$ is given by

$$\begin{cases} \delta q_0 = \dfrac{-a\|\delta\mathfrak{R}\|^2 + f\sqrt{f^2 + (1-a)^2\|\delta\mathfrak{R}\|^2}}{f^2 + \|\delta\mathfrak{R}\|^2} \\ \delta\boldsymbol{\rho} = f^{-1}(a + \delta q_0)\delta\mathfrak{R} \end{cases} \tag{31.13}$$

In RSRUSQUE, the filtering state is defined as

$$\hat{\boldsymbol{X}}_k = \left[\delta\mathfrak{R}_k^T\ \hat{\boldsymbol{X}}_k^{eT}\right]^T \tag{31.14}$$

where $\hat{\boldsymbol{X}}_k^e$ are the components of the state besides the quaternion. In this paper, considering the high precision SINS and LDV adopted and the short alignment time, the other parameters are not estimated to reduce the calculation amount.

Write the measurement model (31.13) as follows:

$$\boldsymbol{y}_k = \boldsymbol{\alpha}(k) - \boldsymbol{C}_{n(k)}^i\boldsymbol{\beta}(k) = \boldsymbol{h}(\boldsymbol{X}_k) + \boldsymbol{v}_k \tag{31.15}$$

where $\boldsymbol{X}_k$ is the ideal value of the filtering state at time instant k, and $\boldsymbol{v}_k$ is the measurement noise.

The state estimation at time instant k is (31.14), and the corresponding covariance is $\boldsymbol{P}_{x,k}$. The RSRUSQUE algorithm for initial alignment is described as follows.

31.3.1 Time Update

The generated sigma points and the weights corresponding to the expectation and covariance matrix are given by

$$\chi_k(i) = \begin{cases} \hat{\boldsymbol{X}}_k, \quad i = 0 \\ \hat{\boldsymbol{X}}_k + \left[\sqrt{(n+\lambda)}\boldsymbol{S}_{x,k}\right]_i, \quad i = 1, 2, \ldots, n \\ \hat{\boldsymbol{X}}_k - \left[\sqrt{(n+\lambda)}\boldsymbol{S}_{x,k}\right]_i, \quad i = n+1, n+2, \ldots, 2n \end{cases} \tag{31.16}$$

$$W^m(i) = \begin{cases} \dfrac{\lambda}{n+\lambda}, \quad i = 0 \\ \dfrac{1}{2(n+\lambda)}, \quad i = 1, \ldots, 2n \end{cases} \tag{31.17}$$

$$W^c(i) = \begin{cases} \dfrac{\lambda}{n+\lambda} + 1 - \sigma^2 + \varsigma, & i = 0 \\ \dfrac{1}{2(n+\lambda)}, & i = 1, \ldots, 2n \end{cases} \tag{31.18}$$

where $\boldsymbol{S}_{x,k}$ is the lower triangular matrix obtained by using the Cholesky factorization of $\boldsymbol{P}_{x,k}$. n is the dimension of the state $\hat{\boldsymbol{X}}_k$, $\lambda = \sigma^2(n+\kappa) - n$. σ is a scale factor, usually a small quantity greater than zero, κ is a tuning parameter which is usually set to 0 and $3 - n$ to capture some higher-order information of the distribution, ς is used to incorporate prior information on the probability density function of the states.

$\boldsymbol{\chi}_k(i)$ can be divided as follows:

$$\boldsymbol{\chi}_k(i) = \left[\boldsymbol{\chi}_k^{\delta\mathfrak{R}}(i)^T \ \boldsymbol{\chi}_k^{e}(i)^T\right]^T \tag{31.19}$$

The quaternion error corresponding to $\boldsymbol{\chi}_k^{\delta\mathfrak{R}}(i)$ is given by

$$\boldsymbol{\chi}_k^{\delta q}(i) = \left[\delta q_{k,0}(i) \ \delta\boldsymbol{\rho}_k(i)^T\right]^T \tag{31.20}$$

which can be calculated by (31.13).

Denote the quaternion-based sigma points by multiplying the error quaternion by the current attitude quaternion

$$\boldsymbol{\chi}_k^{q}(i) = \boldsymbol{\chi}_k^{\delta q}(i) \otimes \boldsymbol{q}_k \tag{31.21}$$

The sigma points obtained by (31.21) propagate forward through process model (31.6), and the propagated quaternion error is calculated by the following equation:

$$\boldsymbol{\chi}_{k+1|k}^{\delta q}(i) = \boldsymbol{\chi}_{k+1|k}^{q}(i) \otimes \left[\boldsymbol{\chi}_{k+1|k}^{q}\right]^{-1} \tag{31.22}$$

The predicted GRP sigma points $\boldsymbol{\chi}_{k+1|k}^{\delta\mathfrak{R}}(i)$ corresponding to $\boldsymbol{\chi}_{k+1|k}^{\delta q}(i)$ can be calculated by (31.12). The propagated sigma points of the state can be determined as

$$\boldsymbol{\chi}_{k+1|k(i)} = \left[\boldsymbol{\chi}_{k+1|k}^{\delta\mathfrak{R}}(i)^T \ \boldsymbol{\chi}_{k+1|k}^{e}(i)^T\right]^T \tag{31.23}$$

The corresponding state prediction and covariance matrix can be calculated, respectively, as

$$\hat{\boldsymbol{X}}_{k+1|k} = \sum_{i=0}^{2n} W^m(i)\boldsymbol{\chi}_{k+1|k}(i) \tag{31.24}$$

$$S_{x,k+1|k} = qr\left\{\left[\sqrt{W^c(1:2n)}\left(\chi_{k+1|k}(1:2n) - \hat{X}_{k+1|k}\right)\ \sqrt{Q_k}\right]^T\right\} \tag{31.25}$$

$$S_{x,k+1|k} = cholupdate\left\{S_{x,k+1|k},\ \sqrt{|W^c(0)|}\left(\chi_{k+1|k}(0) - \hat{X}_{k+1|k}\right),\ \mathrm{sgn}\left(W^c(0)\right)\right\} \tag{31.26}$$

31.3.2 Measurement Update

Similar to (31.16) and (31.19)–(31.21), the new quaternion-based sigma points $\chi^{*q}_{k+1|k}(i)$ can be obtained, and define a new set of sigma points as

$$\gamma^{*}_{k+1|k}(i) = \left[\chi^{*q}_{k+1|k}(i)^T\ \ \chi^{*e}_{k+1|k}(i)^T\right]^T \tag{31.27}$$

The sigma points in (31.27) are propagated directly through the measurement model (31.15), and the predicted sigma points are denoted as $Z_{k+1|k}(i)$.

The predicted mean of measurement, the covariance matrix of measurement and the cross-covariance matrix of the state and measurement are calculated, respectively, as

$$\hat{Y}_{k+1} = \sum_{i=0}^{2n} W^m(i) Z_{k+1|k}(i) \tag{31.28}$$

$$S_{y,k+1} = qr\left\{\left[\sqrt{W^c(1:2n)}\left(Z_{k+1|k}(1:2n) - \hat{Y}_{k+1}\right)\ \sqrt{R_k}\right]^T\right\} \tag{31.29}$$

$$S_{y,k+1} = cholupdate\left\{S_{z,k+1},\ \sqrt{|W^c(0)|}\left(Z_{k+1|k}(0) - \hat{Y}_{k+1}\right),\ \mathrm{sgn}\left(W^c(0)\right)\right\} \tag{31.30}$$

$$P_{xy,k+1} = \sum_{i=0}^{2n} W^c(i)\left\{\left(\chi^{*}_{k+1|k}(i) - \hat{X}_{k+1|k}\right) \times \left(Z_{k+1|k}(i) - \hat{Y}_{k+1}\right)^T\right\} \tag{31.31}$$

The innovation vector is

$$e_{k+1} = y_{k+1} - \hat{Y}_{k+1} \tag{31.32}$$

The measurement noise covariance matrix R_{k+1} can be expressed as

$$R_{k+1} = (1 - \eta_{k+1})R_k + \eta_{k+1}\left(e_{k+1}e_{k+1}^T - S_{z,k+1}S_{z,k+1}^T + R_k\right) \tag{31.33}$$

$$\eta_{k+1} = \frac{\eta_k}{\eta_k + c} \tag{31.34}$$

where $\eta_0 = 1$, and $0 < c < 1$ is called a fading factor, which is usually $c = 0.9 - 0.999$.

Using (31.29)–(31.31), the update equations of the state vector and covariance matrix are determined by

$$\hat{\boldsymbol{X}}_{k+1} = \hat{\boldsymbol{X}}_{k+1|k} + \boldsymbol{P}_{xy,k+1}\left(\boldsymbol{S}_{y,k+1}\boldsymbol{S}_{y,k+1}^T\right)^{-1}\boldsymbol{e}_{k+1} \tag{31.35}$$

$$\boldsymbol{U}_{k+1} = \boldsymbol{P}_{xy,k+1}\left(\boldsymbol{S}_{y,k+1}\boldsymbol{S}_{y,k+1}^T\right)^{-1}\boldsymbol{S}_{y,k+1} \tag{31.36}$$

$$\boldsymbol{S}_{x,k+1} = cholupdate\left(\boldsymbol{S}_{x,k+1|k}, \boldsymbol{U}_{k+1}, -1\right) \tag{31.37}$$

31.3.3 Attitude Update

The updated status vector is expressed as

$$\hat{\boldsymbol{X}}_{k+1} = \left[\delta\mathfrak{R}_{k+1}^T \ \boldsymbol{X}_{k+1}^{eT}\right]^T \tag{31.38}$$

The quaternion error corresponding to $\delta\mathfrak{R}_{k+1}$ is given by

$$\delta\boldsymbol{q}_{k+1} = \left[\delta q_{k+1,0} \ \delta\boldsymbol{\rho}_{k+1}^T\right]^T \tag{31.39}$$

which can be calculated according to Formula (31.13).

The attitude quaternion is updated through

$$\boldsymbol{q}_{k+1} = \delta\boldsymbol{q}_{k+1} \otimes \boldsymbol{\chi}_{k+1|k}^{*q} \tag{31.40}$$

Reset $\delta\mathfrak{R}_{k+1}$ to zeros and go to the next filtering cycle.

31.4 Vehicle-Mounted Field Test

To verify the performance of the proposed RSRUSQUE, the vehicle tests were carried out in Changsha. Figure 31.1 shows the test equipment, which includes a self-developed inertial measurement unit (IMU), a dual-antenna GPS receiver, a

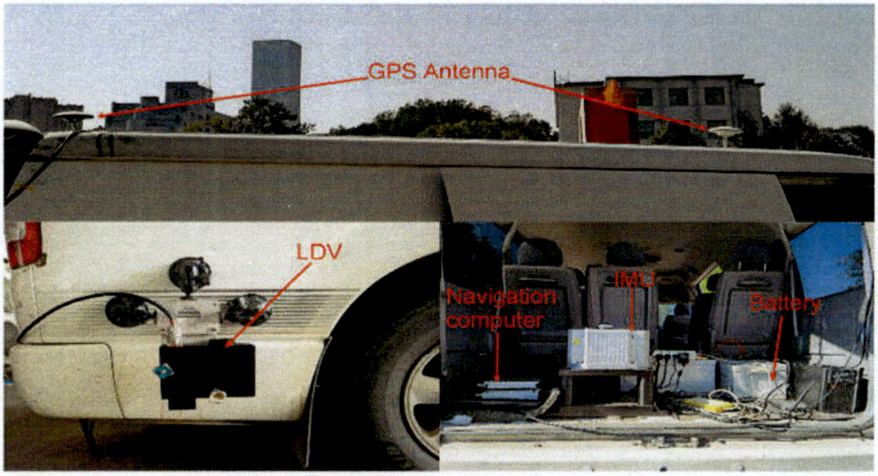

Fig. 31.1 Installation diagram of the experimental system

navigation computer and a self-made LDV. The movement trajectories of vehicles and the outputs of LDV are shown in Figs. 31.2 and 31.3, respectively.

In the vehicle test, the initial value of attitude quaternion is all set as $[1\ 0\ 0\ 0]^T$. For RSRUSQUE, the initial covariance matrix of attitude error is set as diag($[3°\ 3°\ 12°]^T)^2$. The attitude errors are shown in Figs. 31.4, 31.5 and 31.6.

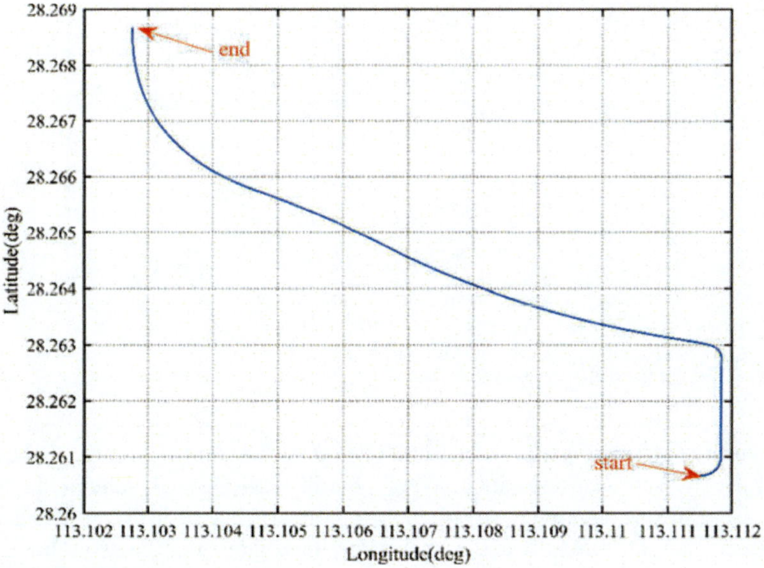

Fig. 31.2 Trajectory of the vehicle in the field test

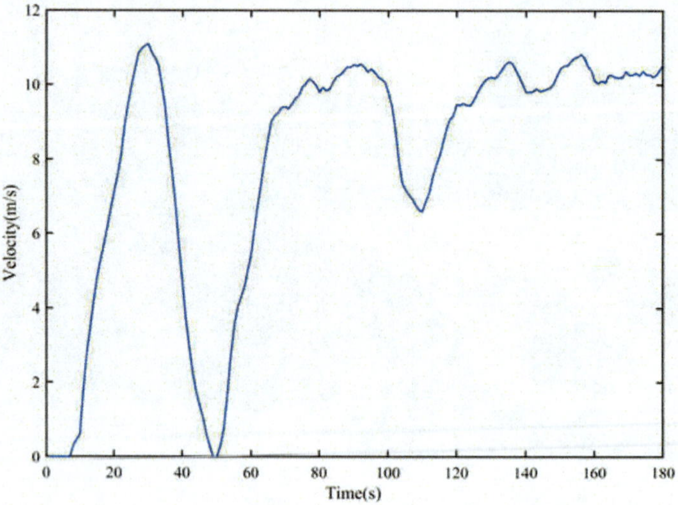

Fig. 31.3 Velocity curve of LDV output

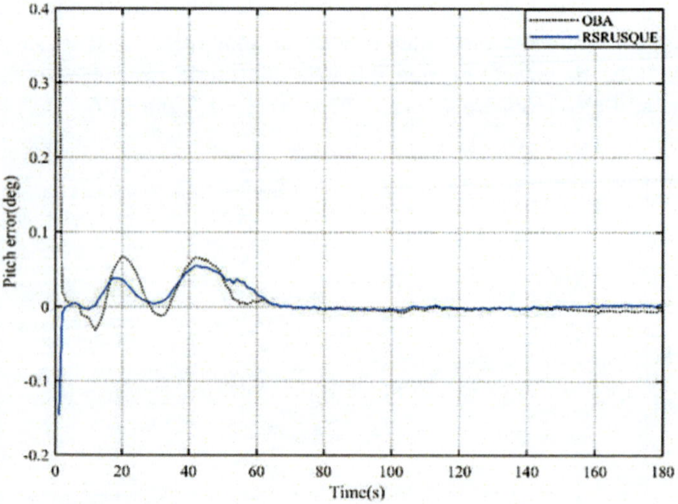

Fig. 31.4 Pitch angle error

As can be seen from Figs. 31.4, 31.5 and 31.6, at the first 60 s of alignment, the pitch error of the two schemes does not converge, but the RSRUSQUE scheme fluctuates less. At 61–180 s of alignment, the pitch error convergence of RSRUSQUE is more gentle than OBA. Different from the pitch error, the roll error of the two schemes has close convergence speed and accuracy. As for heading error, it can be clearly seen from Fig. 31.6 that RSRUSQUE has a faster convergence speed, higher accuracy and

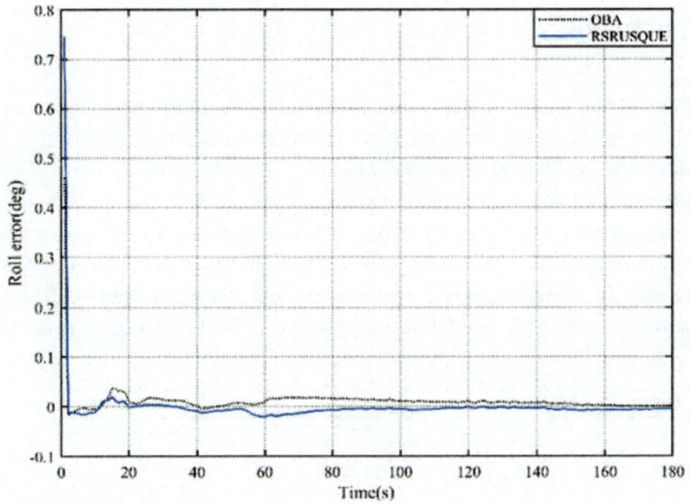

Fig. 31.5 Roll angle error

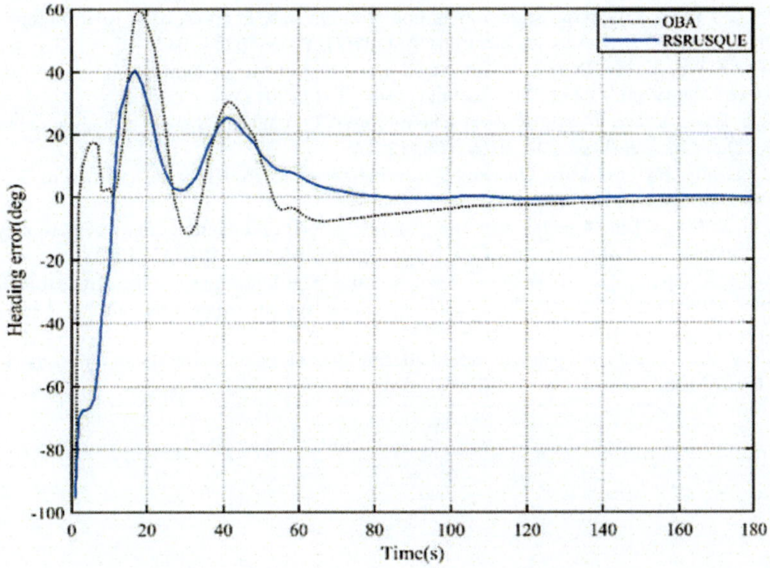

Fig. 31.6 Heading angle error

smaller fluctuation than OBA. In conclusion, compared with the mainstream OBA method, RSRUSQUE has stronger robustness, faster convergence rate and higher alignment accuracy.

31.5 Conclusion

This paper proposes a new fast in-motion initial alignment method for LDV-aided SINS. First, the process model and measurement model for LDV-aided SINS in-motion alignment is derived through skillfully attitude matrix decomposition and velocity kinematic equation reconstruction. In order to improve the robustness of the in-motion alignment process and pursue faster alignment speed and accuracy, the RSRUSQUE is proposed. One group of vehicle field test was carried out to evaluate the performance of RSRUSQUE. Experimental results show that RSRUSQUE has higher alignment accuracy, faster convergence speed and stronger robustness than OBA.

References

1. Q. Wang, C. Gao, J. Zhou, G. Wei, X. Nie, X. Long, Two-dimensional laser Doppler velocimeter and its integrated navigation with a strapdown inertial navigation system. Appl. Opt. **57**(13), 3334–3339 (2018)
2. Y. Wu, X. Pan, Velocity/position integration formula part I: application to in-flight coarse alignment. IEEE Trans. Aerosp. Electron. Syst. **49**(2), 1006–1023 (2013)
3. M. Wu, Y. Wu, X. Hu, D. Hu, Optimization-based alignment for inertial navigation systems: theory and algorithm. Aerosp. Sci. Technol. **15**(1), 1–17 (2011)
4. W. Li, J. Wang, L. Lu, W. Wu, A novel scheme for DVL-aided SINS in-motion alignment using UKF techniques. Sensors **13**(1), 1046–1063 (2013)
5. J.L. Crassidis, F.L. Markley, Unscented filtering for spacecraft attitude estimation. J. Guid. Control Dyn. **26**(4), 536–542 (2003)
6. J. Zhou, X. Nie, J. Lin, A novel laser Doppler velocimeter and its integrated navigation system with strapdown inertial navigation. Opt. Laser Technol. **64**(13), 319–323 (2014)
7. Q. Wang, X. Nie, C. Gao, J. Zhou, G. Wei, X. Long, Calibration of a three-dimensional laser Doppler velocimeter in a land integrated navigation system. Appl. Opt. **57**(29), 8566–8572 (2018)
8. X. Nie, J. Zhou, Pitch independent vehicle-based laser Doppler velocimeter. Opt. Lasers Eng. **131**, 106072 (2020)

Chapter 32
Imaging and Focusing Through Scattering Medium Based on Reflection Matrix Optical Coherence Tomography

Jing Cao and Pinghe Wang

Abstract Multiple scattering inside the random medium limits the imaging depth of optical coherence tomography (OCT) to 1–2 mm, as well as the degree of focus at the deep imaging depth. In this paper, by combining the concept of matrix measurement with a wide-field optical coherence tomography, we have done two aspects of work. The first one is for deeper imaging depth. By reconstructing the huge reflection matrix of the sample and then applying a time-reversal operation to it, we successfully filter out the single scattered light for imaging at the depth of 15 times of the scattering mean free path (SMFP). Since the imaging depth of conventional OCT is 6–7 times of the SMFP, our proposed reflection matrix optical coherence tomography (RMOCT) is about one time deeper than the conventional OCT. The second part of the work is a high-speed wavefront shaping (WFS) method based on a one-time in-and-out complex light field analysis. With the help of a phase-only spatial light modulator, we realize the light focusing through a random medium is ~113 ms. It is about three times faster than the iterative feedback wavefront shaping method. We believe that our work might pave the way to apply WFS to optical imaging methods and open new methods toward deeper imaging through a scattering medium.

32.1 Introduction

Because of multiple scattering, the propagation of light in random media poses a fundamental problem in optical imaging technologies. The challenge becomes much more difficult when considering obtaining mesoscopic-level resolution in a scattering media. Most of the optical imaging methods still rely on the collection of ballistic light to construct the sample information, such as conventional microscopy imaging. They provide diffraction-limited resolution but image penetration is limited to a superficial level. This is because the amount of ballistic (or single scattering) light decreases exponentially with the penetration depth. At present, one method to overcome this is to take advantage of diffuse light reflected back from the layers

J. Cao (✉) · P. Wang
School of Optoelectronic Science and Engineering, University of Electronic Science and Technology of China, Chengdu 610054, China

G. Liu and F. Cen (eds.), *Advances in Precision Instruments and Optical Engineering*, Springer Proceedings in Physics 270, https://doi.org/10.1007/978-981-16-7258-3_32

deep inside the tissue, such as diffuse optical tomography, acousto-optic imaging, and photoacoustic imaging. But their resolutions are severely degraded because they rely on diffused light to form an image. In modern optical society, there are two techniques to discriminate single scattering light from the multiple scattering light for high-resolution imaging capacity. The first option is confocal grating technology, such as confocal microscopy [1] or two-photon microscopy [2–4], which spatially rejects unwanted multiple scattering light based on the spatial filter. Another option is the coherence grating method, such as optical coherence tomography [5, 6], which extracts single scattering light based on the path length they travel in tissue. The combining of these two gates together, optical coherence microscopy [7, 8], can be stronger than either gate individually, achieving greater image penetration in scattering medium without compromising the imaging resolution.

Either for optical imaging or focusing purpose, a lot of remarkable works have been done to break through the limitation caused by multiple scattering. The light manipulation method was proposed [9], known as wavefront shaping, to focus light through a highly scattering medium with the help of a spatial light modulator [10–12]. There have been many remarkable signs of progress such as the investigation of transmission matrix to describe the light propagation in a complex medium [13–16] and the presence of reflection matrix method to imaging through highly scattering tissue [17–20].

In this paper, we have demonstrated a reflection matrix OCT, which has the imaging ability of 15 times of the SMFP. At the same time, by using the lock-in detection instead of the four-step phase-shift method, the acquisition speed of each scanning point has been accelerated to 0.37 s. It is about ten times faster than the previous four-step phase-shift method (4.15 s for each scanning point). Then, we configured the sample arm to a transmitted mode for light focusing purposes. Based on a high-speed and high-sensitivity complex light field reconstruction technology, we realize the light focusing through the sample at the speed of ~113 ms.

32.2 Methods

32.2.1 Experimental Setup

The experimental setup to record the sample's reflection matrix is depicted in Fig. 32.1. A femtosecond laser with center wavelength $\lambda_0 = 790$ nm and spectrum bandwidth $\Delta\lambda = 90$ nm is first collimated and expanded by a pair of lenses. After passing through a beam splitter (BS1), the beam splits into the sample arm and reference arm.

In the reference arm, optical frequency is shifted to $f_0 + f_r$, where f_0 is the center frequency of the laser and f_r is the frequency difference of the two acousto-optic modulators (AOM). In our case, f_r is 40 kHz. After passing through another pair of lenses and pinhole, BS2, PBS2, and a mirror, the reference light is then reflected

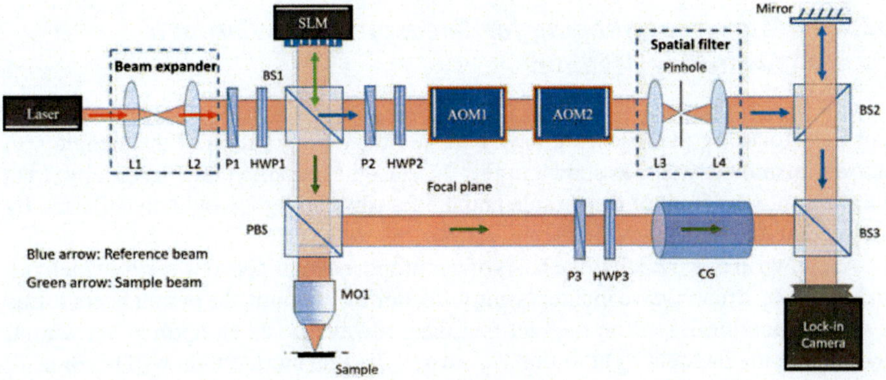

Fig. 32.1 Experimental setup for the measurement of RM

back at BS3. In the sample arm, the phase of the beam is shaped by a spatial light modulator (PLUTO-2, HOLOEYE) operating at phase-only mode, and subsequently transmitted through the BS1 again. After passing through a PBS1 and a QWP oriented at 45°, it focuses onto the sample through a microscope objective (MO). In order to detect the phase of the sample, a piece of compensation glass has been used to match the dispersion introduced by the two AOMs in the reference arm. Finally, the interference signal will be detected by the lock-in camera. The benefit of using this type of camera is that it can directly output two components, in-phase (I) and quadrature (Q), to form the complex light field.

32.2.2 *Time-Reversal Operation for Imaging Through Scattering Medium*

In this setup, we apply a point-to-point scanning strategy to collect the sample information. For each scanning point, there will be a 2D matrix filled with complex light field information. Then, we reshape it into a 1D vector and fill it into one column of the RM. After finishing all the scanning and repeating this filling process, we achieve a huge matrix, the so-called reflection matrix. By using the singular value decomposition and time-reversal operation [21, 22], it is able to restore the imaging of the target beneath the scattering layer.

32.2.3 Wavefront Shaping for Focusing Light Through Scattering Medium

For light focusing purposes, we have changed the configuration of the sample arm into a transmitted mode, as shown in Fig. 32.2. Light from the first microscopy (M1) focused onto the front of the sample and the transmitted light would be collected by the second microscopy (M2).

As shown in Fig. 32.3, it is the basis of our proposed high-speed wavefront method. In Fig. 32.3a, a plane wave incidents onto a scattering medium, the output beam forms a speckle pattern. Based on the fact that there always exists an optimal wavefront, by shaping the incident light to this wavefront, the transmitted light will become all in phase again. Equation (32.1) shows how to calculate this optimal wavefront:

$$O_{in} = E_{in} * E_{out}^{-1} * E_{in} \tag{32.1}$$

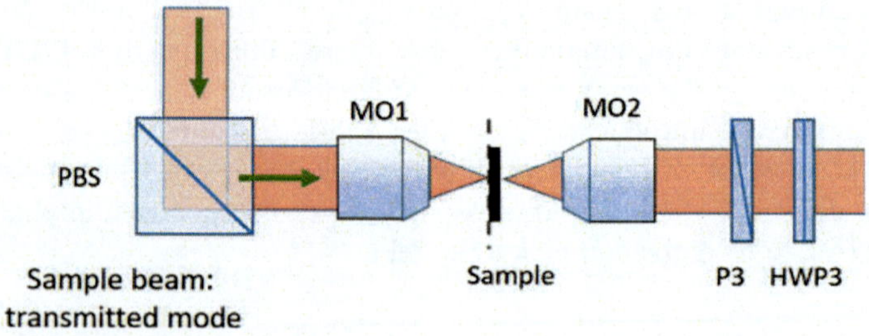

Fig. 32.2 Experimental setup for focusing light

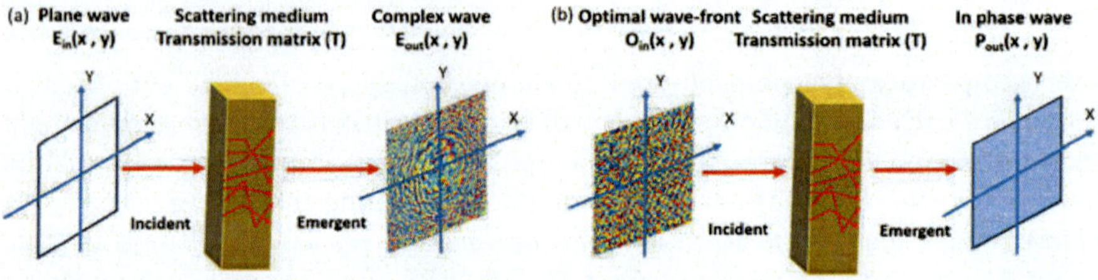

Fig. 32.3 Principle of high-speed wavefront shaping

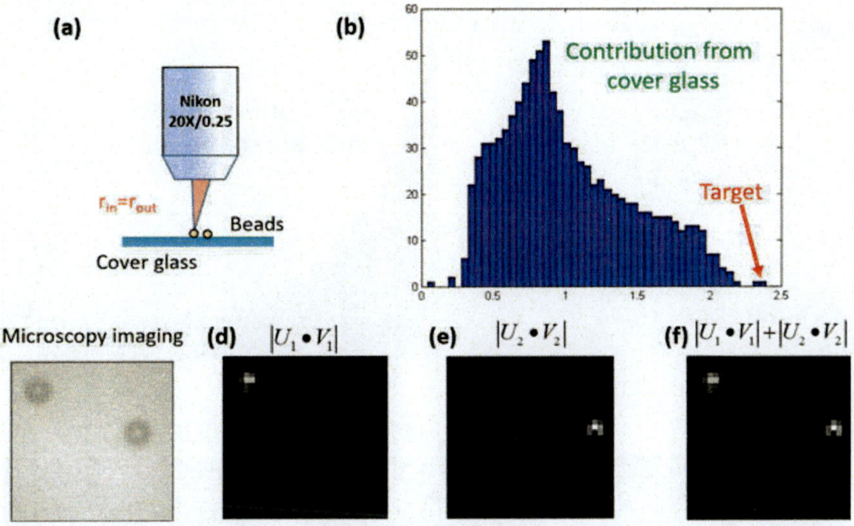

Fig. 32.4 Imaging at the single scattering region

32.3 Results and Discussion

32.3.1 Imaging at the Single Scattering Region

At this step, the imaging target is two beads attached to cover glass and there is no scattering layer (as shown in Fig. 32.4a). Figure 32.4b shows the distribution of singular values. We can see the singular values that have the first two corresponding to the imaging targets. Figure 32.4c is the microscopy image of the target for comparison purposes. Figure 32.4d–f shows the recovered images by using the first singular vector, second singular vector, and the first two singular vectors, respectively.

32.3.2 Imaging at Multiple Scattering Regions

This time, we have inserted a high scattering layer, a piece of A4 paper sheet, between the microscopy and the cover glass (as shown in Fig. 32.5a). The thickness of the sheet is 97 μm and the SMFP is 6.4 μm, which means the optical thickness of the scattering layer is 15.2 SMFP. Figure 32.5b shows the distribution of singular values. We can see the singular values that have the first two corresponding to the imaging targets. Figure 32.5c–e shows the recovered images by using the first singular vector, second singular vector, and the first two singular vectors, respectively. Based on the above results, the RMOCT presents its ability to imaging ability through 15.2 SMFP.

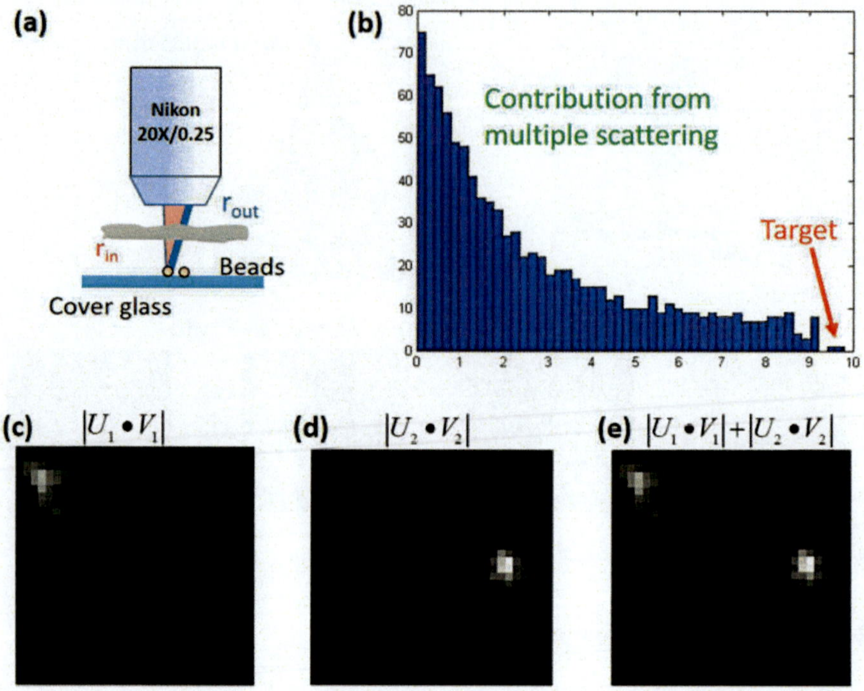

Fig. 32.5 Imaging at multiple scattering regions

32.3.3 Imaging Through a Thick Biological Sample

We have home-made a highly scattering layer (0.8 mm thickness phantom) by adding 20 g titanium dioxide into a 100 g silicone rubber base. The imaging target this time is a polydimethylsiloxane (PDMS) material, which is soft and flexible. Figure 32.6a is the sample setup. We use a water immersion objective with a bigger numerical aperture this time. Figure 32.6b shows the distribution of singular values in decreasing order. Figure 32.6c is the microscopy imaging of the PDMS material. Figure 32.6d and e shows the recovered images by using the first 38 singular vectors at the situation without and with scattering later, respectively.

In order to show the power imaging ability of RMOCT, we use a commercial 1.3 μm swept-source OCT from Thorlabs Inc. to image the same sample. The results are shown in Fig. 32.7. Figure 32.7a is the imaging result of PDMS material without the scattering layer. We can see the boundaries are very clear. Figure 32.7b is the *en-face* imaging of the upper layer. Figure 32.7c is the imaging result of PDMS material beneath the scattering layer, and we can see the boundaries are blurry. Figure 32.7d is the *en-face* imaging of the upper layer of PDMS material. This time, the commercial OCT system lost the ability to recover the image.

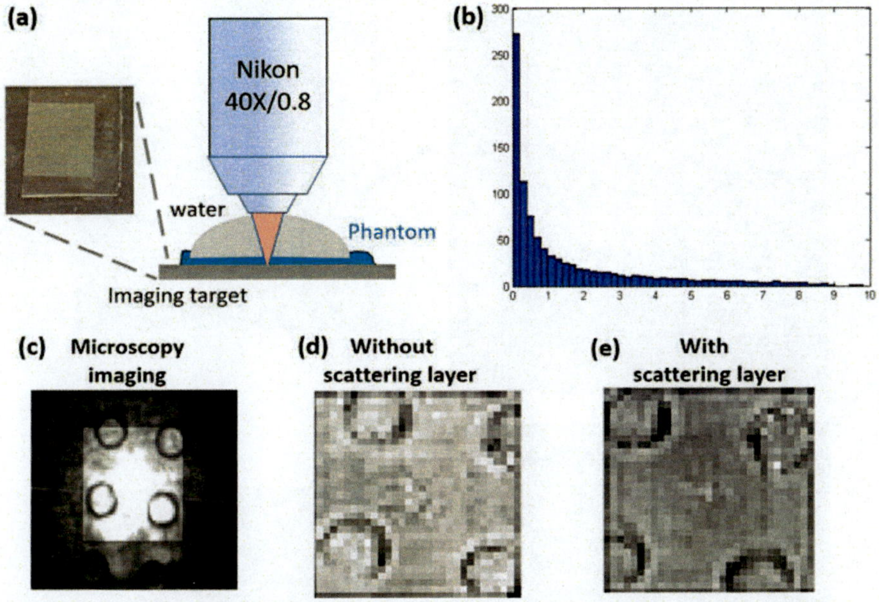

Fig. 32.6 Imaging through a thick biological sample

32.3.4 High-Speed Wavefront Shaping

As for the light focusing experiment, the phase and amplitude before wavefront are all made of chaos. However, the phase and amplitude after wavefront have been shaped to a Gaussian-like distribution. The whole process only needs 133 ms.

32.4 Conclusions

In this paper, we have demonstrated a new type of OCT combined with reflection matrix measurement. The huge reflection matrix of the sample contains abundant information which describes the light propagation process within the sample. We have presented its amazing imaging power at deeper penetration depth, which is 15 times MSFP and two times the conventional OCT's imaging depth. At the same time, we have shown a high-speed WFS method to focus light through a sample in ~133 ms. As a next step, we are now trying to apply WFS technology to OCT to realize deeper imaging depth.

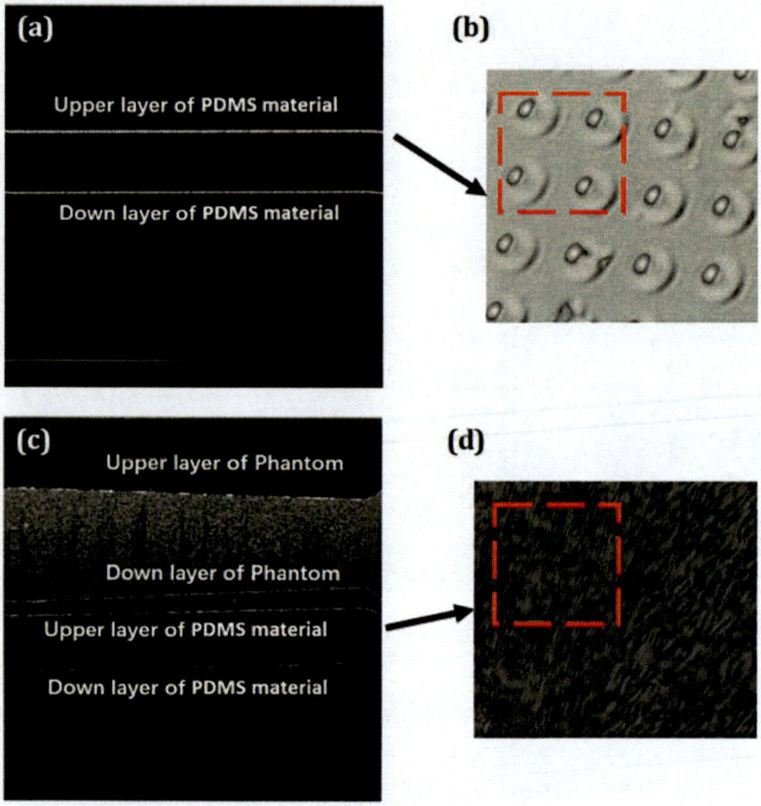

Fig. 32.7 Imaging results of a commercial 1.3 μm swept-source OCT

Acknowledgements This research was financially supported by the National Key R&D Program of China (Grant No. 2016YFF0102003).

References

1. G.J. Tearney, R.H. Webb, B.E. Bouma, Spectrally encoded confocal microscopy. Opt. Lett. **23**(15), 1152–1154 (1998)
2. T. Liu, X. Zhang, Q. Qiao, C. Zou, L. Feng, J. Cui, Z. Xu, A two-photon fluorescent probe for imaging hydrogen sulfide in living cells. Dyes Pigm. **99**(3), 537–542 (2013)
3. P.J. Campagnola, L.M. Loew, Second-harmonic imaging microscopy for visualizing biomolecular arrays in cells, tissues and organisms. Nat. Biotechnol. **21**(11), 1356 (2003)
4. H. Cui, C. Hua, P. Yue, W. Lin, A two-photon fluorescent turn-on probe for palladium imaging in living tissues. Sens. Actuators B Chem. **219**, 232–237 (2015)
5. J. Jang, J. Lim, H. Yu, H. Choi, Y. Park, Complex wavefront shaping for optimal depth-selective focusing in optical coherence tomography. Opt. Express **21**(3), 2890–2902 (2013)

6. H. Yu, J. Jang, J. Lim, J.H. Park, W. Jang, J.Y. Kim et al., Depth-enhanced 2-D optical coherence tomography using complex wavefront shaping. Opt. Express **22**(7), 7514–7523 (2014)
7. J. Liang, L. Zhu, L.V. Wang, Single-shot real-time femtosecond imaging of temporal focusing. Light Sci. Appl. **7**(005), 466–475 (2018)
8. J.H. Park, C. Park, H.S. Yu, J. Park, S. Han, J. Shin et al., Subwavelength light focusing using random nanoparticles. Nat. Photonics **7**(6), 454–458 (2013)
9. I.M. Vellekoop, A.P. Mosk, Focusing coherent light through opaque strongly scattering media. Opt. Lett. **32**(16), 2309 (2007)
10. A.S. Hemphill, Y. Shen, L. Yan, L.V. Wang, High-speed single-shot optical focusing through dynamic scattering media with full-phase wavefront shaping. Appl. Phys. Lett. **111**(22), 221109 (2017)
11. M. Hacker, G. Stobrawa, R. Sauerbrey, T. Buckup, M. Motzkus, M. Wildenhain et al., Micromirror SLM for femtosecond pulse shaping in the ultraviolet. Appl. Phys. B **76**(6), 711–714 (2003)
12. O. Katz, E. Small, Y. Bromberg, Y. Silberberg, Focusing and compression of ultrashort pulses through scattering media. Nat. Photonics **5**(6), 372–377 (2011)
13. M. Cui, A high speed wavefront determination method based on spatial frequency modulations for focusing light through random scattering media. Opt. Express **19**(4), 2989 (2011)
14. H.P. Paudel, C. Stockbridge, J. Mertz, T. Bifano, Focusing polychromatic light through scattering media, in *Mems Adaptive Optics VII* (International Society for Optics and Photonics)
15. J. Aulbach, B. Gjonaj, P. Johnson, A. Lagendijk, Spatiotemporal focusing in opaque scattering media by wave front shaping with nonlinear feedback. Opt. Express **20**(28), 29237 (2012)
16. D. Wang, E.H. Zhou, J. Brake, H. Ruan, M. Jang, C. Yang, Focusing through dynamic tissue with millisecond digital optical phase conjugation. Optica **2**(8), 728–735 (2015)
17. B. Zhuang, C. Xu, Y. Geng, G. Zhao, H. Chen, Z. He, L. Ren, An Early study on imaging 3D objects hidden behind highly scattering media: a round-trip optical transmission matrix method. Appl. Sci. **8**(7), 1036 (2018)
18. A. Aubry, A. Derode, Detection and imaging in a random medium: A matrix method to overcome multiple scattering and aberration. J. Appl. Phys. **106**(4) (2009)
19. C. Xu, B. Zhuang, Y. Geng, H. Chen, L. Ren, Z. Wu, A scanning-free wide-field single-fiber endoscopic image retrieval method based on optical transmission matrix. Laser Phys. **29**(4) (2019)
20. Y. Choi, T.R. Hillman, W. Choi, N. Lue, R.R. Dasari, P.T.C. So et al., Measurement of the time-resolved reflection matrix for enhancing light energy delivery into a scattering medium. Phys. Rev. Lett. **111**(24), 243901 (2013)
21. A. Badon, D. Li, G. Lerosey, A.C. Boccara, M. Fink, A. Aubry, Smart optical coherence tomography for ultra-deep imaging through highly scattering media. Sci. Adv. **2**(11) (2016)
22. Q. Yang, J. Cao, Y. Miao, J. Zhu, Z. Chen, Extended imaging depth of en-face optical coherence tomography based on fast measurement of reflection matrix by widefield heterodyne detection. Opt. Lett. **45**(4) (2019)

Chapter 33
Application of Terahertz Spectroscopy in the Detection of Carbohydrate Isomers

Tonghua Liu, Fang Yan, Junlin Zhang, and Boyang Jin

Abstract In this paper, two isomers of sugars (D-(+)-glucose and D-(−)-fructose) were used as the research object, and the absorption spectra of them in the 0.4–1.9 THz frequency band were obtained by the terahertz time-domain spectroscopy (THz-TDS) system. The results showed that they could be qualitatively identified by two terahertz absorption peaks of 1.43 and 1.64 THz. In order to further explore the formation mechanism of the terahertz absorption peak, the absorption peak of D-(+)-glucose at 1.43 THz was taken as an example to conduct PED analysis on the glucose unit cell. The results showed that the absorption peak of D-(+)-glucose at 1.43 THz was mainly caused by the interaction between molecules. Multiwfn software was used to calculate the RDG of D-(+)-glucose unit cell, and then VMD software was used to visualize the types, positions and intensities of the weak interactions between molecules in D-(+)-glucose unit cell. Terahertz absorption peaks were due to the torsion mode caused by intermolecular hydrogen bonding in the glucose cell. The comparison between the experimental and theoretical results shows that the far-infrared absorption features are highly sensitive to the structure and spatial arrangement of molecules. The research work in this paper provides valuable experimental and theoretical references for the effective detection and identification of sugar isomers and the formation mechanism of D-(+)-glucose terahertz spectral absorption peaks.

33.1 Introduction

Since most of the vibrational and rotational energy levels of macromolecules are in the terahertz band, and macromolecules, especially biological and chemical macromolecules, have their own physical properties, it is possible to analyze and identify the structure and physical properties of substances through characteristic frequencies. The energy of terahertz photon is only a few millielectron volts, so it is not easy

T. Liu (✉) · F. Yan · J. Zhang · B. Jin
Inner Mongolia University of Science & Technology, Inner Mongolia Autonomous Region, Baotou, China
e-mail: liutonghua0613@163.com

G. Liu and F. Cen (eds.), *Advances in Precision Instruments and Optical Engineering*, Springer Proceedings in Physics 270, https://doi.org/10.1007/978-981-16-7258-3_33

to destroy the detected material, so the terahertz spectrum has a broad application prospect in non-destructive testing technology [1]. O. P. Cherkasova et al. found that when the blood sugared concentration of the human body changes, it will cause the change of skin complex refractive index. Therefore, the terahertz wave can be used to predict human blood glucose concentration so as to achieve non-invasive detection of human blood glucose [2]. Zhuanping Zheng et al. studied the spectral characteristics of monohydrous and anhydrous glucose and believed that the spectral characteristics of anhydrous glucose mainly came from the molecular interaction of glucose molecules [3]. Diding Suhandy et al. measured the absorption spectrum and refractive index spectrum of D-(+)-glucose at different concentrations and found that there was a fingerprint spectrum of glucose molecules in the terahertz band, and they could identify glucose molecules and their solutions [4].

In this paper, the terahertz absorption spectra of D-(+)-glucose and D-(−)-fructose were measured by experiments. In order to further explore the formation mechanism of the absorption peak, the absorption peak of D-(+)-glucose at 1.43 THz was taken as an example. Multiwfn and VMD software were used to visually analyze the type and intensity of intermolecular interaction, and the identification of the intermolecular vibration mode at 1.43 THz was completed.

33.2 Experiments and Theory

In this paper, the transmission terahertz time-domain spectroscopy (THz-TDS) system is used to complete the detection of samples to be tested [5]. D-(+)-glucose, D-(−)-fructose and polyethylene (PE) were all purchased from SIGMA-ALDRIDGE. The above samples were all white solid powder. Before determining the sample information, determine that the mass fraction of glucose in the sample is 20%. The terahertz absorption spectra of the three samples were selected to sum and take the average value, that is, the terahertz absorption spectra of the sample. Table 33.1 shows the matching information of sample preparation.

This paper is based on the optical parameter extraction model proposed by Dorney and Duvillarent [6, 7]. Obtain the reference signal $E_{ref}(t)$ and sample signal $E_{sam}(t)$ of the sample in the terahertz optical path, and then the corresponding frequency domain

Table 33.1 Sample mixing information

Sample		Weight (mg)	Thickness (mm)	Proportion (%)
D-(+)-Glucose	Glu1	169.0	1.55	20
	Glu2	170.3	1.57	20
	Glu3	170.5	1.52	20
D-(−)-Fructose	Fru1	168.6	1.55	20
	Fru2	168.4	1.52	20
	Fru3	168.4	1.52	20

signals $E_{ref}(\omega)$ and $E_{sam}(\omega)$ were obtained by Fourier transform. The refractive index $n_s(\omega)$ and absorption coefficient $\alpha(\omega)$ were calculated by formulas (33.1) and (33.2).

$$n_s(\omega) = 1 + \frac{c}{\omega d}\phi(\omega) \tag{33.1}$$

$$\alpha(\omega) = -\frac{2}{d} ln\left\{\frac{|E_{sam}(\omega)|[n_s(\omega)+1]^2}{|E_{ref}(\omega)|4n_s(\omega)}\right\} \tag{33.2}$$

33.3 Results and Discussions

The terahertz time-domain spectroscopy (THz-TDS) system was used to measure the samples of D-(+)-glucose and D-(−)-fructose, respectively. Figure 33.1 shows the characteristic absorption spectra of the two isomers in the band of 0.4–1.9 THz after data processing. As shown in Fig. 33.1, D-(+)-glucose has a characteristic absorption peak at 1.43 THz, and D-(−)-fructose has a characteristic absorption peak at 1.64 THz. The terahertz absorption peaks of the two are significantly different, so it is easy to identify them by the absorption peak position.

Compared with the monomolecular configuration, the intermolecular hydrogen bonds and van der Waals forces can be considered more comprehensively in the crystal cell configuration. The molecular interactions of D-(+)-glucose was studied using Materials Studio 2017 software. The crystal cells were taken from Cambridge Crystallography Data Center (CCDC). As shown in Fig. 33.2, each cell contains four glucose molecules.

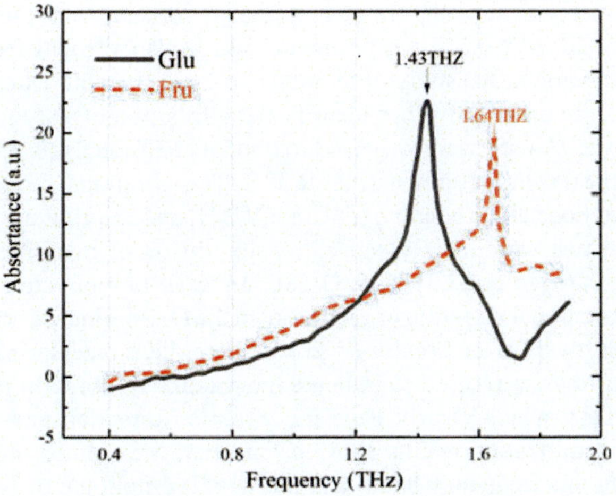

Fig. 33.1 Terahertz absorption spectra of D-(+)-glucose and D-(−)-fructose

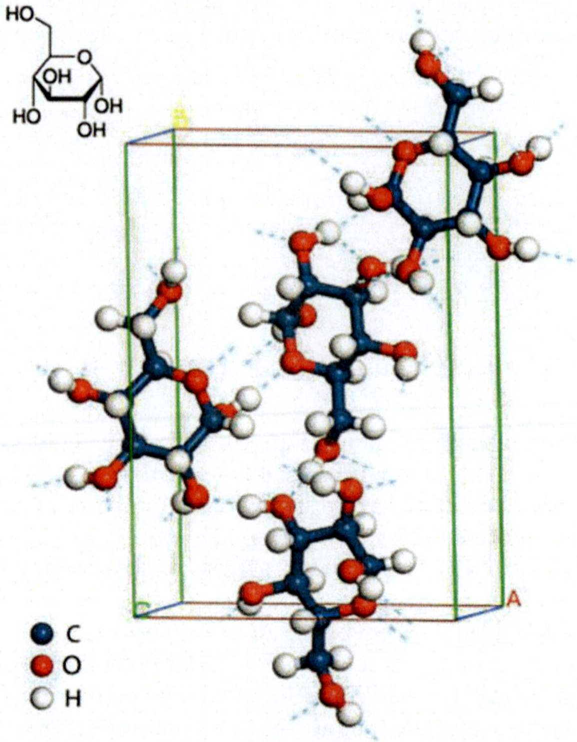

Fig. 33.2 Crystal structure of glucose. The upper left corner is the formula for a single molecule of glucose. The blue spheres are carbon atoms, the red spheres are oxygen atoms and the white spheres are hydrogen atoms

Then, the CASTEP module was used to perform geometric optimization on the glucose cell. The calculated cell simulated spectrum was compared with the experimental spectrum, as shown in Fig. 33.3. The absorption peak of the cell simulated spectrum at 1.48 THz was basically consistent with the absorption peak at 1.43 THz obtained from the experiment, which also verified the feasibility of the CASTEP module in the study of terahertz spectral characteristics of D-(+)-glucose.

In order to further identify the attribution of D-(+)-glucose cell vibration mode, VEDA4 software was used to conduct PED analysis of its absorption spectrum, and the results are shown in Table 33.2. The vibrational modes at 1.48 THz were identified as bond angle bending (58C/83O/31H) and contributed 11% to the vibrational modes, which were mainly caused by the torsion of two glucose molecules connected by hydrogen bond 31H···83O-58C. In order to more clearly and visually represent the rotational vibration of each atom in the D-(+)-glucose crystal cell, Gaussian and VMD software were combined to calculate the molecular vibration vector. In Fig. 33.4, arrows were used to indicate the rotational vibration mode of the molecule at 1.44 THz, which clearly gave the vibration direction and intensity of each atom. The identification results showed that there was almost no vibration of No. 1 glucose in this frequency band, and the wiggle amplitude of (–HO) bond of No. 2, 3 and 4 glucose molecules was the largest.

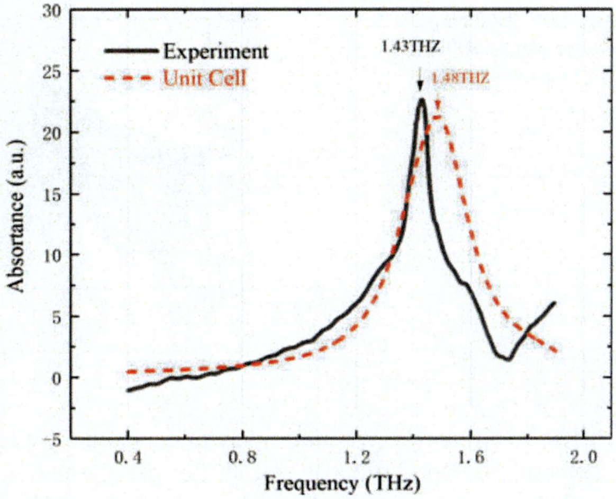

Fig. 33.3 Comparison of D-(+)-glucose experimental spectrum and crystal structure theoretical calculation absorption spectrum

Table 33.2 PED analysis

Sample	Experiment (THz)	Calculation (THz)	PED analysis
D-(+)-Glucose	1.43	1.48	BEND:58C/83O/31H(11)

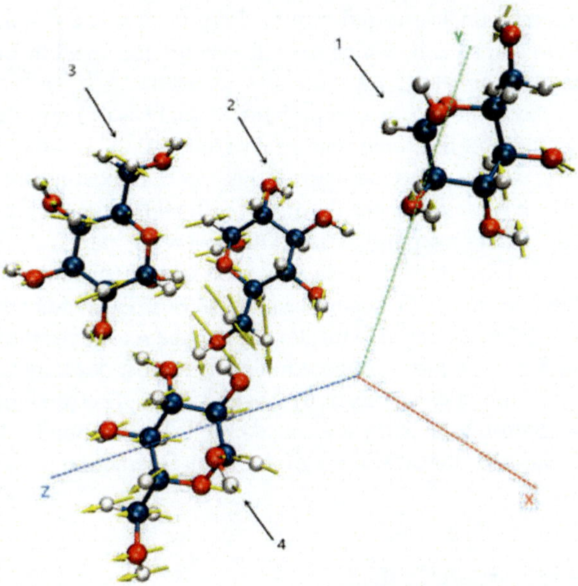

Fig. 33.4 The vibration mode of D-(+)-glucose at 1.48 THz. The yellow arrows show the direction in which the atoms move

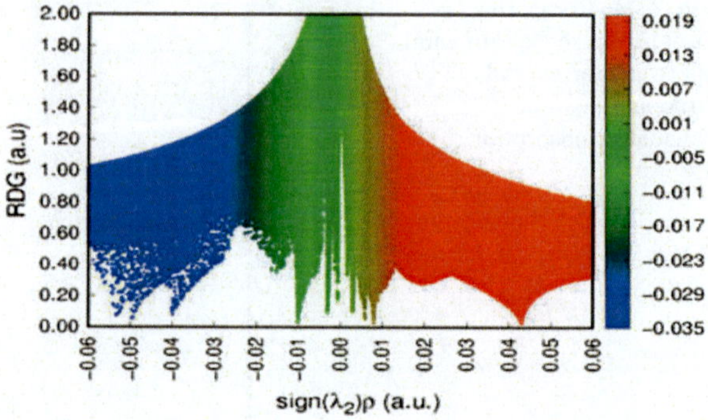

Fig. 33.5 Coloring the scatter plot of RDG for D-(+)-glucose cell

The RDG method can intuitively understand the specific position and type of the interaction in D-(+)-glucose cell. In this paper, the interaction of D-(+)-glucose cell configuration was analyzed based on Multiwfn and VMD [8]. The strength and type of the weak interaction can be clearly shown. Figure 33.5 shows coloring the scatter plot of RDG for D-(+)-glucose cells, in which multiple spikes appear. The spikes in the graph can be divided into three types according to the value of sin(λ2)ρ(r). In Fig. 33.5, the blue region is the hydrogen bonding region. The green region is the van der Waals force action region. Strong steric hindrance of ring or cage molecules is common in the red region. Figure 33.6 is the coloring RDG isosurface of D-(+)-glucose cell, which visually shows the specific location and type of the weak interaction.

According to the comprehensive analysis of Figs. 33.5 and 33.6, the glucose RDG scatter plot has three blue peaks from −0.06 to −0.025, corresponding to the three blue RDG isosurface maps in Fig. 33.6, indicating the strong hydrogen bond interaction between molecules in the cell. Half blue, half red and green spikes at −0.025 to 0. 015 correspond to the RDG isosurface map, indicating weak hydrogen bonds or van der Waals forces. There is a red spike between 0.015 and 0.06, corresponding to the four red RDG isosurface maps in Fig. 33.6, indicating strong repulsion. Through Fig. 33.6, we can also further explain the reason for the torsion of glucose molecules. In Fig. 33.4. No. 1 glucose molecule almost does not torsion, because No. 1 glucose molecule does not form hydrogen bonds. The torsional vibrations of No. 2, 3 and 4 glucose molecules are caused by hydrogen bonds. Table 33.3 shows the specific location of the isosurface of the weak interaction.

33.4 Conclusion

In this paper, the terahertz absorption spectra of sugar D-(+)-glucose and D-(−)-fructose were obtained by the THz-TDS technique. The experimental results show that the absorption peaks of the two samples are significantly different in the band of 0.4–1.9 THz. The theoretical terahertz absorption spectra of D-(+)-glucose cells

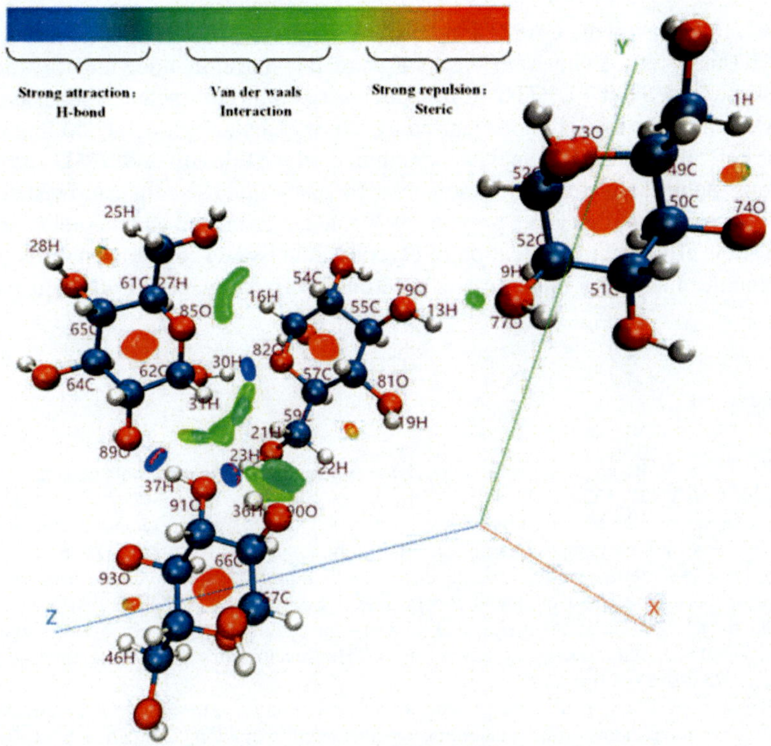

Fig. 33.6 Coloring RDG isosurface of D-(+)-glucose cell

Table 33.3 The specific location of the weak interaction isosurface

Sign(λ2)p(r)	Color	Location of RDG isosurface
−0.06 to −0.025	Blue	82O/30H, 89O/37H, 91O/23H
0.015 to 0.06	Red	94O/68C/69C/70C/66C/67C, 85O/62C/63C/64C/65C/61C, 82O/57C/56C/55C/54C/58C, 73O/49C/50C/51C/52C/53C
−0.025 to 0.015	Half red, half green	1H/74O, 9H/13H, 16H/84O, 16H/61C, 25H/86O, 37H/62C, 23H/87O, 31H/21H, 31H/91O, 46H/93O 59C/62C59C/90O, 59C/66C, 22H/81O, 62C/91O

were calculated using the CASTEP module in Materials Studio software. Combined with PED analysis and Gaussian View software, the vibration mode identification of the absorption peak at 1.43 THz of the glucose cell was completed. The absorption peak was due to the torsion mode caused by intermolecular hydrogen bonding in the glucose cell. Through RDG analysis, combined with Multiwfn and VMD software, the strong, neutral and weak intermolecular forces in glucose cells were visualized. The research work in this paper provides valuable experimental and theoretical references for the effective identification of carbohydrate isomers and the analysis of the formation mechanism of the terahertz spectral absorption peak of D-(+)-glucose.

References

1. L. Wei, L. Yu, H. Jiaoqi, Application of terahertz spectroscopy in biomolecule detection. Front. Lab. Med. **2**(4), 127–133 (2018)
2. O.P. Cherkasova, M.M. Nazarov, A.P. Shkurinov, Application of terahertz time-domain spectroscopy for blood glucose monitoring. Int. Conf. Laser Opt. (LO) **2**, 11 (2016)
3. Z.P. Zheng, Terahertz spectral investigation of anhydrous and monohydrated glucose using terahertz spectroscopy and solid-state theory. J. Mol. Spectrosc. **296**, 9–13 (2014)
4. D. Suhandy, T. Suzuki et al., A quantitative study for determination of glucose concentration using attenuated total reflectance terahertz (ATR-THz) spectroscopy. Eng. Agric. Environ. Food **5**(3), 90–95 (2012)
5. H. Zhang, Z.-H. Zhang et al., Assignment of terahertz vibrational modes of L-glutamine using density functional theory within generalized-gradient approximation. Chin. Phys. B **24**(7), 209–216 (2015)
6. T.D. Dorney et al., Material parameter estimation with terahertz time-domain spectroscopy. J. Opt. Soc. Am. A **18**(7), 1562–1571 (2001)
7. L. Duvillaret, A reliable method for extraction of material parameters in terahertz time-domain spectroscopy. IEEE J. Sel. Top. Quantum Electron. **2**(3), 39–746 (1996)
8. T. Lu, F. Chen, Multiwfn: a multifunctional wavefunction analyzer. J. Comput. Chem. **33**(5), 580–592 (2012)

Chapter 34
Detection of Gas Raman Spectra Based on Double-Cladding Fiber Laser

Chun Feng and Shu-Bo Jiang

Abstract To improve the Raman scattering cross-section of trace gases, a Raman spectrum detection method based on the double-clad fiber laser is designed in this paper. The influence of double-cladding fiber laser as a light source on gas Raman detection is investigated. The higher the output power of the double-clad fiber laser, the larger the gas scattering cross-section and the stronger the Raman effect. Therefore, this paper first conducts a numerical analysis of the factors affecting the output power of the double-clad fiber laser and uses MATLAB to simulate and explore the influence of different pumping modes, pumping power, and rare earth doping concentration on the output power of the fiber laser. Then the gas detection system with ytterbium-doped fiber laser as the light source was used to analyze the material qualitatively and quantitatively. Fiber laser as a Raman detection source has the advantages of high linearity and good stability.

34.1 Introduction

Raman spectrum is a kind of scattering spectrum, which can obtain the information of molecular vibration and rotation through the analysis of the scattering spectrum different from the incident light frequency, and is applied to the study of molecular structure [1]. It can realize the qualitative and quantitative analysis of substances. At the same time, it is a non-contact detection method that can carry on the nondestructive detection to the material. However, Raman scattering is a weak effect. With the development of laser technology, the incident light intensity is improved. The Raman effect is enhanced and the application of Raman spectroscopy is promoted. Therefore, the laser can be used as the Raman detection light source [2].

Since the birth of the first laser, a series of lasers, such as solid-state laser, gas laser, and semiconductor laser, have been developed rapidly. The optical fiber laser instrument has been widely used in optical communication, spectroscopy, aerospace, and

C. Feng · S.-B. Jiang (✉)
College of Electrical Engineering and Control Science, NanJing Tech University, NanJing 211816, China
e-mail: jiangshubo@njtech.edu.cn

G. Liu and F. Cen (eds.), *Advances in Precision Instruments and Optical Engineering*, Springer Proceedings in Physics 270, https://doi.org/10.1007/978-981-16-7258-3_34

laser weapons because of its good beam quality, high conversion efficiency, convenient thermal control and management, large tunable range, and compact structure [3, 4]. In order to further improve the conversion efficiency and output power of conventional fiber laser, double-clad fiber lasers began to appear. It is a kind of optical fiber with a special structure; it has an inner cladding than the conventional optical fiber. The transverse size and numerical aperture of the inner cladding are far larger than that of the fiber core. When rare earth elements are doped in the core, the pump light is reflected in the inner cladding and is absorbed by the doped ions through the core many times, thus being converted into a single-mode laser efficiently. The conversion efficiency of traditional fiber lasers is greatly improved [5].

Based on the principle of double-clad fiber laser, this paper analyzes the fiber laser as a Raman detection source, selects high-power diode-pumped laser, ytterbium-doped double-clad fiber, adopts forward pumping mode and traditional detection optical path, and sets up an oxygen-based Raman gas detection system.

34.2 Basic Theory and Simulation of Double-Clad Fiber Laser

34.2.1 Theoretical Model of Double-Clad Fiber Laser

The theoretical model construction of fiber laser is shown in Fig. 34.1. The fiber laser is composed of fiber combiner, fiber grating, pump source, and double-cladding gain fiber. Instead of discrete optical devices, the optical fiber combiner can effectively make the pump light coupling into the double-clad fiber, reduce the loss and improve the coupling efficiency. At the same time, the output of multiple semiconductor lasers can be connected at the input end of the double-clad fiber to improve the injectable pump power. Fiber Bragg grating FBG1 and FBG2 can be used to achieve fixed wavelength reflection and absorption. The pump source uses semiconductor diodes, double-clad optical fibers [6], and is several meters to dozens of meters in length. The pump light is coupled to the optical fiber through the optical fiber combiner, and the doped rare earth ions in the optical fiber absorb the pump energy, and the energy-level transition occurs, which makes the particles accumulate continuously and form the particle inversion [7]. Then the transition to the lower energy level of the laser produces photons, which are oscillated and amplified in the resonator and

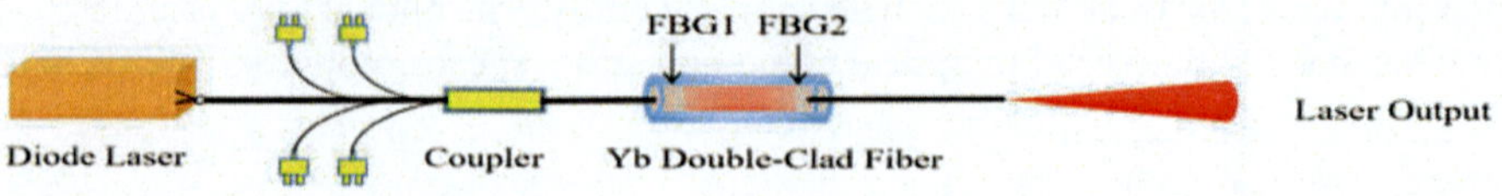

Fig. 34.1 Theoretical model of fiber laser

form the laser output from the grating. The formation of the laser is related to its rate equation. Next, the rate equation of fiber laser is analyzed.

34.2.2 Rate Equation of Double-Clad Fiber Laser

The laser rate equation is used to characterize the differential equations of the number of atoms at each energy level and the number of photons in the cavity changing with time. It is used to describe the interaction between the radiation field and particles. When only pumping light and laser output line width are considered, the rate equations of double-clad fiber lasers can be simplified as follows [7]:

$$\begin{cases} \frac{dP_P^+(z)}{dz} = -\Gamma_P[\sigma_{ap}N - (\sigma_{ap} + \sigma_{ep})N_2(z)]P_p^+(z) - a_p P_P^+(z) \\ \frac{dP_P^-(z)}{dz} = -\Gamma_P[\sigma_{ap}N - (\sigma_{ap} + \sigma_{ep})N_2(z)]P_p^-(z) + a_p P_P^-(z) \\ \frac{dP_s^+(z)}{dz} = -\Gamma_s[(\sigma_{es} + \sigma_{as})N_2(z) - \sigma_{as}]P_s^+(z) - \Gamma_s\sigma_{es}N_2(z)P_0 - a_s P_s^-(z) \\ \frac{dP_s^-(z)}{dz} = -\Gamma_s[(\sigma_{es} + \sigma_{as})N_2(z) - \sigma_{as}]P_s^-(z) - \Gamma_s\sigma_{es}N_2(z)P_0 - a_s P_s^-(z) \end{cases} \tag{34.1}$$

Formula 34.1 describes the relationship between the concentration of upper-level particles in the gain medium at different positions and the pumping power of the forward and backward light, where $N_2(z)$ is the concentration of particles at the upper level. $P_P^+(z)$ and $P_p^-(z)$ are, respectively, the power of pumping light forward and backward; $P_s^+(z)$ and $P_s^-(z)$ refer to the forward and backward transmission power, where N is the concentration of doping medium in the fiber core, A_c is the cross-sectional area of the fiber core, and Γ_P and Γ_s are the power filling factors of double-clad fiber for pumping light and fiber laser, respectively. σ_{ap} and σ_{ep} are the absorption cross-section and emission cross-section of pumped light, respectively, σ_{as} and σ_{es} are the absorption cross-section and emission cross-section of laser, h is Planck constant, υ_s and υ_p are the frequency of laser and pumped light, respectively. τ is the average lifetime of the upper energy level of a particle, where $P_0 = 2h\upsilon_s\Delta\upsilon_s$ is the contribution of spontaneous radiation to the laser within the gain bandwidth, and its value is relatively small. This can be ignored later in the derivation, where $\Delta\upsilon_s$ is gain bandwidth. Under steady-state conditions, the boundary conditions of the linear resonator laser are as follows:

$$\begin{cases} P_p^+(0) = P_p^l \\ P_p^-(L) = P_p^r \\ P_s^+(0) = R_1 P_s^-(0) \\ P_s^+(0) = R_2 P_s^+(L) \end{cases} \tag{34.2}$$

where P_p^l and P_p^r inject the pumping optical power of the inner cladding of the double-clad fiber from the left and right end faces, respectively. Therefore, the output power of the fiber laser is

$$P_{out} = \frac{(1-R_2)\sqrt{R_1}.P_{s,sat}}{(1-R_1)\sqrt{R_2}+(1-R_2)\sqrt{R_1}}.\left[\frac{v_s}{v_p}.(1-\exp(-\Psi))\cdot\frac{P_p^+(0)+P_p^-(L)}{P_{s,sat}} - (N\Gamma_s\sigma_{as}+a_s)L - \ln\left(\frac{1}{\sqrt{R_1R_2}}\right)\right. \quad (34.3)$$

$P_{s,sat} = h\upsilon_s A_c/[\tau\Gamma_s(\sigma_{es}+\sigma_{as})]$ is the output power of the saturated laser and $P_{p,sat} = h\upsilon_p A_c/[\tau\Gamma_p(\sigma_{ep}+\sigma_{ap})]$ is the pumping power of the saturated laser.

34.2.3 *Numerical Simulation of a Double-Clad Fiber Laser*

In view of the above analysis process, this section mainly introduces the results of numerical simulation. The doped fiber selected in this paper is Yb^{3+} particles, because the energy-level structure of Yb^{3+} is simple, and only two multistate expanded energy levels $^2F_{5/2}$ and $^2F_{7/2}$ are related to the wavelength of all light. For Yb^{3+}-doped fiber laser, the central wavelength of pumped LD is generally 915 nm or 975 nm, and the absorption cross-section of 975 nm is about three times that of 915 nm. The quantum efficiency of the laser is also relatively high, so 975 nm LD is used as the pumping source. The basic parameters of the double-clad fiber laser used in the above analysis and simulation are shown in Table 34.1.

Table 34.1 Basic parameters of fiber laser

Symbol	Physical parameters	Numerical value	Unit
λ_p	Pump light center wavelength	975	nm
λ_s	The central wavelength of a fiber laser	1064	nm
τ	Yb^{3+} upper energy level life of a particle	0.8	ms
σ_{ap}	The absorption cross-section of pumped light	2.6×10^{-20}	cm^2
σ_{ep}	The emission cross-section of pumped light	2.6×10^{-20}	cm^2
σ_{as}	The absorption cross-section of a fiber laser	1×10^{-23}	cm^2
σ_{es}	Emission cross-section of optical fiber laser	1.6×10^{-21}	cm^2
A_c	The cross-sectional area of the fiber core	3.1416×10^{-6}	cm^2
a_p	The loss of fiber to the pumped light	2×10^{-5}	cm^{-1}
a_s	Loss of optical fiber to the laser	4×10^{-5}	cm^{-1}
L	Length of optical fiber	20	m
Γ_p	Fill factor for pumping light	0.0024	No unit
Γ_s	Laser power filling factor	0.82	No unit
R_1	The reflectivity of the anterior cavity mirror	0.99	No unit
R_2	Reflectivity of the posterior cavity mirror	0.04	No unit

Pump power under different pumping modes

This section mainly introduces the experimental simulation results under different pumping modes, which simulate the output power of the laser resonator system in three directions: forward pumping, backward pumping, and two-way pumping.

As can be seen from Fig. 34.2, the output power of the laser resonator is not significantly different under the three pumping modes. However, the laser heat dissipation performance will be affected to some extent due to the backward pumping and bidirectional pumping methods. Therefore, these two pumping methods are generally not considered. It can be seen from Fig. 34.2a that the output power of the optical resonator system will increase with the increase of the length of the gain fiber, but it does not increase all the time. The growth rate will become slower and slower with the increase of the gain fiber. When the length of the gain fiber reaches a certain degree, the output power of the laser resonator system reaches a basically stable state. However, when the length of the gain fiber is increased, the output power will decrease. The length of the gain fiber is called the optimal fiber length. As for the output power, the reason is that the excessively long gain fiber will make the pump

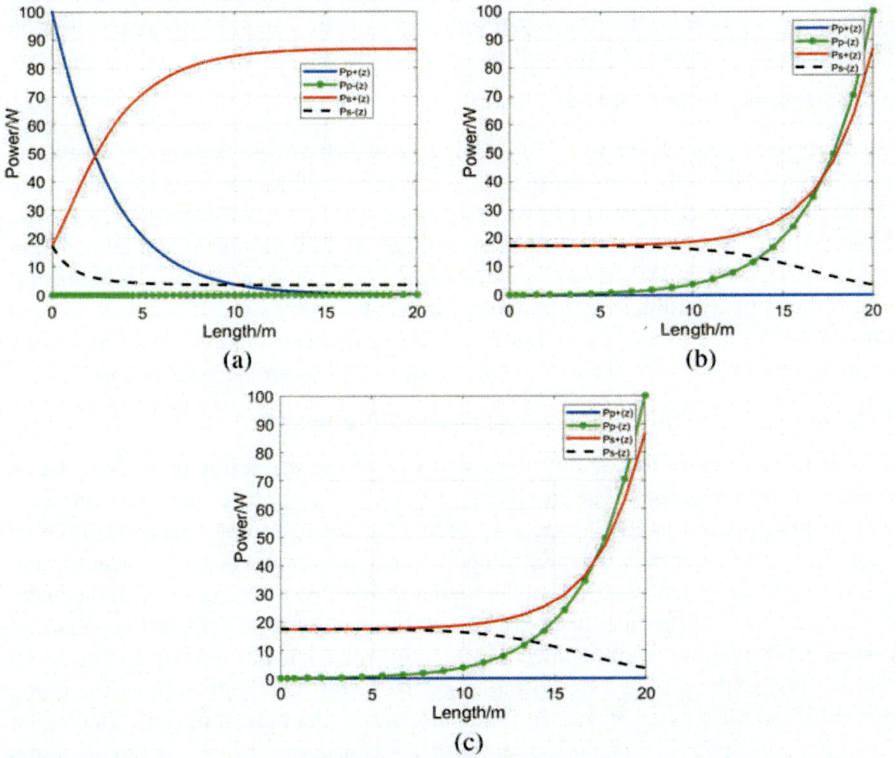

Fig. 34.2 The influence of different pumping modes on the output power of laser resonator system **a** forward pumping mode, **b** backward pumping mode, **c** two-way pumping mode

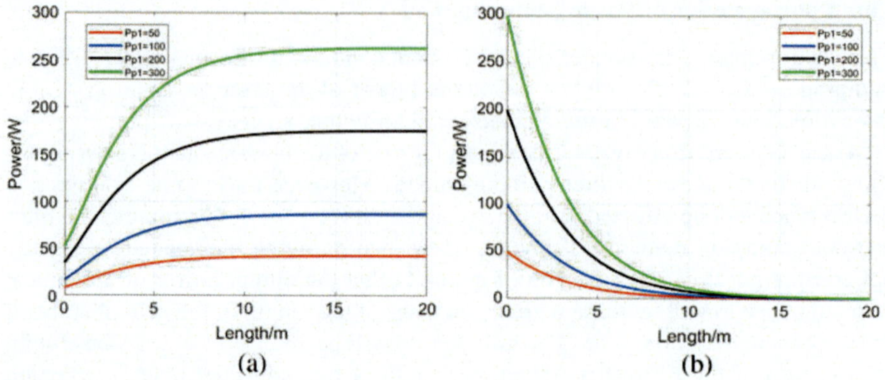

Fig. 34.3 Output power of **a** ytterbium-doped double-clad fiber laser with different pump power on fiber laser performance, **b** pump power

light source be completely absorbed before reaching the end of the fiber, and there is no inversion of particle number, but at this time, the signal light is still in very strong signal power so that the signal light is stimulated and absorbed, resulting in the decrease of the output power of the laser resonator system.

Effect of pump power on laser output

When the pumping power was 50, 100, 200, and 300 W, the doping concentration was 5.535×10^{19} cm^{-3}, and the fiber length was L = 20 m, the laser distribution in the fiber under the condition of forward pumping was simulated numerically. As can be seen from Fig. 34.3a, with the increase of the pump light power, the laser output power also increases. At the same time, the higher the pump power is, the longer the optimal gain fiber length is required to reach the stationary state. It can be seen from Fig. 34.3b that the higher the pump power is, the faster the attenuation rate of the pump is, and the longer the fiber length is required to reach the steady-state.

Influence of core doping concentration on output power

The doped concentration of the fiber core will affect the refractive index, which directly affects the amplification ability of the fiber. Therefore, under the premise that the pump power is 100 W and L = 20 m, change the doping concentration N to obtain the change curve of the output signal optical power and doping concentration. From Fig. 34.4a, it can be seen that the higher the doping concentration is, the faster the output power of the fiber laser grows, and the smaller the fiber length is required to obtain the maximum output power. At the same time, it is observed from Fig. 34.4b that the higher the doping concentration is, the faster the attenuation of the pump power is, and the shorter the transmission distance in the optical fiber is. The higher the doping concentration is, the higher the absorption degree of the fiber to the pump light is, so the maximum signal light output power can be achieved soon. In the remaining length of the fiber, the loss of the fiber will be accelerated, so the output power will be reduced.

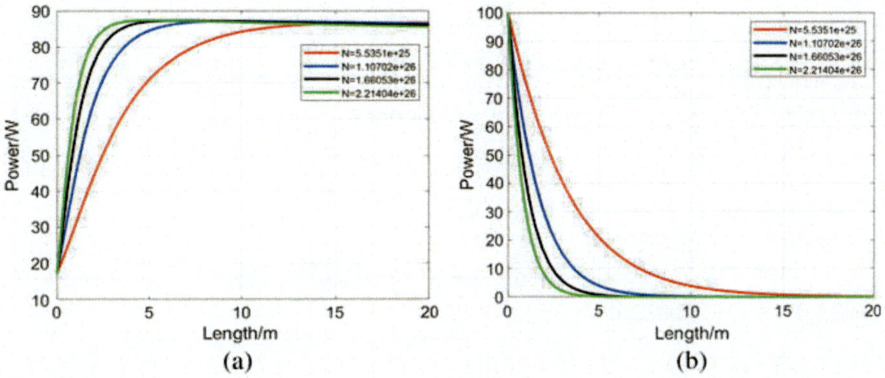

Fig. 34.4 Pump power changes of output power and pump power of ytterbium-doped double-clad fiber laser at different doping concentrations. **a** Output power of ytterbium-doped double-clad fiber laser. **b** Pump power

34.3 Experiment and Result Analysis of Fiber Laser

34.3.1 Experimental Setup

According to the basic structure composition of the double-clad fiber laser introduced in the previous section, it mainly includes pump source, fiber beam combiner, ytterbium-doped double-clad fiber, and fiber grating. So in order to get high power laser output, the pumping efficiency must be very high. The higher the pump power, the higher the output power of the laser. Therefore, the pump source is the diode-pumped laser (DPSSL). It can be seen from the previous Sect. 34.2.3 that forward pumping is the best under three kinds of pumping. The concentration of the doped fiber will affect the output power. Therefore, under the existing conditions, the gain fiber with the doping concentration of 4.0×10^{23} cm^{-3} is selected.

Based on the above analysis and the traditional gas detection system, a double-clad fiber laser gas detection system was built, as shown in Fig. 34.5. The emitted laser passes through a custom connector through KTP (frequency multiplexer) and becomes a central wavelength of 532 nm into a photonic crystal fiber. At the same time, the upper end of the T-interface is used to pass the gas into the photonic crystal optical fiber (HC-580-02, with 580 nm as the central wavelength) with high pressure so that the light interacts with the molecules of the sample to be measured and Raman scattering is generated. At the same time, the light waves without full effect continue to transmit in the photonic crystal fiber, and the FBG4 at one end of the fiber (with a reflectivity of 96% to 580 nm) is reflected back to FBG3 (with a reflectivity of 99% to 580 nm). It keeps oscillating in the optical resonator composed of FBG3 and FBG4, which increases the number of molecules involved in Raman scattering and improves the Raman intensity, finally through the photoelectric acquisition system, and then the data processing.

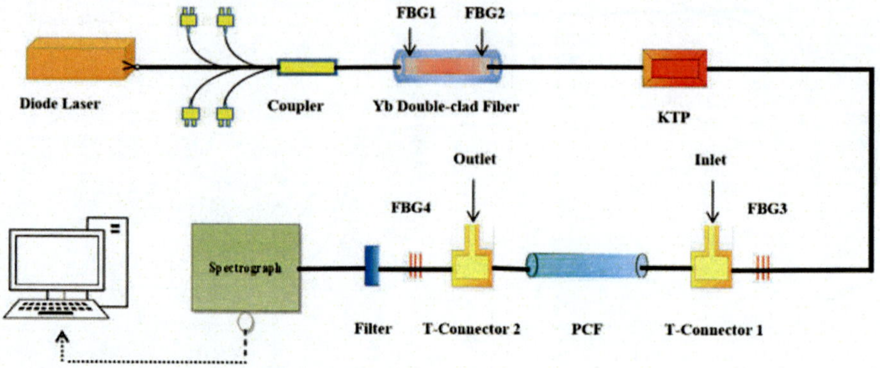

Fig. 34.5 The basic system of Raman gas detection for fiber laser

34.3.2 Test and Analysis of Experimental Equipment

Feasibility test of the experimental device

In optical detection, the laser diode is commonly used as the pump source. Therefore, the light source in Fig. 34.5 is replaced by the laser diode with an output power of 100 MW, forming a common optical detection system. In order to prevent the influence of the air in the optical path on the test results, nitrogen is first supplied for a period of time, and then 5% oxygen is aerated at 3 bar atmospheric pressure. As shown in Fig. 34.6a, it can be clearly seen that the oxygen Raman frequency shift occurs at 1556 cm^{-1}, which is consistent with the standard oxygen Raman frequency shift, proving that this optical detection system is feasible.

In order to explore the detection performance of the double-clad fiber laser as a light source, the above-mentioned laser diode was replaced with a double-clad laser

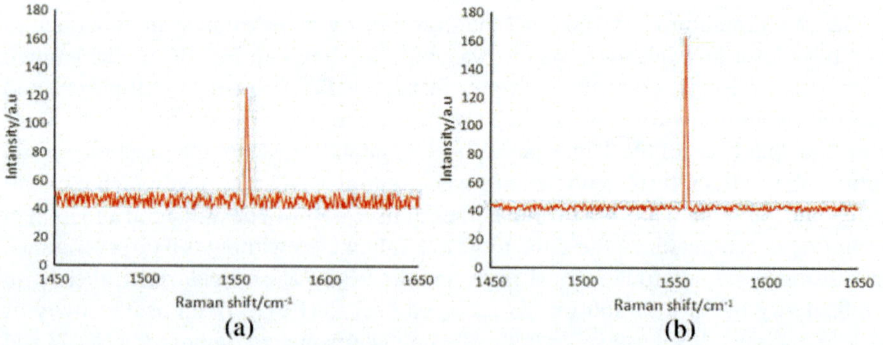

Fig. 34.6 Raman scattering diagram of oxygen molecules under different light sources. **a** Ordinary LD light source as detection light source **b** Ytterbium-doped double-clad fiber laser as detection light source

light source with an output power of 100 MW. As shown in Fig. 34.6b, it can also be seen that there is an oxygen Raman frequency shift at 1556 cm^{-1}, which indicates that neither the use of laser diode source as a Raman light source nor the double-clad fiber laser designed to detect the Raman light source will have any influence on the Raman frequency shift. But when laser diode and double-clad fiber laser are used as the Raman light source, the signal fluctuation is larger and the Raman intensity is smaller. Therefore, a double-clad fiber laser as a Raman detection source has better stability and sensitivity. Therefore, the linearity and stability of the experimental apparatus for the double-clad fiber laser as the light source are then tested.

Linearity test of the experimental device

Linearity is an important index of measuring the precision of a measuring instrument. In order to verify the sensitivity of the system, oxygen calibration experiments were carried out. Using EN4000 distribution device configuration range from 5 to 25% oxygen gas volume, respectively, to test the different concentrations of oxygen gas, oxygen concentration and intensity of Raman scattering is obtained by the experiment shown in Fig. 34.7a. The results of fitting, as shown in Fig. 34.7b, the result of oxygen concentration, and intensity of Raman scattering get an approximately linear relationship. The linear relationship between the two is $f(x) = 36.66591x - 1.85227$, and the linearity reaches 99.94%, indicating good linearity of the system.

Stability test of the experimental device

In order to verify the stability of the system, 10 portions of oxygen with concentrations of 5, 10, 15, 20 and 25% were configured for testing, and the experimental device was rinsed with nitrogen after each concentration of oxygen was tested to ensure the standard of oxygen concentration. The Raman test was carried out with oxygen at a concentration of 5% as an example, and the results were shown in the experimental test figure shown in Fig. 34.8a. The experimental results show that the oxygen concentration obtained from each experiment does not vary much. In the same

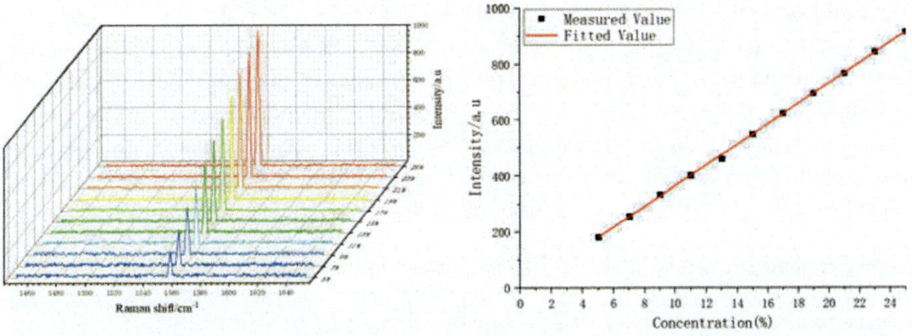

(a) (b)

Fig. 34.7 Sensitivity test results of the experimental device. **a** Detection results under different gas concentrations. **b** Raman scattering intensity fitting of the gas under different concentrations

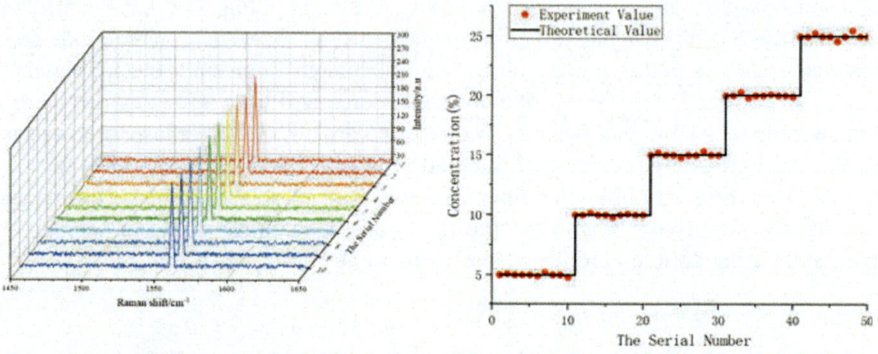

(a) (b)

Fig. 34.8 Stability test results of the experimental setup. **a** Gas with 5% concentration measured multiple times Raman test diagram. **b** Gas repeatability test diagram at different concentrations

way, other oxygen concentrations were measured. In order to verify the stability of the experimental device, the obtained results are summarized to obtain the resulting figure as shown in Fig. 34.8b. The dot represents the measured results and the straight line represents the theoretical concentration. As can be seen from Fig. 34.8b, the experimental results are basically consistent with the standard concentration, and the average repetition rate is 99%, indicating good stability of the system.

34.4 Conclusions

It is found that the double-clad doped laser has good stability and high linearity as the detection light source. Moreover, when the output power of the double-clad laser is large, the scattering cross-section can be effectively changed to enhance the Raman effect and the detection accuracy of low-concentration materials can be improved. Part from the first part of the theoretical study shows different ways of pumping, fiber doping concentration, pump power on the performance of the double-clad fiber laser have influence, therefore can change the way double-clad fiber laser pumping, pump input power, doping concentration, etc. to increase the output power of double-clad fiber laser, in order to enhance Raman optical input power, enhance the Raman effect, and achieve the low concentration gas detection.

Acknowledgements The authors acknowledge the financial support of the National Key R&D program of China (2019YFB1705800), and the 15th Batch of Jiangsu province "six talent peaks" high-level talent project (GDZB-042)

References

1. C.J. Halcrow, C. King, N.S. Manton, Oxygen-16 spectrum from tetrahedral vibrations and their rotational excitations. Int. J. Mod. Phys. **28**(04), 1950026 (2019)
2. R. Salter, J. Chu, M. Hippler, Cavity-enhanced Raman spectroscopy with optical feedback cw diode lasers for gas phase analysis and spectroscopy. Analyst **137**(20), 4669–4676 (2012)
3. L. Zou, Y. Yao, J. Li, High-power, efficient and azimuthally polarized ytterbium-doped fiber laser. Opt. Lett. **40**(2), 229–232 (2015)
4. X. Liu, J. Chen, J. Han et al., Design of large mode area total internal reflection photonic crystal fiber for high power fiber laser. High Power Laser Part. Beams **26**(10), 71–76 (2014)
5. H. Tian, Z. Hou, S. Zhang, G. Zhou et al., Research on ytterbium-doped photonic crystal fiber amplifier for the femtosecond fiber laser. Laser Phys. **26**(1), 1–6 (2016)
6. N. Nishizawa, Development and application of fiber laser. Appl. Opt. **42**(2), 438–445 (2013)
7. D. Mgharaz, N. Rouchdi, A. Boulezhar et al., Double-clad fiber laser design for particle image velocimetry and material science applications. Opt. Lasers Eng. **49**(1), 1–7 (2011)

Chapter 35
All-Fiber-Based Miniaturized Transportable Ultra-stable Laser at 1550 nm

Yafeng Huang, Lingke Wang, Yifei Duan, Yanli Li, Meifeng Ye, Qiuzhi Qu, Liang Liu, and Tang Li

Abstract We demonstrate an ultra-stable miniaturized transportable laser system at 1550 nm by locking it to an optical fiber-delay-line (FDL). To achieve better performance of ultra-stable laser system, a series of necessary measures are designed and implemented which are to restrain the temperature drift of FDL and the excess frequency noise induced by mechanical vibration. The fractional frequency instability of ultra-stable laser system is 3×10^{-15} at 1 s averaging time and below 5×10^{-15} at 1–128 s averaging time, which is the best medium-term frequency stability of a miniaturized transportable FDL-stabilized laser observed to date.

Y. Huang · L. Wang · Y. Duan · Y. Li · M. Ye · Q. Qu · L. Liu · T. Li (✉)
Key Laboratory of Quantum Optics, Shanghai Institute of Optics and Fine Mechanics, Chinese Academy of Sciences, Shanghai 201800, China
e-mail: litang@siom.ac.cn

Y. Huang
e-mail: yfhuang@siom.ac.cn

L. Wang
e-mail: lkwang@siom.ac.cn

Y. Duan
e-mail: duanyifei@siom.ac.cn

Y. Li
e-mail: liyanli@siom.ac.cn

M. Ye
e-mail: yemeifeng@siom.ac.cn

Q. Qu
e-mail: quqiuzhi@siom.ac.cn

L. Liu
e-mail: liang.liu@siom.ac.cn

Y. Huang · Y. Duan · Y. Li
Center of Materials Science and Optoelectronics Engineering, University of Chinese Academy of Sciences, Beijing 100049, China

G. Liu and F. Cen (eds.), *Advances in Precision Instruments and Optical Engineering*, Springer Proceedings in Physics 270, https://doi.org/10.1007/978-981-16-7258-3_35

35.1 Introduction

Ultra-stable lasers with high-frequency stability and phase coherence are indispensable as fundamental tools in the field of precision spectroscopy measurement science, optical atomic clocks, gravitational wave detection, ultralow phase noise optical or microwave synthesis and fundamental physics tests [1–8]. In the last two decades, the development of ultra-stable lasers with high-frequency stability and phase coherence has never been suspended. State-of-the-art ultra-stable lasers are usually realized by stabilizing laser frequency to an ultra-high-finesse Fabry–Pérot (FP) cavity using the Pound–Drever–Hall (PDH) method. Although this approach can achieve a level of frequency stability below 1×10^{-16} [9, 10], it requires precise mode matching, alignment of free-space-optical elements, precise temperature control or even cryocooler, and thus the system is bulky and complex. Therefore, it is difficult to meet the requirements in the transportable applications, such as optical atomic clock for geodesy [6] and gravitational wave detection in space [11–13]. In this letter, as a simple and compact alternative, fiber-delay-line (FDL) stabilized lasers are reported.

35.2 Material and Methods

In Fig. 35.1a, an arm-unbalanced optical fiber interferometer is used as a frequency discriminator to convert the frequency fluctuation ($\Delta\nu$) of the laser into the phase fluctuation ($\Delta\varphi$). The phase fluctuation is then detected and fed to the laser for frequency stabilization. The transfer function of the interferometer can be defined by

$$\Gamma(f) = \frac{\Delta\varphi(f)}{\Delta\nu(f)} = \frac{1 - \mathrm{e}^{-2\pi i f \tau}}{\mathrm{i}f} \tag{35.1}$$

where f is the Fourier frequency, τ is the unbalanced time delay of the interferometer which can be written as $\tau = 2nL/c$, n is the effective refractive index of optical fiber, L is the length difference between the two arms of the interferometer, and c is the speed of light in vacuum. At low frequencies ($f \ll 1/\tau$), the transfer function can be approximately equal to the follows

$$\Gamma(f) = \frac{1 - \mathrm{e}^{-2\pi i f \tau}}{\mathrm{i}f} \approx 2\pi\tau \tag{35.2}$$

The transfer function is plotted in magnitude in Fig. 35.1b. The transfer function has a series of zeros at the frequencies of n/τ ($n = 1, 2, 3\ldots$) and the first one limits the bandwidth of the control loop of the laser frequency stabilization. At low frequencies ($f \ll 1/\tau$), the transfer function is approximately equal to $2\pi\tau$, which means a longer fiber increases the gain of the discriminator while reduces the

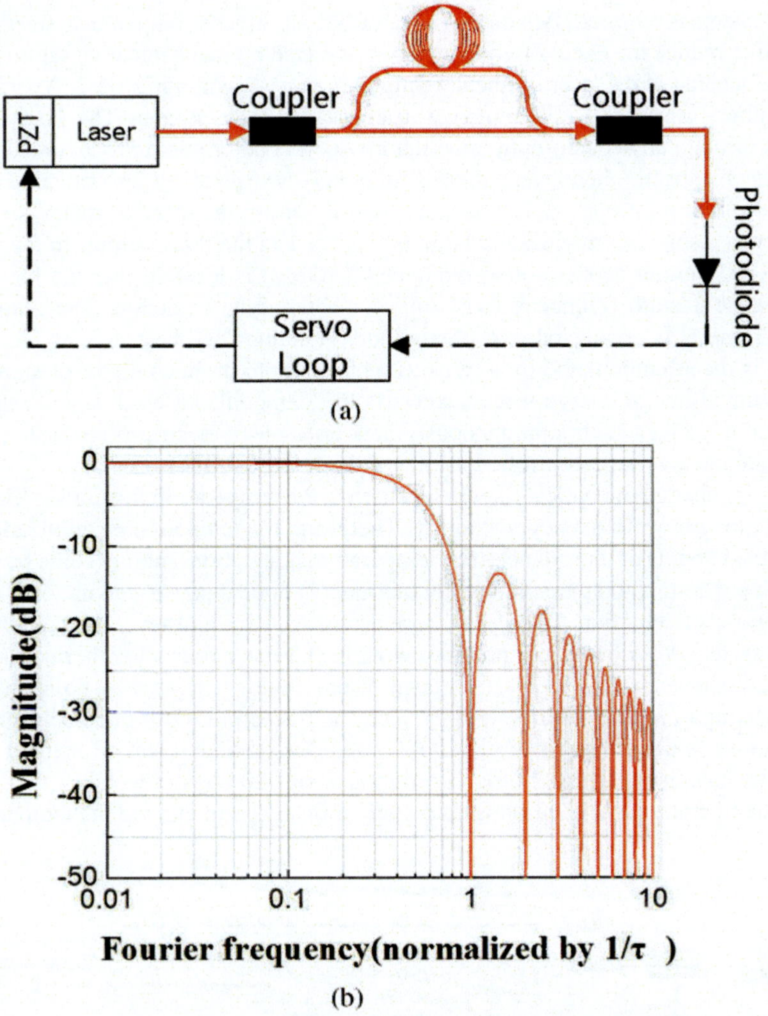

Fig. 35.1 **a** Principle of FDL laser stabilization and **b** transfer function of the interferometer

bandwidth of the control loop. In general, a compromise is required between the bandwidth of the control loop and the gain of the discriminator.

Recently, the FDL laser frequency stabilization has been demonstrated to achieve a sub-hertz linewidth ultra-stable laser [14–16]. Compared to the FP cavity method, this approach can provide not only agile frequency tunability, but also compactness, high reliability, small volume, and light weight [17]. These advantages make it possible to develop a miniaturized transportable ultra-stable laser. However, this approach is sensitive to temperature fluctuation. To migrate this problem, a delicate thermal shield system is required. Besides, an ultralow vibration sensitivity optical fiber spool is used to reduce the excess frequency noise induced by mechanical vibration.

The scheme of the laser frequency stabilization is shown in Fig. 35.2. A 1550 nm DFB fiber laser (NKT photonics) is used as the laser source. The laser beam passes through an acousto-optic modulator (AOM1) before going to an unbalanced Michelson interferometer, and then is split into two parts by an optical coupler with a coupling ratio of 99:1. The small part is sent to the interferometer for laser frequency stabilization while the large one is used as the laser output. In the interferometer, another acousto-optic modulator (AOM2) is inserted into the long arm to produce a radio frequency (RF) shift for heterodyne detection. The output of interferometer is connected to a photodiode, yielding a RF beat note signal. This signal is then demodulated by a tunable synthesizer to produce the error signal. A proportional-integral servo circuit converts the demodulated error signal into the correction signal which simultaneously acts on a piezo-electric transducer (PZT) stretcher and a voltage-controlled oscillator driving the AOM1.

In this scheme, the optical fiber is sensitive to the temperature fluctuation that can cause a long-term frequency instability. Therefore, the temperature stabilization of the optical fiber is of importance. The temperature fluctuations mainly come from (1) cabinet environmental temperature fluctuation; (2) optical power fluctuation; (3) RF signal power fluctuation. To migrate these problems, we adopt several modifications on the system described in our previous work [18]. First, a four-layer thermal shields are used (shown as Fig. 35.3). Each layer of the shields is treated by polishing and gold-plating to reduce their emissivity. The solar absorptivity of the shield surface is about 0.3 while the emissivity is about 0.03. Using a finite-element-method analysis, the calculated thermal time constant is about 20 days. Second, a two-stage active temperature stabilization is employed in this experiment and the temperature

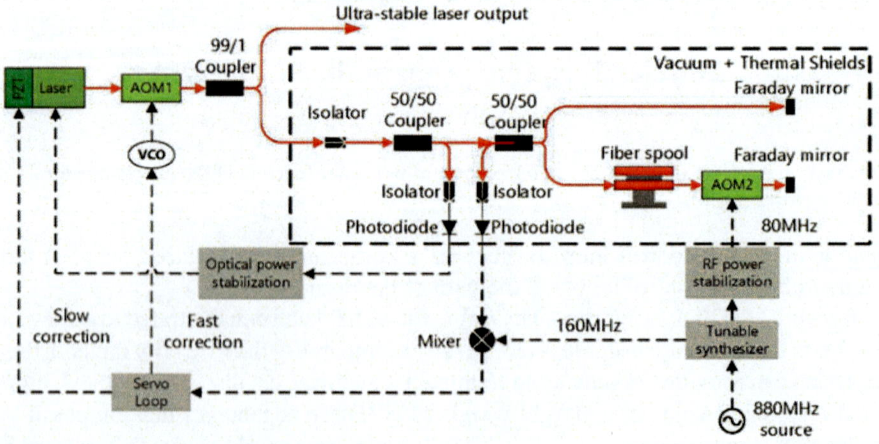

Fig. 35.2 The scheme of the laser frequency stabilization system. PZT: piezo-electric transducer; AOM: acousto-optic modulator; VCO: voltage-controlled oscillator; RF: radio frequency

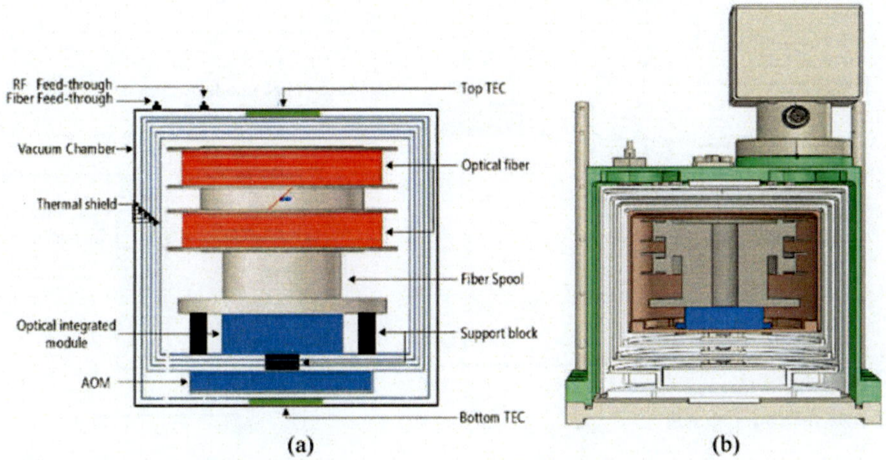

Fig. 35.3 **a** Schematic of the vacuum chamber and **b** its practicality picture

fluctuation of the outer shield is small than 0.5mK over a time of 24 h. Third, both the optical power injecting into the interferometer and the RF power driving the AOM1 are stabilized. Last but not least, to reduce the size of the laser stabilization system, a miniature optical fiber spool with low vibration sensitivity is designed and its volume is small than $1.7L$. The measured vibration sensitivity on the radial and axial direction is about $3 \times 10^{-11}/g$ and $8 \times 10^{-11}/g$, respectively, for a frequency range of 20–200 Hz [19]. The whole volume of the assembled laser stabilization system is about $5.0L$.

35.3 Results and Discussion

We construct two identical laser systems to evaluate the performance of the ultra-stable laser. Each laser is constructed using a separate fiber laser source, a separate fiber interferometer, a separate temperature control system, and separate electronics. Furthermore, both systems are identical and their contributions to the measurements can be considered as the same. The experimental setup used to measure the laser frequency stability is shown in Fig. 35.4. The light beams of the two stabilized lasers are combined by an optical fiber coupler and a heterodyne beat-note signal of approximate 160 MHz is detected by a photodiode. The frequency drift of the beat-note signal is compensated to less than 0.1 Hz/s by applying a small frequency offset to the tunable RF synthesizer [16].

The beat-note signal is then sent to a time interval analyzer (Symmetricom 5125A) for frequency comparison against a reference signal from an active hydrogen maser (iMaser3000, T4Science). The recorded phase data from the time interval analyzer are then used to calculate the frequency stability of the two lasers. For one laser,

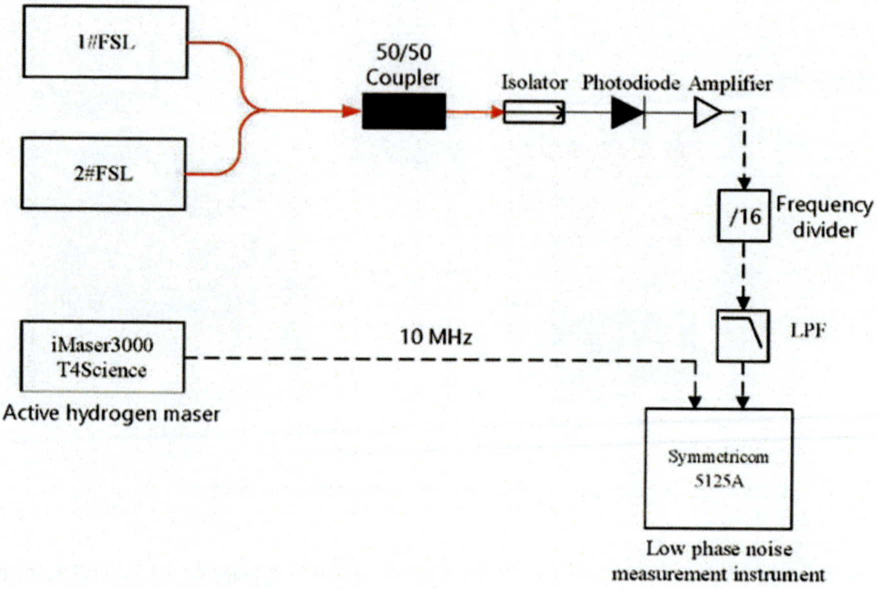

Fig. 35.4 Simplified diagram of the experimental setup. Optical paths are in solid, electrical in dotted line of dark. LPF: Low Pass Filter

the calculated frequency stability should be divided by $\sqrt{2}$. The measured frequency stability for a single system is shown in Fig. 35.5. In the time scale of 1–128 s, the frequency stability is better than 5×10^{-15}. The ascending trend at long-term

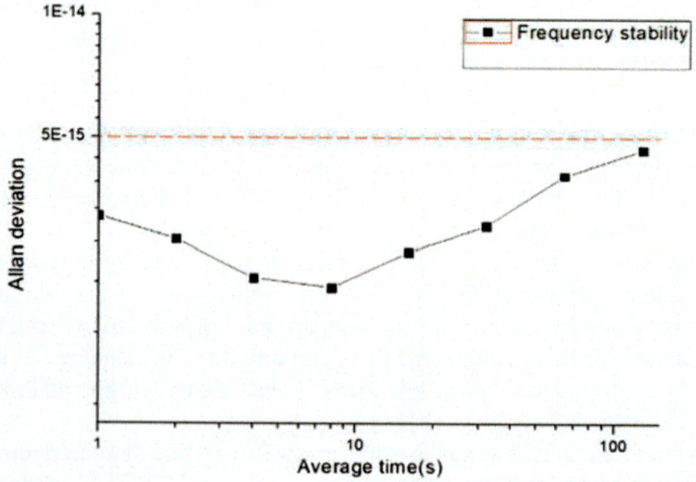

Fig. 35.5 Fractional frequency stability of all-fiber-based ultra-stable laser systems

time scale is mainly due to the residual temperature fluctuations or laser intensity fluctuations.

35.4 Conclusions

In conclusion, we demonstrate an all-fiber-based miniaturized transportable ultra-stable laser with a medium-term frequency stabilization of below 5×10^{-15} in the time scale of 1–128 s. In future work, a more stable temperature control system should be further investigated to improve the long-term frequency stability.

Acknowledgements This work was supported by the National Natural Science Foundation of China (NSFC) (Nos. 11604353, 11274324, and 11704391) and the Key Research Program of the Chinese Academy of Sciences (No. KJZD-EWW02).

References

1. A.D. Ludlow, M.M. Boyd, J. Ye, E. Peik, P.O. Schmidt, Optical atomic clocks. Rev. Mod. Phys. **87**, 637–701 (2015)
2. I. Ushijima, M. Takamoto, M. Das, T. Ohkubo, H. Katori, Cryogenic optical lattice clocks. Nat. Photonics **9**, 185–189 (2015)
3. N. Huntemann, C. Sanner, B. Lipphardt, C. Tamm, E. Peik, Single-ion atomic clock with 3×10^{-18} systematic uncertainty. Phys. Rev. Lett. **116**(063001), 1–1 (2016)
4. S.L. Campbell, R.B. Hutson, G.E. Marti, A. Goban, N.D. Oppong, R.L. McNally, L. Sonderhouse, J.M. Robinson, W. Zhang, B.J. Bloom, J. Ye, A Fermi-degenerate three-dimensional optical lattice clock. Science **358**, 90–94 (2017)
5. Y. Yao, Y.Y. Jiang, L.F. Wu, H.F. Yu, Z.Y. Bi, L.S. Ma, A low noise optical frequency synthesizer at 700–990 nm. Appl. Phys. Lett. **109**(131102), 1–1 (2016)
6. J. Grotti, S. Koller, S. Vogt, S. Häfner, U. Sterr, Ch. Lisdat, H. Denker, C. Voigt, L. Timmen, A. Rolland, F.N. Baynes, H.S. Margolis, M. Zampaolo, P. Thoumany, M. Pizzocaro, B. Rauf, F. Bregolin, A. Tampellini, P. Barbieri, M. Zucco, G.A. Costanzo, C. Clivati, F. Levi, D. Calonico, Geodesy and metrology with a transportable optical clock. Nat. Phys. **14**, 437–441 (2018)
7. W.H. Oskay, W.M. Itano, J.C. Bergquist, Measurement of the $^{199}Hg^{+}$ $5d^{9}6s^{2}$ $^{2}D_{5/2}$ electric quadrupole moment and a constraint on the quadrupole shift. Phys. Rev. Lett. **94**(163001), 1–1 (2005)
8. C. Eisele, A.Y.Nevsky, S.Y. Schiller, Laboratory test of the isotropy of light propagation at the 10^{-17} level. Phys. Rev. Lett. **103**(090401), 1–1 (2009)
9. S. Häfner, S. Falke, C. Grebing, S. Vogt, T. Legero, M. Merimaa, C. Lisdat, U. Sterr, 8×10^{-17} fractional laser frequency instability with a long room-temperature cavity. Opt. Lett. **40**(002112), 1–1 (2015)
10. D.G. Matei, T.Legero, S. Hfner, C. Grebing, R. Weyrich, W. Zhang, L. Sonderhouse, J.M. Robinson, J. Ye, F. Riehle, U. Sterr, 1.5 μm Lasers with Sub-10 mHz Linewidth. Phys. Rev. Lett. **118**(263202), 1–1 (2017)
11. B.S. Sheard, G. Heinzel, K. Danzmann, D.A. Shaddock, W.M. Klipstein, W.M. Folkner, Intersatellite laser ranging instrument for the GRACE follow-on mission. J. Geodesy **86**, 1083–1095 (2012)

12. R.X. Adhikari, Gravitational radiation detection with laser interferometry. Rev. Mod. Phys. **86**, 121–151 (2014)
13. J. Luo, L.S. Chen, H.Z. Duan, Y.G. Gong, S. Hu, J. Ji, Q. Liu, J. Mei, V. Milyukov, M. Sazhin, C.G. Shao, V.T. Toth, H.-B. Tu, Y.M. Wang, Y. Wang, H.C. Yeh, M.S. Zhan, Y. Zhang, Y. Zhang, V. Zharov, Z.B. Zhou, TianQin: a space-borne gravitational wave detector. Class. Quantum Gravity **33**(035010), 1–1 (2016)
14. F. Kéfélian, H.F. Jiang, P. Lemonde, G. Santarelli, Ultralow-frequency-noise stabilization of a laser by locking to an optical fiber-delay line. Opt. Lett. **34**, 914–916 (2009)
15. H.F. Jiang, F. Kéfélian, P. Lemonde, A. Clairon, G. Santarelli, An agile laser with ultra-low frequency noise and high sweep linearity. Opt. Express **18**, 3284–3297 (2010)
16. J. Dong, Y.Q. Hu, J.C. Huang, M.F. Ye, Q.Z. Qu, T. Li, L. Liu, Subhertz linewidth laser by locking to a fiber delay line. Appl Opt. **54**, 1152–1156 (2015)
17. J.C. Huang, L.K. Wang, Y.F. Duan, Y.F. Huang, M.F. Ye, L. Liu, T. Li, All-fiber-based laser with 200 mHz linewidth. Chin. Opt. Lett. **17**(071407), 1–1 (2019)
18. J.C. Huang, L.K. Wang, Y.F. Duan, Y.F. Huang, M.F. Ye, L. Li, L. Liu, T. Li, Vibration-insensitive fiber spool for laser stabilization. Chin. Opt. Lett. **17**(081403), 1–1 (2019)
19. Y.Q. Hu, J. Dong, J.C. Huang, T. Li, L. Liu, An optical fiber spool for laser stabilization with reduced acceleration sensitivity to 10^{-12}/g. Chin. Phys. **24**(104213), 1–1 (2015)

Chapter 36
Research on Dynamic Condition Test of Power Battery Simulation Based on Principal Component Analysis

Hong Pei Li and Guixiong Liu

Abstract In this paper, the vehicle driving conditions used to measure gas emissions and the test conditions of power battery in simulated electric vehicle driving are divided into segments and feature extraction. Through the principal component analysis method, the principal components in the two working conditions are compared and analysed. Select the test condition data US06 of the power battery corresponding to the vehicle driving condition data, compare the characteristic parameters of the two conditions, and analyse the influence of the characteristic parameters of the working conditions on the power battery model. A voltage simulation model is established to detect the key parameters of the test battery. Finally, it is concluded that the state transition frequency is positively correlated with the simulation voltage accuracy of the power battery model in the test condition of the simulated electric vehicle.

36.1 Introduction

At the same time, electric vehicles are driven in ambient temperature changes, power demand changes in a wide range and high frequency of operating conditions, which brings greater difficulties for power battery model construction, parameter estimation and testing. The working condition method is the first experimental method to evaluate the emission status and economy of traditional vehicles [1], and distinguish the test of light vehicles and heavy vehicles. GB/T 38146.1-2019 *China Vehicle Driving Conditions Part 1: Light Duty Vehicles* [2] and GB/T 38146.2-2019 *China Vehicle Driving Conditions Part 2: Heavy Duty Commercial Vehicles* [3] provide the latest standards to suit the driving conditions of domestic vehicles.

GB/T 31467.2-2015 *Lithium-ion Power Battery Packs and Systems for Electric Vehicles Part 2 Test Procedure for High Energy Applications* [4] provides for the simulation of working conditions discharge and charging constant charge/discharge

H. P. Li · G. Liu (✉)
School of Mechanical and Automotive Engineering, South China University of Technology, Tianhe, Guangzhou, Guangdong, China
e-mail: megxliu@scut.edu.cn

G. Liu and F. Cen (eds.), *Advances in Precision Instruments and Optical Engineering*, Springer Proceedings in Physics 270, https://doi.org/10.1007/978-981-16-7258-3_36

test methods for power batteries. The test method is to test the power battery cells for cyclic constant current conditions.

However, in the course of road driving, electric vehicles are affected by the environment and work under the transition of starting, accelerating, braking, maintaining relatively constant speed and idling conditions. The battery modules and cells in the power battery system are regulated by electrical excitation and battery balancing management, and work under cyclic transformations of randomly varying currents for discharge and energy recovery [5]. The model construction and evaluation of key parameters of the power battery under dynamic operating conditions requires the use of simulated operating conditions for relevant tests [6, 7].

Depending on the road conditions and driver habits of each country [8], different driving conditions are constructed, and the variability of the discharge and energy recovery characteristics of the battery cells, modules and systems are also different after transformation by the EV simulation software [9, 10].

36.2 Battery Simulation Data Selection and Segmentation

36.2.1 Battery Simulation Data Selection

The power battery simulated working condition discharge and energy recovery power-time schedule mainly consists of two types: (1) vehicle simulation software, input vehicle driving working condition speed-time schedule, output battery system power-time schedule, such as New Europe Driving Cycle (NEDC), The Federal Urban Driving (FUDS) (2) cycle-shifted multiplier discharge and charging processes at various operating currents, such as Dynamic Stress Test (DST), Beijing Dynamic Stress Test (BJDST). In this paper, we use the working condition test data at different temperatures from the University of Maryland battery test dataset.

Figure 36.1 shows the speed-time diagram for the driving conditions of the electric vehicle and Fig. 36.2 shows the power-time diagram for the power cell under simulated driving conditions of the electric vehicle.

36.2.2 Segmentation

In this paper, we first segment multiple sets of vehicle driving condition speed-time-table data (US06, NEDC, Highway, FTP75), and also segment battery power-time under simulated dynamic conditions (FUDS, US06, DST, BJDST) of the power battery. The condition segmentation transforms the condition engineering into a process of inter-transfer between states and analyses the effect of the frequency of state changes on the battery simulation condition testing. In this regard, the length of time and division of the velocity–time data time segments are given by the following

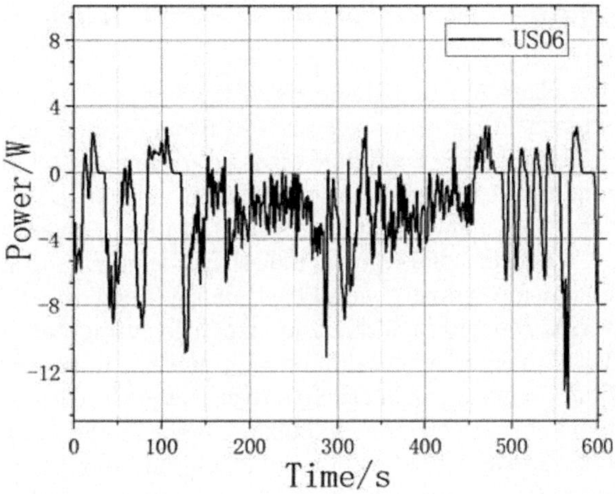

Fig. 36.1 Electric vehicle driving conditions US06 speed-time diagram

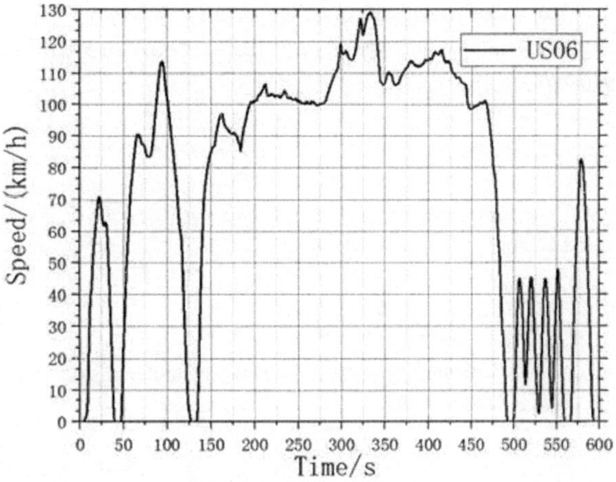

Fig. 36.2 Power battery unit power-time diagram under EV driving simulation

equations.

$$T = t_{end} - t_{start} \tag{36.1}$$

where T is the single state time length, t_{start} is the start time of that state and t_{end} is the end time of that state.

36.2.3 Feature Parameter Construction

The characteristic parameters of the analysed segments were constructed, and the segmented segments were statistically and analytically divided. Table 36.1 shows the selected driving conditions' feature parameters, which are mainly divided into the total time share of the fragment, the fragment speed and state transition frequency acceleration amount and the state transition frequency feature parameters.

Where the total segment time share includes uniform speed Travel time T_1, Acceleration travel time T_2, Braking time T_3, Idling time T_4 and start time T_5. Slice speed versus Amount of acceleration includes average speed $V_{average}$, Maximum velocity V_{max}, maximum acceleration a_{max} and maximum deceleration a_{break}. State transitions Frequency includes Start to accelerate, acceleration to constant speed, uniform to braking, acceleration to braking, brake to idle, braking to standstill. Frequency to time ratio of state transitions includes Start to accelerate f_1, Acceleration to constant speed f_2, Uniform to braking f_3, Acceleration to braking f_4, Brake to idle f_5, Braking to standstill f_6, Frequency to time ratio of state transitions *r.*

The total time share of selected simulated electric vehicle driving condition segments and state transition frequencies are calculated by (36.2) and (36.3) as follows.

$$f_i = \frac{n_i}{\sum_{j=1}^{6} n_j} (i = 1, 2, \ldots, 6) \tag{36.2}$$

$$T_i = \frac{\sum_{k=1}^{m} T_{k,i}}{\sum_{j=1}^{5} \sum_{k=1}^{m} T_{k,j}} (i = 1, 2, \ldots, n) \tag{36.3}$$

Table 36.1 Calculation of battery discharge and energy recovery operating conditions

	US06	BJDST	DST	FUDS
t_1	0.06	0.08	0.12	0.15
t_2	0.91	0.90	0.68	0.62
t_3	0.02	0.01	0.10	0.12
t_4	0.00	0.00	0.10	0.11
$P_{average}$	0.92	0.68	0.88	0.80
P_{max}	14.32	6.34	14.54	14.50
P_{cmax}	11.07	4.18	12.59	15.19
P_{dmax}	−16.38	−5.19	−13.16	−15.19
F_1	0.33	0.33	0.33	0.32
F_2	0.37	0.36	0.35	0.40
F_3	0.00	0.00	0.00	0.00
F_4	0.31	0.30	0.32	0.28
R	0.00	0.00	0.01	0.00

where f_i is the frequency of state transitions in the driving condition of the electric vehicle, n_i is the number of 6 state shifts in the driving condition process; T_i is the total time share of the electric vehicle driving condition state fragment; $T_{k,j}$ indicates the time length of the single state of the kth time slice in state j; m indicates the number of times state j appears. The selected simulated electric vehicle driving condition fragment speed and acceleration quantity calculation formula (36.4) is as follows.

$$V_{average} = \frac{1}{m_{\text{total}}} \sum_{i=1}^{m_{\text{total}}} \left(\frac{1}{T_i} \int_{t_{i,start}}^{t_{i,end}} V_i(t)dt \right) \tag{36.4}$$

where $V_{average}$ is the total average speed of the segments of the electric vehicle driving condition; m_{total} is the total number of time segments; T_i is the time length of a single state of segment i in the condition; t_{start} is the start time of the segment state and t_{end} is the end time of the segment state; V_i is the amount of time variation within the segment.

$$r = \frac{m_{total}}{T_{total}} \times 100\% \tag{36.5}$$

where r is the state transition frequency as a percentage of total time; m_{total} is the total number of time segments; T_{total} is the total length of the driving condition time state end time; V_i is the amount of time change within the segment.

The power battery power state change in the process of simulated power battery discharge and energy recovery is divided and the characteristic quantity is extracted, and the power battery discharge and energy recovery conditions are mainly divided into continuous discharge segment, continuous energy recovery segment and stationary segment.

Where the total segment time share includes Uniform discharge time t_1, Accelerated, Discharge time t_2, energy recovery time t_3 and resting time t_4. Slice power variables include Average output power $P_{average}$, Maximum output Power P_{max}, Maximum power climb rate P_{cmax} and Maximum power drop rate P_{dmax}. State transitions Frequency includes Standstill to acceleration F_1, Acceleration to energy recovery F_2, Uniform discharge to energy recovery F_3, Accelerated discharge to standstill F_4, Frequency to time ratio of state transitions R.

The total time share of the selected simulated power cell discharge and energy recovery segments and the state transition frequency are calculated in (36.5) and (36.6) as follows.

$$F_i = \frac{N_i}{\sum_{j=1}^{3} N_j} (i = 1, 2, 3) \tag{36.6}$$

$$t_i = \frac{\sum_{k=1}^{m} t_{k,i}}{\sum_{j=1}^{5} \sum_{k=1}^{m} t_{k,j}} (i = 1, 2, ..., n) \tag{36.7}$$

where F_i is the power battery state transition frequency, N_i is the power battery working condition power 5 state transfer times; t_i is the power battery working condition power state fragment total time share; $t_{k,j}$ indicates the state j in the kth time piece to maintain a state time length; m indicates the state j appear times. The average output power $P_{average}$ in the simulated power condition of the power cell is calculated as follows.

$$P_{average} = \frac{1}{m_{\text{total}}} \sum_{i=1}^{m_{\text{total}}} \left(\frac{1}{T_i} \int_{t_{i,start}}^{t_{i,end}} P_i(t)dt \right) \tag{36.8}$$

36.3 Principal Component Analysis and Battery Equivalent Circuit Model Construction

Let there be K working conditions, each with p characteristic parameters, denoted as $X = (x_1, x_2, \ldots, x_p)$. Let the mean value of the random variable X be μ and the covariance matrix $\sum$. After normalising the data in the working conditions by $(x_1 - \mu)/\sigma$ and normalising X, the covariance matrix $\sum$ of x is equal to its correlation coefficient matrix. The principal component is the problem of linearly combining p characteristic parameters. A linear transformation of X generates a new composite indicator, the principal component, denoted $y_1, y_2, \ldots, y_p$. The covariance matrix is used to find the eigenvalues $\lambda_1, \lambda_2, \ldots, \lambda_p$ $(\lambda_1 \geq \lambda_2 \geq \ldots \geq \lambda_p)$ and the corresponding eigenvectors A as in (36.9).

$$A = \begin{bmatrix} a_{1,1} & a_{1,2} & \ldots & a_{1,p} \\ a_{2,1} & a_{2,2} & \ldots & a_{2,p} \\ \ldots & \ldots & \ldots & \ldots \\ a_{p,1} & a_{p,1} & \ldots & a_{p,p} \end{bmatrix} \tag{36.9}$$

Then each principal component of the characteristic parameters $x_1, x_2, \ldots, x_p$ after the orthogonal transformation can be expressed as (36.10)

$$\begin{cases} y_1 = a_{1,1}x_1 + a_{1,2}x_2 + \ldots + a_{1,p}x_p \\ y_2 = a_{2,1}x_1 + a_{2,2}x_2 + \ldots + a_{2,p}x_p \\ \ldots \\ y_p = a_{p,1}x_1 + a_{p,2}x_2 + \ldots + a_{p,p}x_p \end{cases} \tag{36.10}$$

where $y_1, y_2, \ldots, y_p$ are called the principal components, respectively.

Fig. 36.3 Thevenin equivalent circuit model of the power cell

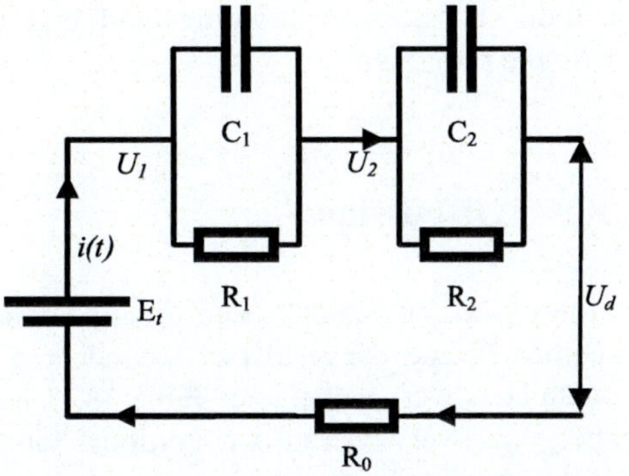

An equivalent circuit model is constructed [11], with the equivalent power cell unit as a second-order equivalent circuit. Figure 36.3 shows the power cell Thevenin second-order equivalent circuit model. R_0 is the internal resistance of the power cell; E_t is the open-circuit voltage potential; R_1 is the electrochemical polarisation internal resistance, R_2 is the concentration polarisation internal resistance; C_1 is the electrochemical polarisation capacitance, C_2 is the concentration polarisation capacitance; U_d is the measurable terminal voltage; U_1, U_2 are the battery polarisation internal resistance voltages and $i(t)$ is the real-time battery operating current.

The power cell unit model is built by Simulink. Figure 36.4 shows a schematic diagram of the Simulink equivalent circuit model of the power cell unit.

The HPPC test is carried out on the power cell, and the OCV-SOC curve and R_0, R_1, R_2, C_1 and C_2 equivalent circuit model parameters of the battery cell are identified by the HPPC condition. Finally, the current in various battery discharging and energy recovery conditions is used as model input to observe the error between the simulated voltage output and the actual observed voltage in the power cell simulation model,

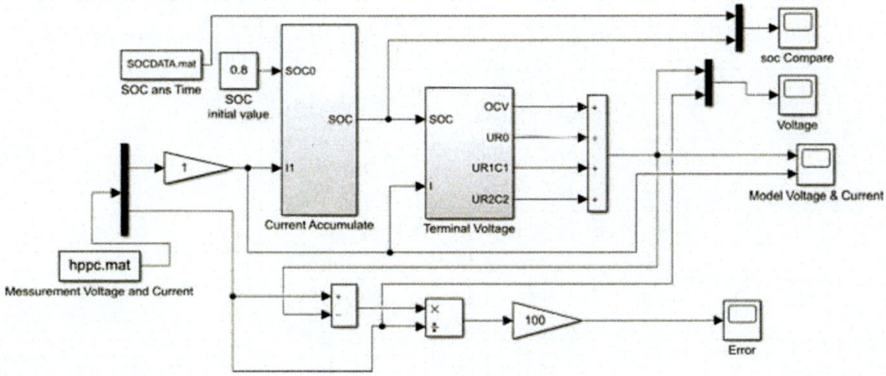

Fig. 36.4 Simulink equivalent circuit model of power battery unit

and the characteristic quantities with high contribution to the battery are compared with the error values.

36.4 Discussion

Multiple sets of vehicle speed-timeline data (US06, NEDC, Highway, FTP75) were segmented and the results of the selected driving condition segmentation feature parameters were calculated. Table 36.1 shows calculation of battery discharge and energy recovery operating conditions. Table 36.2 shows variation of response error of the simulated cell model under different operating conditions. Table 36.3 shows battery discharging and energy recovery working condition principal component score table.

Table 36.2 Variation of response error of the simulated cell model under different operating conditions

Simulated working conditions	Average absolute error of voltage simulation %	Voltage simulation maximum absolute error %	Root mean square error of voltage simulation %
US06	2.41	18.20	3.16
BJDST	2.26	7.06	8.71
DST	1.48	19.48	2.79
FUDS	3.21	19.91	4.70

Table 36.3 Battery discharging and energy recovery working condition principal component score table

Principal component number	Eigenvalue	Contribution	Cumulative contribution
1	6.63	0.51	0.51
2	4.29	0.33	0.84
3	2.08	0.16	1
4	3.87×10^{-16}	2.98×10^{-17}	1
5	3.30×10^{-16}	2.54×10^{-17}	1
6	2.10×10^{-16}	1.63×10^{-17}	1
7	1.06×10^{-16}	8.19×10^{-17}	1
8	-7.47×10^{-17}	-5.74×10^{-17}	1
9	-1.38×10^{-16}	-1.06×10^{-17}	1
10	-2.24×10^{-16}	-1.73×10^{-17}	1
11	-2.45×10^{-16}	-1.89×10^{-17}	1
12	-4.94×10^{-16}	-3.80×10^{-17}	1
13	-1.15×10^{-16}	-8.87×10^{-17}	1

The analysis of the battery discharge and energy recovery working condition principal component score table, with the battery simulation model under this discharge and energy recovery working condition, shows that the prediction error of the power battery model is positively correlated with the high contribution of the characteristic parameters in the principal component analysis of the battery discharge and energy recovery working condition that it is subjected to.

36.5 Conclusion

In this paper, we select and segment the data of various types of electric vehicle driving conditions and the simulated working condition data used to simulate the power battery, extract the relevant characteristic parameters in the working condition data, and analyse the simulated working condition data of the power battery and the electric vehicle driving condition data by using the principal component analysis method. The typical working conditions of the simulated power battery working condition data and the electric vehicle driving working condition US06 are selected to analyse and compare the principal components of the two types of working conditions and to analyse the influence of the working conditions on the construction of the power battery model.

Acknowledgements This paper is one of the stage results of the Guangdong Key Areas R&D Project Power Battery System Testing and Evaluation Technology Research (2019B090908003). The authors also thanks to the Center for Advanced Life Cycle Engineering (CALCE) at the University of Maryland for providing us with an open-source dataset of power cells for our services.

References

1. J. Brady, M. O'Mahony, Development of a driving cycle to evaluate the energy economy of electric vehicles in urban areas. Appl. Energy **177**, 165–178 (2016)
2. GB/T 38146.1-2019, Driving conditions for automobiles in China Part 1: Light-duty vehicles (2019)
3. GB/T 38146.2-2019, Driving conditions for automobiles in China Part 2: Heavy commercial vehicles (2019)
4. China Automotive Technology Research Center, The Eighteenth Research Institute of China Electronics Technology Group Corporation, Tianjin Lishen Battery Co. Lithium-ion power battery packs and systems for electric vehicles Part 2: Test procedures for high-energy applications. J. GB/T 31467.2-2015:24 (2015)
5. X.-Y. Song, C.-L. Wang, Electric vehicle power system modeling based on joint CarSim/Simulink simulation. Agric. Equip. Veh. Eng. **58**, 100–105 (2020)
6. F. Sun, X. Meng, C. Lin, Research on dynamic test conditions of electric vehicle power battery. J. Beijing Univ. Technol. **30**, 297–301 (2010)
7. H. Gong, Y. Zou, Q. Yang, Generation of a driving cycle for battery electric vehicles: a case study of Beijing. Energy **150**, 901–912 (2018)

8. S. Shi, N. Lin, Y. Zhang, Research on Markov property analysis of driving cycles and its application. Transp. Res. Part D-Transp. Environ. **47**, 171–181 (2016)
9. E. Chemali, P.J. Kollmeyer, M. Preindl, State-of-charge estimation of Li-ion batteries using deep neural networks: a machine learning approach. J. Power Sources **400**, 242–255 (2018)
10. F. Wang, Z. Wang, S. Xu, Complex operating conditions testing and simulation of integrated environmental power battery systems. Power Technol. **40**, 527–528+579 (2016)
11. K.X. Wei, C. Qiaoyan, State estimation of lithium-ion power battery based on adaptive traceless Kalman filter algorithm. Chin. J. Electr. Eng. **34**, 445–452 (2014)

Part IV
Optical Remote Sensing and Satellite Image Positioning

Chapter 37
Research on the Application of UAV Remote Sensing Technology in Surveying and Mapping Engineering Survey

Jiabin Yan

Abstract This paper mainly studies the low-altitude UAV (unmanned aerial vehicle) remote sensing digital processing system. The system is mainly a set of digital remote sensing image processing system developed for the data acquired by the system. First, it describes the basic principles of photogrammetry on which the system is based, and mainly explains the advantages of the beam method area network adjustment. This article briefly describes the common methods of surveying technology for power transmission and transformation lines. UAV aerial survey technology and GIS platform are applied to electric power surveys; the UAV aerial survey data are used as the data source to establish a GIS database, and the detailed system design of the structure, interface, and database of the electric power survey geographic information system is carried out. The secondary development realizes the main functions of the system such as data management, spatial query, surface analysis, and spatial analysis, facilitating scientific information management and comprehensive analysis, providing auxiliary decision-making, and bringing new methods to electric power surveys. Secondly, this article points out that the system uses an ordinary digital camera with small image size and large distortion. This article is based on the basic principles of photogrammetry, and it is worthwhile to use navigation GPS. Thirdly, the accuracy of the aerial triangulation results of the system and the accuracy of the digital orthophoto were analyzed through a large amount of data.

37.1 Introduction

With the rapid development of the country's economic construction, most regions already have satellite remote sensing images and traditional aerial image data, and there is a demand for local area real-time, mobility, high-resolution, and high-precision remote sensing image data [1]. Its control method can be divided into remote control, semi-autonomous, program control, or both. It is the future aviation aircraft [2]. UAV has simple structure and low cost. It can not only complete the task

J. Yan (✉)
Engineering College, Yunnan University of Business Management, 296 Haitun Road, Wuhua District, Kunming 650106, Yunnan, China

G. Liu and F. Cen (eds.), *Advances in Precision Instruments and Optical Engineering*, Springer Proceedings in Physics 270, https://doi.org/10.1007/978-981-16-7258-3_37

of manned aircraft, but is more suitable for tasks that are not suitable for manned aircraft, such as reconnaissance and remote sensing monitoring of dangerous areas, and tasks that require long endurance and regular remote sensing monitoring. With the rapid development of the country's economy and society, the demand for new survey technology is becoming more and more urgent. The emergence of new technologies such as drone aerial survey and satellite remote sensing survey provide new technical support for electric power surveys [3].

Since the development and utilization of GPS technology, it has been widely used in engineering applications. In the process of image registration using the method based on the transform domain, a certain transformation of the image is first required, and then a certain area is used as a template for image stitching [4]. Yan [5] proposed to transform the image information into the frequency domain through Fourier transform, and then use the cross power spectrum to calculate the amount of translation between images. The method of extended phase correlation was proposed by Wu [6]. They found that the Fourier transform can be applied to the matching of rotation and translation changes between images. The feature points of the image can be obtained by the wavelet transformation technology, the rotation transformation between the images can be obtained by the iterative method, and the polar coordinate transformation can be used to achieve more accurate and stable matching. It has a smaller workload, faster calculation speed, and can still maintain its performance under the perturbation of multiple parameters such as noise and distortion. Harris corner detection operator is proposed by Zongjian [7]. Some areas of the image have rich texture information. This method can extract a sufficient number of usable feature points. The limitations of this method are also obvious. The feature points that can be extracted from regions with relatively scarce information are very limited. The image stitching model based on the L-Ms algorithm was proposed by Hu [8]. It has a very good effect when stitching a series of images with multiple transformations, and the convergence speed is also very fast. Tziavou et al. [9] used two-dimensional Gaussian fuzzy filtering to obtain the corner model, edge model, and vertex model. The method of extracting corner points by scanning along the arc curve greatly improves the detection accuracy of feature points. Wenquan et al. [10] proposed a method of multi-feature image registration using feature points and straight line segments. The idea is to solve the problem of image matching by finding the most similar parts of the two curves. Once the point with the largest gradient by column can be determined, the position of the image matching can be determined.

This article first introduces the low-altitude UAV remote sensing system that the author has participated in and successfully developed. Figure 37.1 shows a schematic diagram of the hierarchical distribution based on UAV remote sensing technology. First, the composition of the UAV flight platform system, GPS navigation system, flight control system, remote sensing photography system, ground control system, and signal transmission system of the low-altitude UAV remote sensing system is explained. At the same time, it points out that the low-altitude UAV remote sensing system has the characteristics of lightness, flexibility, safety, and easy operation. It does not require complicated application approval procedures like traditional aerial photography, does not require dedicated take-off and landing airports, and does not

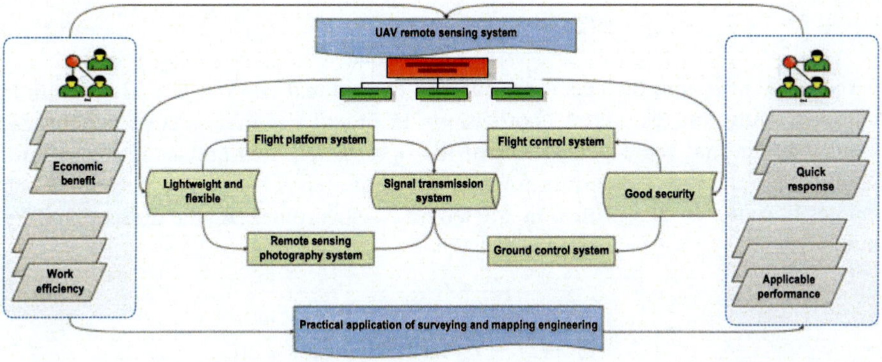

Fig. 37.1 Schematic diagram of hierarchical distribution based on UAV remote sensing technology

need to be expensive. The various navigation and communication equipment can greatly reduce the cost of aerial photography field. In small-scale aerial photography operations, its rapid response capability and economic benefits compared with traditional aerial photography have obvious advantages, and the production process is introduced. Finally, it is pointed out that the system also has its own limitations, and whether it can meet the requirements of the aerial photography field, this article uses a large amount of experimental data to clarify that the image data obtained by the low-altitude UAV remote sensing system meets the requirements of the aerial field industry. Taking a bauxite mining area as the research object to analyze the application of low-altitude drone remote sensing technology in the mining area surveying and mapping, including the acquisition of drone image data, the deployment of field image control points, and the key technology of indoor aerial triangulation. The remote sensing results of low-altitude drones are applied to the land reclamation of mining areas, and the digital orthophotos obtained by the low-altitude drones are used as the basic data for the planning and design of land reclamation, which greatly improves the efficiency of reclamation.

37.2 Construction of Surveying and Mapping Engineering Survey Model Based on UAV Remote Sensing Technology

37.2.1 UAV Remote Sensing Image Acquisition

The UAV is a non-manned aircraft. It is controlled by radio remote control equipment or by its own program. Unmanned Aerial Vehicle Remote Sensing is a combination of unmanned aerial vehicle and remote sensing technology and applied to practical engineering. It can be obtained in a short time.

$$x(1) + x(2) = 0, 0 < x(1) < 1, 0 < x(2) < 1 \tag{37.1}$$

In accordance with the specifications and operational tasks of UAV low-altitude remote sensing images, aerial photography technical parameters such as altitude, photo overlap, and route parameters of the operational area are set in the ground monitoring software. According to the size of the survey area, we determine the number of routes in the entire area, the length of each route, and the direction of the route.

$$y(x) = \sqrt{\frac{x(1) + x(2) + \cdots + x(i)}{x(1) + 2 * x(2) + \cdots + i * x(i)}} \tag{37.2}$$

It uses a high-resolution digital camera carried on the body to collect images of the measured area, so that it can quickly and efficiently obtain relatively high-resolution remote sensing images, and use corresponding internal processing software to quickly process digital image data to produce geographic information products, and related thematic maps that meet the accuracy of the internal industry specifications.

$$\begin{cases} u(x) = a(1) * x(1) + a(2) * x(2) + ...a(i) * x(i) \\ v(x) = \frac{\sum a(i)*x(i)}{\sum a(i)*u(x)} \end{cases} \tag{37.3}$$

When the drone is operating in the survey area, due to its own and external factors, the photo cannot be kept strictly horizontal during photography, and the uneven terrain will also cause the projection center point to shift. The creation of an accurate one-to-one correspondence between the photo point and the actual point in space is the geometric correction of the image, and two coordinate systems, the image side and the object side, are often used in the process. The image coordinate system represents the plane coordinates and space coordinates of the image point, including image plane coordinate system, image space coordinate system, and image space auxiliary coordinate system.

$$f(x, y) = (1 - |\frac{x}{x+1} - 1/2|) * (1 - |\frac{y}{y+1} - 1/2|) \tag{37.4}$$

Both high-resolution remote sensing image data for positioning can be obtained through the UAV low-altitude remote sensing measurement system. At the same time, UAV can obtain multi-spectral, multi-temporal, and multi-resolution remote sensing data by carrying different types of sensors. The high-resolution digital imaging equipment mounted on the UAV platform can not only take upright images, but also have the ability to acquire images in a tilted or even vertical state, covering a wide area, and remote sensing images are obtained in the process of acquiring image data.

37.2.2 Design of Surveying and Mapping Engineering Survey Model

The parameters of several images acquired in the same scene at different times and different lights are determined according to the degree of coincidence of the overlapping parts. The process of obtaining the best matching effect is called image registration, which is the core step of image stitching. There are several methods for image stitching, which can be selected according to the actual situation. (1) Similarity measure: the degree of matching between two images can be expressed by a similarity measure. Selecting appropriate metrics can effectively reduce noise interference, make image registration accurate to a high level, and have stronger anti-interference. The choice of similarity measure should be based on the actual registration situation. (2) Feature space: feature space is mainly based on the collection of some unique features (grayscale and contour, etc.) of the image itself. Whether the selected feature space is suitable is closely related to the accuracy of the image information.

$$\begin{bmatrix} u(x) \\ u(y) \\ u(z) \end{bmatrix} = \begin{bmatrix} 1\,0\,0 \\ 0\,1\,0 \\ 0\,0\,1 \end{bmatrix} * \begin{bmatrix} dx\ dy\ dz \end{bmatrix} \tag{37.5}$$

There are two types of transformation methods: linear and non-linear, and the transformation scope includes three types: global, local, and displacement fields. In order to find the most suitable transformation model, the search strategy adopted needs to be based on spatial search, and the similarity measure is taken as the standard. Using the most suitable algorithm can effectively reduce the workload and greatly improve the speed and accuracy of registration.

Figure 37.2 shows the flow chart of the surveying and mapping engineering survey model based on UAV remote sensing technology. We choose a specific point on the target image as the target matching point, and use the selected target matching point as the center to select a grayscale matrix of pixels as the target area. According to the power spectrum, the position movement, size change, and angular deflection of the two images can be calculated, and then they can be stitched together. In the image matching method based on image features, the image matching speed is fast, the workload is small, and the anti-interference of the gray features is strong. Feature matching refers to matching the retrieval questions that express information needs with the information identifiers stored in the retrieval system, and selecting the same information between the two for output. The overall transformation between images with overlapping regions can be derived from the feature matching relationship, and the model parameters can be further derived. After the geometric model and its parameters are determined, the image can be resampled to unify it in the reference coordinate system to achieve the final matching.

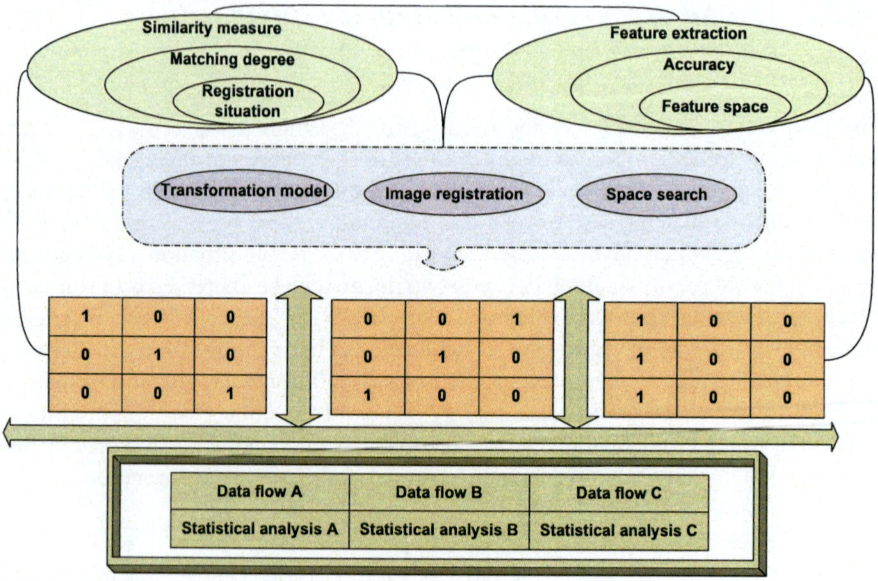

Fig. 37.2 Flow chart of surveying and mapping engineering survey model based on UAV remote sensing technology

37.3 Application and Analysis of Surveying and Mapping Engineering Survey Model Based on UAV Remote Sensing Technology

37.3.1 UAV Remote Sensing Image Classification Data Processing

In order to verify the calculation method, this paper uses two methods of UAV low-altitude remote sensing measurement and GNSS-RTK field measurement to collect data in the same survey area, obtain experimental data, and use CASS9.0 mapping software to process the data. The image point coordinates of each image are the original observations, and the unknowns are the external orientation elements of each image and the ground coordinates of the points to be determined. The error equation directly lists the original observations, which can most conveniently take into account the effects of imaging system errors, and it is most convenient to introduce additional observations, such as navigation data and ground measurement observations. The number of elevation points collected by GNSS-RTK is 372, the sampling interval is 50, and the manual time is 8 h. The amount of point cloud data obtained by using drones is 49013, the sampling interval is 20, and the required time is 2.5 h. Figure 37.3 shows the statistical distribution of remote sensing measurement errors at different sampling points.

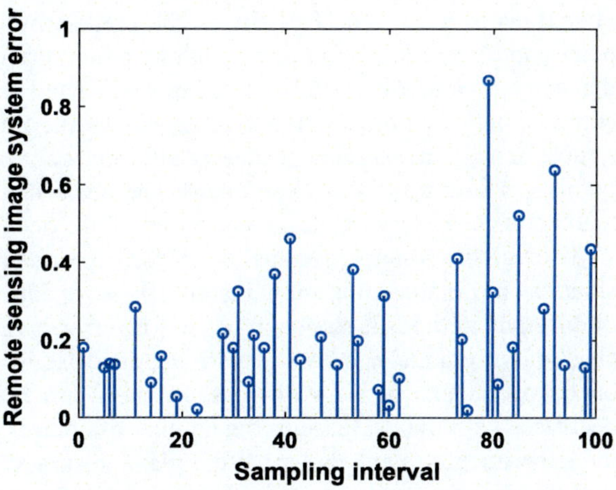

Fig. 37.3 The statistical distribution of remote sensing measurement errors at different sampling points

Through the analysis of the experimental comparison results, it can be seen that for the larger the earthwork calculation area, the advantages of UAV point cloud data are more obvious, the measurement speed is fast, and the accuracy is high. The error of the object coordinate is inversely proportional to the size of the base height ratio (intersection angle) and directly proportional to the measurement error of the image point coordinate. It can be seen that when the error level of the image point coordinates is equivalent, it is mainly related to the proportion of the baseline elevation of the system. Therefore, to further improve the elevation accuracy, the image size must be increased to ensure that the altitude is constant and the existing overlap is maintained. The plane error accuracy of the sparsely distributed control points in the surrounding area is higher than that of the densely distributed points in the surrounding area. The side overlap degree of the system during aerial shooting is generally 60%, which is larger than the traditional side overlap degree of 30–40%. Because of the sparse dots, the theoretical accuracy of the area network will decrease with the increase of the area.

37.3.2 Example Application and Analysis of Surveying and Mapping Engineering Survey Model

In this experiment, the nine-point method, five-point method, and four-point method were used to set up control points for comparative experiments. The low-altitude remote sensing image of UAV is used as the experimental data. The low-altitude remote sensing image data are selected, the aerial camera is Canon IXUS 220 HS, the relative altitude is 165 m, and the ground resolution is 0.05 m/pixel. The set heading overlap rate is 75%, and the side overlap rate is 70%. Before field operations, the route and heading plan must be planned first, and hand-throwing is used in field operations to make the UAV rise to the predetermined aerial height. The control point coordinates are obtained through field measurements using a total station, and the average value of several measurements is used as the experimental data of the control point coordinates. Figure 37.4 shows the three-dimensional histogram of remote sensing measurement data at different control points. A total of 3 control points were set up in this experiment. The beam method area network adjustment method is used to check the data accuracy of the detection points.

Through the comparative analysis of the experimental data, the following conclusions can be obtained: the comparison of the error data in the X-, Y-, and Z-directions of different methods is shown in paper. The error in the X- and Y-coordinates of the check point gradually increases with the decrease in the number of control points, but the overall change is not obvious, and the accuracy of the elevation data gradually decreases with the decrease in the number of control points. The common point of the experimental methods is that the plane accuracy of the check points is higher than the elevation accuracy, especially when the four-point method is used to lay out the photo control points, the elevation accuracy is significantly reduced. Figure 37.5 shows the line graph of the measurement accuracy deviation of UAV remote sensing data.

From the analysis of experimental data and the results, it is known that when arranging control points in the aerial survey area, not only a small number of level points need to be laid around the area network, but also some elevation points need to be added inside it. Using this method for deployment and control can not only

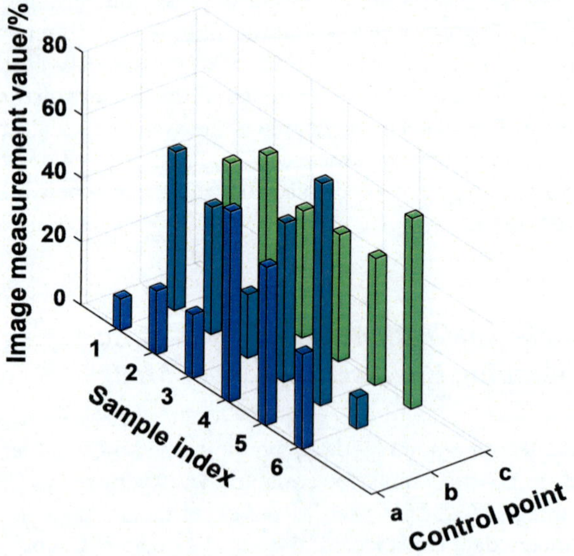

Fig. 37.4 Three-dimensional histogram of remote sensing measurement data at different control points

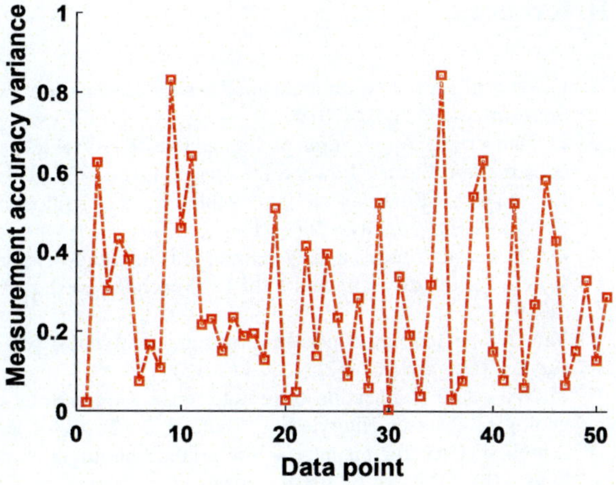

Fig. 37.5 Line graph of measurement accuracy deviation of UAV remote sensing data

obtain higher data accuracy, but also reduce the workload on the site, so as to achieve the purpose of accelerating and controlling the cost.

37.4 Conclusion

This paper studies the influence of different layouts of photo control points on the accuracy of UAV low-altitude remote sensing measurement results. Aiming at the problem of misalignment and color inconsistency caused by image splicing, the splicing and color uniformity of low-altitude remote sensing images of drones have been studied in detail, and a multi-resolution fusion method based on the best stitching line has been proposed, and the feasibility and accuracy of this method are verified through specific experiments. Combining the actual requirements of the survey engineering, we verify that the UAV low-altitude remote sensing technology can be applied to the specific earthwork calculation in the engineering, and the UAV measurement results are applied to the actual earthwork calculation. The calculation speed and calculation results of the earthwork obtained by this method are compared and analyzed with the results, which verifies the feasibility of the UAV in the earthwork calculation and the accuracy of the results. This paper successfully applied the low-altitude UAV remote sensing technology to a bauxite mine survey, and combined it with the actual characteristics of the project to achieve good results. Compared with traditional surveying technology, low-altitude drone remote sensing technology is used for surveying and mapping.

References

1. G. Jin et al., A review on unmanned aerial vehicle remote sensing and its application. Remote Sens. Inf. **2**(9), 88–92 (2019)
2. L. Han et al., Application of unmanned aerial vehicle survey in power grid engineering construction. Electric Power Surv. Des. **2**(1), 62–67 (2020)
3. L. Wen et al., Study on the key technology and application of UAV surveying and mapping data processing. J. Phys. **20**(2), 17–22 (2019)
4. G. Zhou et al., Study on UAV remote sensing image acquiring and visualization management system for the area affected by 5.12 Wenchuan earthquake. J. Remote Sens. **2**(8), 877–884 (2019)
5. Yan, Advantage and application of unmanned aerial vehicle remote sensing in engineering survey. Remote Sens. **2**(2), 12–23 (2018)
6. M. Wu et al., Application of remote sensing technology of UAV in the investigation of the geological disaster. Ningxia Eng. Technol. **2**(12), 133–136 (2020)
7. Zongjian, UAV for mapping—low altitude photogrammetric survey. Photogramm. Remote Sens. **2**(8), 1183–1186 (2019)
8. W. Hu et al., Application of unmanned aerial vehicle remote sensing for geological disaster reconnaissance along transportation lines: a case study. Appl. Mech. Mater. **2**(2), 2376–2379 (2019)
9. P. Tziavou et al., Unmanned Aerial Vehicle (UAV) based mapping in engineering geological surveys: considerations for optimum results. Eng. Geol. **1**(8), 12–21 (2019)
10. Y. Wenquan et al., Primary usages of UAV remote sensing in geological disaster monitoring and rescuing. Geospat. Inf. **2**(11), 5–12 (2019)

Chapter 38
Analysis of Influence Factors on Image Quality of Lensless Fourier Transform Hologram Reconstruction

Tong Xiao, Changhui Tian, and Zhiqiang Gao

Abstract In order to get the accurate mathematical description of lensless Fourier transform hologram and the influencing factors of reconstructed image quality, based on the kind of lensless Fourier transform hologram recording optical path designed, combining with the mathematical expression of lensless Fourier transform hologram recording and reproducing, this paper studied the influence factors of reconstructed image quality of lensless Fourier transform hologram by means of scalar diffraction theory in Fourier optics. The results show that the reconstructed image quality of lensless Fourier transform hologram is related to the ratio of object light and reference light intensity, the position relationship between object and reference light source, and the size of hologram stored in computer. The experimental results prove the correctness of the theoretical analysis.

38.1 Introduction

With the development of computer technology and photoelectric imaging technology, digital holography technology is changing with each day driven by both. Digital holography can record all the information of objects, which has a huge advantage compared with traditional imaging [1]. In addition, it has been widely used in digital holographic interferometry [2–4], digital holographic storage technology [5], three-dimensional topography measurement [5], micro-deformation measurement [6–8], and other aspects. However, digital holography also has some disadvantages, such as small field of view, low resolution, difficulty in filtering speckle noise, and poor quality of reconstructed image.

Compared with off-axis Fresnel holography, image holography and other holograms, lensless Fourier transform holography [9] has the advantages of simple recording optical path, avoiding interference of phase factor brought by lens, and only one Fourier transform is needed to obtain reconstructed image. In recent years, fewer and fewer studies have been conducted on the optimization of imaging quality

T. Xiao · C. Tian (✉) · Z. Gao
Department of Basic Sciences, Air Force Engineering University, Xi'an, China
e-mail: xt415574621@163.com

G. Liu and F. Cen (eds.), *Advances in Precision Instruments and Optical Engineering*, Springer Proceedings in Physics 270, https://doi.org/10.1007/978-981-16-7258-3_38

by using optical methods, and many previous studies by scholars on the analysis of the factors affecting image quality of lensless Fourier transform imaging are partial and not systematic enough. For example, only the effect of recording distance on the reconstructed image quality was studied in literature [10], and only the expression of focal depth in lensless Fourier transform imaging system was studied in literature [11]. In this paper, the scalar diffraction theory is used to systematically analyze the principle of lensless Fourier transform hologram, find out the factors affecting the imaging quality, and verify it through experiments [12].

38.2 Model

38.2.1 *Lensless Fourier Transform Hologram Recording Optical Path*

As shown in Fig. 38.1, the coherent laser reflected by the semi-translucent and semi-reflective mirror BL1 shines on the measured object and reflects on the semi-translucent and semi-reflective mirror BL2 under the effect of beam expansion of the microscopic objective. After a primary reflection, the light passing through the semi-permeable and semi-reflective mirror BL1 expands the beam under the action of the microscopic objective lens and passes through the semi-permeable and semi-reflective mirror BL2. Finally, the two beams of light interferometric imaging on the CCD.

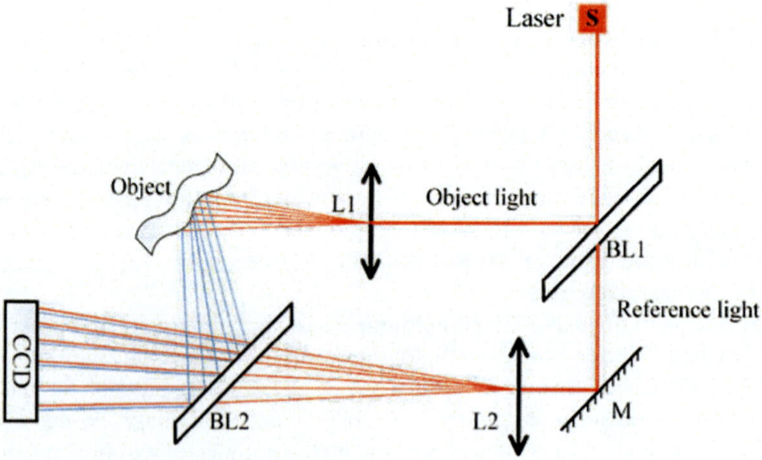

Fig. 38.1 Lensless Fourier transform optical path

38.2.2 Influence of Object Light and Reference Light Intensity Ratio on Imaging Quality

The complex amplitude distribution of the reflected light at the point (x_0, y_0) on the object is $g(x_0, y_0)$, the distance between the object and CCD is d_o, the reference light $R(x, y)$, the distance between $R(x, y)$ and the object is d_r, the distance between the reference light source and the origin of the coordinate on the object in the x-direction is a, and the distance between the reference light source and the object in the y-direction is b. The coordinates of the reference light source are indicated in Fig. 38.2.

The field distribution of object light wave on CCD is

$$O(x, y) = \frac{1}{j\lambda d_o} \exp(jkd_o) \exp\left[\frac{j\pi}{\lambda d_o}(x^2 + y^2)\right] \int_{-\infty}^{\infty} \int_{-\infty}^{\infty} g(x_0, y_0) \exp\left[\frac{jk}{2d_o}(x_0^2 + y_0^2)\right] \times \exp\left[-j\frac{2\pi}{\lambda d_o}(x'x + y'y)\right] dx_0 dy_0 \quad (38.1)$$

Assume that

$$g'(x_0, y_0) = g(x_0, y_0) \exp\left[\frac{jk}{2d_o}(x_0^2 + y_0^2)\right] \quad (38.2)$$

$$G'(f_x, f_y) = F\{g'(x_0, y_0)\} \quad (38.3)$$

The object light wave can be abbreviated as

$$O(x, y) = \frac{1}{j\lambda d_o} \exp(jkd_o) \exp\left[\frac{j\pi}{\lambda d_o}(x^2 + y^2)\right] G'(f_x, f_y) \quad (38.4)$$

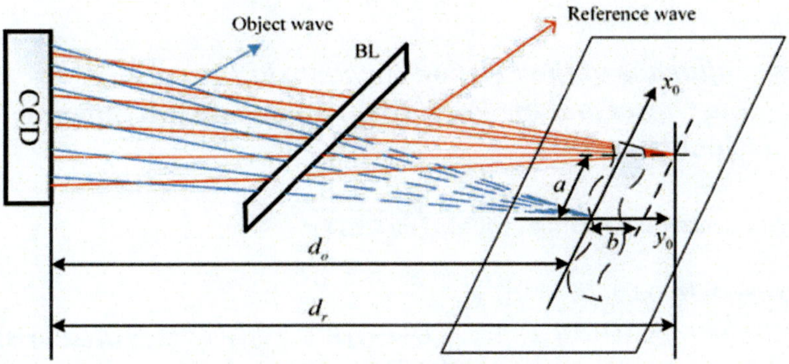

Fig. 38.2 Recording of lensless Fourier transform holograms

Assume that

$$A_o = \frac{\exp(jkd_o)}{j\lambda d_o} \tag{38.5}$$

When the optical path is defined with the laser used, A_o is a constant, representing the amplitude of the object light. The field distribution of the reference light wave on CCD is

$$R(x, y) = \frac{a_0}{d_r}\exp(jkd_r)\exp\left[\frac{j\pi}{\lambda d_r}(x^2 + y^2)\right]\exp\left(j2\pi\frac{xa}{\lambda d_r}\right)\exp\left(j2\pi\frac{yb}{\lambda d_r}\right) \tag{38.6}$$

a_0 is the amplitude per unit distance from the point source. Assume that

$$A_r = \frac{a_0 \exp(jkd_r)}{d_r} \tag{38.7}$$

A_r is also a constant.

$O(x, y)$ and $R(x, y)$ are obtained from the same beam of light through the amplitude division method, satisfying the coherent conditions of the same frequency, constant phase difference, and the same direction of vibration. If the intensity of the reference light is too high, it will drown out the object light. In order to obtain better image quality, the intensity of reference light needs to be controlled. Assume that

$$\gamma = \frac{|R(x, y)|^2}{|O(x, y)|^2} = \frac{(a_0 d_o \lambda)^2}{d_r^2} \tag{38.8}$$

γ depends on the size of the a_0, d_o, d_r, λ. Under the given light path, γ value adjustment can be done through a_0 regulation.

38.2.3 *Influence of the Position Relationship Between the Object and the Point Light Source on the Imaging Quality*

The light intensity distribution recorded on CCD is

$$\begin{aligned} I(x, y) &= |O(x, y) + R(x, y)|^2 \\ &= |O(x, y)|^2 + |R(x, y)|^2 + O(x, y)R^*(x, y) + O^*(x, y)R(x, y) \end{aligned} \tag{38.9}$$

Expand (38.9) to obtain

$$I(x, y) = |O(x, y)|^2 + |R(x, y)|^2 + O(x, y)R^*(x, y) + O^*(x, y)R(x, y) \tag{38.10}$$

Observe the above formula, the third item contains object light information, and expand it

$$\begin{aligned} I_3(x, y) = {} & A_o A_r \exp\left[\frac{j\pi(d_r - d_o)}{\lambda d_o d_r}(x^2 + y^2)\right] G'\left(\frac{x}{\lambda d_o}, \frac{y}{\lambda d_o}\right) \\ & \times \exp\left(-j2\pi\frac{xa}{\lambda d_r}\right) \exp\left(-j2\pi\frac{yb}{\lambda d_r}\right) \end{aligned} \tag{38.11}$$

Since d_o and d_r are not equal, in order to obtain the image of the object through only one inverse Fourier transform during image processing, when designing the optical path, let

$$d_r = d_o = d \tag{38.12}$$

Then $I_3(x, y)$ can be written as

$$I_3(x, y) = A_o A_r G'\left(\frac{x}{\lambda d}, \frac{y}{\lambda d}\right) \exp\left(-j2\pi\frac{xa}{\lambda d}\right) \exp\left(-j2\pi\frac{yb}{\lambda d}\right) \tag{38.13}$$

Inverse Fourier transform of $I_3(x, y)$ is carried out to achieve the optical path as shown in Fig. 38.3, which is obtained

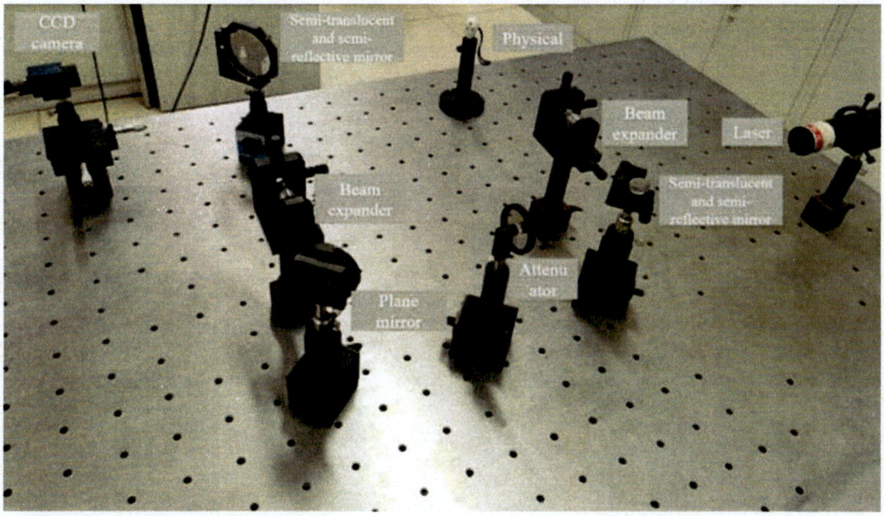

Fig. 38.3 Experimental light path diagram

$$U_3(x', y') = F^{-1}\{I_3(x, y)\} = A_o A_r \int_{-\infty}^{\infty}\int_{-\infty}^{\infty} G'\left(\frac{x}{\lambda d}, \frac{y}{\lambda d}\right)$$

$$\times \exp\left(-j2\pi\frac{xa}{\lambda d}\right)\exp\left(-j2\pi\frac{yb}{\lambda d}\right)\exp\left[j\frac{2\pi}{\lambda f}(x'x + y'y)\right]dxdy$$

$$= A_o A_r (\lambda d)^2 g'\left(\frac{d}{f}x' - a, \frac{d}{f}y' - b\right) \quad (38.14)$$

And similarly, we can solve for the fourth term

$$U_4(x', y') = F^{-1}\{I_4(x, y)\} = A_o A_r (\lambda d)^2 g'^*\left(-\frac{d}{f}x' - a, -\frac{d}{f}y' - b\right) \quad (38.15)$$

At this time, the centres of the two real images are, respectively, located at $\left(\frac{af}{d}, \frac{bf}{d}\right)$ and $\left(-\frac{af}{d}, -\frac{bf}{d}\right)$, and the magnification is $\frac{f}{d}$.

When $d_r \neq d_o$

$$I_3(x, y) = A_o A_r \exp\left[\frac{j\pi(d_r - d_o)}{\lambda d_o d_r}(x^2 + y^2)\right] G'\left(\frac{x}{\lambda d_o}, \frac{y}{\lambda d_o}\right)$$

$$\times \exp\left(-j2\pi\frac{xa}{\lambda d_o}\frac{d_o}{d_r}\right)\exp\left(-j2\pi\frac{yb}{\lambda d_o}\frac{d_o}{d_r}\right) \quad (38.16)$$

With the inverse Fourier transform of $I_3(x, y)$, without considering the phase factor $\exp\left[\frac{j\pi(d_r - d_o)}{\lambda d_o d_r}(x^2 + y^2)\right]$, we get

$$U_3^*(x', y') = F^{-1}\{I_3^*(x, y)\} = A_o A_r \int_{-\infty}^{\infty}\int_{-\infty}^{\infty} G'\left(\frac{x}{\lambda d_o}, \frac{y}{\lambda d_o}\right)\exp\left(-j2\pi\frac{xa}{\lambda d_o}\frac{d_o}{d_r}\right)$$

$$\times \exp\left(-j2\pi\frac{yb}{\lambda d_o}\frac{d_o}{d_r}\right)\exp\left[j\frac{2\pi}{\lambda f}(x'x + y'y)\right]dxdy$$

$$= A_o A_r (\lambda d_o)^2 g'\left(\frac{d_o}{f}x' - \frac{d_o}{d_r}a, \frac{d_o}{f}y' - \frac{d_o}{d_r}b\right) \quad (38.17)$$

Without considering the phase delay of lens constant, the effect of phase factor $\exp\left[\frac{j\pi(d_r - d_o)}{\lambda d_o d_r}(x^2 + y^2)\right]$ is equivalent to adding a lens with focal length $\frac{d_o - d_r}{d_o d_r}$ after $I_3^*(x, y)$, then the frequency spectrum of $I_3^*(x, y)$ at the back focus of the lens is

$$U_3(x', y') = \frac{d_o d_r}{j\lambda(d_o - d_r)}\exp\left[j\frac{d_o d_r}{2(d_o - d_r)}\left(x'^2 + y'^2\right)\right]$$

$$\times A_o A_r (\lambda d_o)^2 g'\left(-\frac{d_o}{f}x' - \frac{d_o}{d_r}a, -\frac{d_o}{f}y' - \frac{d_o}{d_r}b\right) \quad (38.18)$$

By the same token

$$U_4(x', y') = \frac{d_o d_r}{j\lambda(d_o - d_r)} \exp\left[j \frac{d_o d_r}{2(d_o - d_r)} \left(x'^2 + y'^2\right)\right] \times A_o A_r (\lambda d_o)^2 g'\left(\frac{d_o}{f} x' - \frac{d_o}{d_r} a, \frac{d_o}{f} y' - \frac{d_o}{d_r} b\right) \quad (38.19)$$

At this point, the central positions of the object and its conjugate image is $\left(-\frac{af}{d_r}, -\frac{bf}{d_r}\right)$ and $\left(\frac{af}{d_r}, \frac{bf}{d_r}\right)$, and magnification is $\frac{f}{d_o}$. In this paper, the intensity distribution on the observation plane is recorded and measured, and this phase bending has no effect on it. So

$$I_3(x', y') = \left|\frac{A_o A_r d_o^3 d_r \lambda}{j(d_o - d_r)}\right|^2 \left|g'\left(-\frac{d_o}{f} x' - \frac{d_o}{d_r} a, -\frac{d_o}{f} y' - \frac{d_o}{d_r} b\right)\right|^2 \quad (38.20)$$

$$I_4(x', y') = \left|\frac{A_o A_r d_o^3 d_r \lambda}{j(d_o - d_r)}\right|^2 \left|g'\left(\frac{d_o}{f} x' - \frac{d_o}{d_r} a, \frac{d_o}{f} y' - \frac{d_o}{d_r} b\right)\right|^2 \quad (38.21)$$

It is obvious that the intensity distribution of objects and their conjugates on the rear focal plane is their power spectrum.

38.2.4 Influence of Hologram Size on Imaging Quality

After getting the hologram, use a computer to discrete Fourier transform hologram to replace the lens effect; for the size of the $M \times N$ hologram, discrete Fourier transform expression is

$$F(u, v) = \sum_{x=0}^{M-1} \sum_{y=0}^{N-1} f(x, y) \exp[-j2\pi(ux/M + vy/N)] \quad (38.22)$$

And the Fourier transform of the lens can be expressed as

$$F(x', y') = \int_{-\infty}^{\infty} \int_{-\infty}^{\infty} f(x, y) \exp\left[j \frac{2\pi}{\lambda f_x} x' x + j \frac{2\pi}{\lambda f_y} y' y\right] dx dy \quad (38.23)$$

By comparing (38.22) with (38.23), it can be concluded that

$$M = \lambda f_x \quad (38.24)$$

$$N = \lambda f_y \quad (38.25)$$

The lensless Fourier transform hologram can reproduce the real image of the object and its conjugate. According to the expressions of $U_3(x', y')$ and $U_4(x', y')$, it can be obtained that the two real image centres are located at $\left(\frac{aM}{\lambda d}, \frac{bN}{\lambda d}\right)$ and $\left(-\frac{aM}{\lambda d}, -\frac{bN}{\lambda d}\right)$, respectively, with horizontal transverse magnification of $\frac{M}{\lambda d}$ and vertical longitudinal magnification of $\frac{N}{\lambda d}$ [13–17].

38.3 Experiment and Discussion

38.3.1 Experiment

The light path diagram used in the experiment is shown in Fig. 38.3. $He{-}Ne$ laser is used and the working wavelength is 632.8 nm. The object height is 15 mm, the CCD resolution is 2448 × 2048, and the pixel size is 3.45 × 3.45 μm.

The theoretical analysis of influencing factors of lensless Fourier transform hologram reconstruction is verified experimentally by using the optical path diagram. In the experiment, the influence of the ratio of object light to reference light intensity on the imaging quality was verified by changing the intensity of reference light by rotating the attenuator. Change the distance between the object and the camera and adjust the tilt Angle of $BL2$, so as to change the position relationship between the object and the reference light source, and then verify its influence on the reconstruction image quality. CCD is used to obtain holograms of different sizes to verify the effect of the size of holograms on the image quality.

38.3.2 Analysis and Discussion

Figure 38.4a–d are four typical images selected during the process of slowly rotating the attenuator to reduce the reference light intensity. With the weakening of the reference light, the image of the object changes from non-existent to existent, from dark to bright, and then from bright to dark. In Fig. 38.4a, the reference light is too bright to completely drown the object light, and the reconstructed image is only the reference light. In Fig. 38.4b, the image of the object appears, but it can be seen that its brightness is not high and the display effect is not good. In Fig. 38.4c, at this time, the brightness of the object image is high and the imaging effect is good, which is the best among the four images. In Fig. 38.4d, the image quality of the object deteriorates. Thus, it can be concluded that when the intensity of object light is close to that of reference light, the imaging quality will be better. This is consistent with the theoretical analysis.

Figure 38.5a, b, and c are the reconstructed image of the object when $d_r = d_o$, the two-dimensional intensity map, and the one-dimensional pixel intensity value in the y-direction when $x = 150$; Fig. 38.5d, e, and f is the image when $d_r \neq d_o$. By

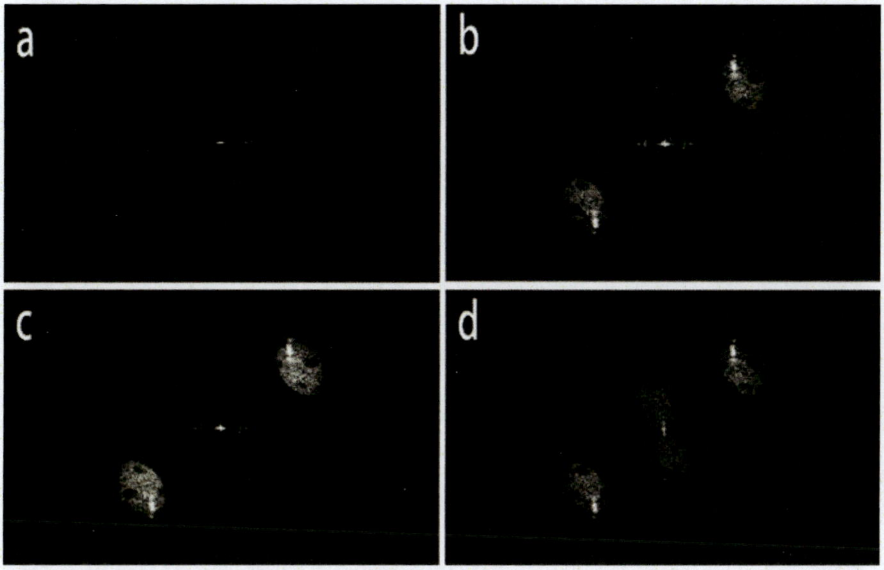

Fig. 38.4 Hologram reconstruction under different ratios of object light and reference light intensity. **a** Point-source of light; **b**, **c**, **d** image of objects

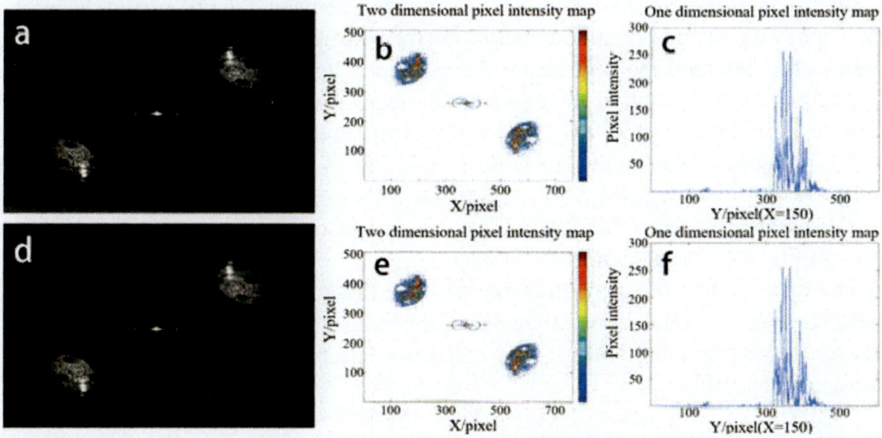

Fig. 38.5 The images when the distance between object, light source, and camera is equal or unequal. **a**, **d** Reconstructed image; **b**, **e** two-dimensional intensity map; **c**, **f** One-dimensional intensity diagram

comparing the three graphs in the two cases, it can be concluded that the intensity distribution diagram of the objects in the two cases is approximately the same, which is consistent with the theoretical analysis.

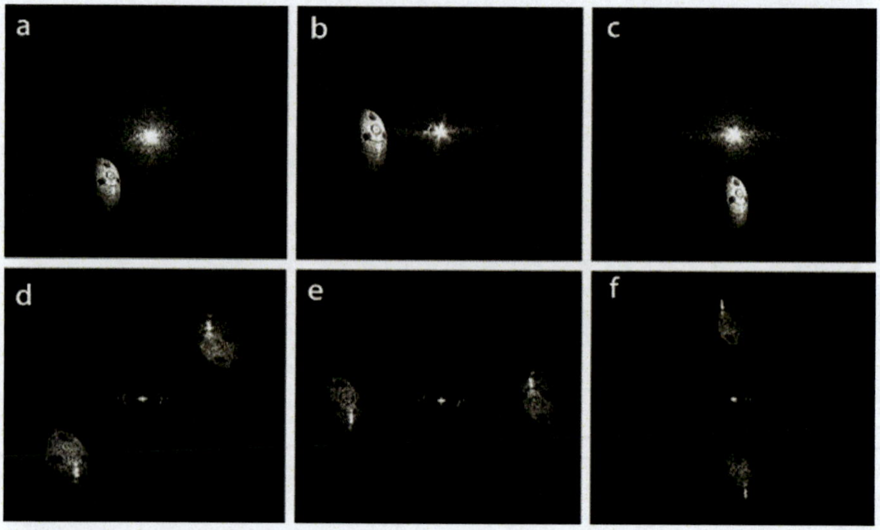

Fig. 38.6 The influence of position relation on reconstructed image. **a–c** The actual position relationship between object and light source; **d–f** reconstruction image under different position relations

Figure 38.6a–c show the positional relationship between the object and the reference light source, and Fig. 38.6d–f show the reconstructed image. When the position relationship between the object and the reference light is oblique, horizontal and longitudinal, the position of the object in the reconstructed image is also consistent with the point light source. This shows that the theoretical analysis is correct.

The hologram sizes (height and width) in Fig. 38.7a–f are 1000×500, 1000×1000, 1000×1500, 1000×2000, 500×1000 and 1500×1000, respectively.

The parameters (approximate range) of its reconstructed image can be obtained through Fig. 38.7, as shown in Table 38.1.

The data in the table are consistent with the theoretical analysis, which verifies the correctness of the theoretical analysis. It can also be found through observation that the larger the hologram, the higher the resolution of the image and the clearer the object [18, 19].

38.4 Conclusions

In this paper, by combining theoretical analysis with experimental verification, the factors affecting the quality of reconstructed images are analyzed by using the mathematic description of the imaging process of lensless Fourier transform hologram. The results show that the intensity ratio of object light and reference light, the position relationship between object light and reference light source, and the size of

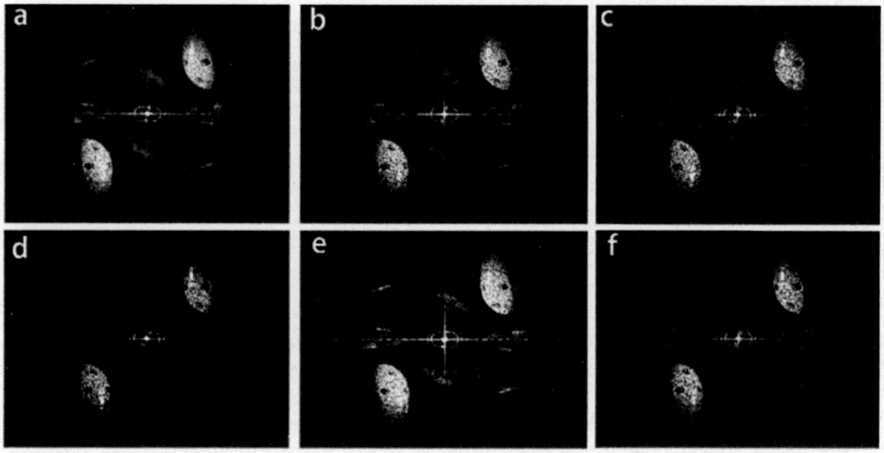

Fig. 38.7 Reconstructed images of holograms of different sizes

Table 38.1 Parameters of different holograms

Size of hologram	Position coordinates of the center of the object	Position coordinates of conjugate image center	Horizontal actual distance/μm	Vertical actual distance/μm
1000 × 500	(800, 150)	(200, 350)	345	1035
1000 × 1000	(800, 300)	(200, 700)	690	1035
1000 × 1500	(800, 450)	(200, 1050)	1035	1035
1000 × 2000	(800, 600)	(200, 1400)	1380	1035
500 × 1000	(400, 300)	(100, 700)	690	517.5
1500 × 1000	(1200, 300)	(300, 700)	690	1552.5

recorded hologram all affect the quality of reconstructed image. In order to obtain high quality holograms, attention should be paid to adjusting the above factors. The results obtained in this paper have some guiding significance for optical path design, experimental realization, and numerical reproduction of lensless Fourier transform hologram.

References

1. K. Harada, Y.A. Ono, Y. Takahashi, Lensless Fourier transform electron holography applied to vortex beam analysis. Microscopy (Oxford, England) **69**(3) (2020)
2. H. Zhou, *Study on Applications of Digital Holographic Imaging* (Soochow University, 2015)
3. B.C. Yin, *The Recording and Reconstruction Technology of Digital Holography* (Changchun University of Science and Technology, 2014)

4. G.J. Wang, *Study on Digital Holography and Its Applications in the Field of Measurement* (Beijing University of Technology, 2011)
5. F. Feng, A.L. Tian, B.C. Liu et al., Full-field three-dimensional test for scratch defects using digital holographic scanning imaging system. Chin. J. Lasers **47**(04), 261–268 (2020)
6. Y.L. Kou, E.P. Li, J.L. Di et al., Surface morphology measurement of tiny object based on dual-wavelength digital holography. Chin. J. Lasers **41**(2), 0209010 (2014)
7. H.J. Wang, F. Li, Digital holographic microscopy applied to measurement of micro devices. J. Xi'an Technol. Univ. **2**, 94–98 (2014)
8. J.C. Zhou, A.L. Gong, Y. Du, Improvement method of measuring object deformation by using double-exposure with gray-level. Laser Technol. **35**(5), 626–628 (2011)
9. R.K. Singh, S. Vyas, Y. Miyamoto, Lensless Fourier transform holography for coherence waves. **19**(11) (2017)
10. Y.P. Wu, H.X. Wang, Z.R. Zhou, Analysis of influence factors of the reconstructed image quality in lensless fourier transform digital holography. Infrared Laser Eng. **S4**, 538–542 (2006)
11. H.Y. Wang, Z.H. Zhang, W. Liao et al., Focal depth of digital lensless Fourier transform micro-holographic system. Acta Phys. Sinica **61**(04), 244–250 (2012)
12. B.S. Duan, *Research on Defect Dection Technology of Rader Absorbing Coating Based on Sherography* (University of Electronic Science and Technology of China, Chengdu, 2018)
13. N.G. Lv, *Fourier Optics*, 3rd edn. (Electronic Industry Press, Beijing, 2019), p. 9
14. R.C. Gonzalez, R.E. Woods, *Digital Image Processing* (Q. Ruan, Y.Z. Ruan, et al., transl.), 6th edn. (Electronic Industry Press, Beijing, 2017), p. 5
15. J.W. Goodman, *Statistical Optics*, 2nd edn. (Wiley, New Jersey, 2015)
16. M.K. Kim, Principles and techniques of digital holographic microscopy. SPIE Rev. **1** (2010)
17. K.J. Gasvik, *Optical Metrology*, 3rd edn. (Wiley, New Jersey, 2002)
18. W.H. Zhang, Y.C. Cao, G.F. Jin, Review on high resolution and large field of view digital holography. Infrared Laser Eng. **48**(06), 104–120 (2019)
19. S. Li, D. Wang, Y.T. Lu, Method for improving imaging resolution of digital holographic adaptive optical system. Chin. J. Lasers **46**(07), 284–291 (2019)

Chapter 39
Error Characteristic Analysis of Tri-axis Rotating Optical Gyro Inertial Navigation System Based on Inertial Frame

Huiying Fan, Kaiwen Tian, Guangxin Cheng, Zhuangzhuang Dong, and Xudong Yu

Abstract Most of the rotation modulation schemes of optical gyro inertial navigation system (INS) are designed based on the navigation frame and implemented in the body frame. However, the inertial sensors are sensitive to the motions in the inertial space, so there are coupling errors between the earth's rotation and the errors of inertial sensors inevitably. In order to solve the problem of continuous deviations of position errors caused by the above coupling errors in the case of long-time navigation, a 16-sequence tri-axis rotation modulation scheme based on inertial frame is proposed. According to the error propagation equations of INS, the characteristics of each inertial sensor error during rotation modulation are analyzed. Simulation and experiments results show that the 16-sequence tri-axis rotation scheme can completely modulate all the drifts, scale-factor errors, and misalignment errors of the inertial sensors. The divergence of navigation position errors of the tri-axis scheme is less than that of the dual-axis scheme, and the position accuracy is improved by 0.32 nm in 5 days, which proves that the proposed scheme can effectively isolate the earth's rotation, and the error modulation effects are better than those of the dual-axis scheme.

39.1 Introduction

In order to offset the impacts of inertial measurement unit (IMU) errors on navigation results, the rotation modulation technique is widely used in high precision inertial navigation system (INS), which is also called system error averaging technique [1–4]. A reasonable rotation modulation scheme should not only avoid the introduction of accumulated errors caused by rotation motions, but also reduce the navigation errors caused by drifts, scale-factor errors, and misalignments of inertial sensors as much as possible [5–7]. The rotating INS is divided into single-axial system, dual-axis

H. Fan · K. Tian · G. Cheng · Z. Dong · X. Yu (✉)
College of Advanced Interdisciplinary Studies, National University of Defense Technology, Changsha 410073, China
e-mail: wind0909@163.com

G. Liu and F. Cen (eds.), *Advances in Precision Instruments and Optical Engineering*, Springer Proceedings in Physics 270, https://doi.org/10.1007/978-981-16-7258-3_39

system, and tri-axis system according to the number of rotating axes. Based on the 8-sequence dual-axis rotation scheme, a 16-sequence dual-axis rotation scheme was proposed in [8]. It not only has the advantages of the 8-sequence rotation scheme, but also solves the problems of scale-factor asymmetry, incomplete error elimination, and the accumulation of calculation errors of the 8-sequence scheme. The 16-sequence scheme is proved theoretically to be a reasonable and practical dual-axis rotation scheme. An improved 16-sequence modulation scheme was proposed in [9], which can not only modulate the constant drifts, misalignments, and scale-factor errors, but also effectively reduce the oscillation amplitudes of velocity and position errors caused by gyro misalignments and significantly improve the navigation accuracy. A 64-sequence rotation scheme was proposed in [7], which divided the 64 rotation sequences into four small cycles, and its modulation effects on mathematical platform misalignment angle errors caused by the symmetry scale-factor errors are better than those of the 16-sequence rotation scheme.

The dual-axis rotation modulation scheme is always implemented on the vehicle carrier. Limited by the body frame, the dual-axis rotation modulation technique cannot eliminate the coupling errors between the IMU errors and earth's rotation. In the case of long-time navigation, the navigation results will be constantly diverged and the navigation accuracy will be affected. In order to solve this problem, a tri-axis rotation modulation scheme based on inertial frame is proposed in this paper. According to the error propagation equations of the INS, the error characteristics of each inertial sensor during the tri-axis rotation modulation are analyzed, and the modulation effects on IMU errors of the proposed scheme will be observed. The navigation results of the proposed tri-axis scheme are compared with those of the dual-axis rotation scheme through simulations and experiments, which proves the effectiveness of the proposed scheme on isolating the earth's rotation. Finally, the conclusion is given.

39.2 Design of Tri-axis Rotation Modulation Scheme

39.2.1 Error Propagation Equations

The phi-angle error equations are used to describe the error propagation model of rotating INS [10]:

$$\begin{aligned} \dot{\boldsymbol{\phi}} &= -\boldsymbol{\omega}_{in}^{n} \times \boldsymbol{\phi} + \delta\boldsymbol{\omega}_{in}^{n} - \boldsymbol{C}_{p}^{n}\delta\boldsymbol{\omega}_{ip}^{p} \\ \delta\dot{\boldsymbol{v}} &= \boldsymbol{f}^{n} \times \boldsymbol{\phi} + \boldsymbol{C}_{p}^{n}\delta\boldsymbol{f}_{ip}^{p} - (2\boldsymbol{\omega}_{ie}^{n} + \boldsymbol{\omega}_{en}^{n}) \times \delta\boldsymbol{v} - (2\delta\boldsymbol{\omega}_{ie}^{n} + \delta\boldsymbol{\omega}_{en}^{n}) \times \boldsymbol{v} + \delta\boldsymbol{g} \end{aligned} \tag{39.1}$$

The meanings of each symbol are shown in Table 39.1.

Table 39.1 The meanings of each symbol

Symbol	Meaning	Symbol	Meaning
n	Navigation frame	p	IMU frame
i	Inertial frame	e	Earth frame
b	Body frame	$\delta \boldsymbol{g}$	Gravity error
$\boldsymbol{\omega}$	Angular rate	$\delta \boldsymbol{\omega}$	Angular rate error
$\boldsymbol{f}$	Specific force	$\delta \boldsymbol{f}$	Specific force error
$\boldsymbol{v}$	Velocity	$\delta \boldsymbol{v}$	Velocity error
$\boldsymbol{C}_p^n$	Transformation matrix from p to n		
$\boldsymbol{\phi}$	Mathematical platform misalignment angle		

The principle of rotation modulation is to change $\boldsymbol{C}_p^n$ periodically by rotation, so that the integrals of the mathematical platform error terms $\boldsymbol{C}_p^n \delta \boldsymbol{\omega}_{ip}^p$ and $\boldsymbol{C}_p^n \delta \boldsymbol{f}_{ip}^p$ in the error propagation model are equal to zero in one or more rotation periods, so as to offset the influences of inertial sensor errors on navigation results. The output error model of optical gyros and accelerometers is:

$$\begin{aligned} \delta \boldsymbol{\omega}_{ip}^p &= (\Delta \boldsymbol{S}_g + \Delta \boldsymbol{M}_g) \boldsymbol{\omega}_{ip}^p + \boldsymbol{\varepsilon} \\ \delta \boldsymbol{f}_{ip}^p &= (\Delta \boldsymbol{S}_a + \Delta \boldsymbol{M}_a) \boldsymbol{f}_{ip}^p + \boldsymbol{\nabla} \end{aligned} \tag{39.2}$$

where $\boldsymbol{\omega}_{ip}^p$, $\boldsymbol{\varepsilon}$, $\Delta \boldsymbol{S}_g$, and $\Delta \boldsymbol{M}_g$ represent the gyro input, drift, scale-factor error, and misalignment, respectively, and $\boldsymbol{f}_{ip}^p$, $\boldsymbol{\nabla}$, $\Delta \boldsymbol{S}_a$, and $\Delta \boldsymbol{M}_a$ are the accelerometer input, drift, scale-factor error, and misalignment, respectively.

It is assumed that the carrier is in a fixed position, without linear or angular motions, and the navigation frame n, the body frame b, and the IMU frame p are coincident at the initial moment, that is $\boldsymbol{C}_p^n = \boldsymbol{C}_b^n = \boldsymbol{C}_p^b = \boldsymbol{I}$, and $\boldsymbol{I}$ is the unit matrix. There is always $\boldsymbol{\omega}_{en} = 0$ and $\boldsymbol{\omega}_{nb} = 0$ during the rotation. Taking the errors of optical gyros as an example, its integral in a rotation cycle is

$$\begin{aligned} \int_0^T \boldsymbol{C}_p^n \delta \boldsymbol{\omega}_{ip}^p dt &= \int_0^T \boldsymbol{C}_p^n (\boldsymbol{S}_g + \Delta \boldsymbol{C}_g) \boldsymbol{\omega}_{ip}^p dt + \int_0^T \boldsymbol{C}_p^n \boldsymbol{\varepsilon} dt \\ &= \int_0^T \boldsymbol{C}_p^n (\boldsymbol{S}_g + \Delta \boldsymbol{C}_g) \boldsymbol{C}_n^p (\boldsymbol{\omega}_{ie}^n + \boldsymbol{\omega}_{np}^n) dt + \int_0^T \boldsymbol{C}_p^n \boldsymbol{\varepsilon} dt \end{aligned} \tag{39.3}$$

where $\boldsymbol{\omega}_{ie}^n$ is the projection of the earth's rotation angular rate in the navigation frame, and $\boldsymbol{\omega}_{np}^n$ is the rotation scheme angular rate in the navigation frame. When a scheme isolates the rotation of the earth, that is, $\boldsymbol{\omega}_{np}^n = \boldsymbol{C}_p^n \boldsymbol{\omega} - \boldsymbol{\omega}_{ie}^n$, where $\boldsymbol{\omega}$ is the equivalent angular rate of the ideal rotation scheme in the inertial frame, (39.3) can be transformed into

$$\int_0^T C_p^n \delta\omega_{ip}^p dt = \int_0^T C_p^n (S_g + \Delta C_g)\omega dt + \int_0^T C_p^n \varepsilon dt \tag{39.4}$$

It can be seen that there is no ω_{ie}^n in the equation anymore. A rotation angular rate which is always equal to the earth's rotation angular rate but in the opposite direction should be added when the tri-axis rotation scheme is adopted. Therefore, the navigation errors caused by the coupling errors between the earth's rotation and the errors of the optical gyros can be eliminated.

According to the above hypothesis, the carrier has no movements relative to the navigation system. Then $C_e^n = \begin{pmatrix} -\sin\lambda & \cos\lambda & 0 \\ -\sin L\cos\lambda & -\sin L\sin\lambda & \cos L \\ \cos L\cos\lambda & \cos L\sin\lambda & \sin L \end{pmatrix}$ is a constant matrix only related to longitude λ and latitude L, which doesn't depend on time t. Then the misalignment angle errors generated by the scale-factor errors are

$$\begin{aligned}
\int C_p^n S_g \omega dt &= C_e^n \bullet \int C_i^e C_p^i dt \bullet S_g \omega \\
&= C_e^n \bullet \int C_i^e C_p^i dt \bullet \begin{pmatrix} S_1 & 0 & 0 \\ 0 & S_2 & 0 \\ 0 & 0 & S_3 \end{pmatrix} \begin{pmatrix} \omega_1 \\ \omega_2 \\ \omega_3 \end{pmatrix} \\
&= C_e^n \int \begin{pmatrix} \cos[(\omega+\omega_{ie})t] & -\sin[(\omega+\omega_{ie})t] & 0 \\ \sin[(\omega+\omega_{ie})t] & \cos[(\omega+\omega_{ie})t] & 0 \\ 0 & 0 & 1 \end{pmatrix} dt \bullet \begin{pmatrix} S_1\omega_1 \\ S_2\omega_2 \\ S_3\omega_3 \end{pmatrix}
\end{aligned} \tag{39.5}$$

When the IMU rotates around the z-axis of inertial frame, there is $\omega = \begin{pmatrix} 0 & 0 & \omega_3 \end{pmatrix}^T$, and the (39.5) is transformed into

$$\begin{aligned}
&\int C_p^n S_g \omega dt \\
&= C_e^n \int \begin{pmatrix} \cos[(\omega+\omega_{ie})t] & -\sin[(\omega+\omega_{ie})t] & 0 \\ \sin[(\omega+\omega_{ie})t] & \cos[(\omega+\omega_{ie})t] & 0 \\ 0 & 0 & 1 \end{pmatrix} dt \bullet \begin{pmatrix} 0 \\ 0 \\ S_3\omega_3 \end{pmatrix} \\
&= C_e^n \int \begin{pmatrix} 0 \\ 0 \\ S_3\omega_3 \end{pmatrix} dt
\end{aligned} \tag{39.6}$$

Compared with the dual-axis rotation in the navigation frame, there is no product of the scale-factor errors and the earth's rotation angular rate in the integral formula. When the IMU is rotated around the z-axis of inertial frame, the cumulative navigation errors caused by the scale-factor errors on the other two axes can be completely eliminated.

39.2.2 Design of Tri-axis Rotation Scheme

Based on the above researches, a tri-axis rotation modulation scheme relative to inertial frame is proposed. This scheme can not only completely modulate the drifts and misalignments of the inertial sensors, but also eliminate the coupling errors between the IMU scale-factor errors and the earth's rotation. The scheme diagram is shown in Fig. 39.1.

Where Zi and Xi are the z-axis and x-axis of the inertial frame, and A, B, C, and D are four stagnation positions. The rotation sequences shown as 1–16 is the same as the rotation sequences of the 16-sequence scheme in [9]. The improvement is rotating based on different frame, that is, rotating the scheme in the inertial frame rather than the navigation frame. The detailed rotation scheme is shown in Table 39.2:

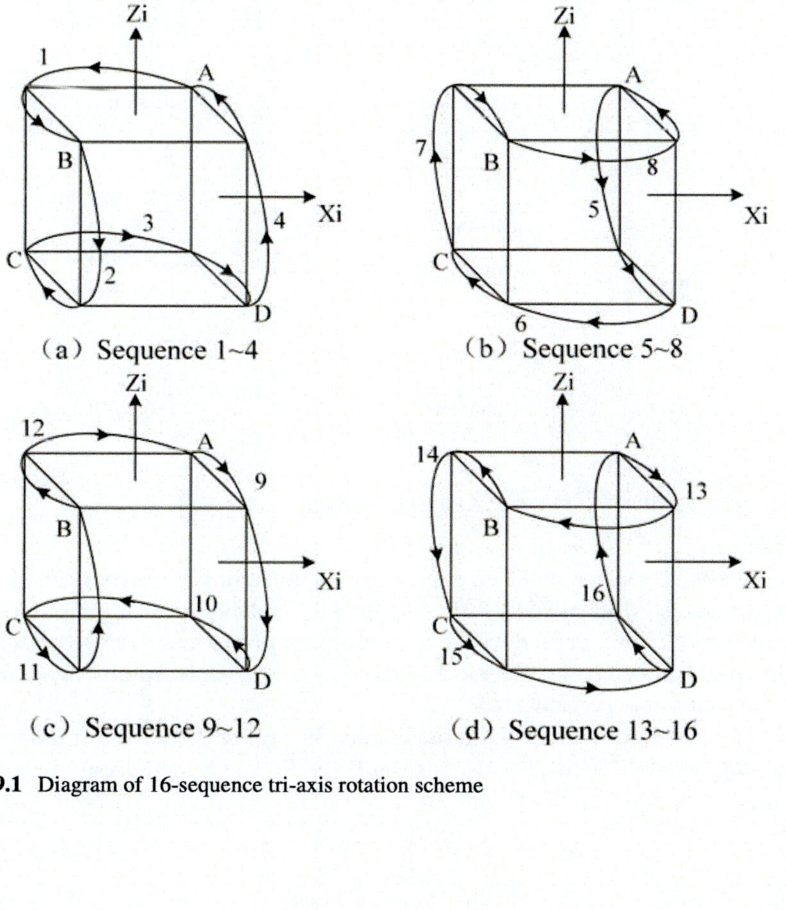

Fig. 39.1 Diagram of 16-sequence tri-axis rotation scheme

Table 39.2 The detailed rotation scheme

Sequence	Rotation path	Rotation angular rate in n frame	Input of optical gyro
1	$A \to B$	$\left(0\ -\omega_{ie}^{N}\ +\omega-\omega_{ie}^{U}\right)^{T}$	$\left(0\ 0\ +\omega\right)^{T}$
2	$B \to C$	$\left(+\omega\ -\omega_{ie}^{N}\ -\omega_{ie}^{U}\right)^{T}$	$\left(+\omega\ 0\ 0\right)^{T}$
3	$C \to D$	$\left(0\ -\omega_{ie}^{N}\ -\omega-\omega_{ie}^{U}\right)^{T}$	$\left(0\ 0\ -\omega\right)^{T}$
4	$D \to A$	$\left(-\omega\ -\omega_{ie}^{N}\ -\omega_{ie}^{U}\right)^{T}$	$\left(-\omega\ 0\ 0\right)^{T}$
5	$A \to D$	$\left(-\omega\ -\omega_{ie}^{N}\ -\omega_{ie}^{U}\right)^{T}$	$\left(-\omega\ 0\ 0\right)^{T}$
6	$D \to C$	$\left(0\ -\omega_{ie}^{N}\ -\omega-\omega_{ie}^{U}\right)^{T}$	$\left(0\ 0\ -\omega\right)^{T}$
7	$C \to B$	$\left(+\omega\ -\omega_{ie}^{N}\ -\omega_{ie}^{U}\right)^{T}$	$\left(+\omega\ 0\ 0\right)^{T}$
8	$B \to A$	$\left(0\ -\omega_{ie}^{N}\ +\omega-\omega_{ie}^{U}\right)^{T}$	$\left(0\ 0\ +\omega\right)^{T}$
9	$A \to D$	$\left(+\omega\ -\omega_{ie}^{N}\ -\omega_{ie}^{U}\right)^{T}$	$\left(+\omega\ 0\ 0\right)^{T}$
10	$D \to C$	$\left(0\ -\omega_{ie}^{N}\ +\omega-\omega_{ie}^{U}\right)^{T}$	$\left(0\ 0\ +\omega\right)^{T}$
11	$C \to B$	$\left(-\omega\ -\omega_{ie}^{N}\ -\omega_{ie}^{U}\right)^{T}$	$\left(-\omega\ 0\ 0\right)^{T}$
12	$B \to A$	$\left(0\ -\omega_{ie}^{N}\ -\omega-\omega_{ie}^{U}\right)^{T}$	$\left(0\ 0\ -\omega\right)^{T}$
13	$A \to B$	$\left(0\ -\omega_{ie}^{N}\ -\omega-\omega_{ie}^{U}\right)^{T}$	$\left(0\ 0\ -\omega\right)^{T}$
14	$B \to C$	$\left(-\omega\ -\omega_{ie}^{N}\ -\omega_{ie}^{U}\right)^{T}$	$\left(-\omega\ 0\ 0\right)^{T}$
15	$C \to D$	$\left(0\ -\omega_{ie}^{N}\ +\omega-\omega_{ie}^{U}\right)^{T}$	$\left(0\ 0\ +\omega\right)^{T}$
16	$D \to A$	$\left(+\omega\ -\omega_{ie}^{N}\ -\omega_{ie}^{U}\right)^{T}$	$\left(+\omega\ 0\ 0\right)^{T}$

39.3 Analysis of Error Characteristics

In this section, according to the error propagation equations of the INS, the rotation modulation model will be established to observe the misalignment angle errors of the mathematical platform caused by the inertial sensor errors in the dual-axis scheme and the tri-axis scheme, and the modulation effects of each rotation scheme on the inertial sensor errors are analyzed.

According to the error propagation equations (39.1) of the INS, if the misalignment angle error $\boldsymbol{\phi}$ caused by a certain error can be periodically modulated to zero, the

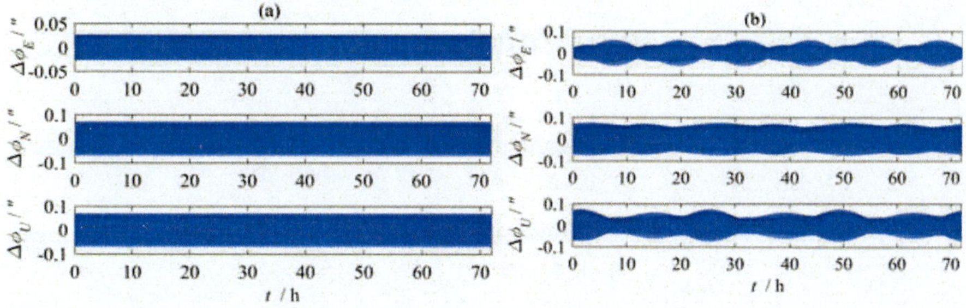

Fig. 39.2 Misalignment angle errors caused by gyro drifts in **a** the 16-sequence dual-axis rotation scheme and in **b** the 16-sequence tri-axis rotation scheme

rotation scheme has a rotation modulation effect on this error. The characteristics of each error in the rotation modulation process are analyzed below.

Figure 39.2 shows the misalignment angle errors caused by optical gyro drifts (0.001°/h) in dual-axis rotation scheme and tri-axis rotation scheme, respectively. It can be seen that, the misalignment angle errors of the two schemes caused by optical gyro drifts are both modulated to zero in a short time, and the average misalignment angle errors of the two schemes is also zero, indicating that both the dual-axis scheme and the tri-axis scheme have modulation effects on the constant drifts.

Figure 39.3 shows the misalignment angle errors caused by the scale-factor errors (1 ppm) in the dual-axis scheme and the tri-axis scheme. It can be seen that the misalignment angle errors caused by scale-factor errors in the dual-axis rotation scheme cannot be modulated to zero, because there are always the coupling errors between the scale-factor errors and the earth's rotation in the dual-axis rotation scheme, leading to the continuous divergence of the misalignment angle errors. In the tri-axis rotation scheme, the misalignment angle errors are modulated to zero in a short time, and does not diverge in a long time, and its average value is zero in one day's modulation cycle. It indicates that the tri-axis rotation can eliminate the coupling errors caused by the scale-factor errors and the earth's rotation, and can

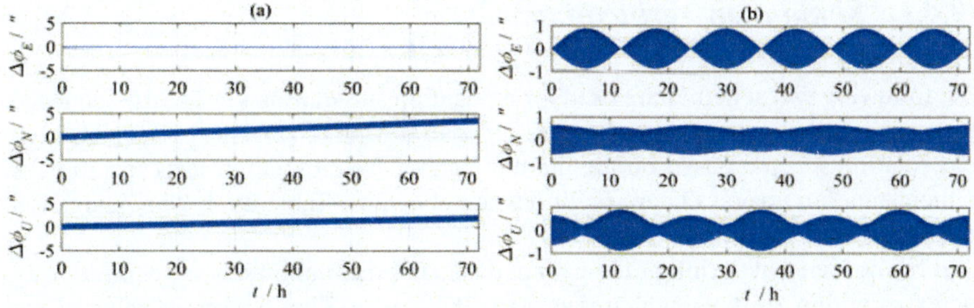

Fig. 39.3 Misalignment angle errors caused by gyro scale-factor errors in **a** the 16-sequence dual-axis rotation scheme and in **b** the 16-sequence tri-axis rotation scheme

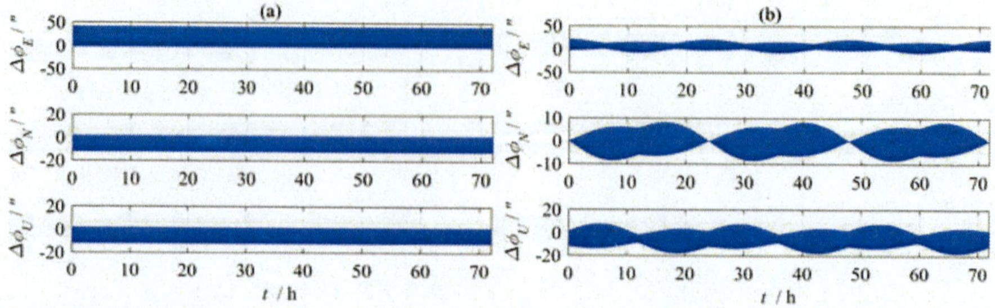

Fig. 39.4 Misalignment angle errors caused by gyro misalignments in **a** the 16-sequence dual-axis rotation scheme and in **b** the 16-sequence tri-axis rotation scheme

restrain the divergence of the navigation errors and improve the navigation accuracy in the case of long-time navigation.

Figure 39.4 shows the misalignment angle errors caused by the optical gyro misalignments (5″) in the dual-axis scheme and tri-axis scheme. It can be seen that the misalignment angle errors in each axis caused by optical gyro misalignments of the two rotation schemes are periodically modulated to zero. The mean errors of the tri-axis scheme are less than those of the dual-axis scheme during one day's modulation cycle.

39.4 Verification of Simulations and Experiments

In this section, the model of the rotation modulation INS will be established. Through simulations and experiments, the navigation results of the dual-axis rotation scheme will be compared with those of the tri-axis rotation scheme to prove the superiority of the tri-axis rotation scheme.

39.4.1 Simulation Verification

The following two schemes are used for navigation simulation: (1) The 16-sequence dual-axis rotation scheme based on the navigation frame; (2) The 16-sequence tri-axis rotation scheme based on the inertial frame. The simulation lasts for 10 days without random errors. Other conditions are shown in Table 39.3. The navigation results are shown in Fig. 39.5.

It can be seen that the longitude error of the dual-axis rotation scheme continuously increases within 10 days, which reaches 0.28″ finally. The maximum value of the longitude error of the tri-axis rotation scheme remains at around 0.1″, proving that the tri-axis rotation scheme can effectively isolate the earth's rotation and restrain

Table 39.3 Parameters used for simulations

Parameter		Value
Gyro	Drift	0.001°/h
	Scale-factor error	1 ppm
	Misalignment	5″
Accelerometer	Drift	10 μg
	Scale-factor error	5 ppm
	Misalignment	2″
	Second-order non-linearity error	5 μg/g²

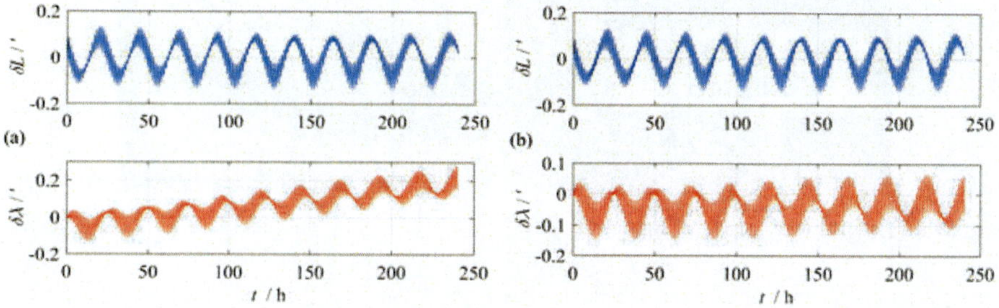

Fig. 39.5 The simulation navigation position errors of the **a** dual-axis 16-sequence rotation scheme and the **b**16-sequence tri-axis rotation scheme

the divergence of position errors. The advantages of tri-axis rotation scheme will be even more obvious in the case of long-time navigation.

39.4.2 Experiment Verification

The experimental system is shown in Fig. 39.6, which consists of a certain type of IMU, a tri-axis turntable, a turntable controller end, and a data acquisition computer. The IMU consists of three optical gyros with bias stability of 0.003°/h and three quartz accelerometers with bias stability of 10 μg.

The 16-sequence dual-axis rotation scheme and tri-axis rotation scheme are used for navigation experiments. The inner and middle axes of the turntable are used to perform the dual-axis scheme. The outer axis is used to isolate the earth's rotation when the tri-axis scheme is performed. The two groups of experiments used the same calibration and initial alignment methods, and the experimental temperature is constant 25 °C, and the rotation angular rates, stagnation interval, and other conditions are completely consistent. The navigation results are shown in Fig. 39.7.

In 5 days of navigation time, the longitude error of the dual-axis rotation scheme keeps diverging and reaches 0.75 nm finally, while that of the tri-axis rotation scheme is always around 0.4 nm and the maximum value is 0.43 nm. It is proved that the

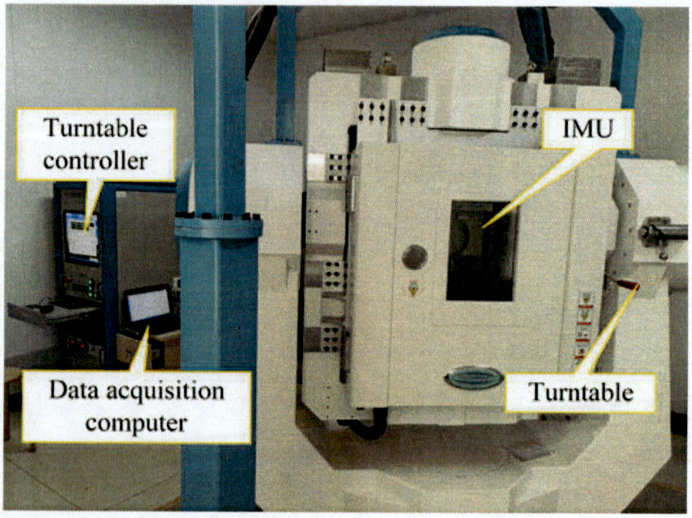

Fig. 39.6 Diagrams of the experimental system

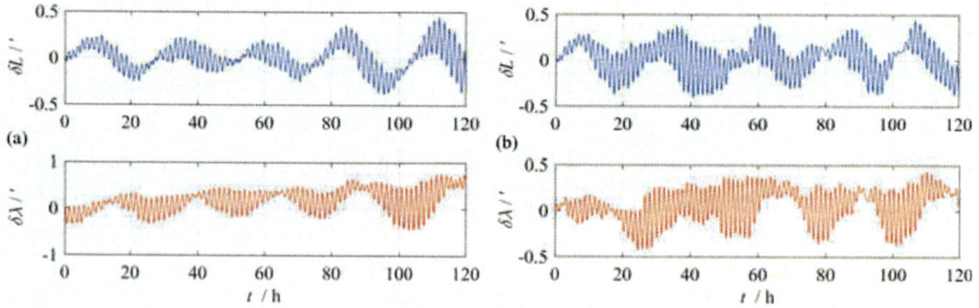

Fig. 39.7 The experiment navigation position errors of the **a** 16-sequence dual-axis rotation scheme and the **b** 16-sequence tri-axis rotation scheme

tri-axis scheme is effective and feasible to isolate the earth rotation, which can reduce the position error by 0.32 nm in 5 days.

39.5 Conclusion

A 16-sequence tri-axis rotation scheme relative to the inertial frame is proposed in this paper based on the error propagation equations of INS. The mathematical platform misalignment angle errors caused by drifts, scale-factor errors, and misalignments of inertial sensors are analyzed, respectively. Compared with the dual-axis rotation scheme based on the navigation frame, the tri-axis scheme can restrain the continuous divergence of the misalignment angle errors caused by the scale-factor errors, indicating that the tri-axis scheme has the effects of eliminating the coupling errors between the scale-factor errors and the earth's rotation. Through simulations and experiments, the tri-axis scheme can reduce the position error by 0.32 nm within 5 days, and effectively restrain the divergence of the position error, indicating that the tri-axis scheme can effectively isolate the earth rotation, which has obvious advantages compared with the dual-axis scheme.

References

1. B. Yuan, G. Rao, On the theory of optical gyro rotating inertial navigation system. J. Natl. Univ. Def. Technol. **28**(6), 76–80 (2006)
2. X. Yu, Y. Wang, P. Zhang et al., Influence of single-axial rotation on INS error characteristics. J. Chin. Inertial Technol. **16**(6), 643–648 (2008)
3. Y. Yang, L. Miao, Fiber-optic strapdown inertial system with sensing cluster continuous rotation. IEEE Trans. Aerosp. Electron. Syst **40**(4), 1173–1178 (2004)
4. E. Levinson, R. Majure, Accuracy enhancement techniques applied to the marine ring laser inertial navigator (MARLIN). Navigation **34**(1), 64–86 (1987)
5. H. Lei, X. Wang, F. Liu, Analysis on error characteristics and key technology of airborne double-axis rotation-modulating RLG INS. Navig. Position. Timing **3**(4), 13–18 (2016)
6. H. Weng, Q. Lu, K. Huang et al., Rotation scheme design for rotary optical gyro SINS. J. Chin. Inertial Technol. **17**(1), 8–14 (2009)
7. Z. Wang, H. Yin, D. Wang, Rotating scheme designing for two-axis rotating laser gyro inertial navigation system. Ship Sci. Technol. **35**(12), 114–120 (2013)
8. B. Yuan, D. Liao, S. Han, Error compensation of an optical gyro INS by multi-axis rotation. Meas. Sci. Technol. **23**(1), 015103 (2012)
9. Z. Ji, C. Liu, S. Cai et al., Improved sixteen-sequence rotation scheme for dual-axis SINS. J. Chin. Inertial Technol. **21**(1), 46–50 (2013)
10. D.H. Titterton, J.L. Weston, *Strapdown Inertial Navigation Technology*, 2nd edn. (The Institution of Electrical Engineers, London, 2004)

Chapter 40
Calculation of the Photon Speed and Photon Energy Discussions

Daocheng Yuan and Qian Liu

Abstract There is an inevitable connection between the wave behavior and the particle behavior of light, which involves a wide range of optical engineering; therefore, in-depth understanding and exploration are required. Based on the hypothesis of the strong interaction between electrons and protons to generate photons, the source and energy of photons are discussed herein. By analyzing and calculating the photon speed, the negative photon speed was found to be in excellent agreement with the existing light speed constant. The Compton effect wavelength shift formula was analyzed and deduced considering the interaction of positive photons and atoms, and new conclusions are drawn, proposing the reason why X-rays are positive photons. Based on existing data, the observed estimation value of the negative photon velocity shift is 0.0295 ppm. Based on the effect of charged spin photons and the electromagnetic potential field, wc explain the reason for the simultaneous existence of reflection and refraction. The relationship between the incident angle and the reflectivity was analyzed, and refractive index formulas were derived to reasonably explain the characteristics of the X-ray refractive index. As a result, edge diffraction is considered a special case of refraction. By analyzing the fluctuation of the potential field inside the atom, the diffraction phenomenon of charged particles is described, and a particle explanation of the phenomenon of light waves and beat frequencies is provided. It is believed that photon energy is kinetic energy and is derived from the potential energy of electrons or protons.

40.1 Introduction

Photons (Solar) are the main source of energy available on Earth and provide hope of clean energy in the future [1, 2]. Optical fiber communication and photolithography have made today's great information age [3, 4], and photons are still the most used tools for human observation of nature [5]. The human body is an energy utilization and information processing system, and photosynthesis is a key link in life activities.

D. Yuan (✉) · Q. Liu
China Academy of Engineering Physics, Mianyang 621000, Sichuan, China
e-mail: yuandaocheng@caep.cn

G. Liu and F. Cen (eds.), *Advances in Precision Instruments and Optical Engineering*, Springer Proceedings in Physics 270, https://doi.org/10.1007/978-981-16-7258-3_40

There is no doubt that photons will bring more human needs, and the propagation of energy by photons at the speed of light is the basis of all photon applications; therefore, a deep understanding of the characteristics of photon energy and its source, as well as photon behavior is important. Human exploration of light has never stopped and will inevitably continue [6–11].

Photons and waves are important controversial topics. The quantum theory of "wave-particle duality" is difficult to understand; it conflicts with common-sense notions derived from observations of the everyday world. Waves are a form of matter movement, and photons are matter (particles) that can travel through a vacuum. Photons behave like particles, and like waves, but are they both? Though we know what photons seem like, we want to know what they are. Therefore, understanding the fluctuations of photons in terms of particle behavior will help to enhance the understanding of photon properties. There are some important questions to consider, for example, why do photons have such a high speed? Is this the limit? Why is the speed of conduction the same as the speed of light? Where does the energy of the electromagnetic waves emitted by electrons come from? How can we understand the relationship between mass and energy?

Nearly 100% of the substances discovered so far comprise electrons and protons (neutrons are a combination of electrons and protons). Hence, there is reason to believe that they are unique and fundamental particles in the material world. There is huge potential energy between electrons and protons (the calculated value of the electron surface potential is greater than 500 kV, and that of the proton surface potential is greater than 1,700 kV) that can result in serious consequences, i.e., an electron–proton explosion, which has exhibited the maximum intensity–volume ratio known to date. This follows the law of Coulomb, Newton's law of motion, and the law of conservation of energy.

It is empirically believed that this intense explosion will produce some smaller broken particles, which are positively or negatively charged, and have the same charge–mass ratio as the parent. Owing to the repelling effect of electricity, these particles can obtain a high speed (kinetic energy). In fact, photons are very important energy carriers. This study assumes that the particles produced by these explosions are photons, including negatively charged negative photons and positively charged positive photons. **Negative photons** are fragments derived from electrons that are negatively charged. Photons of different sizes have different masses and spin magnetic moments, and the charge-to-mass ratio is the same as that of electrons. **Positive photons** are fragments derived from protons that are positively charged. Photons of different sizes have different masses and spin magnetic moments, and the charge-to-mass ratio is the same as that of protons. Based on these assumptions, calculation of the photon speed and photon energy are discussed herein.

40.2 Photon Speed Calculation

The conversion between potential energy and kinetic energy is a common phenomenon. A negative photon leaves the surface of an electron, which is a process of converting potential energy into kinetic energy. Assuming that negative photons are projected from the surface of the electron to the area where the potential is zero, the speed of the negative photons can be calculated according to the conservation of energy.

The potential energy of a system of charges, which is the total work required to assemble the system [12], is given by U_{ext} (40.1). The photon is a tiny part of the charge system, and the assembly of photons and the projection of photons are two reciprocal processes. The kinetic energy $m_{\text{o}}c_{-}^{2}/2$ of the photon obtained by projection is equal to the potential energy required to assemble the photon, and can be calculated using (40.2). c_{-} is the speed at which negative photons leave the surface of the electron to the zero-potential zone (equivalent to in vacuum), q_{o} is the photon charge, m_{o} is the photon mass, e is the electron charge, m_{e} is the electron mass, r_{e} is the electron radius, and $q_{\text{o}}/m_{\text{o}} = e/m_{\text{e}}$.

$$U_{\text{ext}} = \frac{\varepsilon_0}{2}\int_R^{\infty}\left(\frac{Q}{4\pi\varepsilon_0 r^2}\right)^2 4\pi r^2 \mathrm{d}r = \frac{Q^2}{8\pi\varepsilon_0}\int_R^{\infty}\frac{\mathrm{d}r}{r^2} = \frac{Q^2}{8\pi\varepsilon_0 R}. \tag{40.1}$$

$$\frac{1}{2}m_{\text{o}}c_{-}^{2} = \frac{eq_{\text{o}}}{8\pi\varepsilon_0 r_{\text{e}}} \tag{40.2}$$

In another photon velocity calculation method, the electric potential on the surface of the electron is $U_{\text{e}} = e/4\pi\varepsilon_0 r_{\text{e}}$ [12], and the internal force performs the same work on the photon and the electron as the photon leaves the electron. The photon and the electron each gain half of the potential energy $U_{\text{e}}q_{\text{o}}/2$, and the negative photon velocity calculation result is the same as (40.2); thus, the following is obtained (40.3).

$$c_{-} = \left(\frac{eq_{\text{o}}}{4\pi\varepsilon_0 r_{\text{e}} m_{\text{o}}}\right)^{1/2} = \left(\frac{ee}{4\pi\varepsilon_0 r_{\text{e}} m_{\text{e}}}\right)^{1/2}. \tag{40.3}$$

Therefore, the fragmentation of charged particles produces new photons with a certain mass m_{o} (speed of light c), which increases the kinetic energy of the particle system ΔE_{k}, $\Delta E_{\text{k}} = m_{\text{o}}c^2$. This is an understanding of the relationship between mass and energy. Nuclear energy can be understood as part of the potential energy of the particles converted into kinetic energy and released. The photoelectric effect is another case in which the kinetic energy of photons is converted into potential energy.

40.2.1 Negative Photon Velocity

Substituting e and m_e [13], respectively, then $c_- = 2.99792457973037E+8$. Substituting e/m_e [13] into (40.3) as a whole, then $c_- = 2.99792458025253E+8$.

According to the CODATA recommended values of the fundamental constants of physics and chemistry based on the 2014 adjustment [13], the relative standard uncertainty values of e, m_e, and e/m_e are 6.1E−9, 1.2E−8, and 6.2E−9, respectively. The speed of the negative photon was chosen as $c_- = 2.99792458025253E+8$, and the relative error was 8.42E−11, compared with the constant of light speed $c = 2.99792458E+8$. Thus, the calculated negative photon speed has a very good accuracy.

40.2.2 Positive Photon Velocity

The principle of a positive photon obtaining speed c_+ is the same as that of negative photons, (40.4), except that the radius of a proton r_p is smaller, the mass of a proton m_p is larger, and the charge-to-mass ratio is much smaller, taking $r_p = 8.33E-16$ [14].

$$c_+ = \left(\frac{ee}{4\pi\varepsilon_0 r_p m_p}\right)^{1/2}. \tag{40.4}$$

Substituting e and m_p into (40.4), respectively, then $c_+ = 1.286796E+7$. Substituting e/m_p into (40.4) as a whole, then $c_+ = 1.286796E+7$, and $c_-/c_+ \approx 23$.

It can be seen that the calculated speed of the positive photons is significantly smaller than the light speed constant c. Because existing knowledge does not consider photon charging, the speed of positive photons has not been verified, but X-rays exhibit certain properties of positive photons.

The calculation of the speed of photons shows that the speed of light is a normal phenomenon of the strong interaction of charged particles and is the result of the conversion of potential energy into kinetic energy. The calculation model of light speed is not perfect, but it shows that it is possible for photons to achieve such a huge speed. The speed of light particles is related to the electric potential field and can be larger or smaller, but remains fixed in the region where the potential is zero. At this time, the negative photon speed is equal to the light speed constant c, and there is a huge difference between the positive photon velocity and the light velocity constant.

As a charged particle, photons should theoretically participate in the macroscopic conduction process, and responding to changes in the electric field is an inherent property of photons. They silently contribute to the conduction of current, but have not been found, which is consistent with the conduction velocity constant. The photon itself has energy and transfers energy at the speed of light, and hence functions as an energy propagation medium.

Electromagnetic waves also have a speed of light and can propagate in a vacuum. Their electrical properties are remarkable and exhibit all the properties of light. It is reasonable to consider electromagnetic waves as a group of smaller negative photons. Because the photon volume, mass, electricity, and magnetic moment are very small, they have not been detected by current measurement methods.

40.2.3 Compton Effect and Derivation of the Compton Shift

X-ray refraction exhibits positive photon properties (see Sect. 40.3.2 for detailed analysis), so understanding the Compton effect regarding the interaction between positive photons and atoms has certain significance. The Compton effect is the action of X-rays (positive photons) in the electron gravitational field and the nuclear repulsion field. The scattered light can be divided into two parts: One is the scattering (including reflection) of positive photons by the nucleus, with a wavelength shift; the other part is that positive photons escape after accelerating in the gravitational field of electrons, and there is no wavelength shift.

The scattering of positive photons by the nucleus has the same physical mechanism as that of common light reflection. The kinetic energy of the photon enables it to approach the center of the repulsive field very close, so the repulsive force is very large; therefore, the repulsive work is also significant. Photons in the repulsive field continue to lose kinetic energy owing to external work, resulting in a wavelength shift (decline in kinetic energy). Because of the large mass of the nucleus, and the nucleus is not a rigid body, the wavelength shift $\Delta\lambda = \lambda - \lambda_0$ is not significantly affected by the change in the incident wavelength and atomic number, but is greatly influenced by the scattering angle. This is why the Compton shift of an X-ray seems to be independent of the target material. Positive photons escape from the gravitational field of electrons and are regarded as scattered. In fact, they are irregular refractions caused by scattered distribution electrons, and there is no wavelength shift. Compared with scattered photons, the distance between the escaped photons and the center of the gravitational field is much larger. Therefore, the gravitational work is small compared to the repulsive work, and the kinetic energy of refracted (considered as scattered) photons hardly changes.

The absorption, collision, and escape of positive photons coexist in the gravitational field of the electrons. When the atomic number increases, more outer electrons of the atom enhance absorption and refraction, reducing the probability of nuclear scattering, and the intensity of the scattering (with wavelength shift) decreases.

According to the explanation of quantum mechanics, the relation between the shift in wavelength $\Delta\lambda$ and the scattering angle θ is found to be [15]:

$$\Delta\lambda = \frac{h}{m_e c}(1 - \cos\theta) = \lambda_c(1 - \cos\theta), \tag{40.5}$$

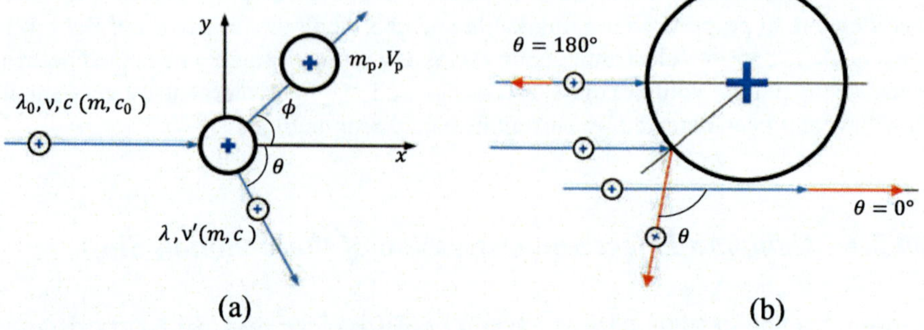

Fig. 40.1 Explanation of the Compton effect. The influence of Lorenz force and angular momentum is not considered. **a** Incident X-ray (positive photons) with energy $E = h\nu$ and momentum $p = 2h\nu/c$ collides with a nucleus. In the collision, positive photons do work on atomic nuclei with repulsive force and lose kinetic energy. **b** Elastic scattering of positive photons and nuclei in the repulsive field

where m_e is electronic mass, c is the light speed constant, and $\lambda_c = h/m_e c$ is the Compton wavelength.

The derivation of the Compton shift is different from the perspective of positive photons. When positive photons hit the nucleus, they are scattered by repulsive fields. Consider the Compton effect as an elastic collision between a photon $h\nu$ and an atomic nucleus m_p, ignoring the influence of the Lorenz force and angular momentum. In the collision, the photon transfers energy and momentum to the nucleus; the scattered X-ray photon thus has a reduced energy $h\nu'$ and a reduced momentum $2h\nu'/c$ ($mc^2/2 = h\nu'$, c should be c_+, use c as usual) (see Fig. 40.1a).

Energy before and after the collision is conserved, so (40.6):

$$h\nu - h\nu' = \frac{1}{2} m_p {V_p}^2. \tag{40.6}$$

For the momentum in the y-direction before and after the collision, we have (40.7):

$$\frac{2h\nu'}{c}\sin\theta - m_p V_p \sin\phi = 0. \tag{40.7}$$

And for the momentum in the x-direction, we have (40.8):

$$\frac{2h\nu}{c} = \frac{2h\nu'}{c}\cos\theta + m_p V_p \cos\phi. \tag{40.8}$$

Substituting $\sin^2\phi + \cos^2\phi = 1$, $\nu\nu' = -c\Delta\nu/\Delta\lambda$, and solving (40.6), (40.7), and (40.8) simultaneously, where $\Delta\nu = \nu - \nu'$ and $\Delta\lambda = \lambda - \lambda_0$, we obtain (40.9):

$$\Delta\lambda = \frac{4hc}{m_p c^2 - 2h\Delta\nu}(1 - \cos\theta) \approx \frac{4h}{m_p c}(1 - \cos\theta). \tag{40.9}$$

This is different from formula (40.5), $\lambda_c = 4hc/\left(m_p c^2 - 2h\Delta\nu\right)$, and requires explanation.

$h\Delta\nu$ is the kinetic energy loss of positive photon scattering, and its value is related to the photon mass and θ, which is the inevitable result of scattering. Figure 40.1b provides a more intuitive explanation. In the figure, the nonlinear trajectory of the photon is simplified as a straight line. When $\theta = 0°$, $h\Delta\nu = \min = 0$, and when $\theta = 180°$, $h\Delta\nu = \max$, hence, it is reasonable to obtain different values of $h\Delta\nu$ with different θ. When the photon mass changes, the same $\theta = 180°$, and $h\Delta\nu = \max$ will have a different magnitude, so λ_c should not be a fixed wavelength shift. The extreme case of elastic collision is the loss of all kinetic energies, which also shows that the wavelength shift is not a fixed value.

The cause of the wavelength shift is the nucleus, which has a large mass, not an electron.

$m_p c^2 \gg 2h\Delta\nu$, $\Delta\lambda \approx 4h(1 - \cos\theta)/m_p c$, and $\lambda_c = 4hc/(m_p c^2 - 2h\Delta\nu)$ is not a constant, but it does not change much.

The kinetic energy of photons decreases due to scattering. Therefore, the essence of the wavelength shift $\Delta\lambda = \lambda - \lambda_0$ is the shift of light speed $\Delta c = c_0 - c$, which corresponds to a reduction in photon kinetic energy $\Delta E = E_0 - E$. By transforming (40.9), we can obtain the formula of light speed shift (40.10), where m is the photon mass.

$$\Delta c = \frac{2mc_0(1 - \cos\theta) + m\Delta c\left(c_0{}^2 - c^2\right)/c_0{}^2}{m_p + 2m(1 - \cos\theta)} \approx \frac{2mc_0}{m_p}(1 - \cos\theta) \tag{40.10}$$

When $\theta = 180°$, $\Delta c = 4mc_0/m_p$, and the incident photon of 1,000 eV can be calculated as

$$\frac{\Delta E}{E_0} = \frac{c_0{}^2 - c^2}{c_0{}^2} = \frac{\Delta c}{c_0}\left(2 + \frac{\Delta c}{c_0}\right) = \frac{9.29\,\text{eV}}{1000\,\text{eV}}.$$

This result deviates significantly from the experimental results in [15]. In the magnetic field outside the nucleus, the Lorenz force enhances the photon scattering and reduces the photon repulsion work to the outside, resulting in a wavelength shift of less than 2.9.

The scattering mass of the nucleus is taken as m_p, because the repulsive force of the photon and the nucleus directly acts on the proton, and it is difficult to determine whether the entire nucleus can be regarded as a rigid body. When the scattering mass of the nucleus is $2.32m_p$, then $\Delta E/E_0 = 4\text{eV}/1{,}000\,\text{eV}$. This result is consistent with the Compton experiment [15], and the value of $2.32m_p$ is reasonable to a certain extent.

When the incident photon energy is 1 meV, the calculated photon mass is greater than the proton mass, which is difficult to explain and verify.

Three points are added to the physical meaning of the Compton effect from the perspective of photon charging:

(1) X-rays are positive photons, as they are significantly repulsed by the nucleus. High-speed positive photons with a mass close to or greater than an electron are impossible to be reflected at $\theta = 180°$ by the electron;
(2) The wavelength shift $\Delta\lambda$ is evidence that the photon loses kinetic energy and the photon velocity changes. The speed shift of light is the result of photons doing work externally, and the scattering angle significantly affects the speed shift of light;
(3) λ_c (corresponding to the speed shift of light) is approximately constant and exhibits a small change with the incident wavelength λ_0.

Inertially confined fusion is an important physical experiment, and the role of X-rays is an important link [16]. If calculated according to the actual wavelength, the X-ray (positive photon) energy is only about 1/23 of the conventional calculation value, and X-rays cannot easily get close to the nucleus. Further, the electrons around the nucleus have an absorption effect on X-rays. These disadvantages clearly increase the difficulty of laser-driven nuclear fusion.

The abilities of positive and negative photons to pass through the Earth's atmosphere are significantly different. The total amount of photon electricity that reaches the earth is not zero, so sunlight may be an influencing factor on the Earth's electromagnetic environment.

40.2.4 Scattering and Speed Shift of Negative Photons

Negative photons also exhibit a similar repulsive field scattering (reflection). When a negative photon moves to an electron, the photon continues to work externally with a repulsive force, resulting in a shift in the speed of light. Equation (40.11) is thus obtained, where m is the mass of the photon and m_e is the mass of the electron.

$$\Delta c = \frac{2mc_0(1-\cos\theta) + m\Delta c\left(c_0{}^2 - c^2\right)/c_0{}^2}{m_e + 2m(1-\cos\theta)} \approx \frac{2mc_0}{m_e}(1-\cos\theta). \tag{40.11}$$

Comparing the speed shifts of negative and positive photons, taking $\lambda_+ = 0.1$ nm and $\lambda_- = 500$ nm photons as an example, the relative speed shift of negative photons is slightly smaller. Negative photons have greater speed, the electron magnetic moment is greater, and the Lorenz force is much larger than for a positive photon. The particularity of the Lorenz force is that it changes the photon momentum (direction) but does not work, and the speed drift of negative photons is small. Equation (40.11) is used only for discussion and comparison with positive photons and has no precise meaning. However, the speed drift of negative photons should exist.

The light velocity shift of negative photons can also be observed, and the magnitude of the light velocity shift is estimated from the experimental data of the beat

Table 40.1 Calculation of light velocity shift of negative photons

λ_1/nm	λ_2/nm	Calculated beat frequency/MHz	Measured beat frequency/MHz
632.991 212 57 (4.7361235361E + 14 Hz)	632.991 354 (4.7361224783E + 14 Hz)	106	(121.33 + 118.59)/2 = 119.96
(119.96–106)/473,612,248 = 0.0295 ppm			

frequency given in [17]. The two stabilized lasers λ_1 and λ_2 participate in the beat frequency, where $\lambda_1 = 632.99121257$ nm (6.3E−11) and $\lambda_2 = 632.991354$ nm (approximately 1E−9); λ_1 undergoes two 90° reflections, and λ_2 undergoes three 90° reflections. Owing to the shift in the speed of light, λ_2 decreases further, resulting in an increase in the beat frequency. Taking the difference between the measured beat frequency and the calculated beat frequency between λ_1 and λ_2 as the contribution of one (3 – 2 = 1) 90° reflection to the light velocity shift, the calculated relative value of the light velocity shift is 0.0295 ppm (see Table 40.1). The extended uncertainty is less than 0.02 ppm, according to the accuracy of the interferometer. 0.0295 ppm is three orders of magnitude smaller than formula (40.11), so it is believed that the Lorenz force considerably influences the reflection of negative photons.

40.3 Photon Reflection and Refraction

Under the effect of the gravitational and repulsive fields in an atom, the movement direction of the photon changes.

Matter is composed of atoms. The potential distribution inside an atom is extremely uneven, and the potential distributions on the surface and inside a substance are extremely uneven at the subatomic scale. The reason why both reflection and transmission coexist is precisely due to the uneven distribution of the electric potential of the material (subatomic scale), the area with gravitational field, and the area with repulsive field, as shown in Fig. 40.3, $U_2 > 0$ or $U_2 < 0$. Negative photons are transmitted (refracted) in the region of positive potential near the center of the atom and reflected (scattered) in the region where the negative potential is large enough around the electron.

The reflection and refraction of positive photons are different. The Compton effect occurs when the angle of incidence is zero. Electrons can cause refraction of positive photons. Because the potential field of electrons in atoms is scattered, the regular refraction pattern is not strong. The nucleus can reflect positive photons, but it is surrounded by electrons, and the reflection phenomenon is not significant.

40.3.1 Photon Reflection

When a negative photon enters a region of negative potential energy equal to its kinetic energy, its velocity drops to zero, and then it is emitted back by the potential energy. This is a negative photon reflection. Positive photons can be reflected by protons, involving complex atomic structures, and electrons are located on the periphery of the atoms, resulting in different reflections of positive photons.

The reflection behavior of the negative photons is the result of their interaction with electrons. Regardless of the influences of magnetic force and Lorentz force, the electric force does work and obeys the law of conservation of energy. Photons are reflected by the conversion of kinetic energy, potential energy, and kinetic energy. Figure 40.3 shows that due to the incident angle θ_i, the kinetic energy movement to the surface of the medium (electrons) is only a component of the total kinetic energy of the photon $m_o c_-{}^2/2$. This component $m_o c_-{}^2\cos^2\theta_i/2$ reduces to make the reflection easier, which is equivalent to the increase in the reflection cross-sectional area of electrons. The coefficient corresponding to the increase in the radius $R(\theta_i)$ of the reflecting circle is $1/\cos^2\theta_i$ (40.13), the coefficient of increasing the reflection area is $1/\cos^4\theta_i$, and the reflectivity $r(\theta_i)$ has a definite relationship with the incident angle θ_i, which is derived as follows.

The speed of light c_- in the zero-potential region is constant (40.12).

$$\frac{1}{2}m_o c_-{}^2 = \frac{eq_o}{8\pi\epsilon_0 r_e}. \tag{40.12}$$

The change in the incidence angle θ_i is related to the electron reflection radius $R(\theta_i)$ (40.13).

$$\frac{1}{2}m_o c_-{}^2\cos^2\theta_i = \frac{eq_o}{8\pi\epsilon_0 R(\theta_i)}. \tag{40.13}$$

The photon's magnetic attraction and kinetic energy (corresponding to the potential-related refractive index) jointly affect the reflection. Let the magnetic influence coefficient equivalent to $\cos^2\theta_i$ be R_m, $R(\theta_i) \sim 1/(\cos^2\theta_i + R_m)$, and $r(\theta_i) \sim 1/\left(\cos^2\theta_i + R_m\right)^2$.

Substituting $r(0) = (n_2 - n_1)^2/(n_2 + n_1)^2$, we obtained:

$$r(\theta_i) = \frac{(1 + R_m)^2(n_2 - n_1)^2}{\left(\cos^2\theta_i + R_m\right)^2(n_2 + n_1)^2}. \tag{40.14}$$

To compare with the existing physical laws, the two reflection coefficients related to the refractive index and incidence angle are given in Fig. 40.2, taking the experience value $R_m = 0.265$ (according to the reflectivity of 632 nm laser at the air–water interface).

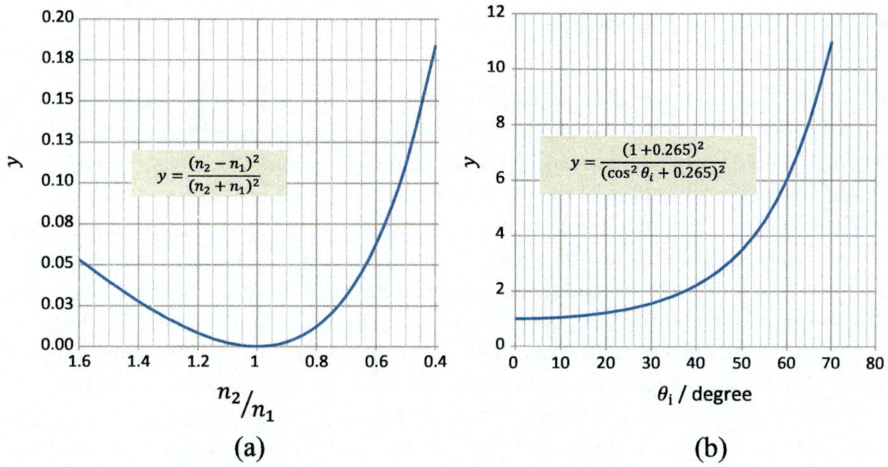

Fig. 40.2 Relationship between reflection coefficient and refractive index and incident angle

Figure 40.2 shows that the effects of the refractive index and incident angle on the reflection coefficient are significantly different. Combined with the results of existing physical experiments, it can be considered that the reflectance of S-light is mainly affected by the incident angle and fixed refractive index, while the reflectance of P-light is affected by the incident angle and changing refractive index. When the photon in a medium of higher index of refraction approaches the other medium, the normal velocity component of the refracted photon will decelerate. As the angle of incidence increases, when the normal velocity tends to zero, the photon flies to the negative potential area, resulting in total reflection. The refracted photon is transformed into a reflected photon, and the photon trajectory has a particularity; total reflection occurs.

The reflection angle θ_r is determined by the ratio of the photon velocity components in the two directions. When the photon enters the repulsive field and returns to the original potential area, regardless of the light speed shift, the velocity component amplitude in both directions remains almost unchanged. The reflection angle is equal to the incident angle (40.15).

$$\theta_r = \theta_i. \tag{40.15}$$

Negative photons use electrons as reflection targets, while positive photons use atomic nuclei as reflection targets. The electron and nucleus have different peripheral potential fields, motion modes, and spatial distributions. Therefore, the reflection characteristics of the positive and negative photons are different. As a reflective target, compared with electrons, atomic nuclei have smaller linear velocity and smaller position change; thus, positive photon reflection optical imaging has advantages.

40.3.2 Photon Refraction

When photons enter the electromagnetic gravitational field and escape, their direction of movement changes with strong consistency, that is, photon refraction. Edge diffraction also occurs due to the change of photon direction, which is clearly affected by the gravitational field, and can be regarded as a special refraction.

Refractive index convention: when a photon passes from one medium into another, if θ_i is the angle of incidence of photons in vacuum and θ_t is the angle of refraction, the refractive index n is defined as the ratio of the sine of the angle of incidence to the sine of the angle of refraction, specifically, $n = \sin\theta_i/\sin\theta_t$. The refractive index is also equal to the velocity v of a photon of a given wavelength in a substance divided by its velocity c in empty space, $n = v/c$ (not $n = c/v$, due to $v > c$), (40.16).

The forces that cause a change in the direction (momentum) of the photon include the electric and magnetic forces. After the photon enters the equipotential field before the incident, although the overall work of the electric force and magnetic force is close to zero, the flight trajectory of the photon changes. Because the photon wavelength determines its magnetic moment, its refraction also changes with the wavelength. The electric force contributes to refraction, without considering the influence of the magnetic force, and the formula for the refractive index of the electric force can be derived.

In the positive electric field near the nucleus, $U_2{>}0$, photon refraction occurs, and positive and negative photons behave differently. In Fig. 40.3, the relative refractive index of the material is n, the corresponding negative photons and positive photons are n_- and n_+, the negative photon incidence speed is c_-, the transmission negative photon speed is c_{-t}, the positive photon incidence speed is c_+, the velocity of the

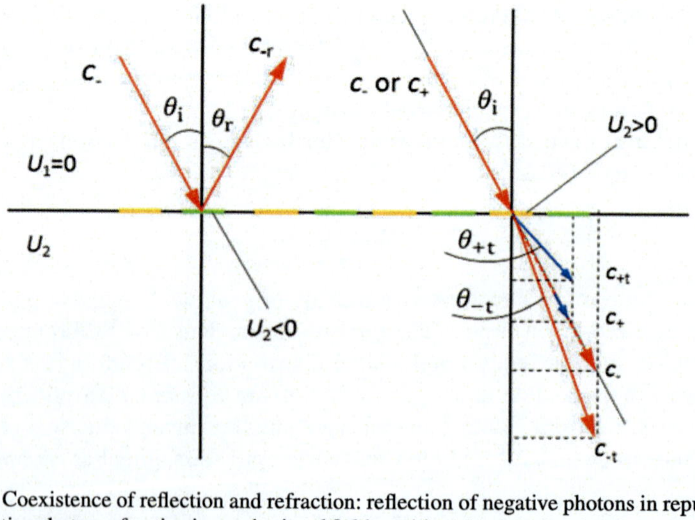

Fig. 40.3 Coexistence of reflection and refraction: reflection of negative photons in repulsive field (left), negative photon refraction in gravitational field, positive photon refraction (actually scattering) in repulsion field (right)

transmitted positive photon is c_{+t} (it is not true refraction for positive photons, but conventional understanding is still used here), the charge of the photon is q_o, the mass is m_o, the electron mass is m_e, the proton mass is m_p, and the positive potential in the transmission area is U (U_2 in Fig. 40.3); thus, $c_{-t} > c_-$ and $c_{+t} < c_+$.

$$n = \frac{\sin\theta_i}{\sin\theta_t} = \frac{v}{c}, \tag{40.16}$$

$$c_{-t} = n_- c_-, \; c_{+t} = n_+ c_+. \tag{40.17}$$

Upon entering the region of positive potential, the kinetic energy of the negative photons increases (40.18).

$$\frac{1}{2} m_o (c_{-t}{}^2 - c_-{}^2) = U q_o. \tag{40.18}$$

When entering the positive potential region, the positive photon kinetic energy decreases (40.19).

$$\frac{1}{2} m_o \left(c_+{}^2 - c_{+t}{}^2\right) = U q_o. \tag{40.19}$$

The electrical refractive indexes of negative photons and positive photons can be obtained as follows, respectively:

$$n_-{}^2 = 1 + \frac{2Uq_o}{m_o c_-{}^2} = 1 + \frac{2Ue}{m_e c_-{}^2}, \tag{40.20}$$

$$n_+{}^2 = 1 - \frac{2Uq_o}{m_o c_+{}^2} = 1 - \frac{2Ue}{m_p c_+{}^2}. \tag{40.21}$$

The electrical refractive index curves of the positive and negative photons were calculated using (40.20) and (40.21), and are shown in Fig. 40.4. Using 0–360 kV as the hypothetical potential, which is lower than the proton surface potential, the calculated value of the proton surface potential is 1,728.649 kV.

The refractive index of X-rays is slightly less than 1.0, due to $c_{+t} < c_+$. The equation $n = v/c$ in this case indicates that the velocity of X-rays is smaller than its velocity in empty space.

The above calculation shows that the electrical refractive indexes of negative and positive photons are very different, and the refractive index of positive photons $n_+ < 1$, which is consistent with the experimental results. This extreme refractive index contrast of positive and negative photons indicates that the photon momentum changes in the opposite direction; therefore, X-rays are positive photons. For the same reason, the total reflection of positive and negative photons appears as distinct external reflection and internal reflection, respectively.

Fig. 40.4 Calculated negative and positive photon electrical refraction indexes under the condition of 0–360 kV potential

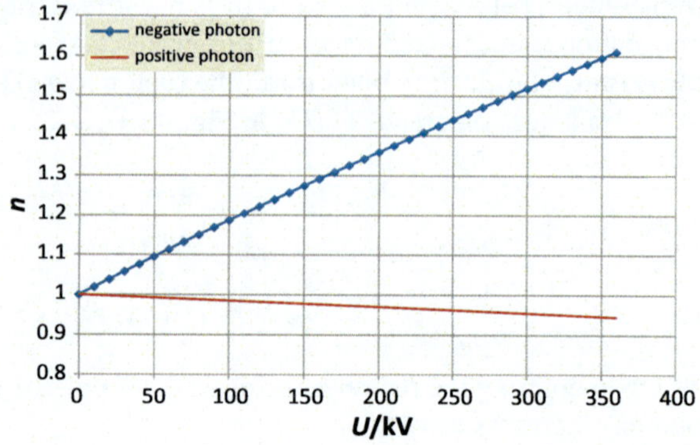

The refractive index of the positive photon and the negative photon will be larger considering the influence of the photon magnetic moment (corresponding to the polarization) force, and negatively relate to the wavelength. Besides refraction, photons can also orbit (photon absorption) in the gravitational field, negative photons fly around the nucleus, and positive photons fly around electrons; the photon orbital radius is positively related to the wavelength. Photon orbital flight is closely related to blackbody radiation and atomic spectrum (Note: Related research content omitted).

40.4 Photon Wave Phenomenon

A nucleus comprises protons and electrons. The density difference between them is approximately 70,000 times. They are combined together, and the center of mass and the center of the overall charge may not coincide. The nucleus has spin motion. In theory, the spin axis passes through the center of mass. When the charge center and spin axis are eccentric, the charge distribution center fluctuates sinusoidally with the spin of the nucleus, resulting in periodic fluctuation of the potential field in the atom.

In addition to the nuclear rotation ω_n, the electric field fluctuation of electrons inside the atom also has a similar effect. Because the orbital electrons rotate relative to the atomic nucleus, as shown in Fig. 40.5, the electromagnetic field caused by the electrons changes with the angular position of the electron rotation ω_e, so the electromagnetic field fluctuates periodically. In general, the periodic fluctuating electromagnetic potential field inside the atom persists over a wide range, and the superposition of various factors produces a fluctuating influence, which is the internal reason for photon wave performance. The influence of the fluctuating electromagnetic field on the reflection and refraction of photons is different, and it is related to optical imaging, hence is worthy of in-depth study.

The wave electromagnetic potential field U inside the atom can be expressed as (40.22), where ω is the angular velocity of the wave potential field.

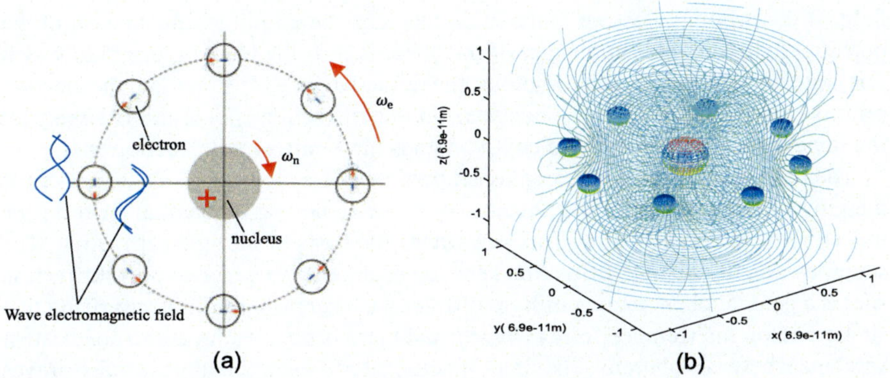

Fig. 40.5 **a** Wave electromagnetic field inside the atom; **b** calculated three-dimensional electric equipotential line

$$U = \sin\omega t. \tag{40.22}$$

As a charged particle, the photon's movement trajectory (refraction) will be affected by this wave potential field U, resulting in a corresponding change in the direction and position $y(U) = y(\sin\omega t)$. To simplify the description, consider $y(U)$ as a simple linear relationship, $y(U) \propto U = \sin\omega t$, which yields (40.23). The time integral $I(y)$ of $y(U)$ is the geometric distribution of a large number of photons, as shown in Fig. 40.6. The light intensity I(y) is a simple fringe distribution that represents the formation mechanism of the photon interference fringes.

$$I(y) \propto \int \sin\omega t d\omega t = 2\sin^2\frac{\omega t}{2}. \tag{40.23}$$

The diffraction spot width is unrelated to the fluctuation period of the potential field, but is affected by the fluctuation amplitude of the electromagnetic potential

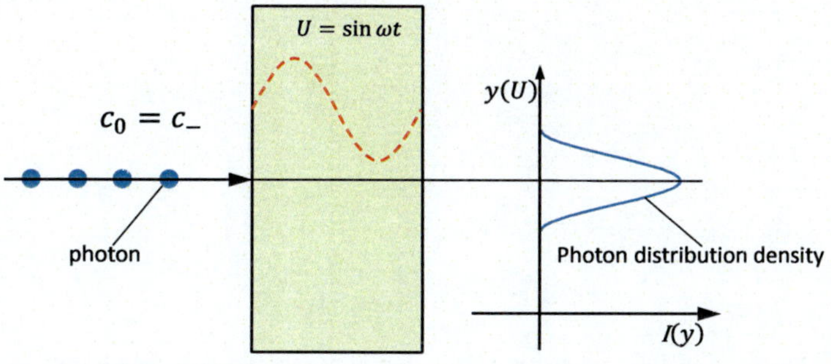

Fig. 40.6 Time-integrated distribution of photons across the wave potential field forms diffraction spot

field. If the orbiting electron is the main factor in the amplitude fluctuation of the potential field, the diffraction spot of the photons near the electron orbit are wider. For large-wavelength negative photons, the refractive index is lower and the electron orbit is closer (farther from the nucleus), so the diffraction spot width is larger; for the same reason, the short-wavelength photons have narrower diffraction spot.

The formation of the abovementioned particle diffraction spot is the result of the merging time integral of the influence of the periodic wave potential field on the trajectory of the photon flow. This constitutes the basis of all light diffraction. The superposition of existing diffraction spots on each other to generate new diffraction spot is a general phenomenon of light diffraction. Therefore, it is not surprising that single-photon interference forms fringes, and there is no need for some nonexistent agreement between photons. The single-photon interference phenomenon also proves that the photon wave originates from the electromagnetic field fluctuation in the optical path. The photon is not a wave, but it shows the characteristics of a wave under the influence of the wave potential field.

The photon (particle) behavior exhibits a wave phenomenon in a statistical sense, which can produce interference fringes, or it can be manifested as wave propagation, which is consistent with the existing wave theory. The photons arranged along the flight direction are like waves propagating in space, exhibiting periodic fluctuations in amplitude and specific wavelengths.

An explanation of the beating phenomenon of dual-frequency lasers is shown in Fig. 40.7, where photons (particles) are represented by the round spots, and light waves are represented by the harmonic curves. Photons are periodically arranged in the flight direction, similar to the propagation of light waves. Different wavelengths λ_1 and λ_2 are mixed together, causing the photon number distribution period (beat frequency) to change. This beat frequency period is much larger than the period of a single wavelength, so that photocells with insufficient time and space resolution can detect it. The beat frequency travels at the speed of light and has a time period

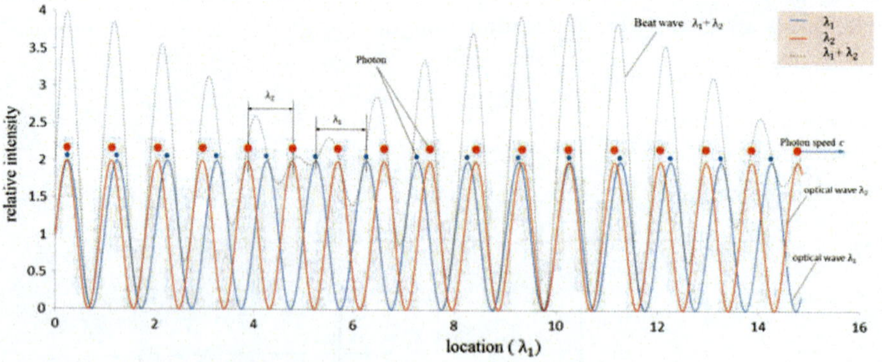

Fig. 40.7 Comparison of light intensity distribution between two-wavelength photon flight and dual-frequency light wave beat frequency

and space period. This is the basis for the use of the Doppler effect in integrating the movement speed to measure the change in length.

For two beams of photons with the same wavelength, the interference fringes are easier to be observed, because their locations do not change over time.

40.5 Conclusion

Considering the widespread existence and interaction of electrons and protons, as well as Coulomb's law, Newton's law of motion, and the law of conservation of energy, the hypothesis that electrons and protons act strongly to generate photons is proposed, and the following conclusions were drawn.

In the zero-potential (equivalent to in vacuum) region, the negative photon velocity is equal to the optical constant c, and there is a huge difference between the positive photon velocity and the light velocity constant. Photons have energy and are also media that transfer energy at the speed of light. Charged photons participate in the macroscopic conduction process, which is why the conduction speed is a constant of the light speed. Electromagnetic waves have the same physical properties as photons and are a group of negative photons with a smaller volume.

The effect of the charged spin photon and the uneven and fluctuating potential field of the atom are the internal mechanisms of all optical phenomena. The main factors affecting these macroscopic optical phenomena include

(1) Positive or negative photons. The behavior of positive and negative photons varies greatly. X-ray is a positive photon;
(2) Gravitational field or repulsion field: scattering (reflection) occurs only in the repulsion field, and refraction, collision, and absorption only occur in the gravitational field. Scattering of the repulsion field causes the photon speed shift;
(3) Nonuniform field and structural characteristics within the atom: electrons are located at the periphery of the atom, with strong reflection laws for the negative photon. The central location of the nucleus increases the significance of the refraction phenomena, and the mechanism of refraction for positive and negative photons remains the same. The scattered electrons in the atoms make the refraction of positive photons less obvious;
(4) Periodic fluctuation of the internal potential field of the atom, which is the internal cause of various diffraction phenomena;
(5) Incident angle of the photons, which significantly affects the reflectivity and photon speed shift.

Very different electrons and protons form atoms, and the microscopic potential field of the material is not uniform at the subatomic scale, resulting in the coexistence of photon reflection and refraction in transparent materials and the Compton effect. The scattering or reflection in the repulsive field causes the loss of photon kinetic energy (light speed shift), and the reflection and refraction behaviors of negative and positive photons are very different. Edge diffraction is a special case of refraction.

Photons are the key carrier and transmission tools of energy. They have kinetic energy that originates from the potential energy of the electrons or protons. It is believed that the charged spin of light particles conforms to the existing physical laws. Understanding the nature of photons will help new developments in energy utilization and measurement technologies.

References

1. B. Viswanathan, *Energy Sources* (Elsevier, Amsterdam, 2017), pp. 139–147
2. N.F.F. Areed, S.M. El Malt, S.S.A., Obayya, Broadband omnidirectional nearly perfect plasmonic absorber for solar energy harvesting. IEEE Photonics J. **8**(5), 1–18 (2016)
3. M. Hirano, Future of transmission fiber. IEEE Photonics J. **3**(2), 316–319 (2011)
4. R. Leuschner, G. Pawlowski, *Photolithography, Materials Science and Technology*, vol. 16 (1996)
5. A. Mcclelland, M. Mankin, *Optical Measurements for Scientists and Engineers (A Practical Guide)* (Cambridge U. Press, 2018)
6. S. Jackson, A.C. Vutha, Magic polarization for cancellation of light shifts in two-photon optical clocks. Phys. Rev. A **99**(6) (2019)
7. S. Ahrens, C.-P. Sun, Spin in Compton scattering with pronounced polarization dynamics. Phys. Rev. A **96**(6) (2017)
8. E. Ikonen, Coherence of radiation as studied by multiple coincidences of photons and particles. Opt. Rev. **17**(3), 239–247 (2010)
9. T. Tashima, H. Morishita, N. Mizuochi, Experimental demonstration of two-photon magnetic resonances in a single-spin system of a solid. Phys. Rev. A **100**(2) (2019)
10. F. Manoocheri et al., Low-loss photon-to-electron conversion. Opt. Rev. **21**(3), 320–324 (2014)
11. M. Tareeva et al., Two-photon excited luminescence in polyethylene and polytetrafluoroethylene. J. Russ. Laser Res. **41**(5), 502–508 (2020)
12. M. Purcell, D.J. Morin, *Electricity and Magnetism*, 3rd edn. (Cambridge U. Press, New York, 2013), pp. 34–706
13. P.J. Mohr, D.B. Newell, B.N. Taylor, CODATA recommended values of the fundamental physical constants: 2014. J. Phys. Chem. Ref. Data **45**, 043102-58-60 (2016)
14. N. Bezginov, T. Valdez et al., A measurement of the atomic hydrogen Lamb shift and the proton charge radius. Science **365**, 1007–1012 (2019)
15. H. Haken, H.C. Wolf, W.D. Brewer, *The Physics of Atoms and Quanta* (Springer, Berlin, Heidelberg, 2005), pp. 60–63
16. S. Nakai, H. Takabe, Principles of inertial confinement fusion–physics of implosion and the concept of inertial fusion energy. Rep. Prog. Phys. **59**(9), 1112–1114 (1996)
17. A. Hapiddin et al., Beat frequency measurement of the stabilized He-Ne laser 633 nm calibration in SNSU-BSN. J. Phys.: Conf. Ser. **1528**, 012003 (2020)

Chapter 41
Research on Simulation of Space-Based Optical Space Debris Images

Yupeng Wang, Jian Huang, Yue Li, Pengyuan Li, and Zhaodong Niu

Abstract A method for imaging simulation of space-based debris based on visible light is proposed. In response to the development needs of space-based space debris detection systems, by analyzing the imaging principles of space-based visible light space debris, the structure and flow of imaging simulation are designed, and the basic simulation of space-based space debris visible light imaging is realized. According to the imaging characteristics of CCD devices, the principle of the Smear tailing effect of super-bright stars is analyzed, and the Smear tailing effect of super-bright stars can be fully reflected in the image generated by the simulation. This method realized the visible light imaging simulation of any orbital platform, any field of view, any time, and any space debris.

41.1 Introduction

With the continuous exploration of space by mankind, increasing space activities have brought hundreds of millions of space debris. The huge number of space debris greatly increases the probability of space collisions. In order to cope with the increasingly severe problem of space debris, countries have developed space target surveillance and detection technologies.

According to the classification of detection methods, space target monitoring and detection technologies are mainly divided into photoelectric detection and radar detection. Compared with radar detection, photoelectric detection has the characteristics of long range and lower cost and has become an important method of space target

Y. Wang · Y. Li · Z. Niu (✉)
National Key Laboratory of Science and Technology On ATR, National University of Defense Technology, Changsha, Hunan, China
e-mail: niuzd@nudt.edu.cn

Y. Wang
e-mail: wangyupeng19@nudt.edu.cn

J. Huang · P. Li
Beijing Institute of Tracking and Telecommunications Technology, Beijing, China

G. Liu and F. Cen (eds.), *Advances in Precision Instruments and Optical Engineering*, Springer Proceedings in Physics 270, https://doi.org/10.1007/978-981-16-7258-3_41

monitoring and detection technology [1]. According to the location of the detection platform, photoelectric detection is divided into ground-based detection and space-based detection. Compared with ground-based detection, space-based detection has incomparable advantages such as being unaffected by the atmosphere, all-time, all-weather, and has become the preferred method of space target monitoring and detection technology. However, in the development process of the space-based space target detection system, it is difficult to obtain the required image data for relevant verification in a short time due to the high cost of the actual installation experiment. Therefore, it is of great significance to study the visible light imaging simulation method of space-based space targets. While providing a large amount of data, it can also verify the performance of the detection and tracking algorithm and reduce the cost in the development process.

Therefore, domestic and foreign scholars have carried out related research on the simulation of visible light imaging of space targets. Literature [2] discussed the visibility of space targets based on geosynchronous orbital observation platforms in combination with the geometrical positions of the sun, space-based observation platforms, and space targets. Literature [3] established a quantitative method for brightness simulation of space targets based on the scattering cross-sectional area and the bidirectional reflectance function. Literature [4] took the development and performance test requirements of real on-orbit optical cameras as the starting point, and designed and realized the imaging simulation software of space-based optical cameras. Literature [5] considered the target's background radiation environment, surface material properties, geometric structure size, orbital elements, etc., through finite element analysis and vector coordinate transformation, using the bidirectional reflection distribution function to establish a mathematical model of the target's optical scattering characteristics. Literature [6] designed a distributed simulation system for space-based surveillance of space targets based on HLA and the extensive application of HLA in the field of aerospace system simulation.

It can be seen from the existing literature that the analysis of the whole process of the simulation of visible light imaging of space targets is relatively weak, and the simulation of the Smear tail phenomenon of super-bright stars is not involved. This paper studies the imaging simulation of space-based debris. The simulation process is divided into three parts. By analyzing the formation mechanism of the super-bright star Smear phenomenon, the super-bright star Smear phenomenon is simulated, forming a complete high fidelity simulation image forming method. The simulation imaging method has a great effect on the development of a space-based space target detection system.

41.2 Simulation Structure and Process Design

The imaging simulation of space debris realized by this method is based on a CCD optical sensor, so the imaging process can be represented by the following figure (Fig. 41.1):

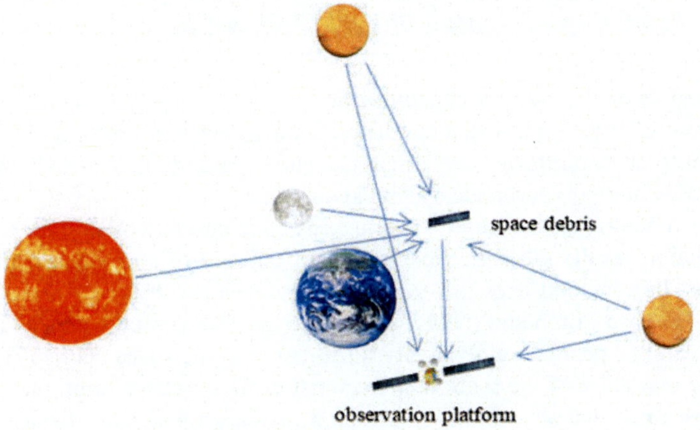

Fig. 41.1 Schematic diagram of space debris imaging process

Space debris is illuminated by space light sources such as sunlight, moonlight, starlight, and diffuse reflected light from the earth. The light reflected on its surface reaches the CCD sensor mounted on the observation platform, and the light from the stars on the starry sky background also reaches the CCD sensor. Because the space debris target and the star are far away from the observation platform, it can be considered as a point light source. It performs photoelectric conversion through the CCD photosensitive unit array and then forms the actual observation image on the CCD imaging plane.

By analyzing the characteristics of the above imaging process, this method divides the imaging process into three parts: space target motion modeling, space debris, and background star optical modeling, and CCD device imaging modeling. The above three parts are interdependent with each other. The overall structure of imaging simulation is as follows (Fig. 41.2):

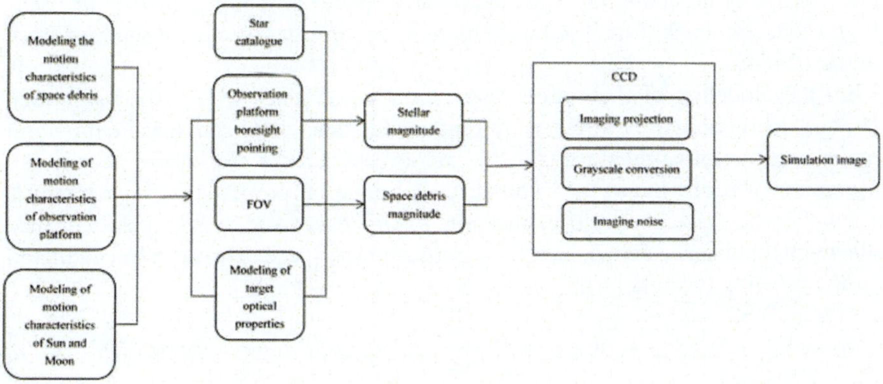

Fig. 41.2 The overall structure of imaging simulation

41.2.1 Modeling of Spatial Target Motion Characteristics

The modeling of the motion characteristics of space targets is mainly based on the relevant information to calculate the position and velocity information of space debris, observation platforms, stellar targets, and solar and moon near-Earth objects in the Earth's inertial coordinate system at any time.

For the modeling of the motion characteristics of space debris, in order to make the simulation results meet the imaging characteristics of space debris under real conditions, this method uses the SGP4 model developed by the North American Joint Air Defense Command (NORAD) to analyze TLE orbital elements issued by it to calculate the position and velocity information of the space debris [7].

For the modeling of the motion characteristics of the observation platform, this method assumes that its attitude is oriented to the ground, and the Kepler orbit elements are used to calculate its position and velocity information.

For the modeling of the motion characteristics of the stellar target, the star catalog selected in this method is the Tycho-2 catalog, which contains information such as the star's right ascension, declination, and right ascension, and the coordinate system used is the earth inertial coordinate system [8]. Taking into account the influence of the star proper motion, precession, and nutation, relevant corrections are made to the star catalog data, and the position information is calculated.

For the modeling of the motion characteristics of the sun and moon celestial bodies, this method uses the JPL DE430 ephemeris issued by the Jet Propulsion Laboratory (JPL) to calculate [9].

41.2.2 Modeling of Optical Properties of Space Debris and Background Stars

The optical property modeling of space debris and background stars is divided into optical property modeling of space debris and optical property modeling of background stars. For background stars, the magnitude information can be obtained from the star catalog.

For the modeling of the optical characteristics of space debris, the magnitude of space debris is mainly affected by ranging (distance between space debris and observation platform), phase angle (the angle between the sun, space debris, and observation platform), and the geometric and material properties of space debris targets. For the space observation platform, the light received by it is mainly diffuse reflected light of space debris, so the magnitude of the space debris can be calculated by the following formula [10]:

$$\mathrm{m} = 1.4 - 2.5\log\gamma - 5\log\mathrm{D} + 5\log\rho - 2.5\lg[\sin\sigma + (\pi - \sigma)\cos\sigma)] \quad (41.1)$$

In the above formula, γ is the diffuse reflection coefficient of the debris surface, D is the diameter of the space debris, ρ is the slant distance from the space debris to the observation platform, and σ is the angle between the sun, space debris, and the observation platform.

41.2.3 CCD Imaging Modeling

The imaging sensor simulated by this method is CCD, and its imaging mechanism is that light is irradiated to the CCD pixel, and photons are injected into the photosensitive unit array for photoelectric conversion to form an actual observation image on the imaging plane. When the CCD performs photoelectric conversion, while generating signal charges, noise signals are inevitably introduced. Therefore, CCD imaging modeling is mainly divided into imaging projection modeling, constellation and image gray conversion modeling, and imaging noise modeling.

For imaging projection modeling, it is mainly divided into space debris imaging projection modeling and background star imaging projection modeling. Space debris imaging projection modeling can be calculated according to the relative position of the space debris and the observation platform and the conversion matrix of the relevant coordinate system. Suppose the position vector of the observation platform in the earth inertial coordinate system at a certain moment is $\boldsymbol{R}_O$, the position vector of space debris in the earth inertial coordinate system is $\boldsymbol{R}_T$, and R_1^* is the rotation matrix from the satellite body coordinate system to the optical camera coordinate system, R_2^* is the rotation matrix from the satellite orbit coordinate system to the satellite body coordinate system, R_3^* is the rotation matrix from the earth inertial coordinate system to the satellite orbit coordinate system, then the position vector P_a of the space debris in the optical camera coordinate system is

$$\boldsymbol{P_a} = R_1^* \cdot R_2^* \cdot R_3^* \cdot (\boldsymbol{R_T} - \boldsymbol{R_O}) \tag{41.2}$$

For the background star imaging projection modeling, in the spatial target motion characteristic modeling part, the rectangular coordinate vector $\mathbf{I}$ in the earth inertial coordinate system is obtained, then the position vector $\boldsymbol{P}_b$ of the star in the optical camera coordinate system is

$$\boldsymbol{P_b} = R_1^* \cdot R_2^* \cdot R_3^* \cdot \mathbf{I} \tag{41.3}$$

Furthermore, according to the principle of small hole imaging, the position vector of the space debris and the background star in the optical camera coordinate system is projected into the imaging plane, and the imaging position in the image is calculated.

For the modeling of magnitude and image grayscale conversion, this method is based on the imaging principle of the CCD optical system to estimate the number of photons received by the CCD, thereby calculating the gray value of the corresponding magnitude [11]. Then the gray value A of the space debris target is

$$A = P/G \tag{41.4}$$

In the above formula, P is the number of photons received by the CDD, and G is the CDD gain parameter. Due to the non-ideal nature of CCD imaging, the imaging point on the image plane will have a diffusion effect. Therefore, this method uses a two-dimensional Gaussian distribution function as the point spread function to process the imaging point for the diffusion effect.

For imaging noise modeling, there are many noise sources for scientific CCD devices. Considering the actual simulation performance requirements, the following three types of noise are mainly considered: photon noise, dark current noise, and readout noise.

41.2.4 Smear Tailing Phenomenon

Through the analysis of the measured images, it is found that in addition to the above-mentioned modeling process, the Smear tailing phenomenon as an inherent characteristic of the CCD image sensor is often ignored. It refers to the phenomenon that striped white bright lines are formed on the image when the CCD is shooting a high-brightness point light source (Fig. 41.3).

For the stellar target, since its position on the imaging plane basically does not change during the exposure time and the readout time, the Smear tail phenomenon of the stellar target always presents the shape of a vertical bright band. Its impact on imaging quality is relatively large, so it is easy to cause the inundation of space

Fig. 41.3 The Smear tail phenomenon of the star in the measured image

debris, and increase the difficulty of detecting space debris, thus it is necessary to simulate it.

The CCD sensor is generally composed of three parts: photosensitive area, imaging area, and register [12]. Take a four-line CCD sensor as an example to analyze the mechanism of the Smear tail phenomenon of stars. Assume that the third row of the CCD sensor is illuminated by a star target. In Figure 41.4, the gray square in the imaging area represents the pixel being imaged, t_1–t_5 is an integration period, and t_6-t_{10} is an integration period. At time t_1, the third row of the CCD sensor generates signal charges. From t_2 to t_5, the signal charge is transferred to the register. As the photosensitive area is continuously exposed during the transfer process, additional charges are continuously generated in the illuminated area. When these charges are transferred to the register, the pixel positions in the first to third rows are occupied by additional charges. At time t_6, the third row of the CCD sensor generates signal charges. After the integration period is over, the signal and the additional charge generated by the charge readout time are transferred to the shift register in order to wait for output and write the image. In this way, the stellar Smear tail phenomenon occurs.

For the Smear tail phenomenon of stars, according to related research, the gray value h is only related to the integration time t_1, the read-out time t_2, the total number of imaging image rows m, and the sum of the gray values of all pixels in the current column [13].

$$h = t_2 \cdot \mathrm{sum}/m/(t_1 + t_2) \tag{41.5}$$

The process of visible light imaging of space-based space debris has been discussed above, and the simulation of visible light imaging of space-based space debris has been realized.

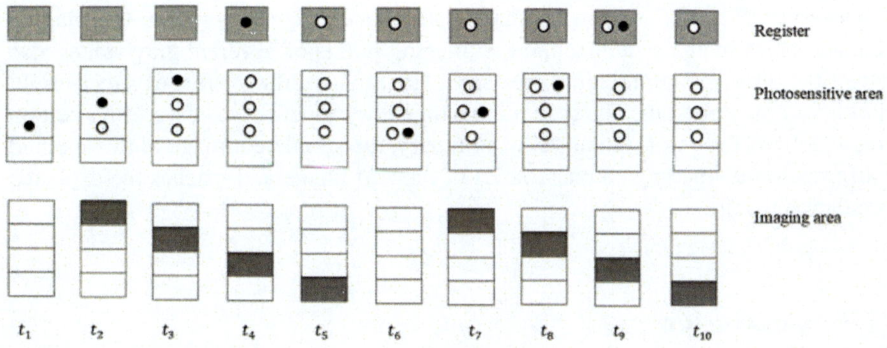

Fig. 41.4 Smear tailing phenomenon generation process

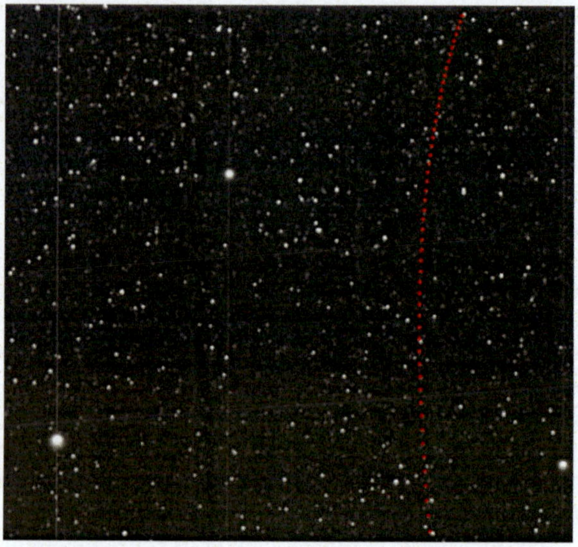

Fig. 41.5 Simulation imaging diagram

41.3 Analysis of Simulation Results

Assuming that the observation platform passes through the ascending node at zero time in the simulation, the space debris orbit is determined by TLE orbital elements. Using a certain model of sensor-related parameters, the number of pixels is 1024 × 1024, and the exposure time is 0.4 s. According to the survey results, the reflectivity is taken as 0.3. For other parameters, suppose the diameter of the space debris is 1 m. The simulation result of a certain frame is as shown in Fig. 41.5, and the space debris track calculated in the following 30 frames is marked in this frame.

Observing Fig. 41.5, we can find that the simulated image truly reflects the imaging characteristics of space debris, and the imaging points of different gray values also reflect the difference in imaging area size. The Smear tailing effect of super-bright stars is naturally obvious. The space debris trajectory appears as a curve with a certain arc, which fits the actual situation. In summary, the simulated image of this method conforms to the imaging characteristics of the real image and reaches the expected simulation result.

41.4 Conclusion

This method divides the imaging process into three parts: modeling of the motion characteristics of space targets, modeling of optical characteristics of space debris and background stars, and modeling of CCD device imaging. Furthermore, the imaging characteristics of the CCD device were analyzed, and the Smear tail phenomenon of super-bright stars was simulated, and the simulation image of space-based debris was obtained. Using this method, an observation simulation image can be obtained at any time, any observation platform, any direction of the visual axis, and any size of the field of view. It is of great significance for the research of space-based space target detection methods and the development of space-based space target detection systems.

Acknowledgements This work was supported by the Youth Science Foundation of China (No.61605243).

References

1. Z. Li, Y. Wang, W. Zheng, Z. Song, Space-based optical observations on space debris via multipoint of view. Int. J. Aerosp. Eng. (2020)
2. K. Zhang, N. Ruan, Fu., Danying et al., Geometry position analysis of space-based detection of space targets in geostationary orbit with visible light. Infrared Laser Eng. **36**(z2), 606–609 (2007)
3. Y. Gao, Z. Lin, J. Li, W. An, Xu., Hui, Research on the starry sky simulation method based on CCD point spread and smear characteristics. Electron. Inf. Warfare Technol. **02**, 58–62 (2008)
4. M. Cong, W. Bao, H. Yu, et a1., Modeling and simulation for optical sensor imaging in space rendezvous and docking, in *IEEE 3rd International Congress on Image and Signal Processing* (2010), pp. 2052–2056.9
5. C. Sun, Y. Yuan, Lu., Qunbo, Modeling and verification of optical scattering characteristics of space-based space targets. Acta Opt. **39**(11), 354–360 (2019)
6. Y. Zhang, Q. Gan, H. Yuan, Fu., Jiangliang, C. Sun, Design of distributed simulation system for space-based surveillance of space targets. J. Syst. Simul. **32**(04), 620–626 (2020)
7. L. Han, L. Chen, B. Zhou, Accuracy analysis of SGP4/SDP4 model for space debris orbit prediction. Chin. Space Sci. Technol. **04**, 67–73 (2004)
8. V.V. Vityazev, A.S. Tsvetkov, S.D. Petrov, D.A. Trofimov, V.I. Kiyaev, Properties of the Tycho-2 catalogue from Gaia data release. Astron. Lett. **43**(11) (2017)
9. M. Li, Q. Wang, T. Chen, Y. Gong, Analysis of the influence of different versions of JPL ephemeris on the transformation of celestial coordinates. J. Hefei Univ. Technol. (Natural Science Edition), 1–9
10. L. Wu, Orbital and detection of artificial satellites and spatial fragments. China Sci. Technol. Press **35** (2011)
11. Q. Zhu, Z. Niu, Y. Duan, Z. Chen, W. Wu, Simulation of space-based observed star map in space radiation environment, in *2016 Progress in Electromagnetics Research Symposium* (2016)
12. K. Powell, D. Chana, D. Fish, C. Thompson, Restoration and frequency analysis of smeared CCD images. Appl. Opt. **38**, 1343–1347 (1999)
13. Y. Duan, Research on low-orbit space target detection technology with large field of view bundled telescope (National University of Defense Technology, 2017)

Chapter 42
Study on Image Processing of Bridge Cable Surface Defect Detection System

Jing Hu, Hongyu He, Guomin Liao, and Guichuan Hu

Abstract The cable-stayed bridge has been widely used because of its excellent bearing capacity, beautiful structure and relatively low construction cost. As one of the most important mechanical components in long-span cable-stayed bridges, the safety of bridge cables is the focus of attention. In order to solve the problems of high risk and low efficiency of traditional manual detection, robots and machine vision equipments are used for automatic detection. This paper aimed at the special situation of strong solar rays and serious light pollution in cable detection environment designed a machine vision inspection system to capture panoramic images of the cable and then preprocessed the image by grayscale transformation and threshold segmentation. Ultimately, the median filtering method is used to reduce the noise and a basis for subsequent defect identification and prediction is provided.

42.1 Introduction

The cable-stayed bridge has been widely used because of its excellent bearing capacity, beautiful structure and relatively low construction cost. As one of the most important mechanical components in long-span cable-stayed bridges, the safety of

J. Hu (✉)
School of Intelligent Manufacturing, Chongqing College of Architecture and Technology, Chongqing, China
e-mail: 278252640@qq.com

H. He · G. Hu
School of Mechanical and Power Engineering, Chongqing University of Science and Technology, Chongqing, China
e-mail: 934437763@qq.com

G. Hu
e-mail: huguichuan@cqust.edu.cn

G. Liao
School of Digital Engineering, Chongqing College of Architecture and Technology, Chongqing, China
e-mail: 1643493318@qq.com

G. Liu and F. Cen (eds.), *Advances in Precision Instruments and Optical Engineering*, Springer Proceedings in Physics 270, https://doi.org/10.1007/978-981-16-7258-3_42

bridge cables is the focus of attention. Due to the complex environment, the load of the cable fluctuates, and there is alternating tensile stress in the live load environment and the protective layer of its surface will harden, crack, peel and other phenomena happen [1]. In addition, the cables have been subjected to wind load and rain erosion in the external environment for a long time, and there are natural disasters such as wind vibration and rain vibration, leading to the extrusion friction of the wires inside the cables, which is prone to corrosion and broken wires.

In this paper, a machine vision system for bridge cable detection was designed, and the image pretreatment of cable surface was carried out by using the methods of gray conversion, threshold segmentation, median filtering and so on, to provide a basis for subsequent defect identification and prediction.

42.2 The Hardware System of Detection

42.2.1 *Image Sensor Configuration*

Image sensor is a device that converts optical images into electronic signals, which is the core component of the image detection system. The most commonly used sensors are CCD and CMOS, and the main advantage of CMOS sensor over CCD sensor is the low power loss. Under the condition that the overall battery capacity of the cable surface detection robot is certain, the small power consumption of the sensor is particularly important [2]. Therefore, the surface scanning camera with CMOS sensor was selected for this project, which can collect the whole image of the cable surface at one time.

For cables with 150 mm od (outer diameter), the camera had a vertical field of view (FOV_{VET}) of up to 140 mm, and the defect identification accuracy (R_{obj}) was less than 0.5 mm [3]. Therefore, the minimum number of elements required to obtain the image was as follows:

$$NE_{\mathrm{VET}} = FOV_{\mathrm{VET}}/R_{\mathrm{obj}} = 140/0.5 = 280 \tag{42.1}$$

$$NE_{\mathrm{HOR}} = FOV_{\mathrm{HOR}}/R_{\mathrm{obj}} = 150/0.5 = 300 \tag{42.2}$$

The distance between the lens and the cable in the section direction was 150 mm. Through analysis and comparison, the system selected a CMOS camera with the model of FIREFLY 6S. The performance parameters of this camera were shown in Table 42.1.

The camera could be powered autonomously or connected to the robot control system using DuPont wire, and the image was transmitted to the robot control system in real time, which was transmitted back to the ground in real time through a wireless adapter for real-time monitoring of the cable surface.

Table 42.1 Camera performance parameters

Index	Parameters
Pixel resolution	720 × 576
AV output system	PAL/NTSC
Compressed format	H264
Aperture	F/2.8, 7 Glass lens
Size	59 × 41 × 21 mm
Working voltage	5 V
Working current	800 mA
Weight	70 g

42.2.2 Data Transmission System

The ground PC transmitted a 2.4 GHz radio through a wireless adapter to transmit signals to each other with the cable robot wireless module. Control the motion of the robot on the cable, and receive the status parameters and video monitoring signals of the cable detecting robot in the process of motion.

42.2.3 Defect Location System

The motor encoder was located at the tail of the motor and integrated with the motor shaft. By automatically judging the rotation direction of the motor and counting it, the motor parameters were transmitted to the slave computer, which fed back to the PC terminal on the ground through its own integrated wireless module. By this method, the motion distance of the cable detection robot on the cable was determined so that the cable detection system could obtain the specific position of the cable surface defect on the cable. The PC terminal on the ground could monitor the mooring time, distance, direction and speed of the robot in real time. By recording the distance and moving time of the robot, the specific position of the defect on the cable could be accurately determined, which provided important guidance information for the subsequent cable maintenance.

42.3 Defect Image Preprocessing

42.3.1 Image Mosaic Based on SIFT Algorithm

The cable detection robot in this paper has three cameras. It needed to splice each frame photo of the three cameras together to unfold the overall picture of the cable surface so as to facilitate the identification and judgment of defects and start the visual detection of cable defects. The SIFT algorithm is used to realize multi-image mosaic fusion [4]. Combined with the actual working conditions, the bilinear interpolation method is used to interpolate the acquired images, and the weighted average method is used to fuse the images.

Bilinear interpolation has a moderate computational amount, and the results are better than nearest-neighbor interpolation, and it does not occupy too much computational resources. The disadvantage is that after bilinear interpolation, the image edge may be blurred due to the smoothing effect of the image edge.

The bilinear interpolation method is to superposition the gray values of the four nearest coordinate points around the target point by weighting the pixel values of the four nearest coordinate points. As shown in Fig. 42.1, points A, B, C and D are the four points closest to the target point (x, y), and each coordinate values are (i, j), $(i, j+1)$, $(i+1, j)$, $(i+1, j+1)$, then the pixel values of E coordinates (i, y) and F coordinates $(i+1, y)$ are as follows:

$$f(E) = (y - j)[f(B) - f(A)] + f(A) \tag{42.3}$$

$$f(F) = (y - j)[f(D) - f(C)] + f(C) \tag{42.4}$$

So the pixel value $f(x, y)$ at the point (x, y) is as follows

$$f(x, y) = (x - i)[f(F) - f(E)] + f(E) \tag{42.5}$$

The weighted average method was used for image fusion, the overlapping regions of the mosaic image $f(x, y)$ and the reference image $g(x, y)$ were given weights ω_f and ω_g, respectively. The expression of the fused image $I(x, y)$ was as follows:

$$I(x, y) = \left\{ \begin{array}{ll} f(x, y) & (x, y) \in f \\ w_f f(x, y) + w_g g(x, y) & (x, y) \in (f \cap g) \\ g(x, y) & (x, y) \in g \end{array} \right\} \tag{42.6}$$

The limiting conditions of weights ω_f and ω_g are as follows:

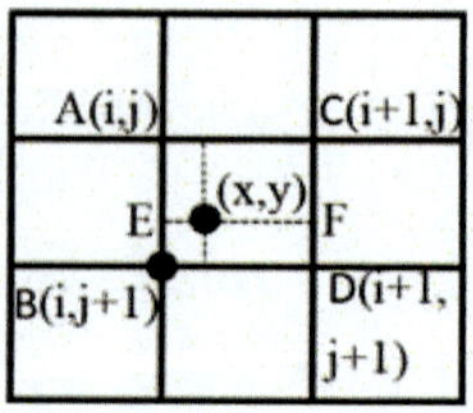

Fig. 42.1 Bilinear interpolation

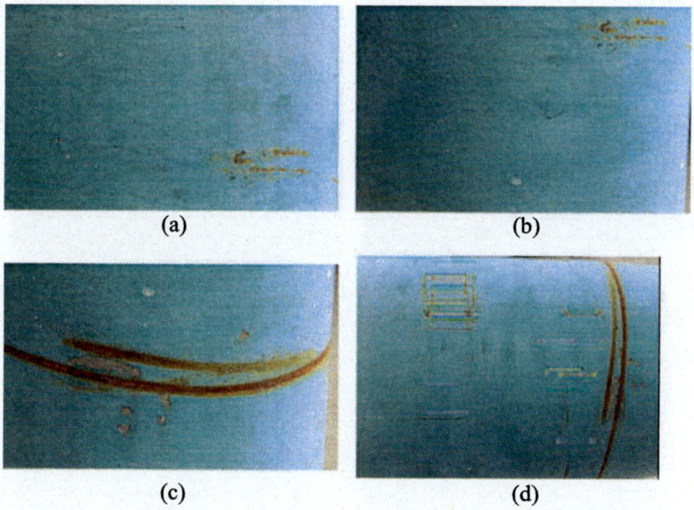

Fig. 42.2 Image mosaic: **a**–**c** three camera images; **d** stitching process

$$\begin{cases} w_f + w_g = 1 \\ 0 < w_f < 1, 0 < w_g < 1 \end{cases} \tag{42.7}$$

In order to make the mosaic image more realistic, the weights of ω_f and ω_g needed to be reasonably valued to achieve seamless stitching and relatively smooth.

As shown in Fig. 42.2, the images taken by the three cameras have overlapping regions, and the image stitching was carried out through the SIFT algorithm's bilinear interpolation and weighted average so that the defect system could obtain the 360° panoramic display image of the cable, which was convenient for subsequent defect processing and analysis.

42.3.2 Grayscale Transformation

Because the camera used in the defect detection system is a color camera, the cable surface is taken as a color image. RGB color image is a common color space, in which a large amount of color information is stored, the amount of data is large, occupies more memory resources and reduces the system computing capacity. Therefore, in order to solve the problem of a large amount of color image data and the occupation of resources, the gray processing of the acquired color image is carried out [5].

From R, G, B three-channel into a single-channel value, accelerate the speed of image processing and analysis. Grayscale usually refers to an 8-bit grayscale map with 256 grayscale levels and the range of pixel values is [0, 255]. When the image is converted from RGB color space to GRAY color space, the processing mode is as

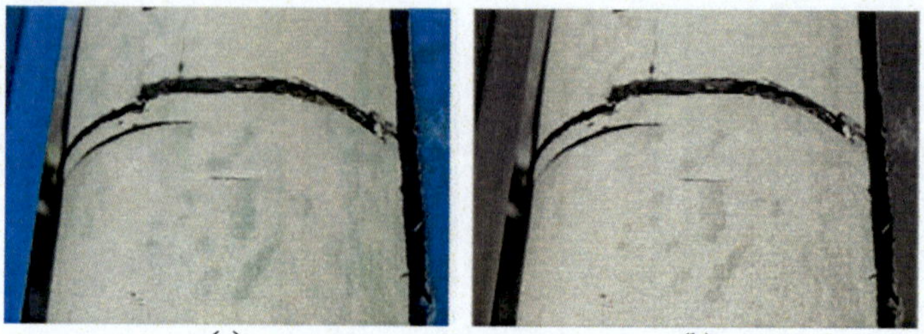

(a) (b)

Fig. 42.3 Grayscale conversion of the defect image **a** Original image; **b** Grayscale image

follows:

$$Gray = 0.299R + 0.587G + 0.114B \tag{42.8}$$

This is the form of the standard conversion and the one used in Python-OpencV. At the same time, the simplified form of the commonly used arithmetic average method is used for grayscale image conversion:

$$Gray = (R + G + B)/3 \tag{42.9}$$

In this way, when the image is converted from RGB color space image to grayscale image, the final value of all channels is the same. Python-OpencV standard conversion was used to transform the cable defect image into grayscale.

As shown in Fig. 42.3, after grayscale conversion, the image became a single-channel image, and the image does not have color information, but the image information was expressed by grayscale.

42.3.3 Threshold Segmentation

After the image was processed into grayscale by the cable surface defect system, threshold processing was needed to display the cable surface image features. Then the gray value of the image is unevenly distributed in light and dark, but because the gray value of the cable surface has little deviation, it is very important to choose the appropriate threshold value to process the cable surface defect image. Different threshold values will result in different defect features.

The image of cable surface defect is processed by anti-binarization threshold [6] (cv2.THRESH_BINARY_INV) and generates a binary image with only two values. Anti-binarization rules for processing pixel values are different from binarization, which are as follows:

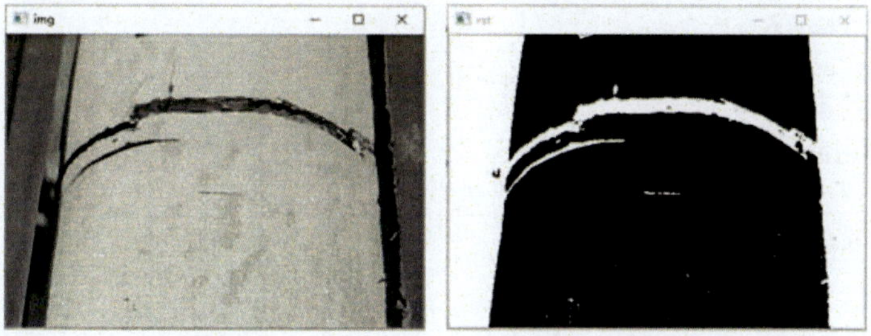

(a) (b)

Fig. 42.4 Threshold processing **a** Grayscale image; **b** Threshold segmentation

$$dst(x, y) = \begin{cases} 0, & src(x, y) > thresh \\ 255, & others \end{cases} \tag{42.10}$$

Thresh is the selected threshold value.

As shown in Fig. 42.4, due to the light color of hdPE on the cable surface, the grayscale value was lower than the ambient disturbance. Through threshold experiments, it was found that when thresh value is 150, anti-binarization threshold processing could extract cable features from the image. Because the gray value of cable image was low, the pixel value of cable image was uniformly converted to 255 after processing, which could clearly observe the features of the cable surface.

42.4 Noise Reduction Processing

42.4.1 Image Noise

When the image was collected, the gray value of the image will fluctuate within a certain range due to the influence of external illumination and motion conditions, which is called noise. According to the noise generated in the image according to the probability density function (PDF), it can be divided into uniform distribution noise, Gaussian noise, pepper and salt noise, and mixed noise [7]. Thc probability density functions and their annotations for different types of noise are shown in Table 42.2.

Due to the defects and impurities on the cable surface, white noise often appears after threshold treatment, which is regarded as salt and pepper noise. Therefore, it is necessary to preprocess the image to reduce and eliminate the interference caused by noise. After modifying the core parameters of several filtering methods, it is found that median filtering has a better denoising ability for cable surface defects. Therefore, the system uses median filtering to de-noise the cable surface image.

Table 42.2 Three types of noise

Noise type	Probability density function	Annotation
Uniform distribution noise	$P(Z) = \begin{cases} \frac{1}{b-a} & a \le z \le b \\ 0 & others \end{cases}$	$\mu = \frac{a+b}{2}, \sigma^2 = \frac{(b-a)^2}{12}$
Gaussian noise	$P(Z) = \frac{1}{\sqrt{2\pi\sigma}} e^{-\sigma^2(z-\mu)/2}$	μ is expectancy value σ^2 is standard deviation σ is variance
Pepper and salt noise	$P(z) = \begin{cases} 0 & f(z) < a \\ 255 & a < f(z) < b \\ f(z) & others \end{cases}$	Noise in the image in the form of a bright or dark point

42.4.2 Median Filtering

Median filtering is by sorting the pixel values in the window from large to small and then taking the pixel values in the middle of the sorting position as the pixel values of the central target point. As shown in Fig. 42.5, the pixel value of the window of 3 * 3 was (97, 95, 94, 93, 78, 90, 66, 91, 101). These values were sorted into (66, 78, 90, 91, 93, 94, 95, 97, 101), and the median value 93 was taken as the pixel value of the window target point.

The window size of median filtering was usually 3 * 3, 5 * 5, 7 * 7 and 9 * 9 for filtering. Generally, the window size was used to filter the image from small to large. The choice of window size is very important to the result of image processing. If the window size is too large, the image denoising ability will be stronger, and the edge information of the image will be eliminated as noise. In contrast, if the window size is too small, the image denoising ability is weaker and the denoising is not complete. For the selection of median filtering window, the filtering effect of 3 * 3, 5 * 5, 7 * 7 and 9 * 9 window sizes on binarization image of defect was compared, as shown in Fig. 42.6.

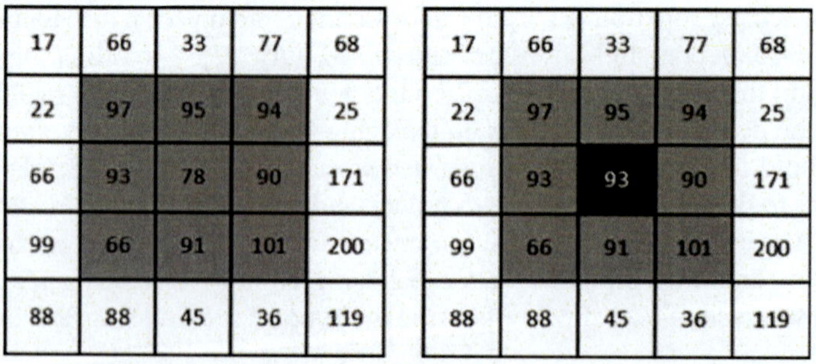

Fig. 42.5 Schematic diagram of median filtering principle

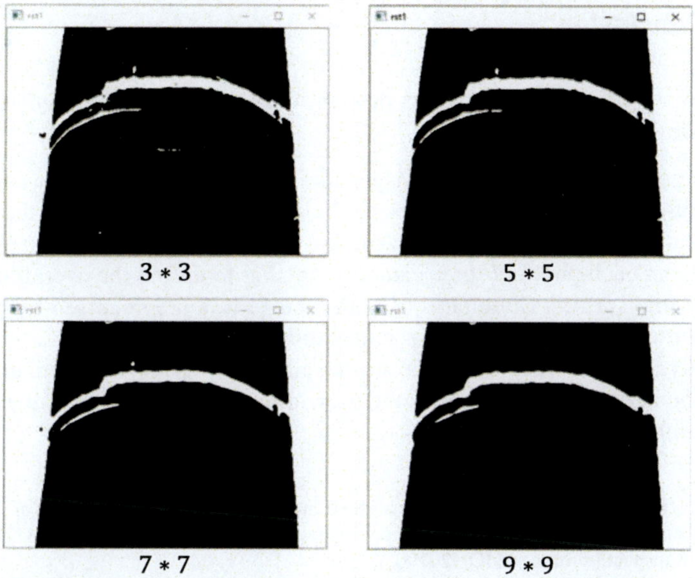

Fig. 42.6 Comparison of median filtering effect

As shown in Fig. 42.7, through comparison, it was found that the window size of 5 * 5 has the best denoising effect on the cable surface defect image. Therefore, in this paper, the window size of 5 * 5 was selected for median filtering of cable surface images.

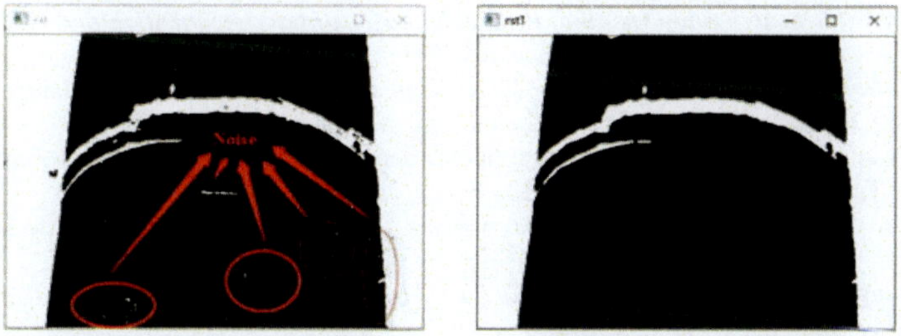

(a) Threshold processing (b) Median filtering 5 * 5

Fig. 42.7 Comparison of median filtering effect

42.5 Conclusion

This paper studies the surface defect detection of a bridge cable, mainly including the following three aspects:

(1) Designing a machine vision inspection system, three cameras were used to capture 360° panoramic images of the cable. In the meantime, the wireless module was used to transmit images and locate the specific location of defects.
(2) Splicing each photo of three cameras together to unfold the overall picture of the cable surface. Simultaneously, the image was preprocessed by grayscale transformation and threshold segmentation.
(3) Analyzing the types of image noise, ultimately using the median filtering method to reduce the noise and providing a basis for subsequent defect identification and prediction.

Acknowledgements This paper is from the science and technology research project of Chongqing Education Commission, Research on cable health monitoring system of cable-stayed bridge based on intelligent inspection robot (KJQN201905203).

References

1. C. BuWen, *Inspection of Cable Surface Defects Design by Image Processing* (Chongqing University, China, 2010)
2. Li., Xinke, *Research on Key Techniques of Surface Defect Image Detection for Cables in Cable-Stayed Bridges* (Chongqing University, China, 2014)
3. W. Shengzhi, *Research and Design of the Cable Surface Image Acquisition and 'Transmission System Based on DSP* (Chongqing University, China, 2010)
4. S. Peng, *Study of Panorama Display Technology Based on SIFT Feature Image Stitching* (University of Electronic Science and Technology of China, China, 2018)
5. Y. Chaoyang, Research on intelligent apparent detection of cable detection robot. Technol. Highway Transp. **36**148(02):90–96 (2020)
6. L. Wenshu, *Digital Image Processing Algorithms and Applications* (Peking University Press, 2012)
7. Y. Shuhong, *Study on the Adaptive and Fast Algrithm of Gray Scale Image Thresholding* (Chongqing University, China, 2014)

Chapter 43
Study on the Structural Performance Degradation of Rigid Airfield Pavements Using HWD

Kunyang Zhao and Panfei Zheng

Abstract At present, most of the airfield pavements are cement concrete pavements. The structural performance of airfield pavement is an important factor affecting operation safety. Under the coupling action of aircraft wheel load and environment, the rigid airfield pavement is prone to structural damage such as void and insufficient capacity. In this paper, based on the heavyweight deflectometer (HWD) deflection of the capital airport runway pavement, we analyze the deterioration laws and structural damage mechanism of three performance indexes, including the pavement structural capacity, the voids beneath rigid pavement slab and the load transfer efficiency at joint. The results indicate that these performance indexes are interrelated and they could reflect the structural performance deterioration of the airfield pavements. The impact load of aircraft has a compaction effect on the pavement foundation, which would slow down the deterioration of pavement structure performance. Grouting reinforcement can effectively improve the performance of pavement structures.

43.1 Introduction

In recent years, China's civil aviation industry has developed rapidly. By 2018, there were 235 transport airports in China [1]. According to the investigation, over 90% of the pavement structure of civil airport is cement concrete pavement [2]. With the increasing growth of air traffic volume and the proportion of large wide-body aircraft, the safety of airfield pavement is seriously challenged. The performance of airfield pavement is mainly divided into structural performance and functional performance [3, 4], which is degraded and attenuated due to the comprehensive effects of aircraft wheel load and environmental factors in the using process. The functional performance degradation is mainly reflected in the decline of the functional indexes such as roughness, friction coefficient and pavement condition index (PCI). The deterioration

K. Zhao · P. Zheng (✉)
China Airport Construction Group Co., Ltd., Beijing 100101, China
e-mail: zhengpf@cacc.com.cn

Beijing Super-Creative Technology Co., Ltd., Beijing 100621, China

G. Liu and F. Cen (eds.), *Advances in Precision Instruments and Optical Engineering*, Springer Proceedings in Physics 270, https://doi.org/10.1007/978-981-16-7258-3_43

of structural performance is manifested in the reduction of the pavement structural capacity and the load transfer efficiency of the joints, which may cause the voids beneath rigid pavement slabs and other diseases. If the structural performance of pavement fails, the pavement function would be unavailable accordingly. Therefore, the structural performance of pavement is directly related to the safety of airport operation.

In order to ensure the safety of airport operation, the regulations on the administration of safe operation of transport airports require that the airport management authority should organize a comprehensive evaluation on the condition of runway, taxiway and apron at least once every five years. At present, the pavement structural performance tests of a civil airport are mainly carried out by the heavyweight deflectometer (HWD) [5, 6]. Through the impact of the drop weight on the pavement, the process of aircraft loading on the pavement is simulated, and the bending response of the pavement under the load is recorded by the sensors so as to evaluate the strength and structure condition of the pavement and the foundation. Then the load transfer and void condition of the pavement can be calculated and analyzed so as to effectively analyze the deterioration of the pavement structural performance. This paper focuses on exploring the pavement structural performance degradation laws of the runway of an airport by using the HWD. We first study and analyze the HWD test data of the runway over the years according to the current research and practical engineering experience. We then discuss the mechanism of performance deterioration and disease. Finally, we summarize the deterioration laws of the pavement structural performance.

43.2 Background

This airport is one of the largest airports in the world with an annual passenger throughput of over 100 million, and its runway undertakes increasingly heavy aircraft take-off and landing tasks. With the continuous growth of the aviation business, the plate breakage diseases of the runway have appeared, and the diseases are getting worse and worse. Since the runway was put into use in 2008, a total of four pavement tests have been carried out up to now, namely in 2013, 2014, 2016 and 2019. The results of the first three tests show that the runway has serious void disease, and the disease is increasing year by year. In order to ensure the safe operation of the runway, the foundation void diseases were treated by grouting from April to June, and then the runway was comprehensively detected in September.

The structure of the runway is as listed below:

(1) At both ends of the runway (1000 m): 44 cm cement concrete, 20 cm cement stabilized gravel and 25 cm cement stabilized gravel.
(2) In the middle of the runway (1000–2800 m): 40 cm cement concrete, 20 cm cement stabilized gravel and 25 cm cement stabilized gravel.

Table 43.1 The detection schemes of the runway

Detection time (year)	The name of the detection line	Detection position	Distance between measuring points (m)	Detection direction	The length of the detection line (m)
2013	2013R-F-W_1	The first panel that locates the west side of runway centerline	40	South to north	3800
2014	2014R-F-W_1		20		2000
2016	2016R-F-W_1		20		3800
2019	2019R-F-W_1		20		3800

43.3 Detection Scheme and Evaluation Method

43.3.1 Detection Scheme

The four deflection detection schemes of the airport runway are shown in Table 43.1.

43.3.2 Detection Equipment

It is stipulated in Specifications for Pavement Evaluation and Management of Civil Airports (MH/T 5024-2019) that the load capacity test of pavement structure should be carried out by the falling weight deflectometer. Therefore, the heavyweight deflectometer (HWD) produced in Denmark (the model of Dynatest 8081) is used as deflection detection equipment. As shown in Fig. 43.1, the maximum load weight of the HWD can reach 24t, which is equivalent to the single wheel load weight of B747-400 aircraft under the maximum take-off full weight state. In addition, the load weight of the HWD can be adjusted according to the pavement structure and the grade of runway. The bearing plate with a diameter of 450 mm is selected for the HWD, and the test load is 238.5 kN. The deflection detection position is the edge and the middle of the pavement slab, as shown in Fig. 43.2.

Fig. 43.1 The heavyweight deflectometer (HWD)

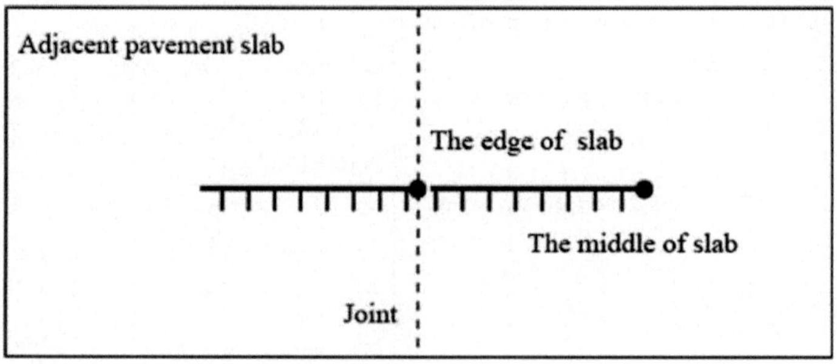

Fig. 43.2 The detection position of the deflection

43.3.3 The Evaluation of Pavement Structural Performance

Based on the pavement deflection detection results, this paper selects the pavement structural capacity, the load transfer efficiency of the joints and the voids beneath rigid pavement slabs as the evaluation index of pavement structure performance [7].

(1) The pavement structural capacity

The capacity of pavement structure is an index to evaluate the comprehensive support effect of pavement and foundation on aircraft operation, which is mainly characterized by impulse stiffness modulus (*ISM*). It is calculated by dividing the load weight of the HWD by the deflection value measured by the sensor under the bearing plate. The larger the *ISM* value in a certain area indicates the stronger the comprehensive support of pavement and foundation. On the contrary, the smaller the ISM value, the weaker is the comprehensive support of pavement and foundation.

(2) The load transfer capacity at joint

The load transfer efficiency at the joint (LTE_δ) is an index to evaluate the load and stress transfer capacity between pavement slabs, which is based on the deflection ratio transfer coefficient.

In the detection, the deflection is measured by two sensors with the same distance from the center of the HWD bearing plate but crossing different plates on both sides of the joint. If the two deflection results are roughly the same, it indicates that the load transfer capacity at the joint is relatively intact. On the contrary, if the deflection results are quite different, it indicates that the load transfer capacity at the joint has been reduced or failed.

(3) The voids beneath the rigid pavement slab

The void beneath the rigid pavement slab is an index to evaluate whether the void exists between the pavement slab and the foundation, which is characterized by the void coefficient (T). Comparing the deflection results in the middle of the pavement slab with the edge of the pavement slab, if the deflection results are roughly the same, it shows that there is almost no void beneath the pavement slab. In contrast, there is void beneath the rigid pavement slab. According to engineering experience, generally, there are the following four situations:

(1) $T < 1.5$ indicates that there is no void beneath the pavement slab.
(2) $1.5 < T < 2$ indicates that there is a slight void beneath the pavement slab.
(3) $2 < T < 3$ indicates that there is a moderate void beneath the pavement slab.
(4) $T > 3$ indicates that there is severe void beneath the pavement slab.

43.4 Detection Results and Discussion

43.4.1 The Evaluation of the Pavement Structural Capacity

The pavement structural capacity distribution of 0–2000 m section at the south end of the runway (the first panel that locates the west side of the runway centerline) is shown in Fig. 43.3, and the average values of *ISM* are shown in Table 43.2. The 800–2000 m section away from the south end of the runway has been foundation grouted.

As illustrated in Fig. 43.3 and Table 43.2, the distributions of the pavement structural capacity in different section tend to change similarly with the ISM. Due to repeated action of aircraft wheel load and structural thickness, the capacity of pavement structure fluctuates significantly.

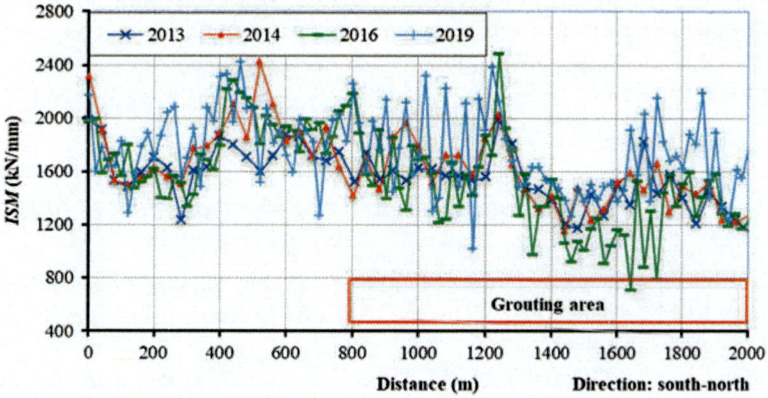

Fig. 43.3 The distribution curve of the pavement structural capacity

Table 43.2 The average values of *ISM*

Section	2013	2014	2016	2019
0–1000 m	1671.56	1793.03	1770.36	1851.27
1000–2000 m	1477.05	1494.22	1364.50	1659.28

Table 43.3 The load transfer efficiency at the joint (LTE_δ)

Detection time (year)	Good (>80%) (%)	Average (80–56%) (%)	Fair (56–31%) (%)	Poor (<31%) (%)
2013	5.6	8.8	36.1	49.5
2014	15.7	17.6	33.3	33.4
2016	41.58	9.91	17.82	30.69
2019	74.0	8.0	12.0	6.0

In the section of 0–1000 m from the south end of the runway, the capacity of pavement structure in 2013 was slightly lower than that of 2014 and 2016. The *ISM* values of the 400–600 m section fluctuate significantly and the floating value is up to 900 kN/mm. This section is the main takeoff and landing area of aircraft. Due to the impact of the aircraft's wheel load, the pavement foundation becomes denser and this would lead to the increase of the *ISM* value. However, because of the uneven impact, the ISM value fluctuates significantly. In the section of 1000–2000 m from the south end of the runway, the average *ISM* in 2013 and 2014 are almost the same.

In 2016, the average *ISM* of this section decreased by about 8.7% compared with 2014. And the average *ISM* value of the 1400–1700 m section decreased significantly, which means the pavement performance of this section has deteriorated significantly.

After grouting, the capacity of the pavement structure has been significantly improved. The average value of *ISM* has increased by about 22.0%. That indicates that the grouting has effectively increased the compactness of the foundation and improved the capacity of the pavement structure. However, from 2013 to 2019, the *ISM* value of pavement fluctuates obviously, which indicates that the pavement structure and basic performance gradually deteriorate with the increase of pavement service time.

43.4.2 The Evaluation of the Load Transfer Capacity at Joint

The load transfer efficiency at joint (LTE_δ) of 0–2000 m section at the south end of the runway (the first panel that locates the west side of runway centerline) is shown in Fig. 43.4 and Table 43.3.

As shown in Fig. 43.4 and Table 43.3, the LTE_δ in 2013 is less than 56%, which indicates that the load transfer capacity at the joint id poor. From 2013 to 2016, the load transfer capacity of pavement has gradually improved and regained. However, the values of LTE_δ fluctuate greatly, which reflects the serious failure of the load transfer capacity in the partial area of the pavement. In the section of 400–600 m from the south end of the runway (main aircraft grounding strap), the pavement load transfer efficiency shows a trend of increasing first and then decreasing. It indicates that the load transfer capacity of the pavement changes irregularly under the impact of aircraft wheel load, but there is still a certain degree of degradation overall. After

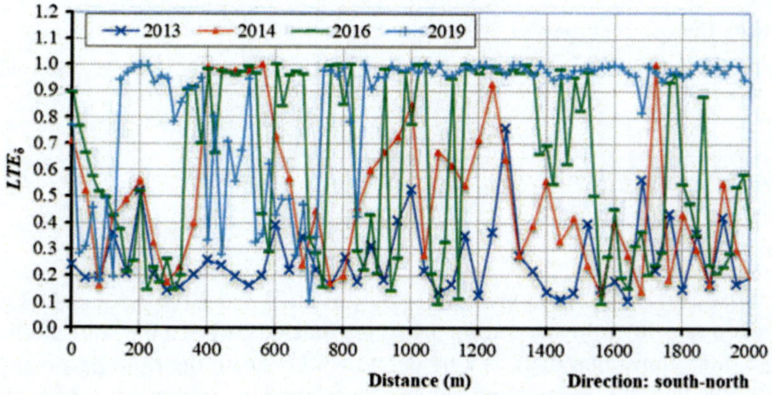

Fig. 43.4 The distribution curve of the load transfer efficiency at joint

grouting in 2019, the load transfer capacity of the road surface in the grouting area has recovered to more than 80%. Therefore, the grouting reinforcement can effectively improve the performance of pavement structures.

43.4.3 The Evaluation of the Voids Beneath Rigid Pavement Slab

The void coefficient (T) of the 0–2000 m section at the south end of the runway (the first panel that locates the west side of the runway centerline) is shown in Fig. 43.5 and Table 43.4.

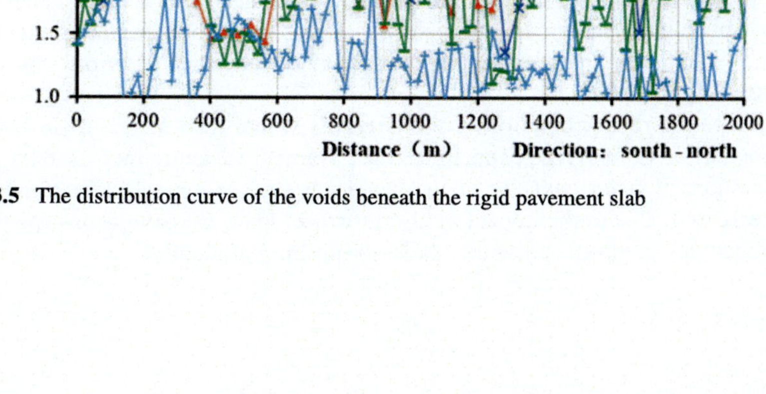

Fig. 43.5 The distribution curve of the voids beneath the rigid pavement slab

Table 43.4 The void coefficient (0–2000 m)

Detection time (year)	Serious void (%)	Moderate void (%)	Slight void (%)	No void (%)
2013	0.0	63.2	33.7	3.1
2014	1	61.8	27.5	9.7
2016	1.98	34.65	46.54	16.83
2019	0	0	22.0	78.0

As illustrated in Fig. 43.5 and Table 43.4, in 2013 and 2014, 63.2 and 61.8% of the detection result of the measuring points are located between the void coefficients of 2.5–3.0. It shows that there is a moderate void beneath the rigid pavement slab. By 2016, the number of detection points of moderate void dropped to 34.65% with a decrease of nearly 30%. The number of detection points of slight and non-void increased by 19% and 7%, respectively. The pavement surface of the 400–600 m section away from the southern end of the runway changed from being moderate void in 2013 to non-void. Under the impact of aircraft wheel load, the pavement structure of this section was effectively compacted. The void basically disappeared, and no obvious change was found in the void situation of the road surface in the remaining areas. Especially, in the section of 400–600 m from the south end of the runway, the voids beneath the rigid pavement slab changed from moderate void to non-void. Due to the impact of aircraft wheel load, the pavement structure is effectively compacted and the voids disappeared gradually.

In Fig. 43.5, we can find that the void coefficient (T) of 1000–2000 m section at the south end of the runway fluctuates significantly in 2014. The absolute values of the void coefficient of some moderate void detection points have increased, but most of them are below 3.0. It shows that the void condition beneath the pavement slab continues to deteriorate, but the deterioration rate is relatively slow. In 2019, the void coefficients of the grouting area were substantially less than 1.5. It indicates that there is no void beneath the pavement slab and grouting reinforcement effectively improves the performance of the pavement structure.

43.5 Conclusions

(1) The three indexes of the pavement structural capacity, the voids beneath rigid pavement slab and the load transfer efficiency at the joint are interrelated and they could effectively reflect the structural performance deterioration of the airfield pavements.

(2) Due to the repeated action of the aircraft's wheel load on the main landing area of the runway, the capacity of the pavement structure increased slightly. The load transfer capacity of the pavement surface increased first and then decreased. Due to the impact of aircraft wheel load, the pavement structure is effectively compacted and the voids disappeared gradually.

(3) With the extension of the use time of the pavement, the fluctuation range of the pavement structure performance index gradually increases. It indicates that the various performance of the pavement deteriorates gradually.
(4) Grouting can effectively improve the capacity of the pavement structure and slow down the performance degradation of the pavement.

References

1. G. Yang, *Quantitative Analysis and Residual Life Prediction of Void beneath Airport Rigid Pavement* (Chang'an University, 2019)
2. Li., Zhufeng, H. Zhiyong, C. Bo, Study on the application of airport pavement bearing capacity evaluation based on nondestructive testing technology. Sci. Technol. Ind. **19**(10), 158–162 (2019)
3. L. Yawei, *The Research of Pavement Structure Analysis System Based on the Deflection* (Civil Aviation University of China, 2015)
4. T. Shenjia, Y. Cong, W. Wang, Q. Linjie., Research on fatigue life prediction of cement pavement with void. J. Guangxi Univ. (NATURAL SCIENCE EDITION) **43**(01), 304–313 (2018)
5. Z. Yuhui, Z. Xianmin, Test methods of airport pavement and subjacent foundation void. J. Traffic Transp. Eng. **16**(06), 1–11 (2016)
6. C. Guoyong, D. Chun, L. Guoguang, Nonzero intercepts of load-deflection curve in load transfer test with airport rigid pavement joint. J. Civ. Aviat. Univ. China **30**(02), 13–16 (2012)
7. Y. Song, Z. Qisen, The dynamic finite analysis of mechanism between foundation void and load transfer capacity in CCP based on FWD. Highway Eng. (06), 70–75 (2007)

Chapter 44
Research on Integrated Navigation and Positioning Technology of Inertial Navigation System and Odometer Based on Factor Graph

Wang Zhisheng, Wang Xingshu, Zhao Yingwei, and Chen Xiaolong

Abstract Research on autonomous navigation technology of vehicle plays a vital role in ensuring vehicle safety and realizing intelligent driving. Integrated navigation technology takes inertial navigation system as the main body and makes full use of other sensor information to improve the accuracy and reliability of navigation system. However, the different error characteristics and operating frequency of different sensors increase the design complexity of navigation system. In order to solve the above problems, this paper proposes the integrated navigation algorithm of inertial navigation system and odometer based on factor graph and constructs the information fusion model of factor graph. The simulation results show that this method can obtain more accurate position information, effectively guarantee the navigation accuracy, and ensure the feasibility of the algorithm for unmanned vehicles. Moreover, it shows the advantages of factor graph algorithm in integrated navigation and information fusion.

44.1 Introduction

Navigation technology is an engineering application technology that provides real-time attitude, speed, and position information for vehicles. With the development of society and technology, navigation technology plays an indispensable role in many

W. Zhisheng · W. Xingshu · Z. Yingwei (✉) · C. Xiaolong
School of Frontier and Interdisciplinary Sciences, National University of Defense Technology, Changsha, China
e-mail: yingweizhao@live.cn

W. Zhisheng
e-mail: wzs0456@163.com

W. Xingshu
e-mail: gfkdwxs@aliyun.com

C. Xiaolong
e-mail: nudtcxl@163.com

G. Liu and F. Cen (eds.), *Advances in Precision Instruments and Optical Engineering*, Springer Proceedings in Physics 270, https://doi.org/10.1007/978-981-16-7258-3_44

application fields. Inertial Navigation System (INS) is a typical autonomous navigation system. Its outstanding advantages are strong autonomy, good concealment, and high short-term accuracy, and it is widely used in military, civil, and aerospace industries [1]. However, due to the influence of initial alignment error and inertial sensor error, the system navigation error gradually diverges with time accumulation. Therefore, in order to improve the navigation accuracy, it is necessary to introduce external reference information to estimate the system error of the INS in real time, so as to meet the requirements of long-endurance and high-precision navigation and positioning. Inertial navigation and odometer (OD) integrated navigation system design because of its high accuracy and adaptability has become an important research direction of land positioning and orientation system in many countries, especially in the military field, land navigation system has formed a series of products in many countries.

At present, most INS/OD integrated navigation algorithms take Kalman filter as the core, and many improvements have been made in order to improve the navigation accuracy and stability. The Kalman filter can only be filtered after all the measured information has arrived. The sensor information of different rates is fused by data synchronization processing method, and the navigation solution is calculated in real time. However, in order to keep the data synchronized, it is often necessary to discard some of the measurement values, which leads to the waste of information. When the measurement information is synchronized, the error is introduced into the filter, which increases the calculation amount [2, 3]. Therefore, Kalman filter has insufficient ability in processing asynchronous information.

Factor graph is a kind of probability graph model. In 2001, Kschischang et al. extracted the concept of factor graph and the sum-product algorithm for the first time, providing a mathematical theoretical basis for future research [4]. The factor graph encodes the probability relationship between the navigation state quantity and the navigation system measurement value and defines the state quantity as variable node and the measurement value of each sensor as factor node. Factor graph model has strong flexibility and can realize the plug and play function of sensor from the algorithm level, which has been widely concerned in the navigation field. Paper [5] applies the factor graph to the attitude estimation of the satellite and obtains better results than the Kalman filter. Paper [6] combined with factor graph to construct a new co-positioning assisted single star positioning method, which improved the positioning accuracy. Paper [7] proposes an IMU/GPS information lag fusion algorithm based on factor graph. By adding the lagging GPS observation information into the factor graph structure, the bad effect of the lagging GPS observation information on the navigation system can be avoided under the condition of ensuring the real-time accuracy.

In this paper, the integrated navigation algorithm based on factor graph is applied to the vehicle INS/OD system, and the factor graph model of the integrated navigation system is constructed. The Gauss–Newton iterative method is used to solve the maximum posteriori estimate of the state, and the reasoning of the factor graph is realized. The real-time performance and positioning accuracy of the inertial integrated navigation system are improved. Therefore, the navigation information can be compatible and the asynchronous problem in the presence of multiple sensors can be avoided.

44.2 Basic Theory of Factor Graph Navigation

Factor graph is a bipartite graph model used to represent factorization. The characteristic of a bipartite graph is that the vertices of the graph can be divided into two non-intersecting nodes so that each edge is connected to two nodes in the non-intersecting set, that is, the vertices in the graph can be divided into two different types. Factor graph can be used to model all kinds of systems. The main purpose of modeling based on factor graph is to factor the complex system. According to the actual problem solved, using the corresponding factor graph model for modeling can improve the efficiency of problem solving. When the nodes of the factor graph represent random variables and probability distributions, the factor graph is also a probability graph model, and most of its applications in navigation are probability graphs.

The factor graph model $G = (F, X, E)$ is composed of variable nodes ($x_i \in X$), factor nodes ($f_i \in F$), and lines $\left(e_{ij} \in E\right)$. When variable nodes are related to factor nodes, a line will be generated between them.

The factor graph G factorizes the function into

$$f(X) = \prod_i f_i(X_i). \tag{44.1}$$

In this equation, X_i is a subset of $\{x_1, x_2, \ldots, x_n\}$ and f_i is a local function.

Suppose $f(X)$ is a function of five variables and can be expressed as follows:

$$\begin{gathered} f(x_1, x_2, x_3, x_4, x_5) = \\ f_1(x_1, x_2) f_2(x_2, x_5) f_3(x_2, x_3) f_4(x_4) f_5(x_1, x_4, x_5). \end{gathered} \tag{44.2}$$

In this equation, $X_1 = \{x_1, x_2\}$, $X_2 = \{x_2, x_5\}$, $X_3 = \{x_2, x_3\}$, $X_4 = \{x_4\}$, and $X_5 = \{x_1, x_4, x_5\}$. Then, the factor graph structure that can be represented by (44.2) is shown in Fig. 44.1.

In Fig. 44.1, the circle represents variable nodes, and the square represents factor nodes. Each factor node generates relationships with different variable nodes, thus forming different local functions.

The factor graph model can transform the solution of maximum posteriori estimation of state variables in navigation system into reasoning of the factor graph. And the reasoning of factor graph is to find the best estimation $\hat{X}$ of X. Therefore, in the factor graph model, the maximum posterior probability estimation algorithm is used to estimate the navigation state.

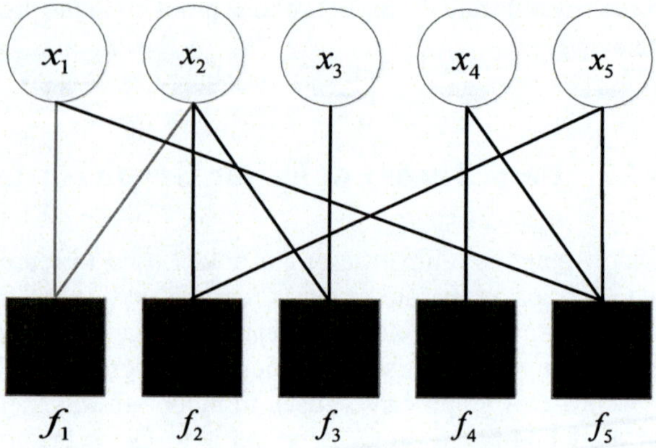

Fig. 44.1 Factor graph example

44.3 IMU and OD Integrated Navigation Factor Graph Modeling

In the application of integrated navigation based on factor graph, inertial navigation is generally regarded as the reference source of integrated navigation system because of its high data generation rate and anti-interference. However, as the errors of INERTIAL navigation will accumulate and diverge over time, it is necessary to introduce other sensor systems for data fusion. Factor graphs therefore provide a very flexible framework for fusing these complementary sources of information.

44.3.1 *Integrated Navigation Framework Based on Factor Graph*

In order to realize autonomous navigation of vehicles, this paper constructs the integrated navigation framework of factor graph as shown in Fig. 44.2. The system adopts inertial navigation system as the main reference source and odometer as the auxiliary.

In Fig. 44.2, the circle represents the state variable node, the black square represents the factor node, x_k represents the navigation state of the system, including

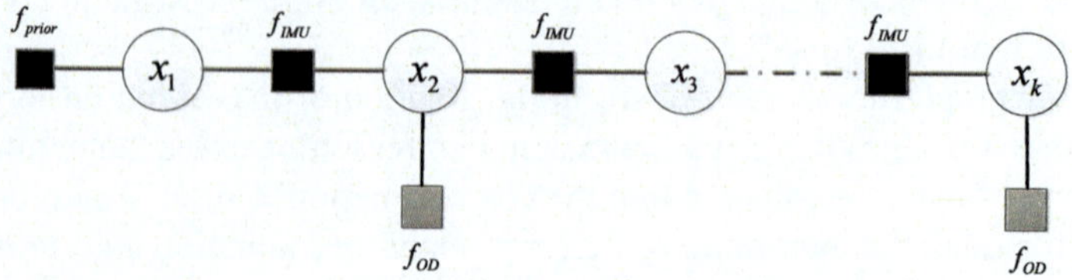

Fig. 44.2 INS/OD fusion framework based on factor graph

attitude, position, velocity, etc., f_{IMU} represents the measurement information from the IMU, and f_{OD} represents the measurement information from the OD.

The integrated navigation system based on factor graph architecture can effectively realize the plug and play of navigation sensors. If the measurement information of other navigation sensors needs to be added to the navigation system, it is only necessary to define the corresponding new factor nodes to expand the factor graph and update the status of the variable edge according to the sensor observation equation and the corresponding cost function. Multi-sensor information fusion method based on factor graph is an effective method to solve the problem of asynchronism of multi-sensor observation data. After receiving the output data of the sensor, the node of the factor graph is extended, and the state of the system is updated quickly and effectively according to the state equation and measurement equation of the system, so as to realize the comprehensive data processing of multiple sensors in the navigation system.

44.3.2 IMU Factor Graph Model

In order to realize autonomous navigation of vehicles, this paper constructs the integrated navigation framework of factor graph as shown in Fig. 44.2. The system adopts inertial navigation system as the main reference source and odometer as the auxiliary.

The discrete form of the IMU equation of state can be abstractly expressed in the following form:

$$x_{k+1} = h\left(x_k, f_k^b, \omega_k^b\right). \tag{44.3}$$

In this equation,x_k represents the navigational state quantity, f_k^b represents the measured value of the accelerometer, and ω_k^b represents the measured value of the gyroscope.

In this case, the factor node can be expressed as follows:

$$f_{INS} = d\left(x_{k+1} - h\left(x_k, f_k^b, \omega_k^b\right)\right). \tag{44.4}$$

In this equation, $d(\cdot)$ is the cost function.

The initial value x_{k+1} is constantly solved by the measurement function $h\left(x_k, f_k^b, \omega_k^b\right)$, and then new factor nodes are added to the factor graph.

44.3.3 OD Factor Graph Model

The odometer provides a pulse signal by installing a speed encoder on the speed drive structure. The speed of the wheel is recorded according to the pulse signal, and then the cumulative distance of the carrier is calculated according to the wheel diameter. Finally, the real-time speed and mileage of the carrier are calculated through the running time to complete the positioning.

The data of OD is expressed as odometer factor, and the measurement equation is expressed as follows:

$$z_k^{OD} = h(x_k) + \upsilon^{OD}. \tag{44.5}$$

In this equation, υ_{OD} represents the OD measurement noise and $h(\cdot)$ represents the corresponding measurement function. Therefore, the OD factor node can be expressed as follows:

$$f_{OD} = d\left(z_k^{OD} - h(x_k)\right). \tag{44.6}$$

44.3.4 *Integrated Navigation Algorithm Based on Factor Graph*

The navigation state is defined as X, the measurement information is Z, and the current time is t; then the maximum posterior probability is estimated as follows:

$$\hat{X} = \underset{X}{\arg\max}\, P(X|Z). \tag{44.7}$$

According to the definition of factors, each factor is represented as an independent term, so

$$P(X|Z) \propto \prod_i f_i(X). \tag{44.8}$$

Each factor has a corresponding error function. Assuming that the error follows a Gaussian distribution, it is defined as follows:

$$f_i(x_i) = \frac{1}{\sqrt{2\pi}} \exp\left(-\frac{1}{2}\|z_i - h_i(x_i)\|^2\right). \tag{44.9}$$

To optimize the maximum posterior probability, the following equation needs to be solved:

$$D(X) = \min \sum_{i=1}^{N} \left((z_i - h(X))^T \cdot R \cdot (z_i - h(X))\right). \tag{44.10}$$

The above equation can be solved by Gauss–Newton iterative method. The specific process is as follows:

$$D(X + \Delta X) \simeq e(X) + J\Delta X. \tag{44.11}$$

In this equation, J is the Jacobi matrix near the initial value of the error function. Further, the following formula can be obtained:

$$\begin{aligned} D(X + \Delta X) &= D(X + \Delta X)^T \cdot R \cdot D(X + \Delta X) \\ &= \left(D\left(\hat{X}\right) + J\Delta X\right)^T R^{-1}\left(D\left(\hat{X}\right) + J\Delta X\right), \\ &= c + 2b^T \Delta X + \Delta X^T M \Delta X \end{aligned} \tag{44.12}$$

$$c = D(X)^T R^{-1} D(X), \tag{44.13}$$

$$b^T = D(X)^T R^{-1} J, \tag{44.14}$$

$$M = J^T R^{-1} J. \tag{44.15}$$

To minimize (44.12), the first derivative has to be 0, so

$$M\Delta X^* = -b. \tag{44.16}$$

Then the solution of the state variable is

$$X^* = \hat{X} + \Delta X^*. \tag{44.17}$$

Incremental ΔX is obtained by iterating $\Delta X^* = 0$. The above is the Gauss–Newton iterative method [8].

44.4 Results and Discussion

In Matlab software, the simulation experiment is carried out. The simulation parameters are set as follows (Table 44.1).

The position errors in the three directions are shown in Fig. 44.3. The errors can be controlled in a small range, and the mean square errors of the three directions are 1.72 m, 1.55 m, and 2.21 m respectively. Simulation results demonstrate the reliability of the factor graph algorithm. Therefore, the integrated navigation algorithm based on factor graph can provide better navigation accuracy for the carrier.

Table 44.1 Simulation parameter setting

Parameters		Value
Basic parameters	IMU frequency	100 Hz
	OD frequency	1 Hz
Initial setup	Attitude	0
	Velocity	0
	Position	28°N, 112°E, 20 m
Gyroscope	Biases	0.03°/h
	Noise	0.004°/h
Accelerometer	Biases	20mGal
	Noise	20mGal
Odometer	Noise	0.1 m

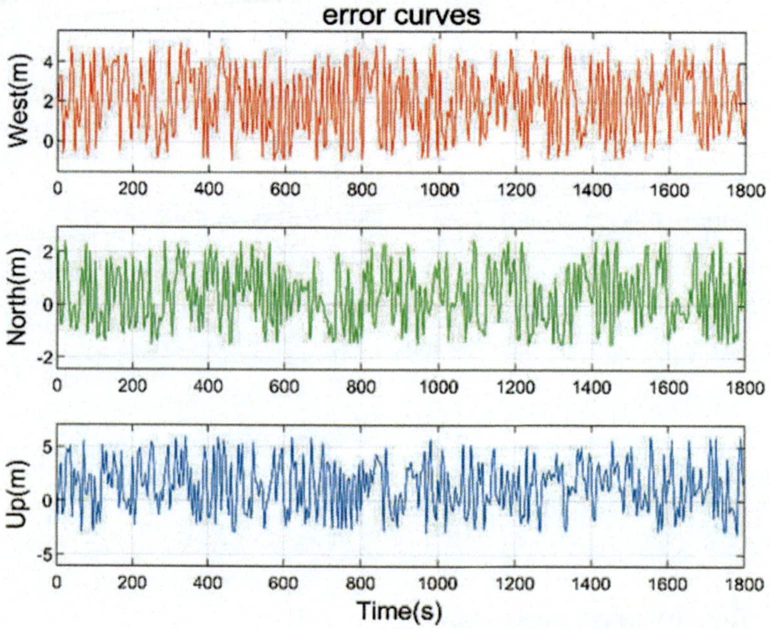

Fig. 44.3 Error curves in three directions

44.5 Conclusions

In order to realize the autonomous and precise navigation of carrier, an INS/OD integrated navigation algorithm based on factor graph is proposed in this paper, and simulation experiments are carried out. The results show that the position of the vehicle can be accurately estimated by the factor graph algorithm. Thus, the feasibility of this algorithm for unmanned vehicle is ensured. In addition, the factor graph algorithm has great advantages in integrated navigation and information fusion, and it is of great significance to carry out relevant research to improve the accuracy and robustness of the system, which is conducive to expanding more kinds of sensors into the system.

References

1. M.S. Grewal, L.R. Weill, A.P. Andrews, *Global Positioning Systems, Inertial Navigation and Integration* (John Wiley & Sons, New York, 2001), pp. 252–261
2. T.L. Xu, P.Y. Cui, H.T. Cui, Design and implementation of multi-sensor integrated navigation system of land vehicle. Syst. Eng. Electron. **30**(4), 686–691 (2008)
3. M. Aftatah, A. Lahrech, A. Abounada et al., GPS/INS/Odometer data fusion for land vehicle localization in GPS denied environment. Mod. Appl. Sci. **11**(1), 62–75 (2017)
4. F.R. Kschischang, B.J. Frey, H.A. Loeliger, Factor graphs and the sum-product algorithm. IEEE Trans. Inf. Theory **47**(2), 498–519 (2001). https://doi.org/10.1109/18.910572.
5. E.Q. Chen, S.Z. Xiao, L. Gao Xin, Estimation of satellite attitude angular velocity based on factor graph. Comput. Simul. **32**(06), 63–66 (2015)
6. C.K. Tang, L.L. Zhang, B.W. Lian, Cooperation factor map of co-location aided single satellite navigation algorithm. Syst. Eng. Electron. **39**(05), 1085–1090 (2017)
7. J.Q. Gao, X.Q. Tang, H. Zhang et al., Processing method of INS/GPS information delay based on factor graph algorithm. J. Comput. Appl. **38**(11), 3342–3347 (2018)
8. C.H. Jiang, S. Chen, Y.M. Bo, et al., Feasibility of GNSS/INS integrated navigation system using factor graph algorithm, in *2018 Inertial Technology Development Trends Seminar Proceedings* (Chinese Society of Inertial Technology, 2018), pp. 43–46

Part V
Optical Communication Electronics and Data Model Recognition

Chapter 45
Research on Influencing Factors and Evaluation Accuracy of Diffuse Reflector Deception Airspace

Jianlu Huang, Tao Shen, and Jing He

Abstract Aiming at the application of diffuse reflector in laser angle deception interference, the influence of interference laser energy, meteorological conditions, weapon threat angle, inclination, and relative azimuth angle of diffuse reflector on the airspace of diffuse reflector interference are studied. It is concluded that when the reflection angle is 0°, the interference laser energy increases to 10 times the original, and the interference distance is 2.62 times the original; when the visibility of the atmosphere is less than 7 km, the meteorological conditions have a greater impact on the interference airspace; the smaller the difference between the angle of diffuse reflector and the threat angle, the larger the interference airspace; when the threat angle is unknown, the diffuse reflector's attitude can be set between 45° and 60° to ensure a better interference effect on weapons from any threat angle; when the relative azimuth angle of the normal of the diffuse reflector is within 30°, a better interference effect can be guaranteed. Considering the weapon threat angle, the distance of the diffuse reflector, and the attitude parameters, an airspace evaluation model for diffuse reflector is established, and the influence of the measurement error of parameter on the interference distance calculation error is analyzed. The research results have certain guiding significance for the layout of diffuse reflectors and the evaluation of interference airspace.

45.1 Introduction

Laser angle deception interference is an effective means to deal with laser-guided weapons. It is usually to place false targets around the protected target, and the false targets reflect the jamming laser signals to deflect the enemy laser guided weapons [1]. At present, diffuse reflectors are often used as false targets to implement angle deception interference. The placement and attitude of the reflectors have a certain

J. Huang (✉) · T. Shen
Rocket Force University of Engineering, Xi'an 710025, Shanxi, China

J. He
Xi'an Technological University, Xi'an 710021, Shanxi, China

G. Liu and F. Cen (eds.), *Advances in Precision Instruments and Optical Engineering*, Springer Proceedings in Physics 270, https://doi.org/10.1007/978-981-16-7258-3_45

impact on the interference airspace, so the research on the deployment of diffuse reflectors is of great significance.

Literature [2] gives the constraint conditions for the placement of false targets on diffuse reflectors. Literature [3] proposed a method to determine the protection angle of false targets through suppression coefficients. Literature [4] proposed a false target placement method for diffuse reflectors without a ceiling area, but the theoretical basis for the method of determining the protection angle by radiation intensity is not rigorous. Literature [5] uses particle swarm algorithm to optimize the layout of false targets with forbidden areas, but the influence of the orientation, inclination, and threat angle of the diffuse reflector panel on the interference airspace was not considered.

In this paper, the interference distance of the diffuse reflector is used as the evaluation index, and the influence of interference laser energy, meteorological conditions, and the posture of the diffuse reflector on the interference distance are studied. A method for measuring the interference airspace of diffuse reflectors is proposed, which can realize the real-time evaluation of the interference airspace of false targets of diffuse reflectors, and finally, the accuracy of the evaluation of the interference airspace is analyzed.

45.2 Laser Energy Transfer Model

Let the output energy of the jamming laser be E_t, the transmittance of the laser emitting system be τ_t, the distance between the jamming laser and the diffuse reflector be R_i, and the laser atmospheric transmittance over the distance from the jamming laser to the diffuse reflector be T_i. Then the light energy of the spot on the diffuse reflector is as follows:

$$E_i = E_t \tau_t T_i \tag{45.1}$$

Assuming that the spot energy is all diffusely reflected into space through the diffuse reflector, then the laser energy density E_r received at the seeker is

$$E_r = \frac{E_i \rho \tau_r T_r \cos\theta_r}{\pi R_r^2} \tag{45.2}$$

where τ_r is the transmittance of the seeker receiving system, ρ is the hemispheric reflectance of the diffuse reflection false target, T_r is the transmittance of the laser atmosphere on the distance between the diffuse reflection plate and the seeker, θ_r is the reflection angle between the seeker and the normal of the diffuse reflection plate, and R_r is the distance between the seeker and the diffuse reflector.

The atmospheric transmittance [6, 6] in (45.1) and (45.2) can be expressed as follows:

$$\begin{cases} T_i = e^{-u_a R_i} \\ T_r = e^{-u_a R_r} \end{cases} \tag{45.3}$$

u_a is the atmospheric attenuation coefficient. Its value can be referenced in the literature [7]. Combined with (45.1), (45.2), and (45.3), the received energy density expression at the seeker can be obtained as follows:

$$E_r = \frac{E_t \rho \tau_t \tau_r e^{-u_a(R_i+R_r)} \cos\theta_r}{\pi R_r^2} \tag{45.4}$$

In (45.4), when the energy density received by the seeker is just the lowest threshold value of the energy density received by the seeker, the distance between the seeker and the diffuse reflector is the interference distance of the diffuse reflector. The expression [8] is as follows:

$$R = \sqrt{\frac{E_t \rho \tau_t \tau_r e^{-u_a(R_i+R_r)} \cos\theta_r}{\pi E_r}} \tag{45.5}$$

45.3 Analysis of Influencing Factors in Interference Airspace

45.3.1 Influence of Jamming Laser Energy on Interference Airspace

Assume that $\tau_t = 0.9$, $\tau_r = 0.8$, $\rho = 0.25$, E_r=1 × 10^{-15} J/mm^2, $u_a = 0.1081$, R_i = 0.1 km, θ_r=0°, 45°, 60°, and 0.1 J ≤ E_t ≤ 1 J. The relationship between R and E_t is obtained through MATLAB simulation, as shown in Fig. 45.1.

In order to more intuitively analyze the relationship between R and E_t, Table 45.1 is obtained according to the MATLAB simulation data.

According to the data in Fig. 45.1 and Table 45.1, when the energy of the jamming laser is constant, the larger the reflection angle is, the smaller the jamming distance is. At the same reflection angle, with the increase of the energy of the jamming laser, the jamming distance gradually increases, and the increasing trend gradually slows down with the increase of the energy of the jamming laser. While θ_r=0° and E_t = 0.1 J, the interference distance R = 2.123 km. If E_t = 1 J, then R = 5.571 km. The energy of the jamming laser is increased by 10 times, and the jamming distance is 2.62 times。By calculating, while θ_r=45° and θ_r= 60°, when the interference energy is increased by 10 times, the interference distance R is 2.68 times and 2.73 times of the original, respectively. Therefore, the interference distance can be increased by increasing the laser energy appropriately.

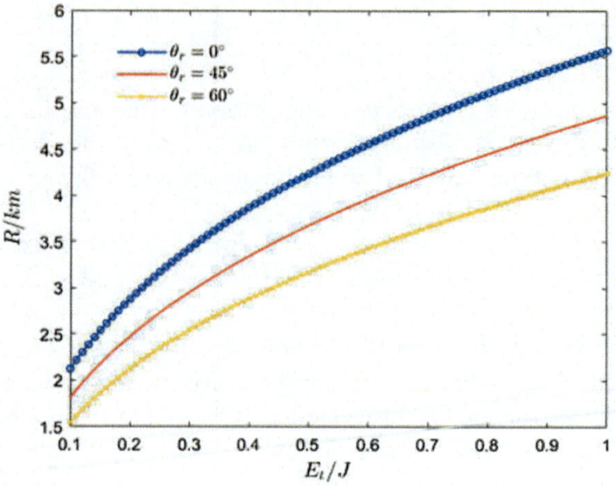

Fig. 45.1 The relationship between R and E_t

Table 45.1 The relationship between R and E_t

R/km	E_t/J									
	0.1	0.2	0.3	0.4	0.5	0.6	0.7	0.8	0.9	1
$\theta_r = 0°$	2.123	2.881	3.426	3.864	4.234	4.558	4.847	5.109	5.349	5.571
$\theta_r = 45°$	1.815	2.477	2.956	3.342	3.671	3.959	4.217	4.452	4.667	4.867
$\theta_r = 60°$	1.548	2.123	2.542	2.881	3.171	3.426	3.655	3.864	4.056	4.234

45.3.2 *Influence of Meteorological Conditions on Interference Airspace*

In order to analyze the influence of meteorological conditions on the false target interference space of diffuse reflector, assume that $\tau_t = 0.9$, $\tau_r = 0.8$, $\rho = 0.25$, E_r=1 × 10^{-15} J/mm^2, $u_a = 0.1081$, $R_i = 0.1$ km, θ_r=0°, 45°, 60°, $E_t = 0.1$ J, and $0 \le V_m \le 50$km. The simulation results obtained are shown in Fig. 45.2.

In order to better analyze the influence of atmospheric visibility on the interference airspace, Table 45.2 is obtained according to the Matlab simulation data.

According to the data in Fig. 45.2 and Table 45.2, at the same reflection angle, the interference distance gradually increases with the increase of atmospheric visibility. When the atmospheric visibility is within the range of 0–7 km, the interference distance increases rapidly with the increase of atmospheric visibility. While $V_m >$ 7km, the interference distance varies little with the increase of atmospheric visibility. While $\theta_r = 0°$, $V_m = 50$km, the interference distance $R = 2.3032$ km. In this case, it is only 1.21 times of the interference distance when $V_m = 7$km, while the interference distance when $V_m = 7$km is 1.38 times that when $V_m = 3$km. It can be concluded that the meteorological conditions have a certain influence on the interference distance, and the worse the meteorological conditions are, the more obvious the influence

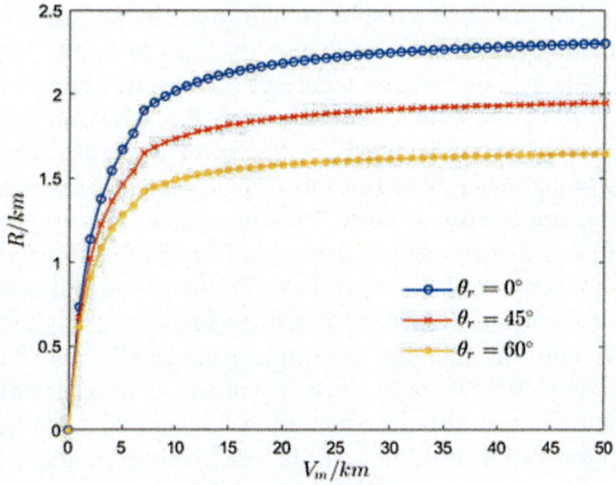

Fig. 45.2 The impact of atmospheric visibility on the interference airspace

Table 45.2 The impact of atmospheric visibility on the interference airspace

R/km	E_t/J									
	1	3	5	7	9	15	25	35	45	50
$\theta_r =$ 0°	0.7365	1.3764	1.671	1.9031	1.9876	2.1253	2.222	2.2674	2.2938	2.3032
$\theta_r =$ 45°	0.6752	1.225	1.4651	1.6478	1.7128	1.8169	1.8885	1.9218	1.941	1.9478
$\theta_r =$ 60°	0.6166	1.0854	1.2793	1.4219	1.4715	1.5498	1.6027	1.6269	1.6409	1.6459

on the interference distance is. When atmospheric visibility is greater than 7 km, meteorological conditions have little effect on disturbance distance.

45.3.3 Influence of Inclination Angle of the Diffuse Reflector on Interference Airspace

The inclination angle of the diffuse reflector referred is the angle between the diffuse reflector plane and the ground normal φ_p. The curves of interference distance with threat angle are analyzed when the inclination of diffuse reflector is 15°, 30°, 45°, 60°, and 75° respectively. For the convenience of analysis, normalization is carried out here, and α represents the ratio of interference distance at different threat angles to the maximum interference distance at the same inclination angle,

$$\alpha = \frac{R}{R_{\max}} \tag{45.6}$$

The simulation results are shown in Fig. 45.3. In order to more intuitively analyze the influence of the inclination of the diffuse reflector on the interference distance, Table 45.3 is obtained according to the simulation data.

It can be seen from the data in Fig. 45.3 and Table 45.3 that the interference coefficient α is symmetric with respect to $\theta = \varphi_p$, and the maximum value is 1 when $\theta = \varphi_p$.That is, when the threat angle is equal to the inclination of the diffuse reflector, the interference distance is the maximum. The value of α depends on $|\theta - \varphi_p|$. As $|\theta - \varphi_p|$ increases, α decreases. That is, the interference distance decreases with the increase of the angle between the seeker's attacking direction and the normal of the diffuse reflector. When the inclination angle of the diffuse reflector is 45° and 60°, its values at 15° threat angle and 90° threat angle are 0.9306, 0.8409 and 0.8409, 0.9306 respectively. Compared with other diffuse reflector, the variation of interference distance is small at 45° and 60°. Therefore, when using the diffuse reflector for protection, if the weapon threat angle is known, the inclination of the diffuse reflector should be adjusted to be equal to the weapon threat angle. At this time, the normal of the diffuse reflector is facing the attacking direction of the laser-guided weapon, so as to maximize the interference airspace of the diffuse reflector. If the threat angle information of the weapon is unknown, the inclination of the diffuse

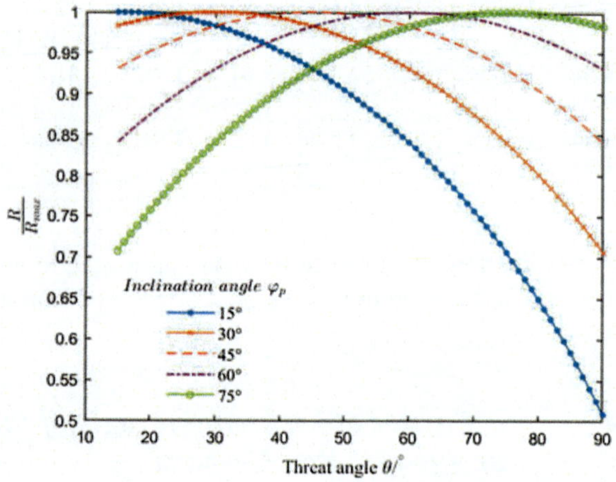

Fig. 45.3 The relationship between the interference coefficient and the threat angle

Table 45.3 The relationship between the interference coefficient and the threat angle

α	$\theta/°$					
	15	30	45	60	75	90
$\varphi_p = 15°$	1	0.9828	0.9306	0.8409	0.7071	0.5087
$\varphi_p = 30°$	0.9828	1	0.9828	0.9306	0.8409	0.7071
$\varphi_p = 45°$	0.9306	0.9828	1	0.9828	0.9306	0.8409
$\varphi_p = 60°$	0.8409	0.9306	0.9828	1	0.9828	0.9306
$\varphi_p = 75°$	0.7071	0.8409	0.9306	0.9828	1	0.9828

reflector can be set between 45° and 60°. Although the optimal jamming effect cannot be guaranteed at this time, when the weapon is attacked at any threat angle, the difference between the jamming distance and the maximum jamming distance can be controlled within 16%, and a good jamming effect can also be obtained.

45.3.4 Influence of Relative Azimuth on Interference Airspace

The horizontal component of the angle between the attack direction of the weapon and the normal of the diffuse reflector is the relative azimuth θ_a. The range of the relative azimuth is [−90°, 90°], and the relation between the interference coefficient is obtained through Matlab simulation, as shown in Fig. 45.4.

In order to better analyze the influence of relative azimuth on interference coefficient, Table 45.4 is obtained according to MATLAB simulation data.

As can be seen from Fig. 45.4 and the odd and even properties of cosine function, the relation curve between the interference coefficient and the relative azimuth angle is symmetric about θ_a=0°. Therefore, only the values of the interference coefficient are listed in Table 45.4. According to the data in Fig. 45.4 and Table 45.4, the maximum value of the jamming coefficient is 1 when $\theta_a = 0°$, that is, the jamming distance is the maximum when the angle between the attack direction of the weapon and the horizontal direction of the normal direction of the diffuse reflector is 0°. As

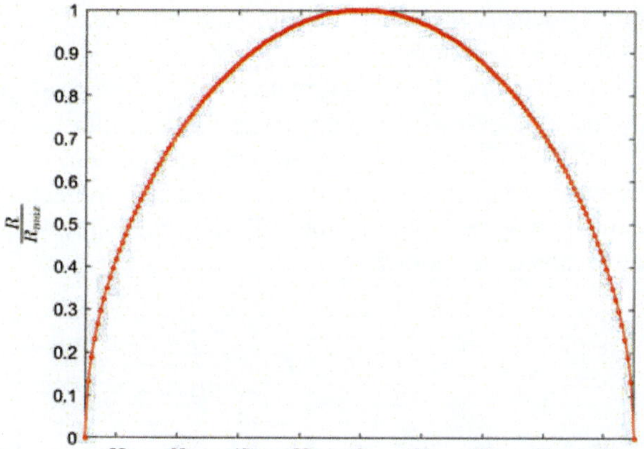

Fig. 45.4 The relationship between interference coefficient and relative azimuth

Table 45.4 The relationship between interference coefficient and relative azimuth

$\theta_a/°$	0	15	30	45	60	75	90
α	1	0.9828	0.9306	0.8409	0.7071	0.5087	0

the absolute value of relative azimuth angle increases, the interference coefficient decreases gradually. When $|\theta_a|$ is small, the interference coefficient decreases at a slower rate. When $|\theta_a| \leq 15°$, the interference coefficient has little change, and when $|\theta_a| = 15°$, the interference distance is only 1.72% lower than the maximum interference distance. When $|\theta_a|$ =30°, 45°, and 60°, the interference distance decreases by 6.94%, 15.91%, and 29.29%, respectively. When = 75°, the interference coefficient decreases to 0.5087, and the interference distance at this time is about half of the maximum interference distance in the normal direction. Therefore, there is no need to adjust the relative azimuth of the diffuse reflector so that it is directly opposite to the attacking weapon in the horizontal direction, and its relative azimuth deviation has little effect on the size of the interference airspace.

45.4 Error Analysis of Diffuse Reflector Interference Airspace

45.4.1 Interference Airspace Evaluation Model

Based on the reflection characteristics of the diffuse reflector, the interference airspace assessment system of the diffuse reflector is constructed. The system consists of laser energy measurement subsystem, the distance and attitude measurement subsystem, and the interference airspace assessment subsystem. The interference space domain of the diffuse reflector can be evaluated only by measuring the laser energy density scattered by the diffuse reflector, the distance and attitude parameters of the diffuse reflector.

Assume that the transmittance of the receiving optical system of the jamming airspace assessment system is the same as that of the seeker, and the measured distance between the system equipment and the false target is L, the included angle between the system equipment and the diffuse reflector is φ, and the receiving energy density of system equipment is E; then the expression of interference distance can be obtained from (45.7),

$$R = L\sqrt{\frac{Ee^{-\mu_a R}\cos\theta_r}{E_r e^{-\mu_a L}\cos\varphi}} \tag{45.7}$$

Reflection angle θ_r has azimuth component θ_{ra} and pitch component θ_{rp} in space. Assuming that the laser is incident parallel to the ground to the false target, then the incident angle θ_i is equal to the inclination ϕ_p of the diffuse reflector. According to the geometric relationship, θ_{rp} is as follows:

$$\theta_{rp} = \begin{cases} \theta - \phi_p\,,\ \theta \geq \phi_p \\ \phi_p - \theta\,,\ \theta < \phi_p \end{cases} \tag{45.8}$$

Combining (45.7) and (45.8) and properties of cosine function, we can get

$$R = L\sqrt{\frac{Ee^{-\mu_a R-L)}\cos\theta-\varphi_p)\cos\theta_{ra}}{E_r\cos\varphi}} \tag{45.9}$$

45.4.2 The Error Analysis

The influence of ranging error on interference distance error

The interference distance error ΔR caused by the ranging error ΔL can be expressed as follows [9]:

$$\Delta R = \frac{\partial R}{\partial L}\cdot\Delta L \tag{45.10}$$

From (45.9), it is easy to know the relationship between R and L,

$$R - L\sqrt{\frac{Ee^{-\mu_a R-L)}\cos\theta-\varphi_p)\cos\theta_{ra}}{E_r\cos\varphi}} = 0 \tag{45.11}$$

Differentiate L on both sides of the equation,

$$\frac{\partial R}{\partial L} - \sqrt{\frac{Ee^{-\mu_a R-L)}\cos\theta-\varphi_p)\cos\theta_{ra}}{E_r\cos\varphi}} - L\frac{\partial\sqrt{\frac{Ee^{-\mu_a R-L)}\cos\theta-\varphi_p)\cos\theta_{ra}}{E_r\cos\varphi}}}{\partial L} = 0 \tag{45.12}$$

Then

$$\frac{\Delta R}{\Delta L} = \frac{\partial R}{\partial L} = \frac{2A+\mu_a AL}{2+\mu_a AL} = \frac{(2+\mu_a L)\sqrt{\frac{Ee^{-\mu_a R-L)}\cos\theta-\varphi_p)\cos\theta_{ra}}{E_r\cos\varphi}}}{2+\mu_a L\sqrt{\frac{Ee^{-\mu_a R-L)}\cos\theta-\varphi_p)\cos\theta_{ra}}{E_r\cos\varphi}}} \tag{45.13}$$

Assuming $\mu_a = 0.108$, $\theta = 45°$, $\varphi_p = \varphi = 20°$, $E_r = 1.0\times10^{-15}$ J/mm², $L \in$ [150 m, 250 m], and $E \in [2.0\times10^{-13}$ J/mm², 3.0×10^{-13} J/mm²], Fig. 45.5 shows the variation curve of the influence of ranging error on interference distance.

As can be seen from Fig. 45.5, both the laser energy density received by the device and the distance from the device to the diffuse reflector have an impact on $\frac{\Delta R}{\Delta L}$. The larger the laser energy density is, the larger $\frac{\Delta R}{\Delta L}$ is. The farther the device is from the diffuse reflector, the smaller the $\frac{\Delta R}{\Delta L}$ is. The laser energy density received by the equipment is also related to the emission energy of the interference laser. When the

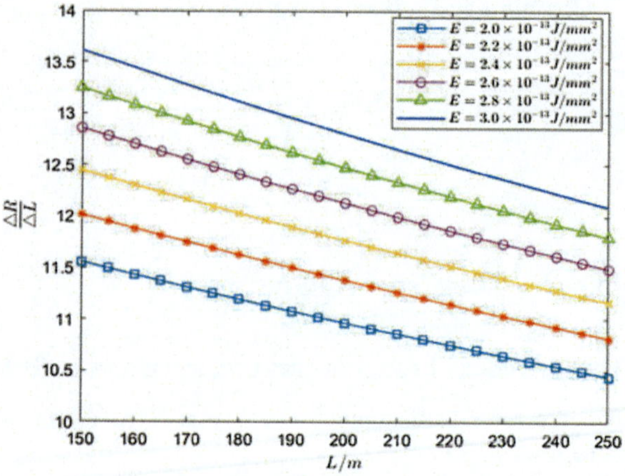

Fig. 45.5 The influence of ranging error on interference distance

relative position of the system equipment and the interference laser is determined, the greater the interference laser energy, the greater the laser energy density received by the equipment will be. Therefore, the greater the emission energy of the interference laser and the closer the evaluation system equipment is to the diffuse reflector, the greater the impact of ranging error on the interference spatial calculation.

The influence of attitude angle error on interference distance error

The interference distance error ΔR caused by the attitude angle calculation error $\Delta\varphi$ can be expressed as follows:

$$\Delta R = \frac{\partial R}{\partial \varphi} \cdot \Delta\varphi \tag{45.14}$$

The horizontal component of φ is assumed to be zero, then $\varphi = \varphi_p$, taking the derivative of both sides of (45.11) with respect to φ, we can get

$$\frac{\partial R}{\partial \varphi} - L\sqrt{\frac{Ee^{u_a L}\cos\theta_{ra}}{E_r}} \cdot \frac{\partial\sqrt{\frac{e^{-\mu_a R}\cos\theta - \varphi)}{\cos\varphi}}}{\partial\varphi} = 0 \tag{45.15}$$

$$\frac{\Delta R}{\Delta\varphi} = \frac{L\sin\theta\sqrt{\frac{Ee^{-(u_a R - L)}\cos\theta_{ra}}{E_r}}}{\cos^2\varphi(2 + u_a L\sqrt{\frac{Ee^{-(u_a R - L)}\cos\theta_{ra}}{E_r}}\sqrt{\cos\theta + \sin\theta\tan\varphi})\sqrt{\cos\theta + \sin\theta\tan\varphi}} \tag{45.16}$$

Assuming $\mu_a = 0.1081$, $E_r = 1.0 \times 10^{-15}$ J/mm^2, $E = 2.0 \times 10^{-13}$ J/mm^2, $L = 200$m, θ=30° 45° 60°, and $\varphi \in$[0°, 60°], the influence of calculation error of φ on interference distance R is shown in Fig. 45.6.

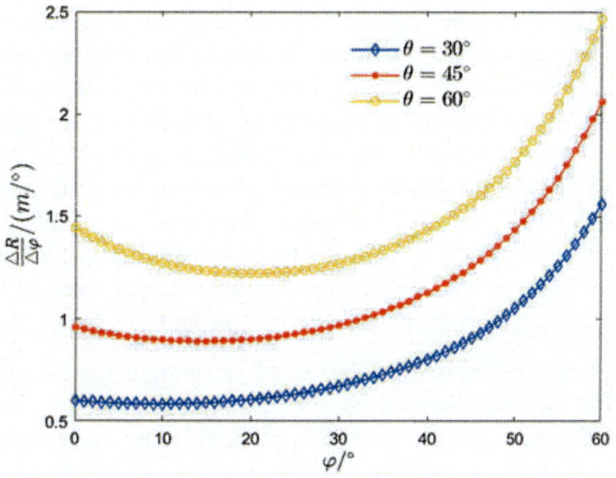

Fig. 45.6 The influence of attitude angle error on interference distance measurement

It can be seen from Fig. 45.6 that the attitude angle calculation error has different effects on the interference distance calculation error under different threat angles of weapons. The larger the threat angle is, the greater the influence of attitude angle calculation error on the interference distance calculation error is. Under the same threat angle, the influence of attitude angle calculation error on interference distance calculation error decreases first and then increases. When attitude angle θ=30° 45° 60° reaches its minimum value of 0.5838 , 0.8914 , and 1.2220 m/° then φ=10°15°, and 21°. The maximum value of $\frac{\Delta R}{\Delta \varphi}$ in the figure does not exceed 2.5 m/°. The interference distance of the diffuse reflector is thousands of meters, so the small attitude angle measurement error has little influence on the interference distance error.

45.5 Conclusion

In this paper, the influence of different parameters on the interference airspace of diffuse reflector in laser angle deception is analyzed by simulation, and the conclusions are as follows:

(1) The interference space can be increased by appropriately increasing the laser energy. However, considering the economic benefits, the laser energy cannot be increased without limit, and the effect of increasing the interference airspace is not obvious when the energy is large.

(2) Meteorological conditions have a certain influence on the interference distance, and the more severe the meteorological conditions, the more obvious the influence on the interference distance. When the atmospheric visibility is greater than 7 km, meteorological conditions have little effect on the interference distance.

(3) When the diffuse reflector is used for protection, if the weapon threat angle is known, the angle of the diffuse reflector should be adjusted to equal the weapon threat angle. If the threat angle information of the weapon is unknown, the inclination angle of the diffuse reflector can be set between 45° and 60°. At this time, the interference airspace of the diffuse reflector of the weapon attacked by any threat angle can be ensured to be large.
(4) It is not necessary to deliberately adjust the relative azimuth angle of the diffuse reflector to make it directly facing the incoming weapon in the horizontal direction, and the interference airspace within the deviation 30° range of its relative azimuth angle is reduced by no more than 6.94%.

According to the false target jamming distance calculation formula, considering the weapon threat angle, the inclination angle of the diffuse reflector, the relative azimuth angle, and other factors, the false target interference airspace evaluation model was established. By measuring the distance and attitude of the diffuse reflector, the measurement of the interference airspace of the diffuse reflector can be realized. Finally, the influence of the distance error and the attitude angle error on the interference distance error was analyzed. It is concluded that small distance and attitude angle measurement errors have little influence on interference airspace evaluation.

References

1. Z. Xiaobao, T. Lipeng, Research on deception jamming technology to laser-guided weapons. Space Electron. Countermeas. **29**(02), 7–8+19 (2013)
2. S. Chunsheng, Z. Xiaohui, R. Jionghui, et al., Layout of diffuse reflection board used for laser decoying. Laser Infrared **43**(03), 252–255 (2013)
3. S. Chunsheng, Z. Xiaohui, Z. Shuang, Guarded angle of laser decoy in laser trickjamming. Laser Infrared **47**(03), 347–351 (2017)
4. L. Song, Y. Wen, Y. Baoqing, et al., Research on flawless deployment of decoys in laser angle cheating jamming. Electro-Opt. Technol. Appl. **14**(05), 38–42 (2016)
5. X. Cheng, W. Yafu, Z. Quan, Research on layout of laser fake targets in complex terrain. Laser Infrared **47**(07), 875–878 (2017)
6. J. Jianzhou, S. De'an, J. Renyao, et al., Attenuation estimation of laser atmospheric transmission. Electron. Inf. Warfare Technol. **25**(04), 73–76+81 (2010)
7. Y. Zang, S. Wang, Z. Liu, et al., On atmospheric attenuating model in HWIL simulation of 1.064 μm laser guided weapons. Electron. Opt. Control **19**(09),17–22 (2012)
8. S. Xiaoquan, L. Yueguang, *Principle and Technology of Laser Counterwork* (The PLA Press, Beijing 2000)
9. D. Jianzhong, L. Zhixing, *Calculation Method* (Xi 'an Jiaotong University Press, Xi'an 2001)

Chapter 46
A Novel Type-Sensitive PageRank Algorithm for Importance Ranking of Heterogeneous Network Nodes

Hang Chen, Jiashun Duan, Yulong Dai, Xiangqian Xu, and Jun Yang

Abstract Systems in the real world are mostly made up of different types of interacting entities. It is of great value to identify the key nodes in a system, which can better grasp the key factors of the system. However, most current studies model systems as homogenous networks without distinguishing the difference information between nodes and edges. In this paper, firstly, the authors introduce the background, challenges, and significance of the research. Secondly, the preliminary of the methods is given. Thirdly, the authors purpose a novel Type-sensitive PageRank algorithm and combine it with substitution rule to solve the importance ranking problem of heterogeneous network nodes and successfully mine important nodes in the social heterogeneous network. Finally, the authors take a social heterogeneous network as an example and use this model to dig out the important nodes.

46.1 Introduction

Recently, information networks are ubiquitous in all aspects of life including citation network [1], media network [2], business network [3], social network [4], and so on. Nodes in networks are used to represent entities, and edges are used to represent relationships between entities. Systems in real life are often complex, with many entities of various types [5–7], and relations between entities are also various. Traditional network model uses homogenous network to model real system, which ignores the difference between entities, resulting in incomplete or missing information. In contrast, heterogeneous network is a graph structure with multiple types of nodes and edges, which can better represent a real system. Due to its universality, heterogeneous network model has been widely applied in deep learning [8], data fusion

H. Chen (✉) · J. Duan · Y. Dai
College of Military Basic Education, National University of Defense Technology, Changsha, Hunan, P. R. China
e-mail: hangchen_nudt@163.com

X. Xu · J. Yang
College of Systems Engineering,, National University of Defense Technology, Changsha, Hunan, P. R. China

G. Liu and F. Cen (eds.), *Advances in Precision Instruments and Optical Engineering*, Springer Proceedings in Physics 270, https://doi.org/10.1007/978-981-16-7258-3_46

[9], military, and other fields. Therefore, the study of heterogeneous network is of great significance.

Identifying the key nodes in heterogeneous networks can effectively understand the operation mode and grasp the key factors of the system. This has importance research value in many application scenarios, such as epidemic prevention and network security. At present, most of the research on the node importance of networks are based on the homogeneous network model. For example, degree centrality [10], betweenness centrality [11], K-Shell method [12] based on network topology structure, PageRank, HITS method based on propagation dynamics, etc. Compared with homogeneous networks, heterogeneous networks contain richer topological structure and semantic information, which also poses new problems and challenges for the identification of important nodes. Therefore, the objective of this study is to propose a node importance ranking method suitable for heterogeneous networks.

In 2016, Zhan et al. [13] reconstructed thc heterogeneous network into a multi-relation network by extracting different information channels in the network and took the importance of nodes in the multi-relation network as the measurement index of their importance. In 2019, Mohammad MehdiKeikha et al. [14] used deep learning to learn feature vectors of nodes in the network. In 2020, Soheila Molae et al. [15] proposed a method based on entropy and meta-path to measure the importance of nodes by integrating local and global information of the network.

In this study, an improved PageRank algorithm based on node type (Type-sensitive PageRank) is used to identify the important nodes in heterogeneous network. The original PageRank algorithm is based on homogeneous network and ignores the difference information between nodes. Based on this deficiency, to solve the problem of richness and semantic complexity of heterogeneous network nodes, the authors improve the original PageRank algorithm and divide the importance of nodes into the importance of each node type. Finally, substitution rule is used to combine the importance of each type into a comprehensive importance index.

The remainder of the paper is structured as follows: Sect. 46.2 is Preliminaries of heterogeneous network model, PageRank algorithm, and substitution rule. Section 46.3 builds the model and provide a detailed description of type-sensitive algorithm. Finally, Sect. 46.4 gives the process of mining the key nodes in the musician influence network to verify the model. Section 46.5 concludes the study with a summarizing discussion.

46.2 Preliminaries

46.2.1 Heterogeneous Network Model

Basic concepts and definitions of heterogeneous network have been illustrated by Shi at el. [16] in 2016 and Yu at el. [17] in 2021.

Fig. 46.1 An example of heterogeneous network on citation network

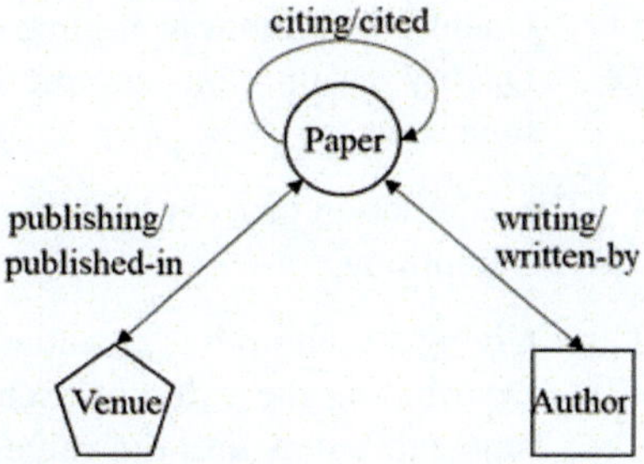

Definition 46.1 A heterogeneous network is defined as a graph $G = (V, E, \varphi, \psi)$. The specific definition is as follows:

V: A set of nodes,
E: A set of edges between nodes,
φ: A node type mapping function $\varphi : V \rightarrow T_V$,
ψ: A edge type mapping function $\psi : E \rightarrow T_E$.

Each node $v \in V$ or edge $e \in E$ belongs to one particular type, that is,

$$\varphi(v) \in T_V,$$

$$\psi(e) \in T_E.$$

In particular, $|T_V| \geq 2$.

Shown in Fig. 46.1, this is a typical heterogeneous network with three types of nodes: paper, venue, and author. For a paper, it can be written by a set of authors and be published in one venue, so a paper node can be connected to a set of author nodes and a venue node.

46.2.2 PageRank Algorithm

PageRank algorithm [18] is used for link analysis and is one of the core algorithms of Google search engine. It can evaluate the importance of each node in the network according to its link relationship. Its idea can be reduced to two assumptions:

(1) Quantity assumption: assume that node with larger indegree is more important.
(2) Quality assumption: assume that node pointed by important nodes is more important.

The calculation process is an iterative process which make full use of the above two assumptions.

(1) Giving a random initial value to each node.
(2) Distributing the value of each node equally to all the nodes it points to and repeat this step until the values of all nodes are stable.
(3) Ranking the nodes by their values.

46.2.3 *Substitution Rule*

Substitution rule is a rule that combines multidimensional indicators. When evaluating an object, it tends to be considered from many aspects. For example, when buying a car, the price, comfort, speed, and other factors will be considered. Therefore, a way to combine multidimensional indicators into a comprehensive indicator is significant. Substitution rules are one such way. It is defined as follows:

Definition 46.2 Assume that there are n evaluation criteria, the evaluation function of each criterion is $u_1, u_2, \ldots, u_n$, and their value interval is [0, 1]. The calculation formula of the substitution rule is as follows:

$$W(u_1, u_2, \ldots, u_n) = 1 - \prod_{i=1}^{n}(1 - u_i). \tag{46.1}$$

From the above formula, it can be found that under the substitution rule, if any index reaches the maximum value, the comprehensive index also reaches the maximum value.

46.3 Model and Method

It is of great research value to identify the important nodes in complex heterogeneous networks. However, previous studies are mostly based on homogeneous networks. According to the characteristics of heterogeneous network, this study improves the PageRank algorithm, proposed a type-sensitive PageRank algorithm, and combines it with substitution rules to establish a model for identifying important nodes of heterogeneous network.

This section describes the process and principle of the method in detail.

Definition 46.3 Let $G = (V, E, \varphi, \psi)$ be a heterogeneous network. Then the adjacent matrix A of G can be expressed as follows:

$$a_{ij} = \begin{cases} 1, if\left(v_i, v_j\right) \in E \\ 0, if\left(v_i, v_j\right) \notin E \end{cases}. \tag{46.2}$$

Definition 46.4 Let $G = (V, E, \varphi, \psi)$ be a heterogeneous network and A be the adjacent matrix of G. Then the state transition matrix P can be expressed as follows:

$$p_{ij} = \frac{a_{ij}}{\sum_{V_i \in V} a_{ij}}. \tag{46.3}$$

Given a heterogeneous network, its adjacencies matrix and state transition matrix are determined accordingly. In the original PageRank algorithm, P will be directly used for the next calculation. But in the Type-sensitive PageRank algorithm, a set of type state vectors s is introduced to distinguish the type of difference between nodes.

Definition 46.5 Let $G = (V, E, \varphi, \psi)$ be a heterogeneous network. For node type t_i, type state vector s_i can be expressed as follows:

$$s_{ij} = \begin{cases} 1, if\varphi\left(v_j\right) = t_i \\ 0, if\varphi\left(v_j\right) \neq t_i \end{cases}. \tag{46.4}$$

As mentioned above, in this model, the nodes importance will be first calculated in each type separately, and then combined. For node type t_i, the calculation process is as follows:

Algorithm 1. Type-sensitive PageRank for node type t_i

Input: heterogeneous network G, damping coefficient q_i, convergence threshold ε_i
Output: importance value vector R_i
1: The state transition matrix P and type state vector s_i are calculated
2: R_i is randomly initialized
3: **while** True **do**
4: $R_i^{'} \leftarrow q_i \times PR_i + (1 - q_i)s_i/|s_i|$
5: **if** $\left|R_i^{'} - R_i\right| > \varepsilon_i$, **then**
6: $R_i \leftarrow R_i^{'}$
7: **else**
8: **return** R_i

Through the above calculation, the importance of each node in node type t_i can be gotten, and then they should be normalized into the interval [0, 1] using Min–Max scaling,

$$U_{ij} = \frac{R_{ij} - \min(R_i)}{\max(R_i) - \min(R_i)}, \tag{46.5}$$

where $\max(R_i)$ and $\min(R_i)$ are maximum and minimum values of R_i.

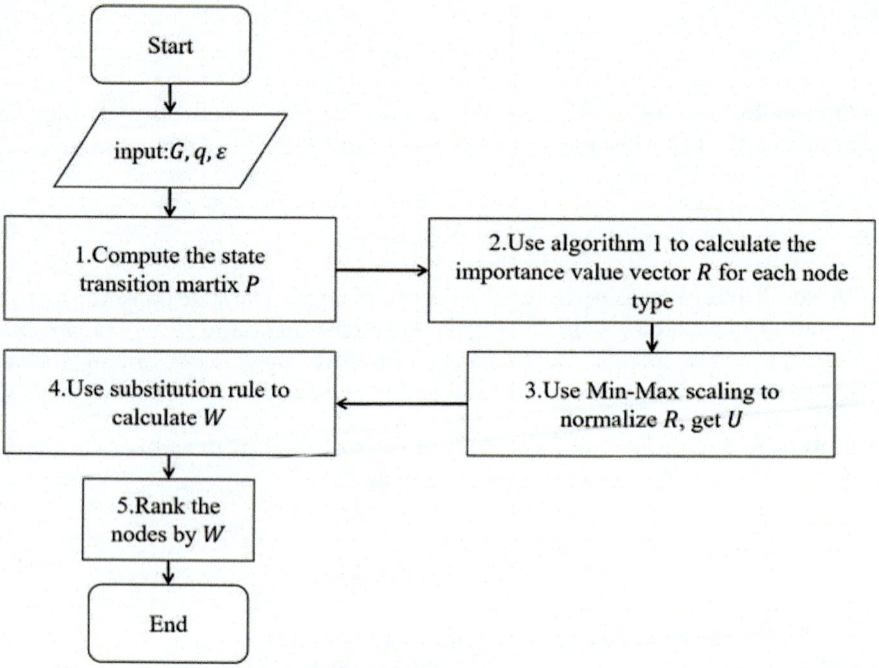

Fig. 46.2 Node importance ranking model

After calculating all node types, the importance matrix U is gotten, where u_{ij} indicates the importance of node v_j in node type t_i.

Obviously, U has the same number of rows as the number of node types, and the same number of columns as the number of nodes.

Then, substitution rule is used to calculate the comprehensive importance of node v_j as follows:

$$w_j = 1 - \prod_{t_i \in T_V} (1 - u_{ij}). \tag{46.6}$$

Finally, the nodes are ranked in descending order by w_j.

Therefore, the steps of node importance ranking model can be described as in Fig. 46.2.

46.4 Experimental Analysis

To further verify the effectiveness of the model, an experimental analysis of the model algorithm is conducted on the musician influence data set.

Table 46.1 Information of the musician heterogeneous network

	Type	Quantity
Nodes	Vocal	162
	Blues	101
	Pop/Rock	2795
	Classical	28
	Comedy/Spoken	46
	Country	403
	R&B	676
	Electronic	208
	Folk	95
	Jazz	405
	Latin	229
Edges	Influence	42,770

46.4.1 Data Declaration

This data set was scraped from *AllMusic.com*, which records the influence relationship of over 5000 musicians since 1920. A musician can be influenced by more than one other musician, and musicians can be divided into 11 genres. Therefore, a heterogeneous network of musician influence is constructed by taking musicians as nodes, differentiating genres as node types, and taking influence relationships among musicians as edges ($follower \rightarrow influencer$). The information of the heterogeneous network is as Table 46.1 .

46.4.2 Experiment Process

Firstly, the state transition matrix P is calculated. Due to space constraints, the specific value of P is not shown. P is a sparse matrix, and it can be seen from P that there is an obvious phenomenon of community division in this network.

Let $\varepsilon_1 = \varepsilon_2 = \cdots = \varepsilon_{11} = 1 \times 10^{-3}$ and $q_1 = q_2 = \cdots = q_{11} = 0.2$. Then, importance matrix U can be calculated by algorithm 1 and formula (46.5). Part of results are shown in Table 46.2.

Then, formula (46.6) is used to calculate comprehensive importance for each node, e.g., In the importance matrix U, the column value of Michael Jackson is

$$[0, 0.32, 0.63, 0, 0.03, 0, 0.91, 0.14, 0, 0, 0.04]^T.$$

So, the comprehensive importance for Michael Jackson is 0.973.

Table 46.2 Top 3 musicians in Pop/Rock, R&B and Country

Genre	Musician	Importance
Pop/Rock	The Beatles	1.00
	Bob Dylan	0.84
	The Rolling Stones	0.81
R&B	Marvin Gaye	1.00
	James Brown	0.93
	Michael Jackson	0.91
Country	Hank Williams	1.00
	Johnny Cash	0.92
	Merle Haggard	0.86

The result is shown in Table 46.3. The basic idea of PageRank algorithm is reflected. In the network, The Beatles has the largest indegree and it is one of the most influential musicians. There are more nodes that point to Michael Jackson than Marvin Gaye, but musicians influenced by Marvin Gaye are more influential, so Marvin Gaye is more influential than Michael Jackson. Substitution rules and comprehensiveness are also reflected in the result. Michael Jackson is not only influential in R&B, but also in many other fields including Pop/Rock, Blues, etc. Therefore, he is more influential than James Brown, who is more influential in R&B than him.

The distribution of node importance values is shown in Fig. 46.3 below. As can be seen from Fig. 46.3, there are few nodes with very high ([0.95, 1]) or very low ([0, 0.05]) importance, most of which are in the range of high ([0.6, 0.75]) or low ([0.1,

Table 46.3 Experimental result

Importance interval	Nodes	Description
1	The Beatles (Pop/Rock), Marvin Gaye (R&B), Hank Williams (Country), John Coltrane (Jazz)… (11 in total)	The most influential musicians in a particular field
[0.75, 1)	The Rolling Stones (Pop/Rock), Michael Jackson (R&B), Tito Puente (Latin), Johnny Cash (Country)… (623 in total)	Musicians with strong influence in a particular field
[0.5, 0.75)	Indigo Girls (Pop/Rock), Pat Martino (Jazz)… (2192 in total)	Musicians with moderate influence in a particular field
[0.25, 0.5)	Melba Moore (R&B), Kruder&Dorfmeister (Electronic), (1521 in total)	Musicians with a week influence in a particular field
[0, 0.25)	King Krule (Pop/Rock), Madeleine Peyroux (Vocal), (1659 in total)	Musician with little influence

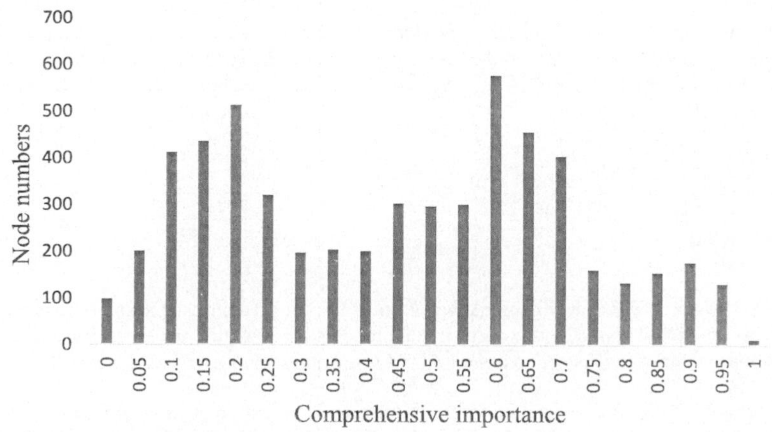

Fig. 46.3 Distribution of node importance values

0.25]) importance, and the number of nodes with medium ([0.3, 0.55]) importance is not many. Therefore, this experiment successfully distinguishes the important nodes from the non-important nodes.

46.5 Conclusions

The importance ranking of heterogeneous network nodes has important research value and practical application value. However, most of the researches are conducted based on homogeneous networks, which will cause the loss of differential information in the network and fail to achieve accurate calculation.

To deal with the problem described above, the authors purpose a novel Type-sensitive PageRank algorithm and combine it with substitution rule, which makes full use of differential information among nodes. It provides an efficient, accurate, and concise solution to the problem. Through experiments, the model has achieved good results in social heterogeneous networks.

The innovation of this study is firstly that Type-sensitive PageRank algorithm is proposed to solve the problem of computing the importance of nodes in A single node type in A heterogeneous network. Secondly, the comprehensive importance index is obtained by substitution rules.

The shortcoming of this study is that the differential information of edges is not fully utilized, and the model does not work as well in heterogeneous networks in other domains as in social networks.

Therefore, in the future work, the model should be expanded and perfected by adding the differential information of edges and using other merge rules to suit different domains.

Acknowledgements We are thankful to the Editor and the reviewers for their valuable comments and detailed suggestions to improve the presentation of the paper. Further, we also acknowledge the support in part by the National Natural Science Foundation of China under grant No. 71690233 and No. 71671186.

References

1. M. Yasunaga, J. Kasai, R. Zhang, A.R. Fabbri, I. Li, D. Friedman, D.R. Radev, ScisummNet: a large annotated corpus and content-impact models for scientific paper summarization with citation networks, in *Proceedings of the 33rd AAAI Conference on Artificial Intelligence* (2019), pp. 7386–7393
2. L. Yongjun, Z. Su, A comment on ãcross-platform identification of anonymous identical users in multiple social media networks, IEEE Trans. Knowl. Data Eng. **30**(7), 1409–1410R (2018); F. Bruni, A. Cesarone, A. Scozzari, F. Tardella, On exact and approximate stochastic dominance strategies for portfolio selection. Eur. J. Oper. Res. **259** (2017)
3. Y. Zheng, R. Hu, S.-f. Fung, C. Yu, G. Long, T. Guo, S. Pan, Clustering social audiences in business information networks. Pattern Recognit. **100**, 107126 (2020)
4. E.G. Tajeuna, M. Bouguessa, S. Wang, Modeling and predicting community structure changes in time-evolving social networks. IEEE Trans. Knowl. Data Eng. **31**(6), 1166–1180 (2019)
5. J. Tang, M. Qu, M. Wang, M. Zhang, J. Yan, Q. Mei, LINE: large-scale information network embedding, in *Proceedings of the 24th International Conference on World Wide Web* (2015), pp. 1067–1077.
6. M. Gong, C. Yao, Y. Xie, M. Xu, Semi-supervised network embedding with text information. Pattern Recognit. **104**, 107347 (2020)
7. E.L. Martelot, C. Hankin, Fast multi-scale detection of relevant communities in large-scale networks, Comput. J. 1136–1150 (2018)
8. H. Ji, X. Wang, C. Shi, B. Wang, P. Yu, Heterogeneous graph propagation network. IEEE Trans. Knowl. Data Eng. https://doi.org/10.1109/TKDE.2021.3079239
9. Z. Yan, J. Liu, L.T. Yang, W. Pedrycz, Data fusion in heterogeneous networks. Inform. Fusion **53**, 1–3 (2020)
10. G. Sabidussi, The centrality index of a graph. Psychometrika **31**(4), 581–603 (1966)
11. L.C. Freeman, Centrality in social networks conceptual clarification. Soc. Netw. **1**(3), 215–239 (1979)
12. M. Kitsak, L.K. Gallos, S. Havlin et al., Identification of influential spreaders in complex networks. Nat. Phys. **6**(11), 888–893 (2012)
13. Q. Zhan, J. Zhang, P.S. Yu, S. Emery, J. Xie, Discover tipping users for cross network influencing (Invited Paper), in *2016 IEEE 17th International Conference on Information Reuse and Integration (IRI)* (2016), pp. 67–76. https://doi.org/10.1109/IRI.2016.17
14. M.M. Keikha, M. Rahgozar, M. Asadpour, M.F. Abdollahi, Influence maximization across heterogeneous interconnected networks based on deep learning. Expert Syst. Appl. **140**, 112905 (2020). ISSN 0957-4174
15. S. Molaei, R. Farahbakhsh, M. Salehi, N. Crespi, Identifying influential nodes in heterogeneous networks. Expert Syst. Appl. **160**, 113580 (2020). ISSN 0957-4174
16. C. Shi, Y. Li, J. Zhang, Y. Sun, S.Y. Philip, A survey of heterogeneous information network analysis. IEEE Trans. Knowl. Data Eng. **29**(1), 17–37 (2016)
17. Y. Xie, B. Yu, S. Lv, C. Zhang, G. Wang, M. Gong, A survey on heterogeneous network representation learning. Pattern Recogn. **116**, 107936 (2021). ISSN 0031-3203
18. J. Lee, M. Olvera-Cravioto, PageRank on inhomogeneous random digraphs. Stochast. Process. Appl. **130**(4), 2312–2348 (2020). ISSN 0304-4149

Chapter 47
Trapping and Storing Photons via a Dynamically-Formed Nanocavity

Jia-Hui Chen, Chao Li, and Jun-Fang Wu

Abstract In this paper, we demonstrate numerically that we can create a potential-well-like ultrahigh-Q nanocavity to trap a traveling signal pulse with a long storing time. This approach has potential application in photon memories and quantum computing.

47.1 Introduction

At present, there are already many methods to constructing a high-Q microcavity, which are basically achieved by partially modifying the mode gap or by constructing a whispering gallery [1–8]. However, we cannot design our structure which need to achieve light movement based on these traditional methods because the high-Q microcavities formed by these methods are fixed in one place and cannot be moved [9–13]. For this reason, we designed a method based on local refractive index increasing on a photonic slab to form a high-Q nanocavity realized by local mode gap modification. Our structure is based on the W1 line-defect waveguide in a 2D triangular air-hole dielectric photonic crystal slab shown in Fig. 47.1a. We increase the refractive index of the dielectric material in the rectangular shaded area in Fig. 47.1b, while the refractive index of the air hole remains unchanged. Through this process, the area refractive index increased forms a potential well, which can confine the light so that a high-Q microcavity is formed. The high-Q nanocavity formed by this method is reversible because we have not made any permanent change on the structure like move the air holes or fill some air holes, and the refractive index increasing which can be receive by fast optical nonlinear processes like Kerr nonlinearity is reversible so that we can use this method of dynamic modulation to make high-Q nanocavity very flexible and can be constructed at any time and at any position on the waveguide. When we do not modulate, the whole structure is a simple

J.-H. Chen · C. Li · J.-F. Wu (✉)
School of Physics and Optoelectronic Technology, South China University of Technology, Guangzhou 510640, China
e-mail: wujf@scut.edu.cn

G. Liu and F. Cen (eds.), *Advances in Precision Instruments and Optical Engineering*, Springer Proceedings in Physics 270, https://doi.org/10.1007/978-981-16-7258-3_47

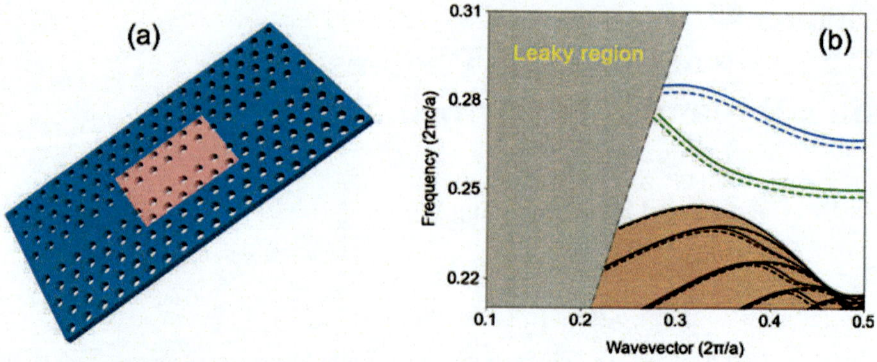

Fig. 47.1 **a** 2D triangular-lattice PC slab with a W1 waveguide. **b** Calculated band structure for the W1 waveguide in **a**

waveguide structure, the waveguide mode introduced by line defects can propagate freely in the waveguide.

In this paper, we propose a simple method for realizing ultrahigh-Q nanocavities which can be formed instantly at any time and at any position in a PC waveguide and can be moved on demand. We also apply this method to light storage.

47.2 Model

We start by considering a triangular-lattice PC slab as shown in Fig. 47.1a. The refractive index of this silicon slab is $n_0 = 3.4$. The lattice constant a, the radii of air holes r, and the thickness of the slab h are 420 nm, 110 nm, and 210 nm, respectively. A W1 line-defect waveguide is formed by removing a row of air holes, as shown by Fig. 47.1b. If we increase the refractive index of the waveguide region slightly, the mode curve will descend a little, as shown by the dashed lines in Fig. 47.1b. Thus, we can form a potential well to trap the photons through altering the refractive index in waveguide local region. With the help of the mode gap, the walls of the formed potential well can act as perfect mirrors to form an ultrahigh-Q nanocavity. When no modulation, the whole system is a simple PC waveguide structure, and the waveguide modes introduced by line defects can propagate freely along the waveguide.

To trap and store signal photons, we suppose in the beginning, a signal light pulse is injected into a W1 PC waveguide from the input end. Since at present this simple PC slab has no resonator, the injected signal pulse will travel down the waveguide freely. When we want to trap this signal light, just swiftly increasing the refractive index of the local region in the waveguide to form a potential-well-like cavity. In this way, the signal photons will be trapped and stored by the dynamically-formed nanocavity.

47.3 Application

We perform finite-difference time-domain (FDTD) simulations [14] on the PC structure shown in Fig. 47.2a. The length of the PC waveguide is set to be 87*a*. When a sudden index variation of $\Delta n/n_0 = 0.8\%$ is added to the dielectric material of the modulation area, a potential-well-like nanocavity will be dynamically formed. FDTD simulations show that the resonance wavelength of the formed cavity is $\lambda_0 =$ 1683.14 nm, and the Q factor is calculated to be ~12,000,000 by an exponential fit of the electromagnetic energy decay. These results confirm that the mechanism for dynamically generating an ultrahigh-Q nanocavity is feasible and efficiency, without any spatial structure tuning.

This dynamically formed ultrahigh-Q nanocavity offers us a new opportunity to make the manipulation on photons flexible. The signal light is a long light pulse with a wavelength of 1668 nm which is close to the edge of the wavelength of the mode induced by the line defect. The light pulse injects our structure from the left side of the waveguide, which is shown as the red arrow in Fig. 47.2a, and propagates along the waveguide to the right end. At $t = 13$ ps, the signal light pulse reaches the center of the waveguide, and we begin to modulate the rectangular shaded area to increase the refractive index of dielectric material to 1.008*n*. It can be seen from field distribution snapshots in Fig. 47.2b–e that the signal light is captured by the region increased refractive index, stays at the center of the waveguide, and the light density can be well reserved even after a long time. The frequency of the captured light in the refractive index modulated area is 1683.14 nm, which is exactly the resonant

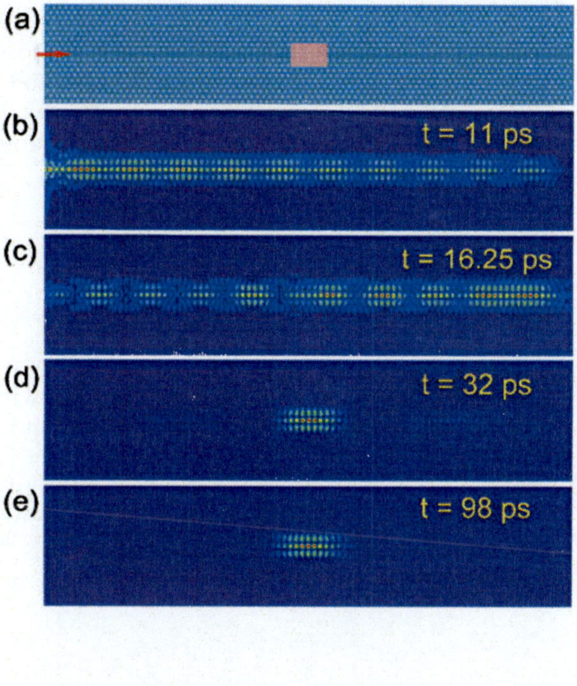

Fig. 47.2 **a** A Gaussian pulse with center wavelength at 1668 nm is incident. **b–e** Field distribution at different moments during optical trapping and storing process

frequency of the microcavity, proving that the captured light intensity exists as the resonant mode of the microcavity, so its energy decay rate is very small, and it can stay in this region for a very long time. This can also be confirmed from Fig. 47.2, which shows the magnetic field strength monitored in the center of the waveguide as a function of time. When we start to increase the refractive index of the rectangular shaded area at $t = 13$ ps, the magnetic field strength begins to fluctuate sharply, and at a segment of time after this, a large part of the light intensity captured by the refractive index increased region and finally exist as the resonant mode of the formed high-Q nanocavity. It can be seen that the decay speed of the captured signal light intensity is very slow, which make the captured light can be held for a long time.

Above we showed how to capture signal light through dynamic modulating the refractive index on a rectangular area of a 2D triangular air-hole dielectric photonic crystal slab with W1 line-defect waveguide. In the following, we will show how to achieve controllable light movement through dynamic refractive index modulation on our structure, using the flexibility of the high-Q microcavity we formed. After the light capture process, the signal light has been fixed at the center of the structure, where the rectangular shaded area is located. Then, at $t = 50$ ps, we start to move the rectangular refractive index modulation region (the refractive index of the dielectric material on area newly covered by the moving rectangular area increase to $1.008n$, and the refractive index on area no longer been covered return to n) with a constant speed of $0.15c$. The region finally reaches a place, which is $26a$ away from the starting position.

We found the captured signal light originally stayed at the center of the whole structure which is exactly the position the rectangular shaded region is located. Then, when $t = 50$ ps, the refractive index modulation region begins to move, and the captured signal light move along with the movement of the region, that is, toward waveguide moves to the right, with a speed almost the same as the speed of the movement of the region, and finally reaches and stays where the rectangular shaded area stays. As a result, we realized the controllable light movement by using the flexibility of the high-Q microcavity we formed, which used a dynamic refractive index modulation method.

47.4 Conclusions

In summary, we have demonstrated numerically that we can create a potential-well-like ultrahigh-Q nanocavity to trap a traveling signal pulse with a long storing time. This approach has potential application in photon memories and quantum computing.

Acknowledgements This research was financially supported by the National Natural Science Foundation of China (11774098); Guangdong Natural Science Foundation (2017A030313016); Science and Technology Program of Guangzhou (202002030500).

References

1. K. Nozaki, A. Shinya, S. Matsuo, Y. Suzaki, T. Segawa, T. Sato, Y. Kawaguchi, R. Takahashi, M. Notomi, Ultralow-power all-optical RAM based on nanocavities. Nat. Photon. **6**, 248 (2012)
2. L. Fan, J. Wang, L.T. Varghese, H. Shen, B. Niu, Y. Xuan, A.M. Weiner, M.H. Qi, An all-silicon passive optical diode. Science **335**, 447 (2012)
3. D.L. Sounas, J. Soric, A. Alù, Broadband passive isolators based on coupled nonlinear resonances. Nat. Electron. **1**, 113
4. K. Hennessy, A. Badolato, M. Winger, D. Gerace, M. Atatüre, S. Gulde, S. Fält, E.L. Hu, A. Imamoğlu, Quantum nature of a strongly coupled single quantum dot–cavity system. Nature **445**, 896 (2007)
5. B.-S. Song, S. Noda, T. Asano, Y. Akahane, Ultra-high-Q photonic double-heterostructure nanocavity. Nat. Mat. **4**, 207–210 (2005)
6. B.-S. Song, T. Asano, S. Noda, Physical origin of the small modal volume of ultra-high-Q photonic double-heterostructure nanocavities. New J. Phys. **8**, 209 (2006)
7. Y. Akahane, T. Asano, B.-S. Song, S. Noda, High-Q photonic nanocavity in a two-dimensional photonic crystal. Nature **425**, 944 (2003)
8. K. Ashida, M. Okano, M. Ohtsuka, M. Seki, N. Yokoyama, K. Koshino, M. Mori, T. Asano, S. Noda, Y. Takahashi, Ultrahigh-Q photonic crystal nanocavities fabricated by CMOS process technologies. Opt. Express **25**, 18165 (2017)
9. Y. Lai, S. Pirotta, G. Urbinati, D. Gerace, M. Minkov, V. Savona, A. Badolato, M. Galli, Genetically designed L3 photonic crystal nanocavities with measured quality factor exceeding one million. Appl. Phys. Lett. **104**, 241101 (2014)
10. Z.-M. Xu, C. Li, J.-F. Wu, Dynamic trapping and releasing photonics beyond delay-bandwidth limit in cascaded photonic crystal nanocavities New J. Phys. **22**, 063030 (2020)
11. Y. Tanaka, J. Upham, T. Nagashima, T. Sugiya, T. Asano, S. Noda, Dynamic control of the Q factor in a photonic crystal nanocavity. Nat. Mater. **6**, 862 (2007)
12. J. Upham, Y. Tanaka, T. Asano, S. Noda, Dynamic increase and decrease of photonic crystal nanocavity Q factors for optical pulse control. Opt. Express **16**, 21721 (2008)
13. B. Wang, J.-F. Wu, C. Li, Z.-Y. Li, Dynamic tuning of the Q factor in a photonic crystal nanocavity through photonic transitions. Opt. Lett. **43**, 3945 (2018)
14. A. Taflove, S.C. Hagness, *Computational Electrodynamics* (Artech House, 2000).

Chapter 48
Breaking the Delay-Bandwidth Limit in a Dynamically Tuned Nanocavity

Shuang Liu, Jun-Fang Wu, and Chao Li

Abstract In this paper, we present a novel approach to trap, store, and release a signal pulse and eventually break the delay-bandwidth limit, based on a dynamically tuned nanocavity. Using this mechanism, we design a compact silicon photonic crystal system with long storing time, and the trapped signal pulse can be released at will.

48.1 Introduction

Achieving high-quality (Q) factor nanocavities, which can enhance the interaction of light and matter since the photons can be strongly confined in a tiny space, are a promising fundamental component in physics and engineering, including integrated photonics, nonlinear optics, cavity quantum electrodynamics, and quantum information processing [1–4]. Unfortunately, ultrahigh-Q nanocavity resulting ultra-narrow linewidth of resonance spectra dictates that the optical pulse cannot be effectively coupled into/out of the cavity. It is not conductive for all optical signal processing in high speed communication. Actually, the delay-bandwidth product, characterizing the storage capacity of the any type of resonator system, is a fundamental limit in physics and engineering [5]. However, if the dynamic tuning of the optical structure can be used, it can theoretically demonstrate that breaking the fundamental limit can be achieved [6–8].

This dynamic process has the following characteristics: trapping light with broadband of resonance from the input waveguide and storing light with ultra-narrow linewidth of resonance in the resonator and releasing light with broadband of resonance. So far, various methods have been proposed to realize dynamic tuning of optical structure [9–12]. By constructing an analogue of EIT structure and modulated the refractive index of the resonator, the photon bandwidth can be compressed

S. Liu · J.-F. Wu · C. Li (✉)
School of Physics and Optoelectronic Technology, South China University of Technology, Guangzhou 510640, China
e-mail: lichao@scut.edu.cn

G. Liu and F. Cen (eds.), *Advances in Precision Instruments and Optical Engineering*, Springer Proceedings in Physics 270, https://doi.org/10.1007/978-981-16-7258-3_48

to zero and it realizes a new mechanism for stopping light [13]. Breaking the delay-bandwidth limit by controlling the coherent interference between two low-Q ring-resonators is a method to realize photonic memory and release [3]. In order to increase the storage time, it is necessary to continuously refresh the control pulses within the carrier lifetime. In addition, dynamic release of photon can be realized in photonic-crystal nanocavity with highly Q factor, through adiabatic frequency tuning[5] which makes a blue shift for the frequency of the cavity resonance and decreases the Q factor. In fact, this technique allows to change the quality factor from high-Q to low-Q, conversely it is not. For this reason, it can release the photon from a high-Q cavity, but it is not conductive for the light to efficiently couple into the microcavity before photonic memory. Therefore, finding a way to achieve broadband processing, high efficiency trapping light, long storage time, and arbitrary timing of releasing light is still a challenge. In this paper, we will propose a new method to realize dynamic process that controls the arbitrary timing of storing light and releasing light through photonic transition in photonic-crystal nanocavity and eventually break the delay-bandwidth limit.

48.2 Theory

In order to demonstrate this photonic transition process, we consider a silicon cavity coupling to a waveguide. Here, the cavity supports several defect modes which have different modal profile. Assuming only one of the defect modes of this cavity is excited (mode ω_1), photonic transitions to another mode ω_2 can be induced by a modulated cavity with time dependent perturbation of permittivity (for simplicity, assuming the presence of the photonic bandgap can isolate related two modes, even though for several defect modes),

$$\epsilon(\mathrm{r},\mathrm{t}) = \epsilon_0 + \Delta\epsilon(r)\cos(\Omega t), \tag{48.1}$$

where Ω is the modulation frequency such that $\Omega = \omega_2 - \omega_1$. Here, ϵ_0 is dielectric distribution of cavity which is time-independent, and modulation amplitude distribution is described by $\Delta\epsilon(r)$.

By substituting (48.1) into Maxwell's equation,

$$\nabla^2 E - \mu\frac{\delta^2\epsilon}{\delta t^2}E - 2\mu\frac{\delta\epsilon}{\delta t}\cdot\frac{\delta E}{\delta t} - \mu\epsilon\frac{\delta^2 E}{\delta t^2} = 0, \tag{48.2}$$

where μ is the magnetic permeability, under slowly varying envelope approximation and assuming $\Omega \ll \omega_1, \omega_2$, and only two modes of perturbation need to be considered, we obtain the period of two modes going through mutual conversion,

$$T = \frac{8\pi}{\sqrt{\omega_1 \omega_2 K_1 K_2}}. \quad (48.3)$$

It means that a photon in defect mode at ω_1 will make a complete transition to the defect mode at ω_2 at a time point where it is coherence time $L_t = \frac{T}{2}$.

For cavity-waveguide system, it means the cavity is introduced a decay mechanism. If the time-modulated system is not applied to cavity, the attenuation factors between different modes do not interact with each other. However, in the modulated cavity, this process, where photonic transitions happen between two defect modes, is mode coupling and has a property of attenuation. From (48.2), we can know that $e^{-\frac{(\upsilon_2-\upsilon_1)t}{2}} e^{-\upsilon_1 t} = e^{\frac{-(\upsilon_2+\upsilon_1)t}{2}}$ and $e^{-\frac{(\upsilon_1-\upsilon_2)t}{2}} e^{-\upsilon_2 t} = e^{\frac{-(\upsilon_2+\upsilon_1)t}{2}}$ are attenuation factor of mode 1 and mode 2 in modulated system, respectively. It means that these two modes have the same attenuation factor during mutual conversion process even though they have the dramatically different attenuation factor when no modulation.

48.3 Application

We demonstrate the photonic transition between defect modes and manipulating bandwidth concept as discussed above by using a finite-difference time-domain method. The structure consisting of waveguides and cavity is a direct-coupled two-dimensional photonic crystal which is triangular lattice of air hole with a radius $r = 0.275\,a$ (a is a lattice constant) (Fig. 48.1a). The cavity supports several defect modes, one of which exhibits a large quality factor Q and extremely small mode volume. Importantly, the interval between frequencies of various modes is not the same so that photonic transitions can eliminate cascaded processes. Here, photonic transitions are induced between defect mode 1 at a frequency $\omega_1 = 0.2704(2\pi c/a)$ with Q =

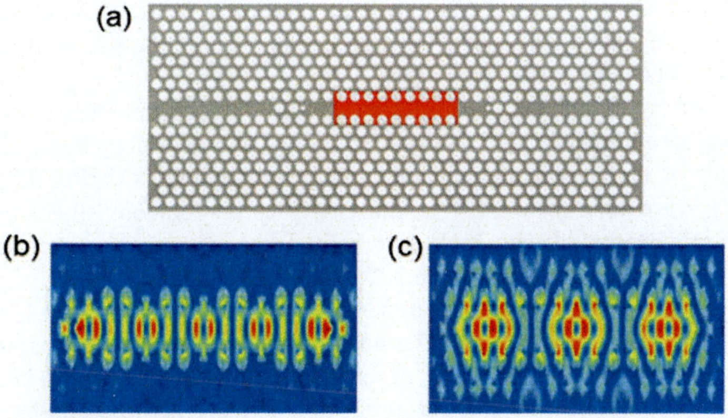

Fig. 48.1 **a** Sketch of a dynamically tuned nanocavity. **b**, **c** Field patterns for the high-Q mode and low-Q mode, respectively

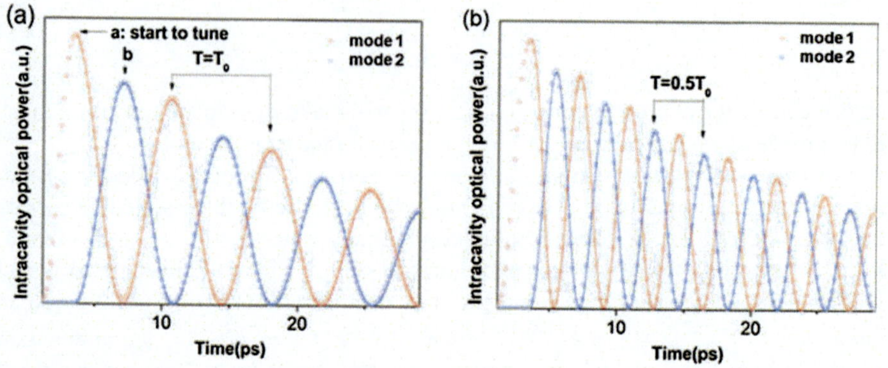

Fig. 48.2 Photon transitions between an ultrahigh-Q mode and low-Q mode in time domain. **a** $\varepsilon/\varepsilon_0 = 7.2 \times 10^{-3}$. **b** $\delta\varepsilon/\varepsilon_0 = 3.6 \times 10^{-3}$

17,821 and defect mode 2 at a frequency $\omega_2 = 0.2608(2\pi c/a)$ with $Q = 146{,}225$, by additional perturbation of permittivity $\Delta\epsilon(r, t) = 7.194 \times 10^{-3}\epsilon_0\cos(\Omega t)$, where $\Omega = \omega_2 - \omega_1$. The modal profile of these two defect modes is shown in Fig. 48.1b, c. For maximize the transition efficiency and considering $K_1 \neq 0$ (or $K_2 \neq 0$), the modulated region covers only a part of this cavity, as shown in Fig. 48.1a.

In the simulation, when an incident signal light pulse centered at ω_1 is launched at the left end of the direct-coupled structure shown in Fig. 48.1a, a defect mode with ω_1 in the cavity will be excited. In the time-modulated system, the two defect modes will go through a periodic process of mutual conversion, where they oscillated periodically along the time. In Fig. 48.2a, when defect mode 1 reaches peak in the cavity (as point a), modulation occurs and photons start to make a transition to defect mode 2 until it reaches peak. In point b, defect mode 1 is completely converted to defect mode 2 with ω_2 except little part of the energy leak. When defect mode 2 reaches peak, it starts to convert to defect mode 1.

Figure 48.2a shows that there is excellent agreement between numerical simulation and the theory. In the theory, through the analysis of the model structure, all parameters can be determined accurately.

Next, we show dynamic stored light by photonic transition. Figure 48.3 shows the temporal evolutions in three stages of storing and releasing light. Under the requirements of high speed optical information processing, a defect mode $\omega_1 = 0.2704(2\pi c/a)$ with $Q = 17{,}821$ in the cavity as an incident signal light with relatively high bandwidth can high efficiently enter into cavity, comparing with $\omega_2 = 0.2608(2\pi c/a)$ with $Q = 146{,}225$. The incident light is a 45 ps gaussian pulse. Figure 48.3a show that if no dynamic tuning of cavity happens, mode ω_1 follows an exponential delay curve after energy of mode ω_1 reaches peak in cavity.

The first tuning occurs when energy of resonance mode 1 reaches peak in cavity, as shown Fig. 48.2a. At this moment, the energy of resonance mode ω_2 start to increase from mode ω_1. To store the light, it can stop to tune when energy of resonance mode 2 and mode 1 reaches peak and bottom respectively. Since the long photon lifetime

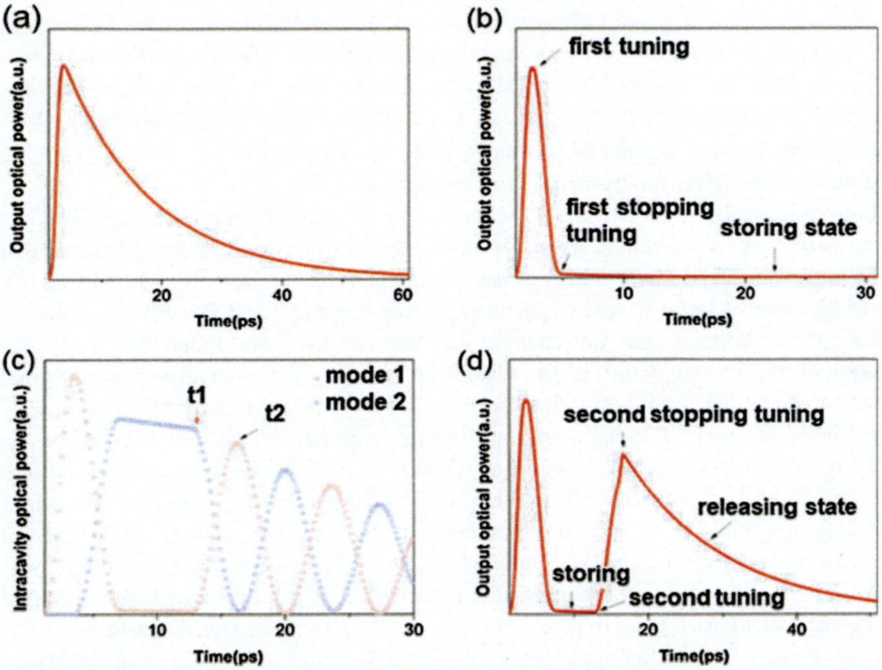

Fig. 48.3 **a** Output optical power without dynamic tuning. **b** First turning and then stopping to realize storing. **c, d** Releasing at t_2 after a given delay

of high Q cavity is an efficient way to store energy, this state of the cavity is used as a storing state. The time domain spectra of output of this state is shown in Fig. 48.3b, where the timing of stopping tune is marked by an arrow. It shows that output optical power (resonance mode ω_1) drop to zero once the tune stops since the energy of mode ω_1 is completely converted to the mode ω_2 except a bit of leaking, which is stored in the cavity. Comparing with [3], where it is sensitive to the intensity of the modulation and need to refresh the control pulse if a longer storage time is required, here this method of storing is determined by mode ω_2, which plays a key role in storage time. The higher Q factor of the mode ω_2, the longer storage time will be.

To release the energy from cavity, we can change the cavity from stored state to releasing state by using time modulated system once again. In order to release light, for example, we can choose t_1 as second tuning time and then the energy of mode ω_2 start to convert to that of mode ω_1, as shown Fig. 48.3c, where the timing of second tuning is marked by a red arrow. When the optical power of mode ω_2 reaches peak, cavity can become releasing state by stopping the second modulation (the timing of stopping is t_2 which is marked by black arrow as shown Fig. 48.3c). Figure 48.3d shows that output optical power of a released pulse appears after the second stopping modulation occurs. It can be seen that the timing of released pulse is completely determined by the timing of stopping modulation. The time interval, between the second tuning and second stopping modulation, is determined by the modulation strength $\frac{\Delta\epsilon(r)}{\epsilon_0}$. From (48.3), we can know that when the $\frac{\Delta\epsilon(r)}{\epsilon_0}$ double, the oscillation period T will be cut down a half, as shown Fig. 48.2b. Note that in Fig. 48.3d, the peak power of released light drops due to quality factor Q of the resonance mode ω_2 is not large enough and decay factor of mutual conversion ($e^{\frac{-(\upsilon_2+\upsilon_1)}{2}}$). Note that these losses can be overcome by optimizing the optical structure.

In fact, quality factor $Q = \frac{\omega_0}{\Delta\upsilon}$, where ω_0 is the system resonance and $\Delta\upsilon$ is the bandwidth of resonance, is inversely proportional to the bandwidth. It means that the narrower the bandwidth of resonance, the longer the storage time. In (48.2), (6), and (7), we can know that when the modulation stops, C_1 and C_2 will be a constant since the amplitude does not change over time. It has been understood that, after modulation, the bandwidth of the resonance is same as the corresponding original bandwidth before modulation. In numerical calculation, the bandwidth of ω_1 and ω_2 is 0.068 THz and 0.008 THz without tuning, respectively. We find that, when the cavity is stored state, the linewidth of light ω_1 with 0.068 THz is compressed into that of light ω_2 with 0.008 THz. When the cavity is releasing state, the line width of light ω_2 with 0.008 THz is expanded into that of light ω_1 with 0.068 THz. It means the storage time can overcome the traditional delay-bandwidth product limit[3] by photonic transition. In other word, it is a method of manipulating bandwidth that linewidth of light can be compressed and expanded by photonic transitions.

In recent years, inducing photonic transitions is achieved between discrete modes in a silicon cavity, by using ultrafast tuning of the refractive index in a time[10]. The time is comparable to inverse of the frequency interval of adjacent modes. It is an optical controlling method that free carriers are produced by illuminating the top of the cavity with pump pulse. In addition, there is an electrically controlling optical waveguide consisting of pn-junction, to realize nonreciprocal transmission under temporal refractive-index modulation by applying sinusoidal voltage V. With fabricating p-i-n junction integrated photonic crystal nanocavity and applying electrical signal, refractive index modulation of nanocavity can be achieved [14–17]. Our theory can be realized in a similar system. Appropriately designing nanocavity can meet the modulation frequency that can be achieved in electro-optic modulation which is usually on the order of 10 GHz [18].

48.4 Conclusions

In summary, we have shown that in modulated photonic crystal cavity, mutual conversion occurs between defect mode1 with low Q and defect mode 2 with high Q, which can be used to releasing state and stored state. It is an efficient way to realize stored light dynamically. When the cavity is stored state, the storage time is determined by the linewidth of light being compressed. The degree of being compressed is determined by linewidth of defect mode 2. If the resonator can be designed appropriately, we can obtain ultrahigh-Q defect mode and low-Q defect mode as stored state and releasing state, respectively. It means that this structure can realize ultralong storage time and ultrafast releasing, such that meeting high speed optical information processing. For this work, it points to propose a new method to dynamically control Q factor and manipulate the output spectrum from the nanocavity.

Acknowledgements This research was financially supported by the National Natural Science Foundation of China (11774098); Guangdong Natural Science Foundation (2017A030313016); and Science and Technology Program of Guangzhou (202002030500).

References

1. B.-S. Song, S. Noda, T. Asano, Y. Akahane, Ultra-high-Q photonic double-heterostructure nanocavity. Nat. Mater. **4**, 207–210 (2005)
2. C. Li, M. Wang, J.-F. Wu, Broad-bandwidth, reversible, and high-contrast-ratio optical diode. Opt. Lett. **42**, 334 (2017)
3. Xu., Qianfan, Po., Dong, M. Lipson, Breaking the delay-bandwidth limit in a photonic structure. Nat. Phys. **3**, 406–410 (2007)
4. C. Li, S.-Y. Wu, J.-F. Wu, Broad-bandwidth, reversible nonreciprocal light transmission based on a single nanocavity. Opt. Express **27**, 16530 (2019)
5. T. Tanabe, M. Notomi, H. Taniyama, E. Kuramochi, Dynamic release of trapped light from an ultrahigh-Q nanocavity via adiabatic frequency tuning. Phys. Rev. Lett. **102**, 043907 (2009)
6. Q. Xu, B. Schmidt, S. Pradhan, M. Lipson, Micrometre-scale silicon electro-optic modulator. Nature **435**, 325–327 (2005)
7. C.T. Phare, Y.H. Daniel Lee, J. Cardenas, M. Lipson, Graphene electro-optic modulator with 30 GHz bandwidth. Nat. Photon. **9,** 511–514 (2015)
8. Z. Yu, S. Fan, Complete optical isolation created by indirect interband photonic transitions. Nat. Photon. **3**, 91–94 (2009)
9. H. Lira, Z. Yu, S. Fan, M. Lipson, Electrically driven nonreciprocity induced by interband photonic transition on a silicon chip. Phys. Rev. Lett. **109**, 033901 (2012)
10. P. Dong, S.F. Preble, J.T. Robinson, S. Manipatruni, M. Lipson, Inducing photonic transitions between discrete modes in a silicon optical microcavity. Phys. Rev. Lett. **100**, 033904 (2008)
11. K. Fang, Yu., Zongfu, S. Fan, Realizing effective magnetic field for photons by controlling the phase of dynamic modulation. Nat. Photon. **6**, 782–787 (2012)
12. L.D. Tzuang, K. Fang, P. Nussenzveig, S. Fan, M. Lipson, Non-reciprocal phase shift induced by an effective magnetic flux for light. Nat. Photon. **8**, 701–705 (2014)
13. T. Tanabe, K. Nishiguchi, E. Kuramochi, M. Notomi, Opt. Express **17**, 22505 (2009)
14. T. Tanabe, E. Kuramochi, H. Taniyama, M. Notomi, Electro-optic adiabatic wavelength shifting and Q switching demonstrated using a p-i-n integrated photonic crystal nanocavity. Opt. Lett. **35**, 3895–3897 (2010)
15. K.L. Tsakmakidis, L. Shen, S.A. Schulz, X. Zheng, J. Upham, X. Deng, H. Altug, A.F. Vakakis, R.W. Boyd, Breaking Lorentz reciprocity to overcome the time-bandwidth limit in physics and engineering. Science **356**, 1260–1264 (2017)
16. M.F. Yanik, W. Suh, Z. Wang, S. Fan, Stopping light in a waveguide with an all-optical analog of electromagnetically induced transparency. Phys. Rev. Lett. **93**, 233903 (2004)
17. M.F. Yanik, S. Fan, Stopping light all-optically. Phys. Rev. Lett. **92**, 083901 (2004)
18. G.T. Reed, G.Z. Mashanovich, F.Y. Gardes, M. Nedeljkovic, Y. Hu, D.J. Thomson, K. Li, P.R. Wilson, S.-W. Chen, S.S. Hsu, Recent breakthroughs in carrier depletion based silicon optical modulators. Nanophotonics **3**(4), 229–245 (2013)

Chapter 49
Investigation About Large Capacity Optical Transmission System Based on Wavelength Division Multiplexing and Multiplexing Division Multiplexing

Zhuolun Song

Abstract With the rapid development of internet and the emergence of new internet applications such as clouding computing, big data, and 5G, the network capacity of optical fiber communication system is increasing exponentially. It is necessary to study large optical transmission technology to satisfy the capacity requirements. In this paper, the key technologies of large-capacity optical fiber communication systems based on single-mode fiber (SMF) and few-mode fiber (FMF) are studied. First, wavelength division multiplexing (WDM) transmission technology based on the optical frequency comb in SMF is introduced, and four implementation schemes of the optical frequency comb are discussed. Next, mode division multiplexing (MDM) based on FMF is introduced to overcome the nonlinear Shannon limit of SMF. The key technologies to realize mode multiplexing and demultiplexing are introduced, including free-space optics scheme, the all-fiber mode multiplexing scheme and the silicon-based integrated mode multiplexing scheme.

49.1 Introduction

With the emergence of new internet services such as enterprise networks, data centers, cloud computing, high-definition video, virtual reality, mobile communications, and the Internet of Things, the demand for optical fiber communication network capacity continues to grow exponentially [1]. According to a research report [2] released by Bell Labs in 2015, the actual growth rate of optical fiber capacity has been roughly maintained at around 80% since the 1990s. After 2002, the capacity growth rate has slowed down and stabilized at around 20%. At present, although the optical communication system based on single mode fiber (SMF) can still provide bandwidth far exceeding actual requirements, the bandwidth margin provided by the SMF-based transmission system gradually decreases, and the capacity crisis gradually appears [1].

Z. Song (✉)
Qingdao Foreign Language School, Qingdao 266012, Shandong, China
e-mail: jwwthu@163.com; 1901110384@pku.edu.cn

G. Liu and F. Cen (eds.), *Advances in Precision Instruments and Optical Engineering*, Springer Proceedings in Physics 270, https://doi.org/10.1007/978-981-16-7258-3_49

In the past 20 years, wavelength division multiplexing (WDM) technology, high-order modulation formats, polarization division multiplexing, coherent detection, and digital coherent receiving algorithms have been utilized to increase the capacity of optical fiber communication systems [3, 4]. At present, the maximum capacity of SMF has reached 100 Tb/s, and the spectral efficiency is as high as 11b/s/Hz [5], approaching the nonlinear Shannon limit of SMF. For SMF, multiple degrees of freedom such as wavelength, amplitude, and phase have been fully utilized in WDM coherent optical communication systems, and it cannot satisfy the growing demand for traffic. Therefore, it is necessary to explore other degrees of freedom to further increase the communication capacity of a single optical fiber.

Space division multiplexing is a multiplexing technology with a new degree of freedom (mode and core), to further increase transmission capacity [6, 7]. Space division multiplexing has been discussed since the 1980s. However, due to the limitations of early technology, the project has only gradually been paid attention in recent years. Mode division multiplexing (MDM) is one of the main implementations of space division multiplexing. The orthogonal modes in the few-mode fiber (FMF) or multi-mode fiber (MMF) can be used as an independent transmission channel for data transmission so that the transmission capacity of the fiber can be doubled.

In this paper, the key technologies of large-capacity optical fiber communication systems are studied. First, the WDM transmission system based on the optical frequency comb is introduced, including the system architecture and the implementation scheme of the optical frequency comb. Next, the key technologies to realize mode multiplexing and demultiplexing are discussed.

49.2 Optical Frequency Comb-Based WDM System

In optical fiber communication systems, WDM is a technology in which a group of light waves of different wavelengths carrying data information are coupled into the same optical fiber through a multiplexer so that the transmission capacity of a single optical fiber can be increased. For common WDM systems, many independent lasers must be used to provide optical carriers. At the same time, a complete WDM system must meet the following points: one is to ensure that the center frequency and wavelength space of the system are stable; the other is to ensure that the number of channels in the system can be expanded as the number of end users increases; third is to ensure that the power flatness of all channel carriers is better to ensure the sensitivity of the system receiver; fourth is to ensure that the channels of the system can be dynamically adjusted as the state of mutual interference between channels changes Interval.

In traditional WDM transmission systems, distributed feedback laser (DFB) laser arrays are often used as laser carrier sources. However, the laser array will have wavelength drift due to temperature, and wavelength management will be complicated. The optical frequency comb is composed of a series of spectral lines with constant mode spacing, ultra-low phase noise, and equidistant frequency space. A

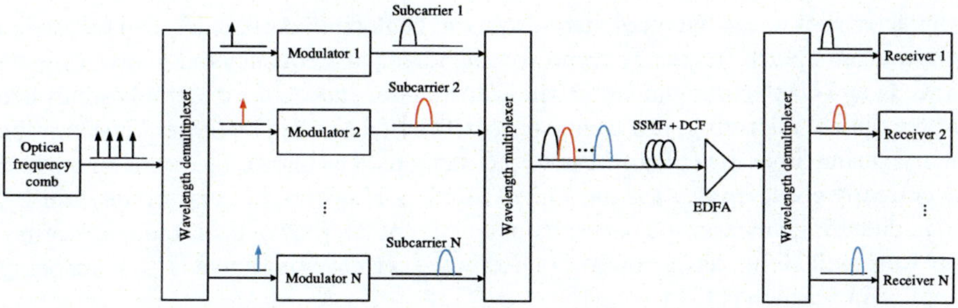

Fig. 49.1 Schematic diagram of WDM transmission system based on optical frequency comb

stable optical frequency comb can not only meet the requirements of frequency stability between various optical carriers but also minimize the required guard interval between adjacent transmission optical wavelengths, which cannot be met by independent lasers. Therefore, the optical frequency comb can be used as an ideal light source for high-speed optical transmission systems. The schematic diagram of WDM system with optical frequency comb is illustrated in Fig. 49.1. In the coherent communication system, the optical frequency comb can be used not only as lasers at the transmitting end but also as local oscillators at the receiving end for signal coherent detection. Next, the typical methods to generate optical frequency comb are discussed.

49.2.1 Mode-Locked Laser Method

In mode-locked laser method, the optical frequency comb was originally produced by a mode-locked laser. The phase relationship of the different modes is locked in the laser resonant cavity. The result observed from the time axis is a periodic pulse sequence [8], and the distance in the frequency domain can be obtained by Fourier transform. The advantage of this scheme is that the structure is simple and the number of generated carriers is large. However, the volume and cost of traditional femtosecond-level mode-locked lasers are very high. Moreover, it is difficult to control the number of carriers, poor flatness and carrier spacing, which limits the wide application in optical communication networks [9].

49.2.2 Microring Resonator Method

In this method, Kerr effect of the optical microring resonator is used to generate an optical frequency comb [10]. Kerr effect is a third-order nonlinear effect in optical media, which can change the transmission characteristics of light in the media. The achievable bandwidth of the Kerr optical frequency comb depends on the dispersion characteristics of the microresonator, so it can be flexibly designed. It can be used to generate an optical frequency comb with hundreds of frequency tones, covering the S, C, L, and U telecommunication bands. In addition, due to the extremely small size of the optical microring, if it can be successfully fabricated, its power consumption will be many times lower than the previous methods. The microring resonator method is currently considered as the most ideal solution. However, because of the stability, manufacturing process, and noise characteristics of the Kerr optical frequency comb, the Kerr optical frequency comb obtained in the current experiment is far from being applied in practice [11].

49.2.3 Cyclic Frequency Shift Method

The basic principle of cyclic frequency shift method is to use a Recirculating Frequency Shifter (RFS) to generate an optical frequency comb. The common RFS includes single-sideband RFS and multiple-path RFS. In addition, RFS can be combined with phase modulator (PM) to generate optical frequency combs. As shown in Fig. 49.2, the RFS contains a closed optical fiber loop, a tunable optical bandpass filter, an IQ modulator, and two optical amplifiers. The band pass filter is used to control the number of sidebands generated, and the optical amplifier is used to compensate for the loss of frequency conversion. Start by inputting an optical sideband with a center frequency of f0. After N cycles, optical sidebands with center frequencies of f1, f2, … fn are generated. Each cycle, the coupler divides the fi sideband into two parts, one part outputs RFS, and the other part returns to the input of the optical IQ modulator [12]. The advantage of using RFS to generate an optical frequency comb is that the flatness is better, and the more the number of cycles, the more spectral lines can be generated, but this method makes the optical frequency comb have the disadvantages of insignificant carrier phase relationship and greater noise.

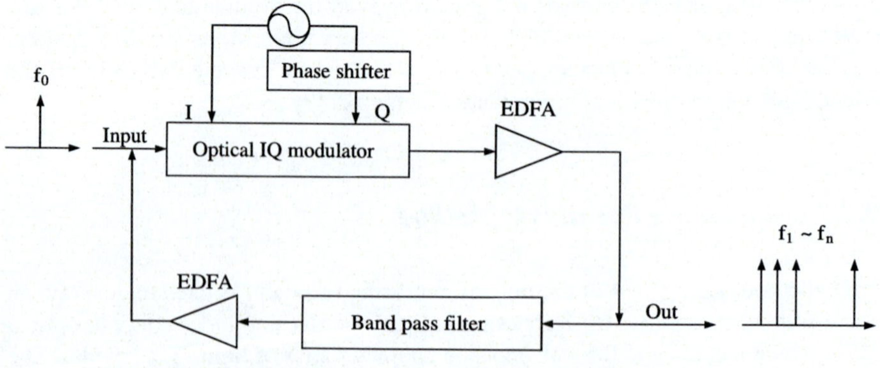

Fig. 49.2 The schematic diagram of cyclic frequency shift method

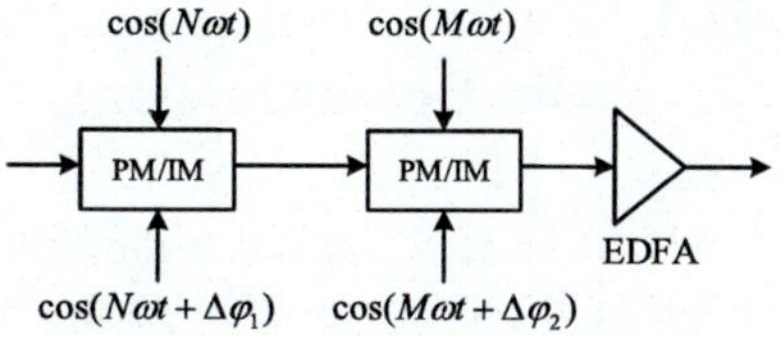

Fig. 49.3 The schematic diagram of cascaded PM and IM to generate optical frequency comb

49.2.4 Electro-Optical Modulator Method

This method is relatively easy to implement and is currently a widely used optical frequency comb generation scheme. Common schemes used for optical frequency combs include an intensity modulator (IM), a PM, and a cascade scheme of them. The advantage of using a phase modulator is that it can output a stable frequency comb without controlling the modulation bias voltage. The advantage of using an intensity modulator is that the bias voltage can be controlled to generate 2 or 3 flat photonic carriers for the optical comb. The signal is further refined. By cascading the two (shown in Fig. 49.3), multiple flat optical carriers can be generated.

49.2.5 Research Progress of WDM Transmission System Based on Optical Frequency Comb

D. Hillerkuss et al. reported a mode-locked fiber laser to generate a frequency comb with 325 combs, based on which, a 16QAM-OFDM signal with a transmission rate of 26Tbit/s was transmitted 50 km in a SMF [13]. In the case of no dispersion compensation, the 16QAM Nyquist-WDM signal with a transmission rate of 32.5Tbit/s was transmitted 227 km in a SMF, and the spectral efficiency reaches 6.4bit/s/Hz [14]. Yi et al. used an optical frequency comb generated by a conventional electro-optical modulator to seamlessly transmit an OFDM 16QAM signal with an optical bandwidth of 318 GHz and 1.008 Tb/s [15]. Puttnam et al. demonstrated the generation of 2.15Pb/s PDM-64QAM signal utilizing the optical frequency comb generated by the electro-optical modulator [16]. In terms of chip-level optical frequency combs, Weimann et al. based on Silicon-organic hybrid (SOH) generated 9 optical frequency combs with a channel spacing of 25 GHz, and transmitted 1.008 Tb/s Nyquist-WDM signals for 300 km in optical fibers [17]. T. Shao et al. used Gain-switched laser to generate 24 optical frequency tones with a channel spacing of 12.5 GHz, and transmitted 1.876 Tb/s Nyquist-WDM signals over 300 km in optical fiber [18] (Vidak et al.) The Quantum-dash Passively Mode-Locked Lasers method was used for the first time to generate 80 optical frequency combs with a channel spacing of 22.7 GHz, and the 2.256 Tb/s 16-QAM OFDM signal was transmitted in the optical fiber for 3 km.

49.3 Key Technologies of Mode Multiplexing and Demultiplexing

In MDM system, the mode multiplexer/demultiplexer is the most critical part. The mode multiplexer can convert the fundamental mode in the single-mode fiber and couple it to a plurality of different high-order modes in the few-mode fiber to realize mode multiplexing. The demultiplexer has a dual function and converts different modes into the fundamental mode in a single-mode fiber. This article gives a detailed introduction to the current mainstream mode multiplexer/demultiplexer implementation schemes.

49.3.1 Free Space Optics Scheme

Free space optics mode multiplexer is a mode multiplexer based on space matching. It is generally composed of lens group, optical splitter/combiner, and mode converter. It has high mode isolation and good mode control, which is widely used in various MDM experimental systems. According to the different key equipment for mode conversion, free space optical multiplexers are mainly divided into spatial light modulator (SLM)-based mode multiplexer and phase plate-based mode multiplexer [19–21].

Figure 49.4 illustrates the schematic diagram of phase plate-based mode multiplexer. Spatial optical elements such as lenses, phase plates, and light splitting prisms are adopted. Phase plates with different mask modes are used to convert the fundamental mode to different higher-order modes. These modes are combined together by optical beam splitters. However, due to the transflective nature of the optical beam splitter, a 3 dB loss will occur every time the beam passes through the beam splitter, and the input light of port 0 and port 1 will lose 6 dB of optical power.

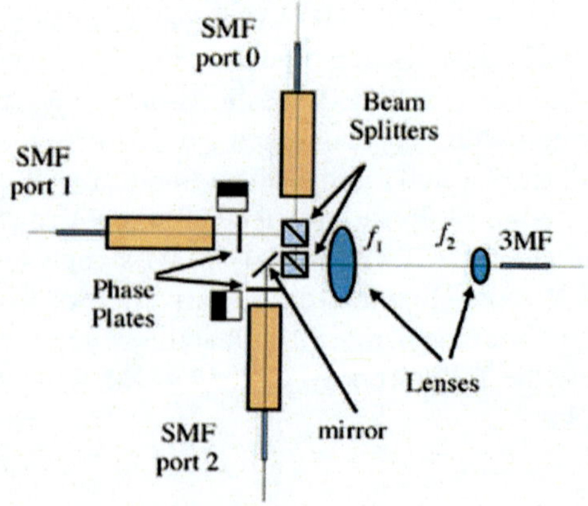

Fig. 49.4 Schematic setup of the six-mode FMF mode multiplexer based on phase plates

The mode multiplexer and demultiplexer based on the SLM can control the performance of the LCOS element through computer programming. It can produce different phase modulations on the fundamental mode beam, thereby obtaining different high-order mode fields. It can also be superimposed with blazed grating phase plate as a diffraction device [21]. Different modes can be generated at the same time through one LCOS, and the system recombination after LCOS greatly reduces the complexity of the system and improves the crosstalk.

49.3.2 All-Fiber Scheme

All-fiber mode multiplexing and demultiplexing are mainly divided into fiber coupler type and photon lantern [22, 23]. There are two main implementation schemes for fiber coupler-type mode multiplexing and demultiplexing: one is few-mode coupler, and the other is mode-selective coupler (MSC). Both of these couplers can be prepared by the fusion taper process. The difference is that the few-mode coupler cannot directly realize mode conversion and a mode converter, such as a long-period fiber grating (LPFBG) is required to be added to the front. While the MSC can directly perform mode conversion. Both couplers can multiplex multiple modes by cascading. Ismaee et al. made MSC and used it in MDM system to achieve selective excitation of high-order modes. The efficiency and purity of the mode conversion are relatively high.

The photon lantern type multiplexer/demultiplexer has been extensively studied because of its simple structure and high efficiency in realizing mode conversion and mode coupling and can greatly improve the integration of the device. In real life, the common photon lantern multiplexer is made of optical fiber fusion taper. The structure of the photon lantern is shown in Fig. 49.5. The photon lantern can also be used as a demultiplexer. After the multiplexed light in the few-mode fiber passes through the photon lantern again, the different high-order modes are converted into the fundamental mode and coupled to the different receivers.

49.3.3 Silicon-Based Integration Scheme

Common silicon-based integrated mode multiplexing schemes are divided into mode multiplexers based on directional coupling, multimode interference coupling mode multiplexers, Y-branch directional couplers, and mode multiplexers based on microring resonator (MRR) [23–26]. The asymmetric directional coupler (ADC) and MRR are mostly investigated in the experiment. ADC has asymmetric waveguide structure. The two coupled waveguides of the coupler are required to meet the phase matching condition. When the effective refractive index of the fundamental mode in the single-mode waveguide and the higher-order mode in the multi-mode waveguide are equal, the phase matching condition is satisfied, and the mode conversion

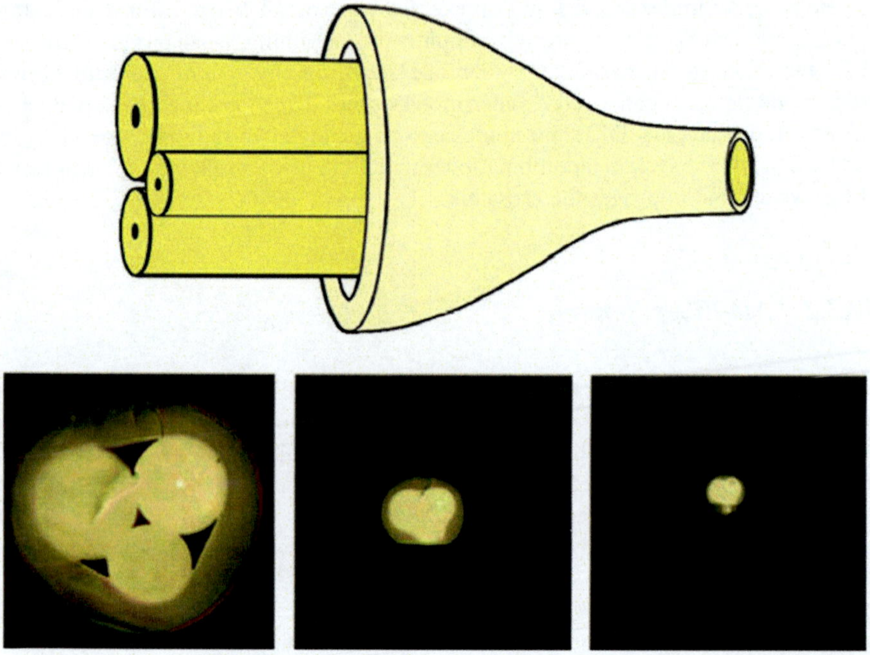

Fig. 49.5 Schematic diagram of photon lantern

from the fundamental mode to the higher-order mode is realized. The MRR structure introduces a micro-ring structure in the asymmetric waveguide, which is equivalent to adding a resonance mechanism. Such a change increases not only the flexibility of the device but also because the MRR's requirements for the coupling length are not as accurate as the ADC, which improves the complexity of the process and is more suitable for WDM systems.

49.3.4 Research Progress of MDM Transmission System

The optical frequency comb-based WDM transmission technology can also be applied to MDM systems. Recently, there are also some reports about large capacity MDM transmission experiment using optical frequency comb. G. Rademacher combined MDM and PDM through the PDM-16QAM modulation format and realized $3 \times 348 \times 24.5$G bandwidth signal transmission on 1045 km of three-mode fiber, with a maximum rate of 159Tbit/s [27]; Daiki Soma, Yuta Wakayama et al. achieved a transmission rate of 10.16 Pb/s on 11.3 km of 6-mode 19-core few-mode fiber. This amazing rate is also the highest transmission rate reported for a MDM system. Rate [28]; Shibahara, Mizuno et al. conducted theoretical research and transmission experiments on a 3-mode 12-core MC-FMF with a length of 2500 km and a 3-mode few-mode fiber with a length of 6300 km. The results show the loss between modes (Mode-dependent loss, MDL) In the weakly coupled fiber that grows linearly with the fiber length, through the new transmission scheme and hierarchical MIMO design, the transmission distance of the mode division multiplexing system based on the few-mode fiber is sufficient to realize transoceanic transmission [29]; Rademacher et al. achieved single-mode 93.34 Tb/s transmission on a 30 km three-mode graded few-mode fiber, which further proved the huge expansion potential of mode multiplexing technology [30].

49.4 Summary

In this paper, the key technologies of large-capacity optical fiber communication systems are studied. First, the WDM transmission system based on the optical frequency comb is introduced and four implementation schemes of the optical frequency comb are discussed, including the mode-locked laser method, the micro-ring resonator method, the cyclic frequency shift method, and the electro-optic modulator method. The electro-optic modulator method is simple to implement and is widely used in large-capacity WDM transmission systems. In the future, integrated optical frequency combs are a promising development direction. Next, the mode multiplexing and demultiplexing technology used to realize the mode division multiplexing is introduced, including free-space optics scheme, the all-fiber mode multiplexing scheme and the silicon-based integrated mode multiplexing scheme. Both the free-space optics scheme and the all-fiber mode multiplexing scheme can be applied to the multiplexing system of multimode fiber, and also used in the multiplexing system of few mode fiber. However, free-space optics scheme requires complex optical platforms and high-precision optical equipment. Although free-space optical multiplexers have advantages in mode selectivity, mode purity, and mode crosstalk, they are difficult to use in non-laboratory environments. Due to the mismatch between the mode field in the silicon-based waveguide and the mode field in the optical fiber, the silicon-based integrated mode multiplexing scheme has not yet been applied to the optical fiber transmission system, and further research is needed.

References

1. Chralyvy, Plenary paper: the coming capacity crunch. ECOC 1–2 (2009)
2. P.J. Winzer, Scaling optical fiber networks: challenges and solutions. Opt. Photonics News **26**(3), 28–35 (2015)
3. T. Li, The impact of optical amplifiers on long-distance lightwave telecommunications. Proc. IEEE **81**(11), 1568–1579 (1993)
4. R. Giles et al., Optical amplifiers transform long-distance lightwave telecommunications. Proc. IEEE **84**(6), 870–883 (1996)

5. D. Qian, et al., 101.7-Tb/s (370×294-Gb/s) PDM-128QAM-OFDM transmission over 3 × 55-km SSMF using pilot-based phase noise mitigation. OFC 1–3 (2011)
6. G. Li et al., Space-division multiplexing: the next frontier in optical communication. Adv. Opt. Photonics **6**(4), 5041–5046 (2014)
7. D.J. Richardson et al., Space-division multiplexing in optical fibres. Nat. Photonics **7**(5), 354–362 (2013)
8. T. Hellwig et al., Toward an all-optically stabilized frequency comb based on a mode-locked fiber laser. Opt. Lett. **39**(3), 525–527 (2014)
9. J. Davilarodriguez et al., Frequency stability of a 10 GHz optical fre-quency comb from a semiconductor-based mode-locked laser with an intracavity 10,000 finesse etalon. Opt. Lett. **38**(18), 3665–3668 (2013)
10. P. Del'Haye, et al., Octave spanning frequency comb on a chip. Physics (2009)
11. A.B. Matsko, et al., Optical hyperparametric oscillations in a whispering-gallery-mode resonator: threshold and phase diffusion. Phys. Rev. A **71**(3), 033804 (2005)
12. J. Li, et al., Analysis of the stability and operation of the single-side-band modulator based on re-circulating frequency shifter used for the T-bit/s optical communication transmission. Opt. Exress **18**(17), 17597–17609 (2010)
13. D. Hillerkuss et al., 26 Tbit s-1 line-rate super-channel transmission utilizing all-optical fast Fourier transform processing. Nat. Photonics **5**(6), 364–371 (2011)
14. D. Hillerkuss, et al., Single-laser 32.5 Tbit/s Nyquist WDM transmission. J. Opt. Communi. Netw. **4**(10), 715–723 (2012)
15. X. Yi et al., Tb/s coherent optical OFDM systems enabled by optical frequency combs. J. Lightwave Technol. **28**(14), 2054–2061 (2010)
16. B. Puttnam, et al., 2.15 Pb/s transmission using a 22 core homogeneous single-mode multi-core fiber and wideband optical comb. ECOC 1–3 (2015)
17. C. Weimann et al., Silicon-organic hybrid (SOH) frequency comb sources for terabit/s data transmission. Opt. Express **22**(3), 3629–3637 (2014)
18. T. Shao et al., Integrated gain switched comb source for 100 Gb/s WDM-SSB-DD-OFDM system. J. Lightwave Technol. **33**(17), 3525–3532 (2015)
19. R. Ryf et al., Mode-division multiplexing over 96 km of few-mode fiber using coherent 6×6 MIMO processing. J. Lightwave Technol. **30**(4), 521–531 (2011)
20. B. Franz, et al., Mode group multiplexing over graded-index multimode fiber. ICTON (2012)
21. H. Chen, et al. Single multi-mode mask for multi-channel mode division demultiplexing. OFC (2013)
22. C.P. Tsekrekos et al., Symmetric few-mode fiber couplers as the key component for broadband mode multiplexing. J. Lightwave Technol. **32**(14), 2461–2467 (2014)
23. H. Chen et al., Design constraints of photonic-lantern spatial multiplexer based on laser-inscribed 3-D waveguide technology. J. Lightwave Technol. **33**(6), 1147–1154 (2014)
24. M. Greenberg et al., Multimode add-drop multiplexing by adiabatic linearly tapered coupling. Opt. Express **13**(23), 9381–9387 (2005)
25. Y. Ding et al., On-chip two-mode division multiplexing using tapered directional coupler-based mode multiplexer and demultiplexer. Opt. Express **21**(8), 10376–10382 (2013)
26. D. Dai et al., Silicon mode (de)multiplexer enabling high capacity photonic networks-on-chip with a single-wavelength-carrier light. Opt. Lett. **38**(9), 1422–1424 (2013)
27. G. Rademacher, et al., 159 Tbit/s C+L Band Transmission over 1045 km 3-Mode Graded-Inde Few-Mode Fiber. OFC (2018), paper Th4C.4
28. D. Soma et al., 10.16-Peta-B/s dense SDM/WDM transmission over 6-mode 19-core fiber across the C+L band. J. Lightwave Technol. **36**(6), 1362–1368 (2018)
29. K. Shibahara, et al., DMD-unmanaged long-haul SDM transmission over 2500-km 12-core × 3-mode MC-FMF and 6300-km 3-mode FMF employing intermodal interference cancelling technique. OFC (2018), paper Th4C.6
30. G. Rademacher, et al., 93.34 Tbit/s/mode (280 Tbit/s) transmission in a 3-mode graded-index few-mode fiber. OFC (2018), paper W4C.3

Chapter 50
Manipulating Photons with a Dynamic Nanocavity

Yuan-Bao Zhang, Jia-Hui Chen, Chao Li, and Jun-Fang Wu

Abstract We present a simple yet effective approach to realize a movable nanocavity based on dynamically tuning the local refractive index of a photonic crystal (PC) waveguide. Using this movable cavity, we demonstrate that it can trap, store, and transfer signal photons. The approach proposed here provides a new way for manipulating photons.

50.1 Introduction

At present, there are already many methods to form high-quality factor (Q) microcavities which can strongly confine photons in a volume in optical wavelength scale, and their implementation principles are basically achieved by partially modifying the mode gap to achieve the purpose of light confinement. These traditional high-Q microcavities are mostly realized by structure modifying like locally changing the crystal lattice [1, 2] or moving some air holes to modify the width of the line defects [3–7]. These irreversible realization methods determine that the position of their microcavities are fixed. We proposed a new method for realizing high-Q microcavities which can be formed instantly at any time and at any position in the waveguide, have a high Q, and can be moved at will. We also applicate this method to optical storage. Traditional optical storage structures based on high-Q microcavities are passive that the process of light entering the microcavity requires complex Q-switching technology (first enter when the structure is adjusted to low Q, and then store when adjusted to high Q). In addition, when these structures try to release stored signal light, the light will travel along the waveguide at both ends of the microcavity, which is said that the direction cannot be controlled. The optical storage structure we designed base on our new method of high-Q microcavity implementation is active. It can grab any part of signal light at any time during the light travel process and achieve efficient storage without Q-switching. At the same time, in the light release process,

Y.-B. Zhang · J.-H. Chen · C. Li · J.-F. Wu (✉)
School of Physics and Optoelectronic Technology, South China University of Technology, Guangzhou 510640, China
e-mail: wujf@scut.edu.cn

G. Liu and F. Cen (eds.), *Advances in Precision Instruments and Optical Engineering*, Springer Proceedings in Physics 270, https://doi.org/10.1007/978-981-16-7258-3_50

our structure can make the stored signal light transmit in the specified direction with only a simple operation other that a complex Q-switching one. In addition, unlike the traditional optical storage structure that light is trapped in a static microcavity, we can move the captured signal light to any desired position by moving the high-Q microcavity, and the moving speed can be arbitrary controlled, which is more flexible than traditional slow light waveguides [8–11].

In order to realize light storage and light release, we designed a new structure. Our structure is based on the W1 line-defect waveguide in a 2D triangular air-hole dielectric photonic crystal slab (PCS) which is shown in Fig. 50.1a. We reduce the refractive index of dielectric in the two rectangular shaded areas shown in Fig. 50.1b, while the refractive index of the air hole remains unchanged. By this way, we formed two potential barriers like what is shown in Fig. 50.1d through locally modify the mode gap so that the waveguide segment clamped by these two tuned areas, due to the presence of mirrors at both ends, forms an Fabry–Pérot (F–P) cavity. We can further understand the realization principle of forming a microcavity through two refractive index reduced regions through the calculated band structure plotted in Fig. 50.1c. Light waves that can originally propagate in the waveguide will stay in the waveguide segment clamped by the two refractive index reduced regions for the reason that the local refractive index reducing locally modified the mode gap and formed two potential barriers, therefore the light wave in the waveguide segment can exist as the resonant mode of the formed FP cavity.

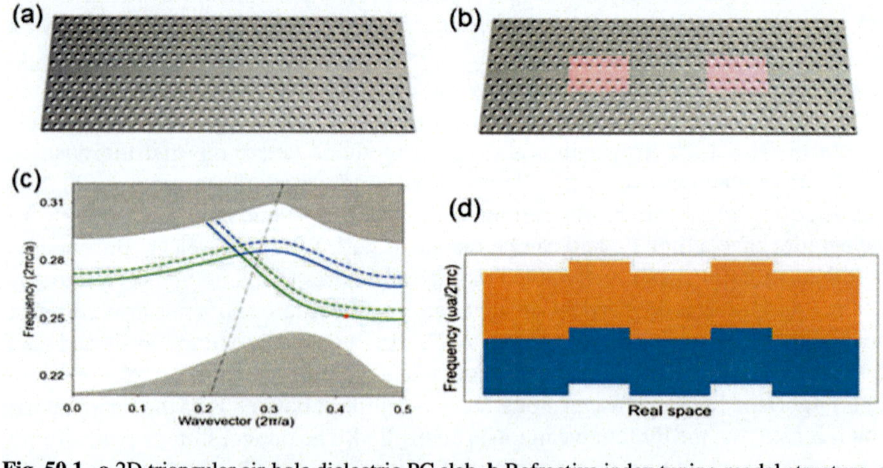

Fig. 50.1 **a** 2D triangular air-hole dielectric PC slab. **b** Refractive index tuning model structure. **c** Calculated band structure. **d** Schematic diagram along the waveguide direction

50.2 Model and Analysis

To calculate the Q and resonant wavelength of formed microcavity of our structure, we use finite-difference time-domain (FDTD) [12] to simulate our structure. The index of dielectric slab is 3.46, and the effective refractive index n used in our 2D-FDTD is 2.8. Lattice constant a, air hole radius r, and thickness t of the slab are 420 nm, 110 nm, and 210 nm respectively. For the W1 line defect used, the central distance of the nearest-neighbor air holes at both sides of line defect is $\sqrt{3}a$. The refractive index of the dielectric material in two rectangular shaded areas is $0.98n$, the length and width are 6a and 4a respectively, and the central distance of the two areas is $14a$. From the numerical calculation, we get the Q of the microcavity is 8×10^5, and the resonant wavelength is 1.67014 μm.

Compared with other mentioned methods of forming high-Q microcavities, our method does not change the structure, which means that this modulation is reversible, suggesting that we can not only stick to a static modulation approach, a dynamic one also can be used, that is, the refractive index of the dielectric material in the two shaded areas changes as a function of time, rather than constantly decreasing. We will show later that through such dynamic refractive index modulation, the structure we designed will realize the storage, release, and movement of light, which are all important parts of optical information processing.

We already know that by reducing the refractive index of dielectric material in two areas to clamp a segment of waveguide, a high-Q nanocavity can be formed between the two tuned areas. Now, we hope to make use of this, to capture light waves through dynamic tuning, and to use the formed high-Q microcavity to achieve optical storage. Figure 50.2a is the schematic diagram of our study in this section. In order to clearly see the process of a light pulse propagating and storing in our structure, we used a much longer structure in this study. The length of the structure is $87a$, and other parameters are exactly the same as the previous model structure used in Q and resonant frequency calculation. A long light pulse with a wavelength of 1.67014 μm, which is exactly the resonant frequency of the formed microcavity measured before, injected from the left end of the waveguide and propagates toward the right end. When $t = 14$ ps, the light pulse arrives the center of the waveguide, also the center of the waveguide segment clamped by these two rectangular shaded areas, the two shaded areas are modulated to decrease the refractive index of their dielectric material by 2% steeply so that their refractive index becomes $0.98n$. The Hy distribution snapshot shown in Fig. 50.2b–h shows the complete process, which begins with the injection of the light pulse and ends with storing of the light pulse. After the two shaded areas are modulated to decrease the refractive index, most of the light intensity is captured by the two modulation areas, staying at the center of the waveguide. After a long time, we can still see the light staying here. It clearly shows that we have achieved optical storage by dynamically reducing the refractive index of the two regions to clamp a segment of waveguide.

We monitored the magnetic field intensity in the center of the waveguide and plotted as in Fig. 50.3, where $t = 14$ ps is the moment we start to reduce the refractive

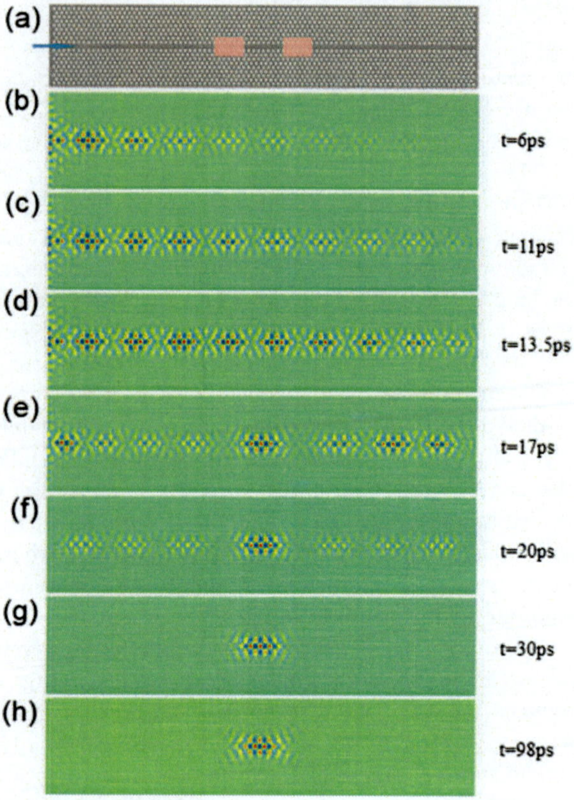

Fig. 50.2 Snapshots of field distribution in optical storage process

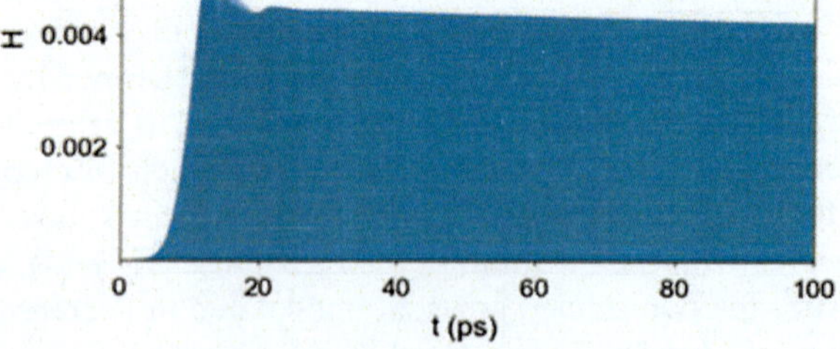

Fig. 50.3 Magnetic field in dynamic cavity as a function of time

index of the two shaded areas, which can be found in Fig. 50.3 as the moment the magnetic field intensity fluctuates sharply. It can be seen that most of the intensity of injected light is captured by the two regions and continually stay at the center of the waveguide, and for the reason that the captured light intensity exists as the resonant mode of the high-Q microcavity, its decay rate in the microcavity is very slow.

We have shown the optical storage process of our structure above. It is hard to applicate a storage structure if what been stored have not a way to be controllably released. So, in the following, we study the light release process of our structure. Continuing the optical storage process above, which we have performed dynamic modulation of the refractive index reduction on two rectangular regions at $t = 14$ ps, so that the light intensity stays between the two regions. Next, at $t = 60$ ps, we remove the refractive index modulation on the right shadow area, that is, make the refractive index of the dielectric material in the right rectangular shadow area recovers from $0.98n$ to n. Meantime, refractive index of the dielectric material in the shaded area on the left remains unrecovered and is still $0.98n$. From what have been discussed above, we already know that the light intensity can be confined between the two regions with reduction of refractive index. Then, when one of the two regions (here is the right side one) which have been reduced refractive index recover from tuning, the constraint on the light wave that originally existed on the right side of the waveguide segment will disappear so that the light wave will propagate toward the direction which the constraint is disappeared, and finally leave structure from the end of waveguide.

From the field distribution snapshots in Fig. 50.4a–e, it can be seen that the light wave that originally stored between the two tuned areas moves to the right along the waveguide after the refractive index modulation in right side shadow area is removed at $t = 60$ ps and finally leaves waveguide from the end of right side. We also monitored the magnetic field intensity at the right end of the waveguide and plot as in Fig. 50.5. It can be seen from Fig. 50.5 that there is an obvious peak at about

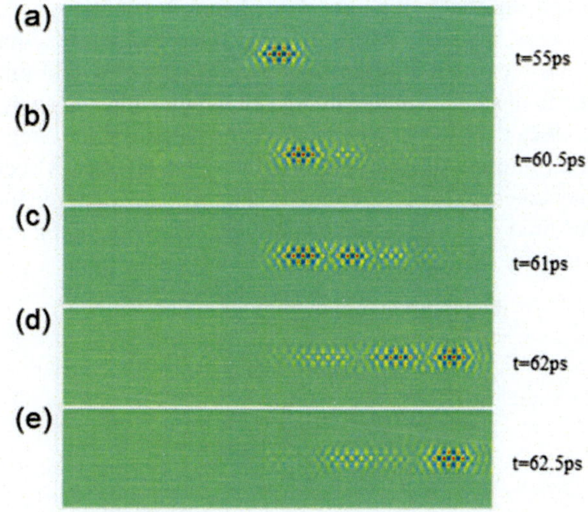

Fig. 50.4 Snapshots of field distribution in optical release process

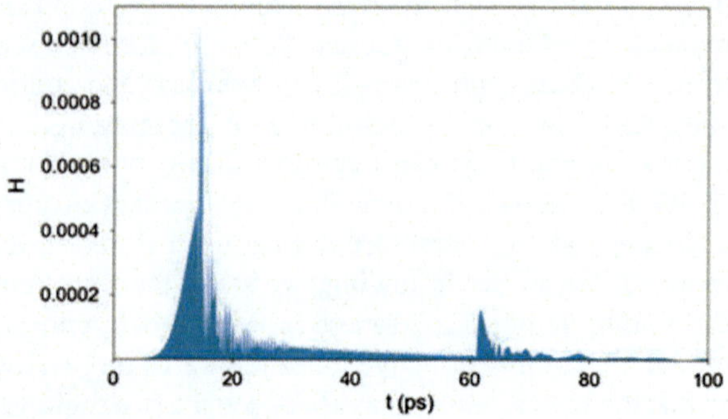

Fig. 50.5 Magnetic field at the right end of waveguide as a function of time

$t = 61.8$ ps, which is formed when the light wave leaves the right end of the waveguide after removing the modulation on the right refractive index modulated area. So far, we have realized the light release process by dynamically tuning the refractive index to remove refractive index reduction on one of two rectangular shadow areas.

Above we showed how our model structure achieve optical storage and release process through dynamical tuning. In this section, we are going to study the controllable movement of light intensity through dynamical tuning on our structure. Figure 50.6a is a schematic diagram of our study. The two dark rectangular shaded areas are the same as the previous optical storage process which will reduce their refractive index to $0.98n$ at $t = 14$ ps to complete the storage process. After the optical storage process, at a certain moment after the light intensity can stably stay between the two modulation regions, we move.

the two modulation regions toward right side which is shown by the blue arrow in Fig. 50.6a with same speed, that is, refractive index of the dielectric material in the place covered by the rectangular shaded area becomes $0.98n$, and the refractive index of the place no longer be covered recovers n. The two modulation areas finally reach and stay in the light shaded area shown in Fig. 50.6a. After the optical storage process, when $t = 40$ ps, the two refractive index reduced regions start to move to the right with a constant speed of $0.012c$. The total time of movement is 3 ps, and the final stay position for these two regions is $26a$ away from the starting position.

It can be clearly seen from the field distribution snapshots in Fig. 50.6b–f that the light intensity captured by the two refractive index reduced regions move with the movement of the two regions and finally stay between the two regions which reached the final position. And it can be seen from Fig. 50.6g that after a long time, the light intensity still stays between the two regions.

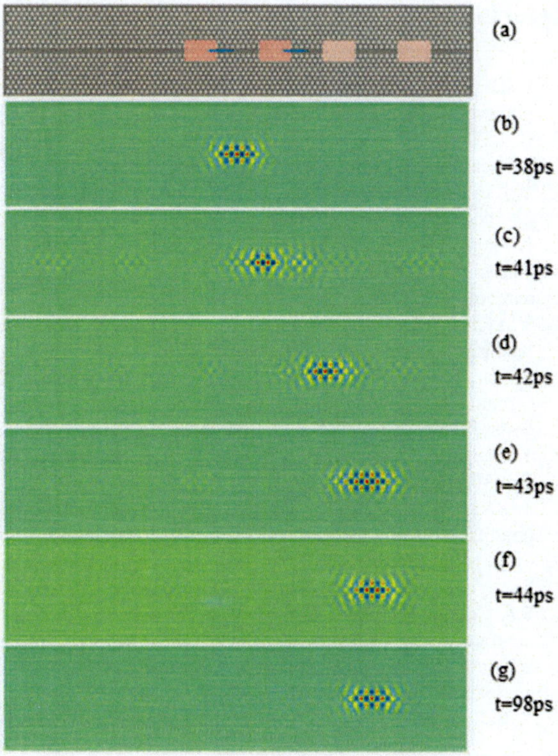

Fig. 50.6 Snapshots of field distribution in optical move process

50.3 Conclusions

In this paper, we present a simple yet effective approach to realize a movable nanocavity based on dynamically tuning the local refractive index of a photonic crystal (PC) waveguide. Using this movable cavity, we demonstrate that it can trap, store, and transfer signal photons. The approach proposed here provides a new way for manipulating photons and is promising in the fields of advanced photonic circuits, all-optical information processing, and optical communications.

Acknowledgements This research was financially supported by the National Natural Science Foundation of China (11774098); Guangdong Natural Science Foundation (2017A030313016); and Science and Technology Program of Guangzhou (202002030500).

References

1. B.-S. Song, T. Asano, S. Noda, Physical origin of the small modal volume of ultra-high-Q photonic double-heterostructure nanocavities. New J. Phys. **8**, 209 (2006)
2. B.-S. Song, S. Noda, T. Asano, Y. Akahane, Ultra-high-Q photonic double-heterostructure nanocavity. Nat. Mat. **4**, 207–210 (2005)
3. Y. Akahane, T. Asano, B.-S. Song, S. Noda, High-Q photonic nanocavity in a two-dimensional photonic crystal. Nature **425**, 944 (2003)
4. K. Ashida, M. Okano, M. Ohtsuka, M. Seki, N. Yokoyama, K. Koshino, M. Mori, T. Asano, S. Noda, Y. Takahashi, Ultrahigh-Q photonic crystal nanocavities fabricated by CMOS process technologies. Opt. Express **25**, 18165 (2017)
5. T. Asano, Y. Ochi, Y. Takahashi, K. Kishimoto, S. Noda, Photonic crystal nanocavity with a Q factor exceeding eleven million. Opt. Express **25**, 1769 (2017)
6. Y. Lai, S. Pirotta, G. Urbinati, D. Gerace, M. Minkov, V. Savona, A. Badolato, M. Galli, Genetically designed L3 photonic crystal nanocavities with measured quality factor exceeding one million. Appl. Phys. Lett. **104**, 241101 (2014)
7. Y. Akahane, T. Asano, B.-S. Song, S. Noda, Fine-tuned high-Q photonic-crystal nanocavity. Opt. Express **13**, 1202 (2005)
8. Y. Zhao, Y.N. Zhang, Q. Wang, H. Hu, Review on the optimization methods of slow light in photonic crystal waveguide. IEEE Trans. Nanotechnol. **14**, 407 (2015)
9. M.K. Moghaddam, R. Fleury, Slow light engineering in resonant photonic crystal line-defect waveguides. Opt. Express **27**, 26229 (2019)
10. S.A. Schulz, J. Upham, L. O'Faolain, R.W. Boyd, Photonic crystal slow light waveguides in a kagome lattice. Opt. Lett. **42**, 3243 (2017)
11. S. Serna, P. Colman, W. Zhang, X. Le Roux, C. Caer, L. Vivien, E. Cassan, Experimental GVD engineering in slow light slot photonic crystal waveguides. Sci. Rep. **6**, 26956 (2016)
12. A. Taflove, S.C. Hagness, *Computational Electrodynamics* (Artech House, 2000)

Chapter 51
Pig Pose Recognition Method Based on Openpose

Guangchang Li, Qing Jv, Feifei Liu, and Zhibo Yao

Abstract Early detection and treatment of disease in pigs can increase pork production. The changes of pig posture information have high diagnostic value, but these subtle changes are not easy to be quantified and need to be observed and analyzed for a long time, which is difficult to achieve in the process of large-scale breeding. In this paper, a pig pose recognition method based on Openpose is proposed for pig pose recognition. Firstly, the collected images were input into the OpenPose pose estimation model to obtain the key point location information and key point connection information of pigs. The two groups of information were converted into joint angle features and joint spacing features, which were input into the KNN algorithm to classify pig poses. The experimental results show that the accuracy of the model for pig attitude recognition is more than 93%, and the FPS is between 10 and 12, which is a method with both accuracy and speed.

51.1 Introduction

In recent years, affected by African Swine fever and other infectious diseases of pigs, the price of pork skyrocketed, seriously affecting people's daily life. Therefore, early detection of damage to the health of pigs and timely treatment can increase pork production so as to improve people's livelihood.

G. Li · Q. Jv · F. Liu (✉) · Z. Yao
School of Mechanical and Electrical Engineering, Jiangxi University of Science and Technology, Ganzhou, China
e-mail: gzlff@126.com

G. Li
e-mail: 2359795949@QQ.COM

Q. Jv
e-mail: 940680576@QQ.COM

Z. Yao
e-mail: 1406394534@QQ.COM

G. Liu and F. Cen (eds.), *Advances in Precision Instruments and Optical Engineering*, Springer Proceedings in Physics 270, https://doi.org/10.1007/978-981-16-7258-3_51

As a kind of easily accessible biological information, pig posture information is an important indicator to measure the health status of pigs. However, the changes of posture information is not easy to be quantified and requires long-term observation by staff, which is difficult to achieve in the process of large-scale breeding. Therefore, the objective monitoring of the change of attitude information through sensors has attracted wide attention [1]. Based on the analysis of pixel difference between continuous images, a method of automatic detection of aggressive behavior between pigs by image analysis was proposed [2]. The Kinect depth camera was used to extract the behavior information of pigs, and the support vector machine was used to detect attack behavior. This method was cost-effective and effective in terms of accuracy [3].

In this paper, the OpenPose pose estimation algorithm [4] is used to identify the key points of pigs, the position and connection information of the key points are obtained, then the two groups of information are converted into the bone joint angle and joint spacing of pigs, and then the KNN [5] algorithm was used classification and recognition of pig posture.

51.2 Materials and Methods

51.2.1 Data Preparation

The experimental data came from a pig breeding base in Ganzhou city, Jiangxi Province. Cameras were used to shoot the video at different angles and with different degrees of occlusion. And 4000 images were selected as the training and testing sample data. The image size was adjusted to 360 pixels $\times$ 520 pixels to improve the training speed of the model.

Labelme image annotation tool was used to annotate the key points and pose categories of pigs. Finally, Json files with information were used. Key points and pose category comments are shown in Fig. 51.1.

See Fig. 51.2, the selected image is rotated, mirrored, and scaled to enhance and expand the sample image.

51.2.2 Pig Pose Estimation

In actual breeding, multiple pigs are raised in a pig pen, so the OpenPose pose estimation algorithm is adopted in this paper to carry out real-time pose estimation for multiple pigs. The OpenPose algorithm adopts bottom-up approach and has achieved good results in real-time pose estimation [4].

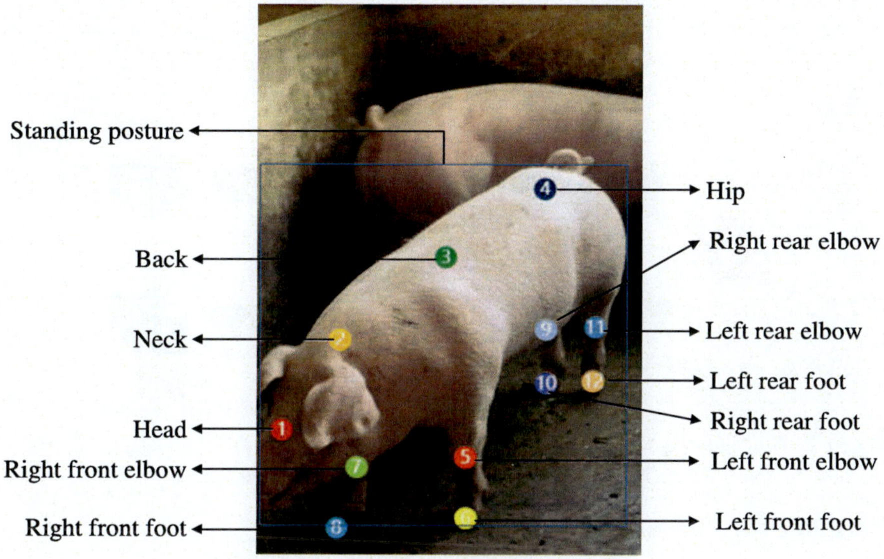

Fig. 51.1 Picture annotation schematic diagram

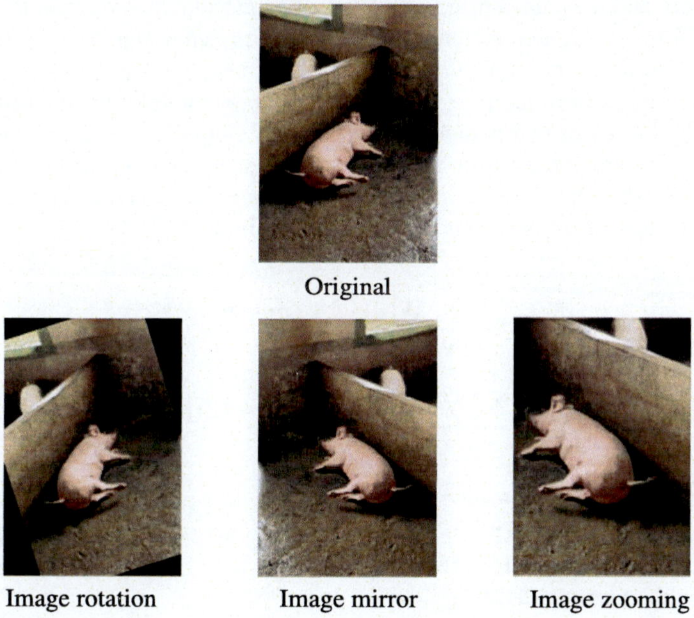

Fig. 51.2 Data to enhance

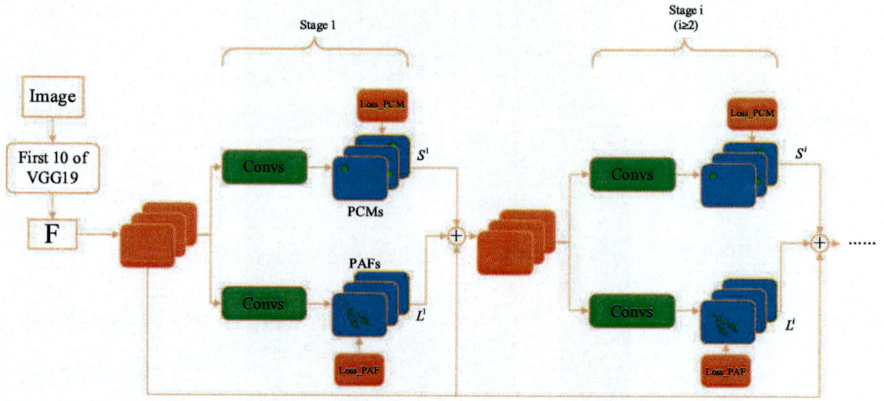

Fig. 51.3 The OpenPose Network architecture

OpenPose Network Architecture

The OpenPose network generates Part confidence maps (PCMs) and Part affinity fields (PAFs) based on the input images. The former is used to predict the location of key points, and the latter is used to group key points and finally obtain the attitude skeleton. The network architecture of OpenPose is shown in Fig. 51.3.

The first 10 layers of VGG19 [6] network were used for feature extraction, and the feature images were input to the subsequent Stage module for processing. The Stage module contains two branches, S and L. The S branch is used to generate the confidence diagram of the key point, and the L branch is used to generate the affinity domain of the key point. L2 Loss is used to calculate the Loss of each branch, and the Loss function of the branch is shown in the formula.

$$f_{\mathbf{S}}^{i} = \sum_{j=1}^{J} \sum_{\mathbf{p}} \mathbf{W}(\mathbf{p}) \cdot \left\| \mathbf{S}_{j}^{i}(\mathbf{p}) - \mathbf{S}_{j}^{*}(\mathbf{p}) \right\|_{2}^{2},$$
$$f_{\mathbf{L}}^{i} = \sum_{c=1}^{C} \sum_{\mathbf{p}} \mathbf{W}(\mathbf{p}) \cdot \left\| \mathbf{L}_{c}^{i}(\mathbf{p}) - \mathbf{L}_{c}^{*}(\mathbf{p}) \right\|_{2}^{2}. \quad (51.1)$$

Among them, $W(p)$ represents the weight of the key point. If the position is not marked, 0; otherwise, 1;$\mathbf{L}_{c}^{*}(\mathbf{p})$ and $\mathbf{S}_{j}^{*}(p)$ represents the predicted value; $\mathbf{L}_{c}^{i}(\mathbf{p})$ and $\mathbf{S}_{j}^{i}(p)$ represents the true value. The total loss is the sum of the losses of each branch, as shown in the formula (51.2),

$$f = \sum_{t=1}^{T} \left(f_{\mathbf{S}}^{i} + f_{\mathbf{L}}^{i} \right) \quad (51.2)$$

Starting from the second Stage module, the input of the module consists of three parts, the feature diagram F, the output of S^{i-1}, and the output of L^{i-1}, as shown in the formula (51.3).

$$\mathbf{S}^i = \rho^i\left(\mathbf{F}, \mathbf{S}^{i-1}, \mathbf{L}^{i-1}\right), \forall i \geq 2,$$
$$\mathbf{L}^i = \phi^i\left(\mathbf{F}, \mathbf{S}^{i-1}, \mathbf{L}^{i-1}\right), \forall i \geq 2. \tag{51.3}$$

OpenPose components

Part confidence maps are used to represent the position of the key point. The Gaussian function is used to create the confidence graph of the key point. The value of each pixel in the confidence graph represents the confidence degree of the position, which is calculated by the formula (51.4).

$$g(x, y) = \frac{1}{2\pi\sigma^2} e^{-\frac{\left((x-x')^2+(y-y')^2\right)}{2\sigma^2}}. \tag{51.4}$$

In this paper, the PCMs output contains 12 porcine key point location information and one containing background information. The background information serves as input to the next Stage, where better semantic information can be obtained. When σ set to 0.008, the output looks like Fig. 51.4.

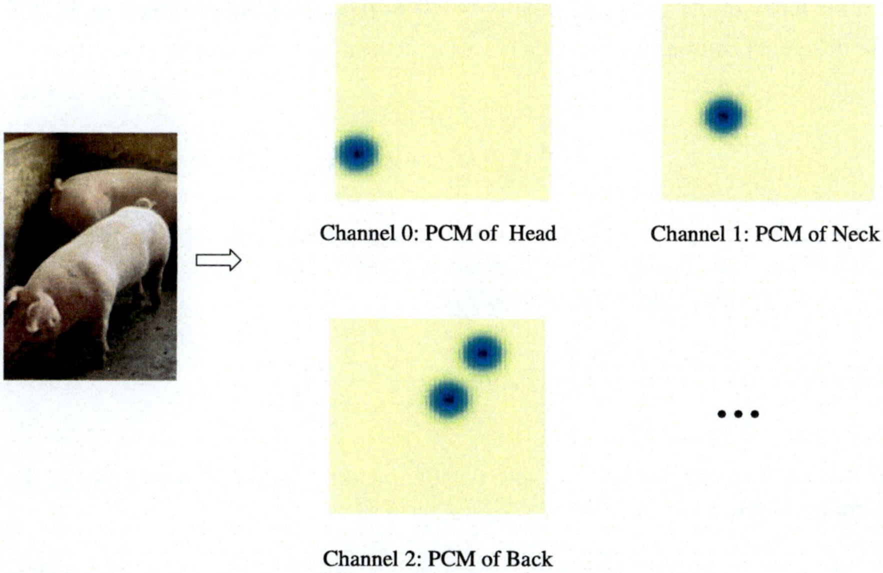

Fig. 51.4 PCMs output. There are two regions with high confidence in the channel 2, so it can be judged that there are two pigs in the input image

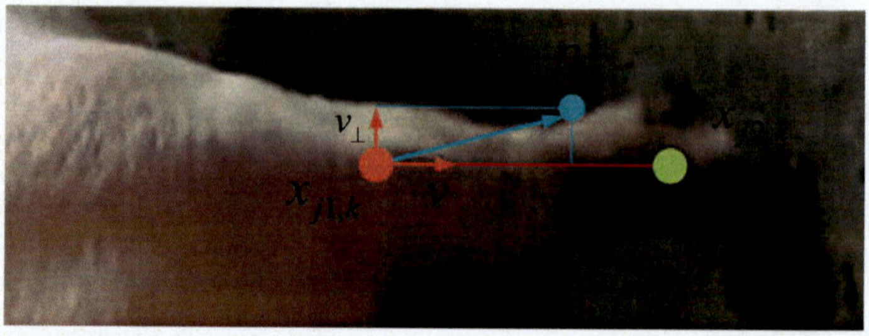

Fig. 51.5 Left front leg of pig

Part affinity fields are used to judge the affinity between different key points. Each pixel in the PAF region holds a two-dimensional vector, which is used to represent the directional information of the bone. Take the left front leg as an example, as shown in Fig. 51.5. $x_{j1,k}$ and $x_{j2,k}$ respectively represent the position coordinates of the left front elbow and the left front hoof.

For position p, its vector is expressed by the formula (51.5).

$$L^*_{c,k}(\mathbf{p}) = \begin{cases} \mathbf{v}, \ \textbf{if } p \textbf{ on limb } c, k \\ \mathbf{0}, \ \textbf{otherwise} \end{cases}. \tag{51.5}$$

c is the index of the limb, and $\mathrm{v} = \frac{(x_{j2,k} - x_{j1,k})}{\left\| x_{j2,k} - x_{j1,k} \right\|_2}$ represents the pointing unit vector. For different positions use formula (51.6) to determine whether it is on the limb.

$$\begin{aligned} & 0 \le \mathbf{v} \cdot (p - x_{j1,k}) \le l_{c,k} \\ & \textbf{and } \left| \mathbf{v} \times (p - x_{j1,k}) \right| \sigma_l, \end{aligned} \tag{51.6}$$

where $l_{c,k} = \left\| x_{j2,k} - x_{j1,k} \right\|_2$ indicates the length of the limb, and σ_l indicates the width of the limb. For any two key points $\mathbf{d}_{j_1}$ and $\mathbf{d}_{j_2}$, the affinity between the key points can be calculated using formula (51.7),

$$E = \int_{u=0}^{u=1} L_c(\mathbf{p}(u)) \cdot \frac{\mathbf{d}_{j_2} - \mathbf{d}_{j_1}}{\left\| \mathbf{d}_{j_2} - \mathbf{d}_{j_1} \right\|_2} du. \tag{51.7}$$

$\mathbf{p}(u)$ is obtained by position and interpolation, and the calculation is shown in the formula (51.8)

$$\mathbf{p}(u) = (1 - u)\mathbf{d}_{j_1} + u\mathbf{d}_{j_2}. \tag{51.8}$$

11 limbs are defined in this paper, as shown in Fig. 51.6.

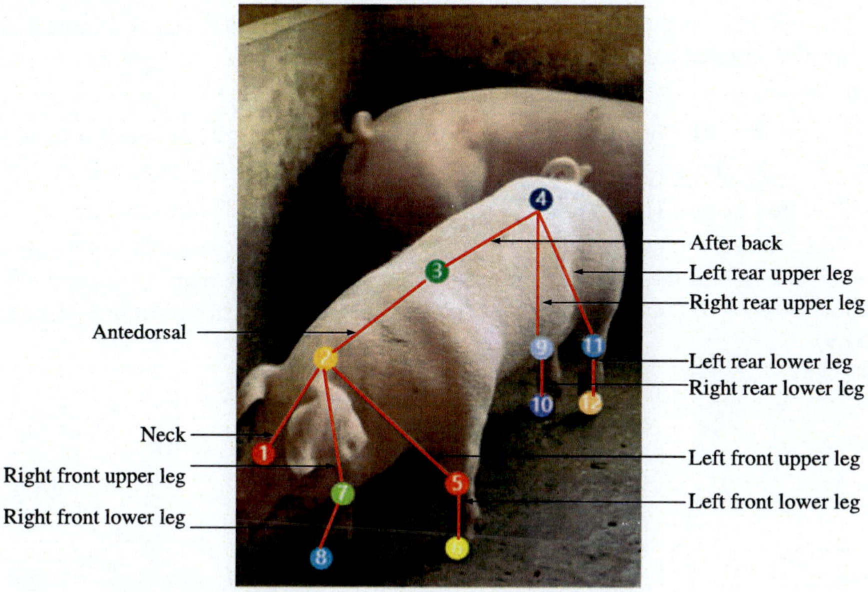

Fig. 51.6 Definition of pig limbs

In this paper, there are 22 PAFs output channels, including 11 limb direction vectors. When σ_l set to 0.0015, the key affinity domains of 11 limbs are obtained, as shown in Fig. 51.7.

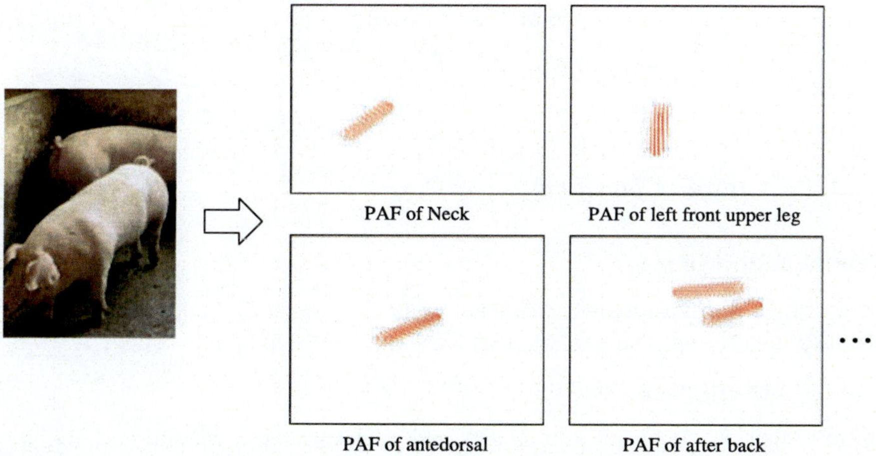

Fig. 51.7 PAFs output. The arrow direction of the output of PAF of the limb represents the direction of vectors in the affinity domain of key points

A set of discrete candidate positions of key points can be obtained from the confidence graph of key points. All candidate key points constitute a set $D_J = \left\{d_j^m : \text{for } j \in \{1 \ldots J\}, m \in \left\{1 \ldots N_j\right\}\right\}$, define variables $z_{j_1 j_2}^{mn} \in \{0, 1\}$ to indicate candidate key points and whether $d_{j_1}^m$ and $d_{j_2}^n$ can be joined to get the collection. Therefore, get the set $Z = \left\{z_{j_1 j_2}^{mn} : \text{for } j_1, j_2 \in \{1 \ldots J\}, m \in \left\{1 \ldots N_{j_1}\right\}, n \in \left\{1 \ldots N_{j_2}\right\}\right\}$ when only the limb c is considered, the two key points corresponding to it are respectively represented by j_1 and j_2, and the candidate set corresponding to the key point is respectively represented by D_{j_1} and D_{j_2}, then the objective function is shown in the formula (51.9),

$$\begin{aligned} &\max_{Z_c} E_c = \max_{Z_c} \sum_{m \in D_{j_1}} \sum_{n \in D_{j_2}} E_{mn} \cdot z_{j_1 j_2}^{mn}, \\ &s.t. \;\; \forall m \in D_{j_1}, \sum_{n \in D_{j_2}} z_{j_1 j_2}^{mn} \le 1, \\ &\forall n \in D_{j_2}, \sum_{m \in D_{j_1}} z_{j_1 j_2}^{mn} \le 1. \end{aligned} \tag{51.9}$$

E_c represents the sum of weights corresponding to the trunk c, Z_c is the subset Z corresponding to the trunk c, and E_{mn} is the affinity of the key point $d_{j_1}^m$ and $d_{j_2}^n$. For all the torsos, the optimization objective function is transformed into the formula (51.10),

$$\max_{Z} E = \sum_{c=1}^{C} \max_{Z_c} E_c. \tag{51.10}$$

51.2.3 Attitude Classification of Pigs

Typical posture of pigs

Four typical pig postures were defined, including crawling, lying on one's side, standing, and sitting. The four attitudes are shown in Table 51.1.

Attitude feature extraction

In this paper, a mixture of joint angles and joint spacing is used to describe the posture of pigs. Take any two segments of bone in an individual pig as an example, the coordinates of the corresponding key points are $A(x_1, y_1)$, $B(x_2, y_2)$, $C(x_3, y_3)$, $D(x_4, y_4)$, then the joint angle θ can be calculated by formula (51.11), and the joint spacing L can be calculated by formula (51.12),

Table 51.1 Schematic diagram of four kinds of pigs

Type	Schematic diagram
Prone position	
Lateral position	
Standing position	
Sitting-dog position	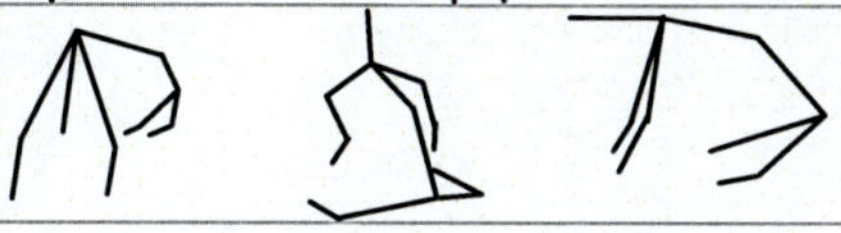

$$\cos\theta = \frac{\overrightarrow{AB} \cdot \overrightarrow{CD}}{(|\overrightarrow{AB}| \cdot |\overrightarrow{CD}|)}, \tag{51.11}$$

$$L = \frac{L_{MN}}{d}\sqrt{(x_1 - x_2)^2 + (y_1 - y_2)^2}. \tag{51.12}$$

L_{MN} is the head height of the individual, and d is the standard head height.

In this paper, 10 groups of joint angle characteristics and 15 groups of joint spacing characteristics were defined. The angle position of each joint was shown in Fig. 51.8, and the distance position of each joint was shown in Fig. 51.9. 10 groups of joint angles.

Pig attitude classification method based on KNN algorithm

When an unknown sample is input, among the K samples closest to the unknown sample, the category that accounts for more determines the category of the unknown

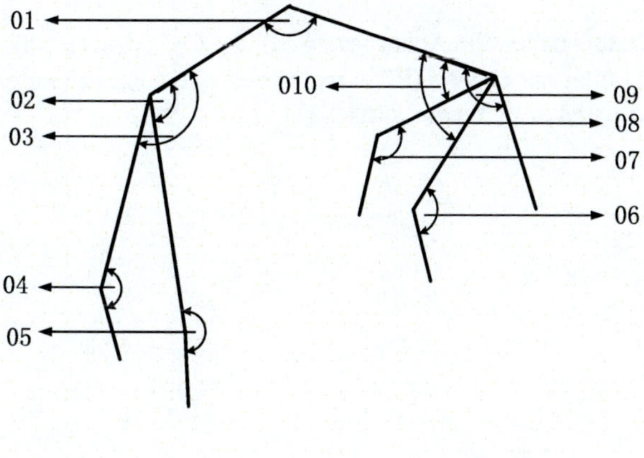

Fig. 51.8 10 groups of joint angles

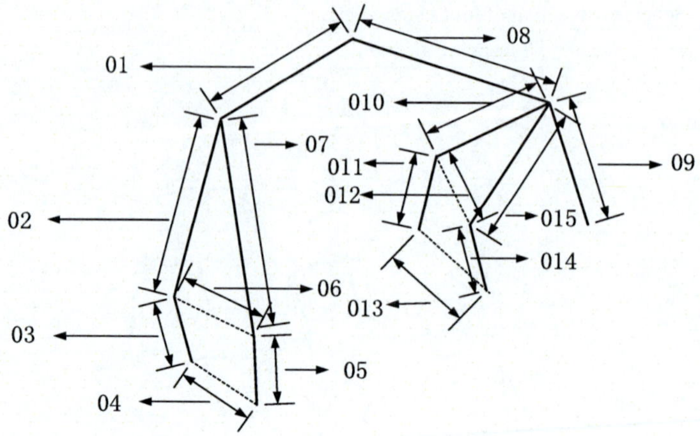

Fig. 51.9 15 groups of joint spacing

sample [5]. In this paper, the Euclidean distance is used to represent the distance between the unknown sample and the sample in the feature space. The Euclidean distance between two points $x_1(x_{11}, x_{12} \ldots, x_{1n})$ and $x_2(x_{21}, x_{22} \ldots, x_{2n})$ in the n dimensional feature space can be calculated by the formula (51.13),

$$L_2 = \sqrt{\sum_{i=1}^{n} (x_{1i} - x_{2i})^2}. \tag{51.13}$$

In the process of experiment, this paper adopts normalization treatment for the values of all sample features and determines the value of K through cross validation. The algorithm process is shown in Fig. 51.10.

51.2.4 Evaluation Indicators

In this paper, the Accuracy (Acc), the Precision (P), the Recall (R), the F_1 score, and Frames per second (FPS) were used as evaluation indexes of the model, which were calculated by formula (51.14),

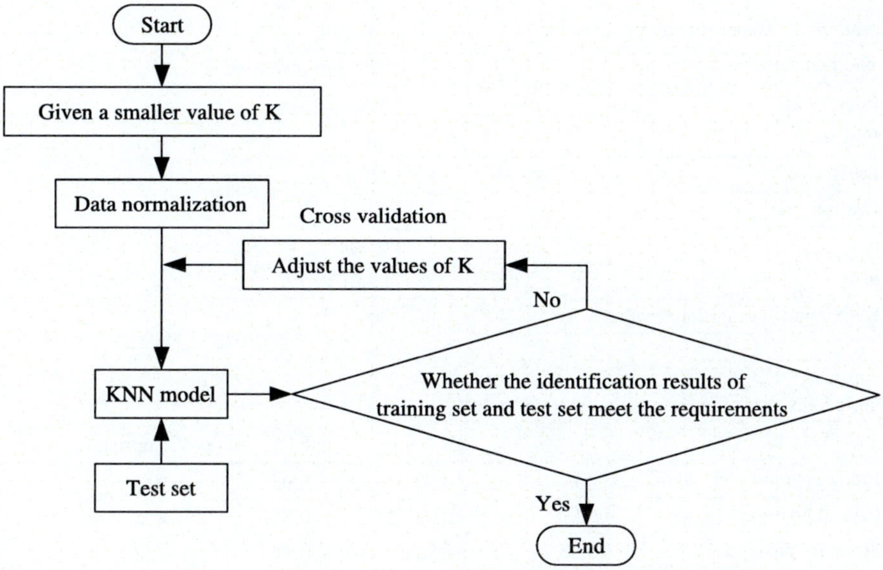

Fig. 51.10 Pose recognition training and testing process based on KNN algorithm

$$
\begin{aligned}
Acc &= \frac{TP+TN}{TP+TN+FP+FN}, \\
P &= \frac{TP}{TP+FP}, \\
R &= \frac{TP}{TP+FN}, \\
F_1 &= 2\cdot\frac{\mathrm{P}\cdot\mathrm{R}}{\mathrm{P}+\mathrm{R}}, \\
\mathrm{FPS} &= \frac{\mathrm{frameNum}}{elapsedTime}.
\end{aligned}
\tag{51.14}
$$

TP represents the true example, TN represents the true counter example, FP resents the false positive example, FN represents the false counter example, frameNum represents the image frame number, and elpsedTime represents the time interval.

51.3 Results and Discussion

The experimental platform GPU adopts NVIDIA GTX 1080TI, 11 GB video memory, 16G installation memory, TensorFlow version 2.1.0, Python version 3.6 Related experimental parameters are shown in Table 51.2.

The image data of the test set was input into the pose recognition model, and the pose category of pigs was predicted. The recognition results were shown in Table 51.3.

Table 51.2 Experimental parameters

The parameter name	The parameter value
Image size	360 × 520
Epoch	50
Batch size	10
Learning rate	0.001
σ	0.008
σ_l	0.0015
KNN algorithm initial K value	20

Table 51.3 Attitude recognition results

	P	R	F1	Support
Prone position	0.94	0.92	0.93	912
Lateral position	0.92	0.94	0.93	875
Standing position	0.93	0.92	0.92	922
Sitting-dog position	0.93	0.92	0.92	875
Accuracy	0.93	0.93	0.93	0.93
Macro avg	0.93	0.93	0.93	3584
Weighted avg	0.93	0.93	0.93	3584

It can be seen from Table 51.3 that the average accuracy of pig attitude classification algorithm based on KNN algorithm, which takes joint angle and joint spacing as input features, is more than 93% for pig attitude recognition. Among them, the recognition accuracy of prone position is the highest, reaching 94%, while the recognition accuracy of lateral position is lower, reaching 92%. The reason for the low recognition accuracy of lateral position may be that there are more key points on the body of pigs when they are lying on their side, which leads to less characteristics of input joint angle and joint spacing, and thus reduces the recognition accuracy of posture.

During the experiment, the FPS of the model is stable between 10 and 12, and the real-time performance is normal. It may be because the first ten layers of VGG19 network are used as the feature extractor, which makes the model a large amount of computation and takes a long time. However, through the observation of the daily behavior of pigs, it is found that the posture of pigs is relatively fixed within a period of time. The model can meet the requirements of real-time recognition of pig posture.

51.4 Conclusions

In this paper, a pig pose recognition method based on deep learning is proposed for pig pose recognition. Firstly, the collected images were input into the OpenPose pose estimation model to obtain the coordinate information of the key points and the connection information of the key points. The two groups of information were converted into joint angle features and joint spacing features, which were input into the KNN algorithm to classify the poses of the pigs. The experimental results show that the accuracy of the model for pig attitude recognition is more than 93%, and the FPS is between 10 and 12, which is a method with both accuracy and speed.

However, this method also has some limitations: (1) the data are collected in the same breeding base and the scene is single; (2) the recognition speed is slow. In view of the above problems, the following research will further expand the data set and optimize the attitude estimation model and attitude classification algorithm to improve the detection efficiency under the condition of ensuring the accuracy.

Acknowledgements Foundation items: National Natural Science Foundation of China (61364014); Key Laboratory of Machine Vision and Intelligent System, No.20181BCD40009.

References

1. Q. Yang, D. Xiao, A review of video-based pig behavior recognition. J. Sci. Appl. Animal Behav. Sci. **2020**, 233 (2020)
2. S. Viazzi, G. Ismayilova, M. Oczak et al., Image feature extraction for classification of aggressive interactions among pigs. J. Comput. Electron. Agricult. **104**, 57–62 (2014)
3. L. Jonguk, L. Jin, P. Daihee et al., Automatic recognition of aggressive behavior in pigs using a kinect depth sensor. J. Sensors **16**(5), 631 (2016)
4. C. Zhe, T. Simon, S.E. Wei, et al., Realtime multi-person 2D pose estimation using part affinity fields, in *2017 IEEE Conference on Computer Vision and Pattern Recognition*, (Hawaii, 2017), pp. 172–186
5. T. Abeywickrama, M.A. Cheema, D. Taniar, K-nearest neighbors on road networks: a journey in experimentation and in-memory implementation. J. Proc. VLDB Endowment **9**(6) (2016)
6. K. Simonyan, A. Zisserman, Very deep convolutional networks for large-scale image recognition. J. Comput. Sci. (2014)

Chapter 52
Research on Data Management Technology of Construction Project Planning Completion Survey

Lijun Wei

Abstract As the process of urban modernization is accelerating, the scale of the urban construction industry is expanding. At the same time, rapid economic development has promoted the increase in construction demand, which has led to a gradual increase in construction planning projects. Planning completion measurement is an important environment in a construction project, and the measurement level is directly related to the construction quality of the construction project. In addition to data measurement, data management also occupies an important proportion of planned completion measurement. Therefore, it is very important to integrate data management technology to optimize the planning of the completion measurement process. To this end, this article analyzes the construction project planning completion measurement accuracy evaluation method, further explores the data management technology, and constructs the corresponding data model, aiming to provide some useful references, improve the measurement process, and improve the measurement accuracy.

52.1 Introduction

With the rapid social and economic development, the country is also paying more and more attention to urban modernization and development. As an important part of social development, cities are important carriers and signs of social modernization [1, 2]. Accelerating urban modernization is an effective way to promote the sustainable development of the national economy and the harmonious development of society. In this macro policy environment, local governments have increased their policy support for construction projects and effectively promoted the innovative development of the construction industry. As the scale of the construction industry expands, the corresponding construction engineering planning projects also increase [3, 4]. As part of the construction project, the planned completion measurement can effectively speed up the measurement data processing efficiency by applying data management

L. Wei (✉)
Engineering College, Yunnan University Of Bussiness Management, KunMing 650106, YunNan, China

G. Liu and F. Cen (eds.), *Advances in Precision Instruments and Optical Engineering*, Springer Proceedings in Physics 270, https://doi.org/10.1007/978-981-16-7258-3_52

technology, ensure that the measurement process is more standardized and scientific, and improve the accuracy of the measurement data. Based on this, analysis and exploration of construction project planning completion measurement data management technology is an important content of this paper [5–7].

52.2 Completion Measurement and Model Construction of Architectural Engineering Planning

52.2.1 *Evaluation Method of Construction Project Planning Completion Measurement Accuracy*

In the process of construction project planning and completion measurement, CORS technology and three-dimensional laser scanning technology, two surveying and mapping technologies, have more in-depth applications. CORS is a new surveying and mapping technology based on network technology. It has a variety of characteristics. It can measure and plan completed project data in real time, and dynamically track spatial attribute data, providing real and reliable data basis for relevant departments. The core of 3D laser scanning technology is different from the former. It measures the horizontal and vertical angles of the laser beam and the distance from the target to the center of the instrument, and then calculates the 3D coordinates of the measuring point according to the measured distance, as follows:

$$\begin{bmatrix} x \\ y \\ z \end{bmatrix} = \begin{bmatrix} r\cos\theta\cos\phi \\ r\cos\theta\sin\phi \\ r\sin\theta \end{bmatrix} \leftrightarrow \begin{bmatrix} r \\ \theta \\ \phi \end{bmatrix} = \begin{bmatrix} \sqrt{x^2+y^2+z^2} \\ \tan^{-1}\left(\frac{z}{\sqrt{x^2+y^2}}\right) \\ \tan^{-1}\left(\frac{y}{x}\right) \end{bmatrix} \tag{52.1}$$

52.2.2 *Construction of the Construction Project Planning Completion Measurement Accuracy Evaluation Model*

Accuracy evaluation model of planning completion measurement based on CORS technology

In the actual measurement process, the planning and completion measurement accuracy assessment based on CORS technology can start from two perspectives, the first perspective is internal measurement, and the second perspective is external measurement.

The specific method of internal accuracy measurement is as follows: select the measurement area, select a certain number of points in the area, and use the selected points as the measurement points. Use CORS technology to perform interval testing on these points, and collect the coordinate information of the test points in turn. The coordinate data and average value are compared with the test results, The internal measurement accuracy can be judged by observing the error value between the two. The specific calculation formula is as follows:

$$M = \pm\sqrt{\frac{vv}{n-1}} \tag{52.2}$$

In the calculation formula, M represents the internal measurement accuracy of CORS technology in the three directions of x, y, and h, v represents the deviation of each measurement result from the arithmetic average in the three directions, and n represents the number of measurements. By observing the change of the M value, you can clearly grasp whether the measurement result is stable. If the value of M is smaller, it means that the measurement result is more stable, which means that the degree of stability of the completion measurement data of the construction project is also higher.

Different from the internal accuracy measurement method, the external accuracy measurement can introduce external data for testing. The specific process is as follows: select a fixed area, which includes the CORS service area and the surrounding area. In the selected area, select a certain number of known points and use these known points as test points. Continuously observe the test points, record the observation results, count the coordinate information of each test point in the three directions x, y, and h, and calculate the final difference. The specific calculation formula is as follows:

$$M = \pm\sqrt{\frac{\Delta\Delta}{n-1}} \tag{52.3}$$

In the calculation formula, M represents the external measurement accuracy of CORS technology in the three directions of x, y, and h, n represents the number of measurements, and Δ represents the deviation between each measurement result and the arithmetic average in three directions. By observing the change of the M value, we can clearly grasp whether the measurement result is accurate. If the value of M is smaller, it means that the measurement result is more accurate, which means that the accuracy of the completion measurement data of construction project planning is also higher.

Planning the completion measurement accuracy evaluation model based on 3D laser scanning technology

In the process of construction project planning and completion measurement, the geometric form of the tested building will affect the quality of the scanned point cloud to a certain extent, and then affect the actual scanning effect. To this end, in order to further explore the influence of different geometric conditions on the quality of the scanned point cloud, the next step will be an in-depth discussion from the angle of incidence.

In a sense, the scanning geometric conditions of the 3D laser scanning technology are usually determined by its own geometric erection position and the scanned object. Here, the vector is expressed as

$$P_i = [x_i, y_i, z_i]_{i=1,\cdots,n}$$

In the vector, *P* represents the direction vector of the laser beam vector emitted by the scanner to the surface of the scanned object, *n* represents the number of points in the scanned point cloud, and *x*, *y*, and *z* represent three different dimensions, respectively. On the basis of the vector, the incident angle is taken as the angle between the direction vector of the laser beam and the normal vector of the scanned object surface, and the following formula is obtained:

$$\alpha_i = \cos^{-1}\left(\frac{P_i * N}{|P_i||N|}\right) \tag{52.4}$$

According to the incident angle formula, determine the specific range of the incident angle $\left[0 < \alpha_i < \frac{\pi}{2}\right]$. Due to the divergent characteristics of the laser beam, when it hits the surface of the object, a circular mark will appear. When the incident angle is zero, the laser beam energy presents a circular state. At this time, if the distance between the scanner and the object is farther, will be larger, indicating that the signal reflection intensity of the three-dimensional scanner is also lower. If the incident angle is not zero, the circular mark will transform into an elliptical shape. At this time, if the distance between the scanner and the object is farther, the elliptical mark will become longer, indicating the signal reflection intensity of the three-dimensional scanner is also higher.

52.3 Construction Project Planning Completion Measurement Data Management Technology and Model Construction

In addition to data measurement, data processing is also very important in the completion measurement process of construction engineering planning. In particular, different building completion projects have different requirements for measurement data types, scope, and accuracy, and higher standards are also set for data measurement and processing technology. Therefore, in order to meet the requirements of different projects, the staff tried to apply the data processing technique of scale transformation to manage the measurement data and constructed the corresponding model.

The scaling conversion is to convert the corresponding parameter value into a measured engineering quantity value that can directly display the dimension, that is, to convert the standard parameter value into a value that people can intuitively understand. In this way, it is convenient for the staff to be more clear about the size of the measured parameters when carrying out the construction project planning and completion measurement, so as to more accurately measure the data and improve the quality of work. From a principle point of view, the scale conversion is to transfer the parameter value from the sensor to the AD terminal through the amplifier, and then from the AD terminal to the computer terminal, and translate the parameter value into the value you want, and finally "translate" the value Send to the digital tube for display.

52.3.1 Linear Scale Transformation Method

There are two types of scaling methods: linear and non-linear. Among them, linear scaling is the most commonly used method of scaling. When there is a linear relationship between the sensor output signal of the measuring instrument and the measured parameter, then the staff can use the linear scale transformation method to process the data. The specific formula is

$$A_x = (A_m + A_0)\frac{N_x - N_0}{N_m - N_0} + A_0 \tag{52.5}$$

In the linear scale conversion formula, A_x represents the actual measured value, A_m indicates the upper limit of a measuring instrument (the maximum value of the measuring range), A_0 indicates the lower limit (the minimum value of the measuring range) of a measuring instrument at a time, N_x represents the digital quantity corresponding to the actual measured value, N_m indicates the digital quantity corresponding to the upper limit of the instrument, and N_0 indicates the digital quantity corresponding to the lower limit of the instrument.A_0, A_m, N_0, and N_m are all constants. In order to further simplify the procedure, the lower limit of many measuring instruments A_0 is usually set to 0; the simplified formula is

$$A_x = A_m\frac{N_x}{N_m} \tag{52.6}$$

From the simplified linear scale conversion formula, it can be found that when the upper limit value of a measuring instrument and the digital quantity corresponding to the upper limit of the meter are both constant values, the actual measured value is directly related to the digital quantity corresponding to the actual measured value. The larger the actual measurement value, the larger the digital quantity corresponding to the actual measurement value. Therefore, the co-workers can directly observe the actual measured values if they want to know the size relationship between the measured values when they perform the completion measurement of the construction project planning. Moreover, by comparing the actual measured value with the required parameter value, the staff can also grasp the gap between the actual situation of the construction project and the established goal in time, so as to take targeted measures to optimize and improve the project implementation plan to ensure the overall construction project quality of work.

52.3.2 Non-Linear Scaling Method

The prerequisite for the staff to use the non-linear scale transformation method for the completion measurement of the construction project planning is that there is a non-linear relationship between the sensor output signal of the measuring instrument and the measured parameter. Therefore, when the staff chooses the non-linear scaling transformation method as the data processing technology, the differential pressure method is used as an example to measure the flow rate. The relationship between the flow rate and the differential pressure is shown in (52.7), where Q represents the fluid flow rate, and K represents the flow rate. Scale factor ΔP represents the differential pressure before and after the throttling device.

$$Q = K\sqrt{\Delta P} \tag{52.7}$$

It can be seen from the relational formula that the fluid flow rate and the square root of the pressure difference generated before and after the fluid to be measured flows through the throttling device is a proportional functional relationship. The scale conversion formula is

$$Q_x = (Q_m - Q_0)\sqrt{\frac{N_x - N_0}{N_m - N_0}} + Q_0 \tag{52.8}$$

In the non-linear scale transformation formula, Q_x represents the measured value of the flow rate of the liquid being measured, Q_m represents the upper limit of the differential pressure flow meter, Q_0 represents the lower limit of the differential pressure flow meter, N_x represents the digital quantity corresponding to the differential pressure measured by the differential pressure flow meter, N_m represents the digital quantity corresponding to the upper limit of the differential pressure flow meter, and N_0 represents the digital quantity corresponding to the lower limit of the differential pressure flow meter. When Q_0 is set to 0, the non-linear scale transformation formula can be simplified as

$$Q_x = Q_m\sqrt{\frac{N_x}{N_m}} \tag{52.9}$$

From the simplified non-linear scaling conversion formula, it can be found that there is a close relationship between the measured flow rate of the liquid being measured and the digital quantity corresponding to the differential pressure measured by the differential pressure flow meter. The differential pressure is measured by the differential pressure flow meter. The larger the digital value corresponding to the value, the larger the measured flow rate of the liquid being measured.

52.4 Conclusion

In the process of construction project planning and completion measurement, measurement accuracy is easily affected by factors such as measurement range and incident angle. In addition, data management is also closely related to data measurement. For this reason, the staff should make full use of data processing techniques when making measurements, such as scale conversion, to transform parameter values into their desired values, so as to more clearly define the measurement objectives, standardize the measurement process, and improve the accuracy of the measurement results. The study provides scientific data basis for subsequent construction project planning and project decision-making. Due to the limitation of the length of the article, the research on the completion measurement data management technology of construction engineering planning in this article is not in-depth and perfect enough. In the future, we should continue to pay attention to the research trends of the completion measurement data management technology of construction engineering planning, enrich the research experience, and make up for the lack of research in this article.

Completion survey is a surveying and mapping work. Compared with quality management, the surveying and mapping engineering quality control system is more institutionalized and systematic. Quality management is the basis for the establishment of surveying and mapping engineering quality control systems. Due to the large number of surveying and mapping enterprises in China, it is necessary to conduct on-site investigation before capital investment, understand the details of each enterprise, and formulate the distribution plan to ensure the quality of surveying and mapping products, improve the level of science and technology, and ensure the integrity of data. In the process of recruiting talents, it is necessary to conduct a comprehensive and reasonable review of talents, improve the staff's sense of teamwork while ensuring the professional knowledge and scientific quality of talents, and establish an all-round and high-quality talent team.

References

1. C. Chuanshuang, Analysis of construction project planning completion measurement technology and method. Green Environ. Protect. Build. Mater. **06**, 140–141 (2021)

2. S. Hening, T. Chunxiao, The application of HERON backpack SLAM laser scanning system in architectural planning measurement. Beijing Survey. Mapping **35**(04), 485–488 (2021)
3. T. Xindong, Discussion on quality control in the whole process of construction project completion measurement. China Build. Mater. News (2021)
4. Q. Wei, Analysis of the role of planning completion survey in urban planning. Real Estate World **03**, 114–115 (2021)
5. G. Jinhua, Research on the key process of construction project planning, acceptance and completion measurement[J]. Decision Exploration (Medium) **12**, 29–30 (2020)
6. T. Xu, Z. Xianze, Z. Guofeng, Analysis of construction project planning completion survey technology and method. Survey. Spatial Geographic Inform. **43**(10), 198–200 (2020)
7. L. Ruixia, Analysis of the key process of construction project planning acceptance completion measurement. China New Technol. New Prod. **15**, 111–112 (2020)

Chapter 53
Research on Active Suspension System of Heavy Commercial Vehicle Controlled by PID Controller Based on Genetic Algorithm

Yitong Wang, Ke Chen, and Mingming Dong

Abstract Heavy commercial vehicles have high requirements for ride smoothness due to their objective requirements. In order to study the characteristics of their suspension system, 1/4 passive suspension and 1/4 active suspension models were established, respectively. Firstly, the front axle data and road excitation model of the experimental sample vehicle are substituted into 1/4 passive suspension; the reliability of the simulation parameters is verified by comparing the vertical acceleration of the body with the data collected from the real vehicle vibration test. MATLAB Simulink platform was used to compare and analyze the performance of the passive suspension, traditional PID control active suspension, and genetic algorithm PID control active suspension with the body vertical acceleration, suspension dynamic travel, and suspension dynamic load as indexes. The simulation results show that the genetic algorithm PID active suspension system can improve the ride smoothness of heavy commercial vehicles, which lays a theoretical foundation for the subsequent bench test of active suspension of the front axle of heavy commercial vehicles to a certain extent.

53.1 Introduction

Suspension is a general term for all the power transmission connection devices between the frame and the wheel (axle). Its performance directly affects the ride smoothness, controllability, and road condition adaptability of the car. According to different principles, suspension is currently mainly divided into passive suspension, semi-active suspension, and active suspension [1, 2].

Y. Wang · M. Dong (✉)
Beijing Institute of Technology, Beijing, China
e-mail: vdmm@bit.edu.cn

Y. Wang
e-mail: 657038907@qq.com

K. Chen
Inner Mongolia First Machinery Group Co.,Ltd, Inner Mongolia, China

G. Liu and F. Cen (eds.), *Advances in Precision Instruments and Optical Engineering*, Springer Proceedings in Physics 270, https://doi.org/10.1007/978-981-16-7258-3_53

Traditional passive suspension consists of a spring-damping structure, and its stiffness and damping are fixed, so the adaptability to different road conditions is poor. Based on the traditional suspension structure, active suspension can actively adjust the stiffness and damping according to the motion state of the vehicle and the size of road feedback excitation [3]. The essence of active suspension is to continuously apply regulating force to improve ride smoothness, controllability, and comfort of the vehicle.

If the active suspension is adopted, it is necessary to design the system control strategy. At present, the relevant control strategies mainly include neural network fuzzy control, immune control, PID control, and fuzzy PID control [4]. Traditional PID has been widely used due to its high control accuracy and good robustness, but its parameter determination has some blindness. Due to the popularization of genetic algorithm in recent years, genetic algorithm has become an effective approach for PID parameter tuning [5].

Heavy commercial vehicles have high requirements for ride smoothness due to their objective requirements. This paper takes the vertical acceleration of the body, suspension travel, and dynamic load as the research targets. The simulation parameters are determined by comparing the real vehicle test and the simulation test of passive suspension. Finally, the performance index obtained by the active suspension simulation design is compared with the performance index after the passive suspension simulation, which verifies that the ride smoothness of heavy commercial vehicles is greatly improved after the genetic algorithm PID control.

53.2 Basic Model Building

53.2.1 Establish a ¼ Suspension Model

The two-degree-of-freedom suspension system model has the basic characteristics of solving practical problems, which can better express the mechanical characteristics of the suspension, and the simple system structure greatly reduces the description of the system parameters and makes calculation easier. Therefore, the ¼ suspension system model is widely used to study the vertical vibration of vehicle body caused by road roughness. The function of passive suspension is to highlight the control effect of active suspension system, dynamic models of 1/4 passive suspension and 1/4 active suspension are established, respectively, based on Newton's second law, as shown in Figs. 53.1 and 53.2.

In Fig. 53.1, m_a is the unsprung mass, m_b is the sprung mass, K_t is the tire stiffness, K_s is the suspension system stiffness, C_s is the suspension system damping coefficient, x_q is the road surface displacement function, x_a is the vertical displacement of the equivalent unsprung mass, and x_b is the vertical displacement of the suspension sprung mass. The dynamic equation is shown in (1):

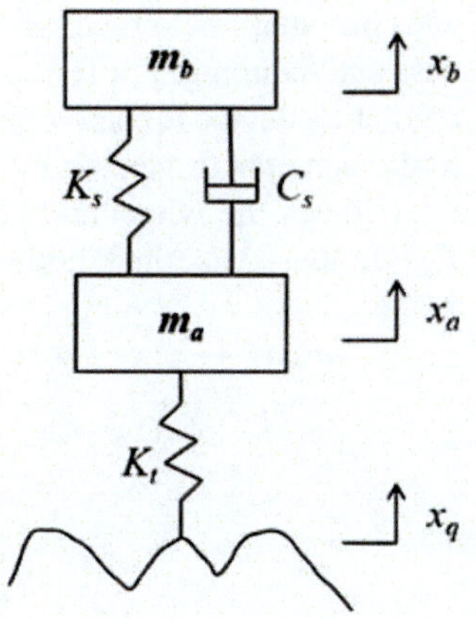

Fig. 53.1 1/4 passive suspension model

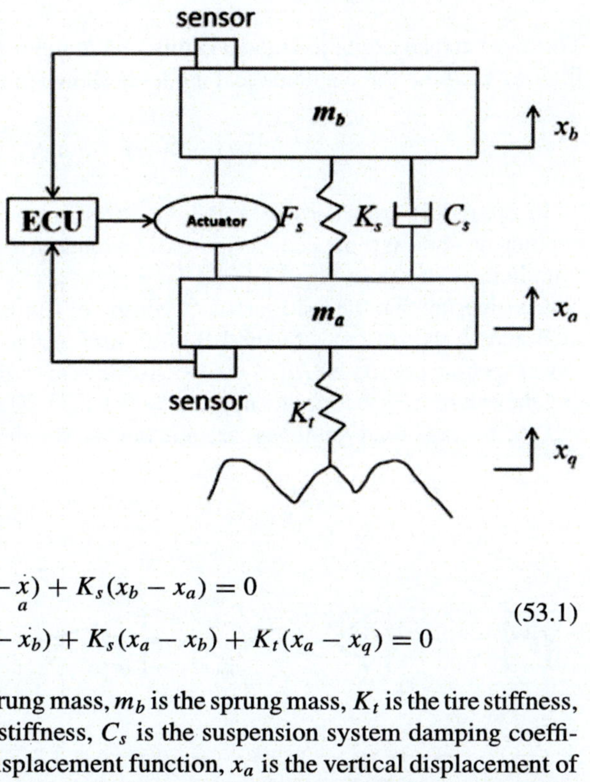

Fig. 53.2 1/4 active suspension model

$$
\begin{aligned}
& m_b \ddot{x}_b + C_s(\dot{x}_b - \dot{x}_a) + K_s(x_b - x_a) = 0 \\
& m_a \ddot{x}_a + C_s(\dot{x}_a - \dot{x}_b) + K_s(x_a - x_b) + K_t(x_a - x_q) = 0
\end{aligned}
\tag{53.1}
$$

In Fig. 53.2, m_a is the unsprung mass, m_b is the sprung mass, K_t is the tire stiffness, K_s is the suspension system stiffness, C_s is the suspension system damping coefficient, x_q is the road surface displacement function, x_a is the vertical displacement of the equivalent unsprung mass, and x_b is the vertical displacement of the suspension sprung mass. The main principle of the active suspension control is to collect the vertical acceleration information of the sprung mass and the unsprung mass by the sensor, and transmit the acquired analog signal to the ECU. The control signal is transmitted to the active suspension system through the analog-to-digital conversion and computational processing of the ECU to control the dynamic changes of its stiffness and damping. For the convenience of research, this paper idealized it as a force regulating device for active suspension, and transmitted the signal processed by ECU to the actuator to generate the reverse regulating force adapted to the acceleration direction of the wheel and body [6]. The output regulating force of the actuator is F_s. On this basis, the kinetic equation can be established as shown in (53.2):

$$\begin{aligned} m_b \ddot{x}_b + C_s(\dot{x}_b - \dot{x}_a) + K_s(x_b - x_a) + F_s = 0 \\ m_a \ddot{x}_a + C_s(\dot{x}_a - \dot{x}_b) + K_s(x_a - x_b) + K_t(x_a - x_q) - F_s = 0 \end{aligned} \quad (53.2)$$

53.2.2 Build the Road Excitation Model

The random road excitation model in this simulation is established by the white noise filtering method; the mathematical model is shown in (53.3):

$$\dot{q}(t) = W(t) - avq(t) \quad (53.3)$$

In Eq. (53.3), $q(t)$ is road excitation; $W(t)$ is white noise; a is road roughness coefficient; and v is the speed. On this basis, a random road model is built in MATLAB Simulink, as shown in Fig. 53.3.

Considering that the road surface spectrum of China is basically within the range of B and C, the national standard B-class road surface is selected, where the road power spectral density is $Gq(n) = 6.4 \times 10^{-5} \text{m}^2/\text{m}^{-1}$, the speed is selected as 20 m/s, and the reference space frequency is $n = 0.1\ \text{m}^{-1}$. The time domain curve and the PSD of B-grade road roughness are obtained as shown in Figs. 53.4 and 53.5:

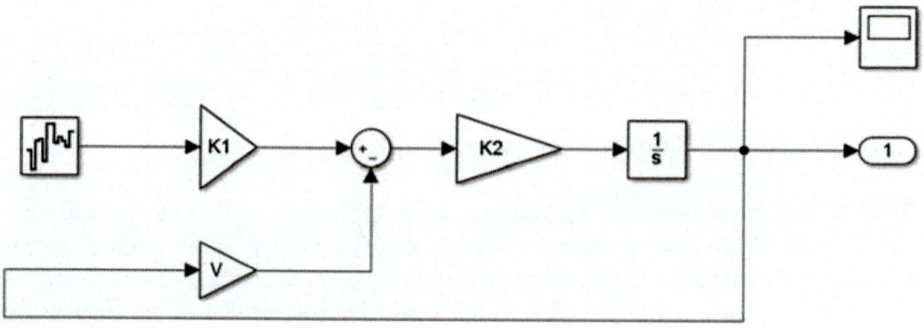

Fig. 53.3 White noise filtering random road model

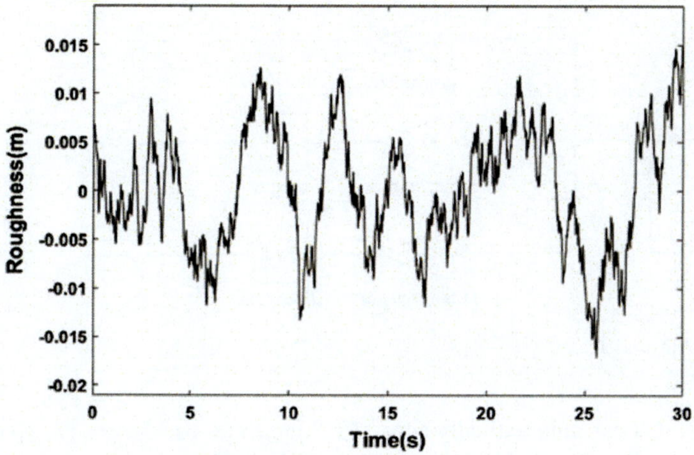

Fig. 53.4 Time domain curve of class B road roughness

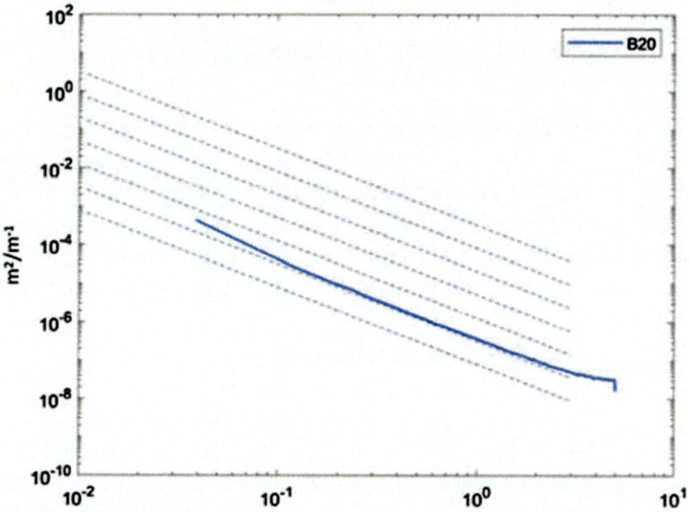

Fig. 53.5 PSD of class B road surface

53.3 Establish the Genetic Algorithm PID Optimal Control Model

For a different control system, the traditional PID setting parameters are often not the most ideal control parameters, resulting in the performance of the process control system not achieving the desired effect, and the control effect of the PID controller cannot get into full play. Therefore, it has great practical application value to design a

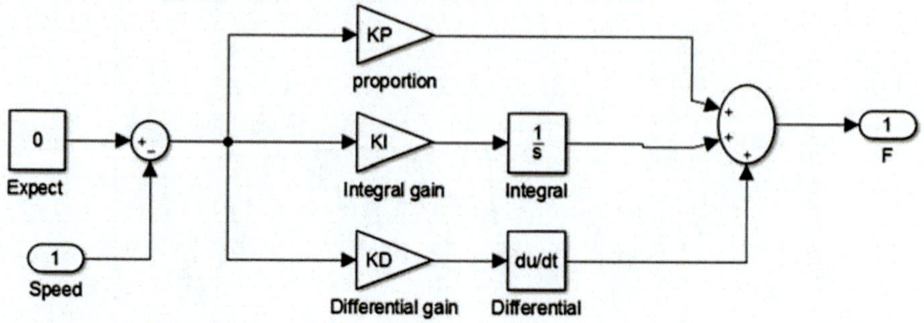

Fig. 53.6 PID model

technology that can automatically adjust PID controller parameters [7, 8]. Establish the PID controller Simulink model as shown in Fig. 53.6.

The input is the vertical acceleration of the body, the input reference value is 0 m/s, and the output is the actuator control force F. The formula is (53.4).

$$F = k_p e(t) + k_i \int_0^T e(t)dt + k_d \frac{de(t)}{dt} \tag{53.4}$$

In order to solve the above problems existing in traditional PID parameter tuning, this paper adopts a genetic algorithm to optimize PID parameters. The design flowchart of PID control based on genetic algorithm is shown in Fig. 53.7.

Firstly, the basic idea is to select K_p, K_i, and K_d as design variables and expect the minimum vertical acceleration of the body. Therefore, the minimum root mean square value of the vertical acceleration of the body is taken as the objective function.

$$\min \; J = \sqrt{\frac{\Sigma_{i=1}^{N}(\ddot{\underset{b}{z}}(i)^2)}{N}} \tag{53.5}$$

In Eq. (53.5), N is the number of data points; and $i = 1, 2, \ldots, N$; $\ddot{\underset{b}{z}}(i)$ is the vertical acceleration of the car body.

The design process of the genetic algorithm is as follows:

The initial population is generated first: *the Binary coding is used for the initial population coding, and each chromosome is a real vector.*

Assign values to K_p, K_i, and K_d *of the encoded chromosomes, and then substitute them into the PID control model to solve the regulating force F, and run the whole system model to get the fitness function value.*

To calculate the population fitness function value, the optimization objective of the fitness function value L of the genetic algorithm is set as follows:

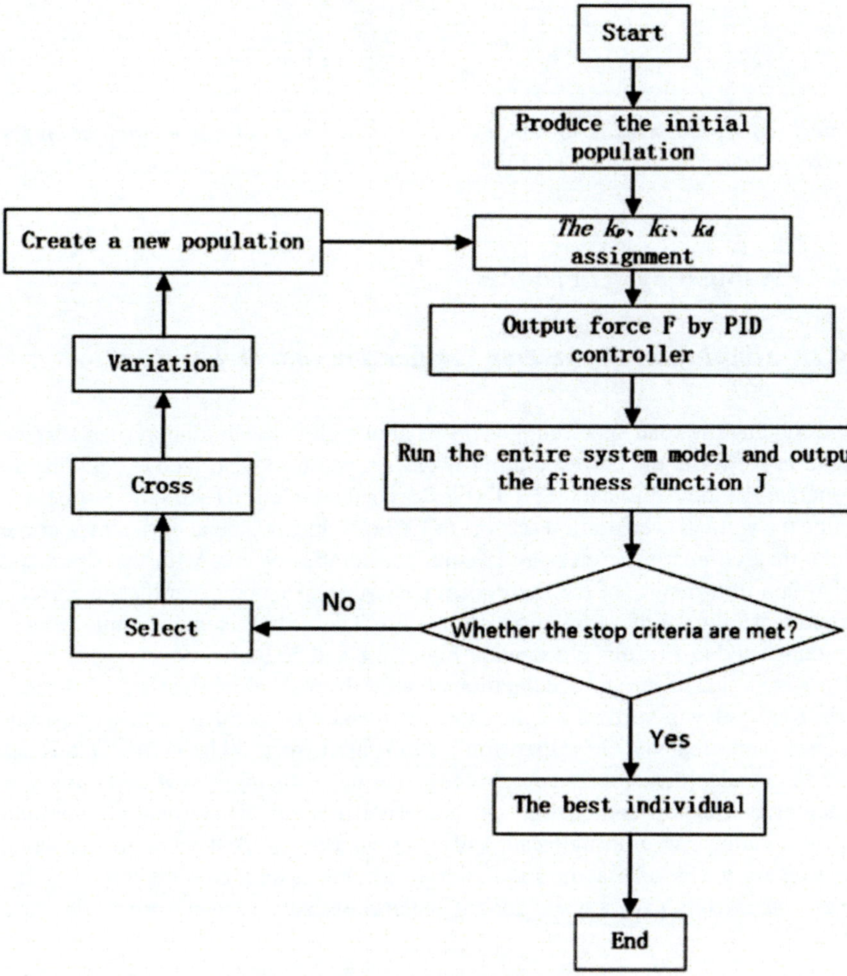

Fig. 53.7 Design process of the Genetic algorithm PID controller

$$L = \frac{1}{RMS\left[\ddot{z_b}\right]} \tag{53.6}$$

In order to ensure the improvement of the vertical acceleration condition of the body and improve the controllability of the car, this design restricts the dynamic deflection of the suspension, and the limiting conditions are shown in (53.7) *and* (53.8):

$$\max(z_b(i) - z_a(i)) \le 0.3 \tag{53.7}$$

$$\sqrt{\frac{\sum_{i=1}^{N}(z_b(i) - z_a(i))^2}{N}} \leq 0.1 \tag{53.8}$$

Select, cross, and mutate to retain the best target and get a new population for the next round of the genetic algorithm.

53.4 Establish System Model

53.4.1 Simulation Parameter Verification and Determination

By introducing the road spectrum generated above into the simulation road surface generation software, the corresponding B-class random road surface excitation can be generated in the American MTS 320 series road simulation hydraulic test bench. Arrange the vertical acceleration sensors at 1/4 body and axle head of the test sample vehicle, then compare the measured actual acceleration values with the simulated acceleration information of the theoretical passive suspension to verify the correctness and feasibility of the simulation parameters. The comparison of sample vehicle test pictures and test results is shown in Figs. 53.8 and 53.9.

The RMS value of vertical acceleration obtained by real vehicle test is 0.6571 m/s^2, and the RMS value of vertical acceleration obtained by 1/4 passive suspension simulation based on a real vehicle suspension system parameters is 0.6191 m/s^2. Consider that in the actual test exist some complex engineering problems such as asynchrony of data acquisition and differences between actual and theoretical parameters, combined with acceleration curve fitting trend, and by comparing the root mean square value of parameters, we found that the simulation results are close to the actual test results, so these parameters can be used for the subsequent active suspension simulation

Fig. 53.8 Sample vehicle test

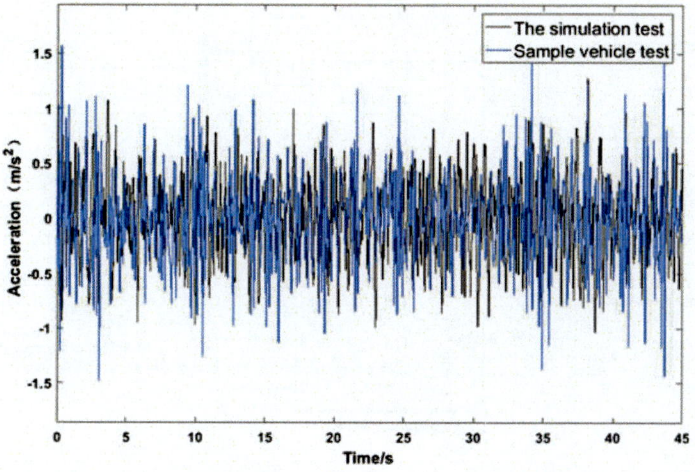

Fig. 53.9 Body vertical acceleration contrast curve

Table 53.1 The simulation parameters

Parameter	Numerical
m_a/kg	150
m_b/kg	2500
$K_s/(N \cdot m^{-1})$	192,000
$K_t/(N \cdot m^{-1})$	50,000
$C_s/(N \cdot s \cdot m^{-1})$	10,000

design. Table 53.1 shows some parameters of the front axle suspension based on the test sample vehicle.

53.4.2 Build SIMULINK System Model

The GA-PID control model of the 1/4 passive suspension system and 1/4 active suspension system of the vehicle was established in MATLAB/Simulink software, as shown in Fig. 53.10. It is mainly composed of the ¼ active and passive suspension model and genetic algorithm PID model. The input signal is the road excitation model mentioned above, and the output signal is the body vertical acceleration, suspension dynamic travel, and suspension dynamic load of 1/4 active and passive suspension, respectively.

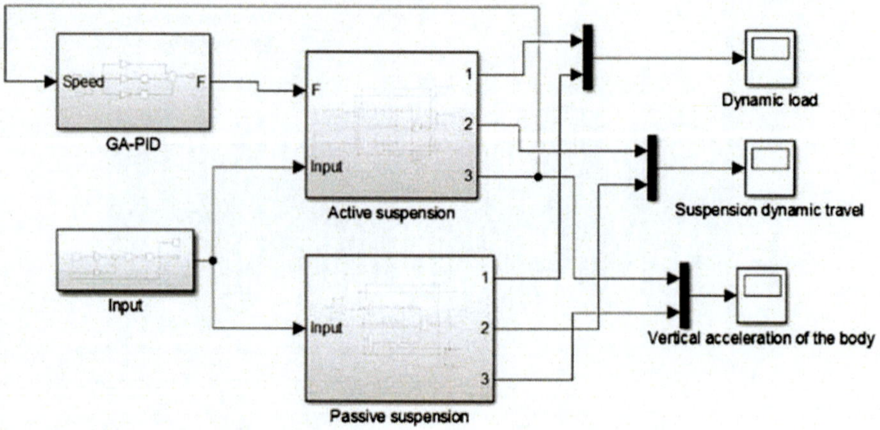

Fig. 53.10 System simulation model

53.5 Simulation Experiment and Result Analysis

The initial conditions for setting the genetic algorithm are as follows: the search range of K_p, K_i, and K_d parameters is [1,2000], with the binary code, the population number is 100, the maximum evolutionary algebra is 230 generations, the crossover probability is 0.85, and the mutation probability is 0.1. The initial population is randomly generated by the computer within the range of variables. The optimized results are obtained by substituting them into the above model: $K_p = 1001.2$, $K_i = 496.15$, and $K_d = 1.25$. Figures 53.11, 53.12, and 53.13 are the curves of simulation results of body vertical acceleration, suspension travel, and dynamic load before and after PID control and genetic algorithm optimization. The root mean square values of each performance are shown in Table 53.2. As you can see from the chart, the active suspension under the PID control compared to the passive suspension in the body

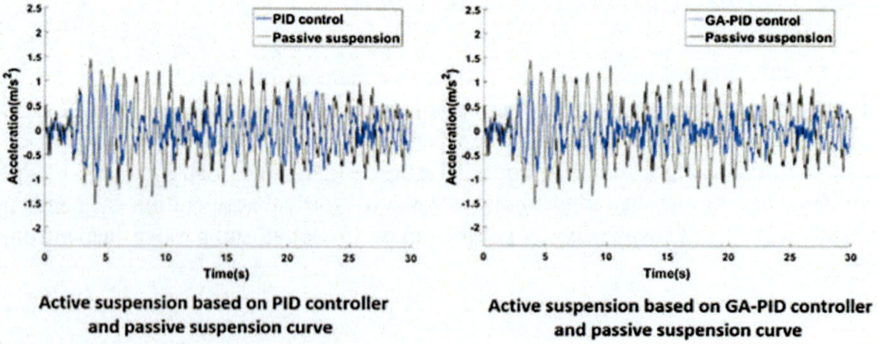

Fig. 53.11 Body vertical acceleration curve

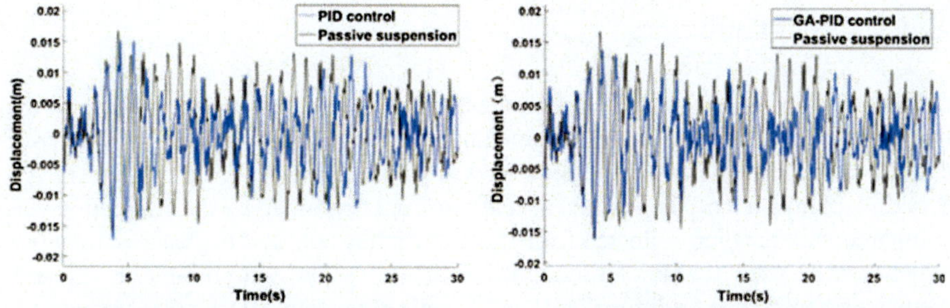

Active suspension based on PID controller and passive suspension curve

Active suspension based on GA-PID controller and passive suspension curve

Fig. 53.12 Suspension dynamic travel

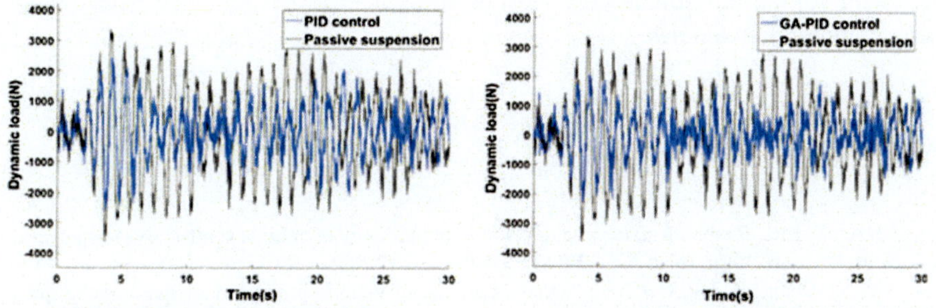

Active suspension based on PID controller and passive suspension curve

Active suspension based on GA-PID controller and passive suspension curve

Fig. 53.13 Dynamic load

Table 53.2 Comparison of suspension performance

Root mean square value of evaluation index	Passive suspension	PID control	Genetic algorithm-PID control
Vertical acceleration of the body (m/s^2)	0.6191	0.3686	0.2897
Dynamic load (N)	1.4225×10^3	825.7994	687.7302
Suspension dynamic travel (m)	0.0068	0.0053	0.0046

vertical acceleration, suspension travel, and dynamic load has large improvement. After a genetic algorithm to optimize PID controller, compared with the ordinary PID control of active suspension in the body vertical acceleration comfort increased by 21.4%, the ride smoothness of suspension dynamic travel is improved by 16.72%, and the ride smoothness of suspension dynamic load is improved by 13.21%. Therefore, it can be concluded that the PID control optimized by a genetic algorithm plays a great role in improving the ride smoothness of heavy commercial vehicles.

53.6 Conclusion

Based on the front axle parameters of the test sample vehicle, this paper set up the two degree-of-freedom quarter master model of passive suspension and the road excitation model. 1/4 passive suspension was simulated with the help of MATLAB Simulink platform, and the vertical acceleration data obtained by the simulation was compared with the data collected from the real vehicle test, and the basic parameters which can be used for the subsequent 1/4 active suspension simulation research were determined. The traditional PID and genetic algorithm PID control are used for further simulation, and the simulation results were compared with the three indexes of passive suspension, which verifies its role in improving ride smoothness of heavy commercial vehicles, and also proved the advantage of the active suspension system optimized by genetic algorithm PID control. To a certain extent, it laid a theoretical foundation for the subsequent bench test of the active suspension of the front axle of heavy commercial vehicles.

References

1. Q. Wang, X. Mei, Research status and development tendency of vehicle semi-active suspension. J. Hefei Univ. Technol. (Nat. Sci.) **36**(11), 1289–1294 (2013)
2. J. Nie, X. Zhang, B. Hu, L. Chen, A new approach of vehicle passive suspension techniques. Vehicle Power Technol. **128**(02), 59–64 (2012)
3. K. Zhou, Z. Han, Design of LQG controller of automobile active suspension based on simulink. Automotive Technol. **2**, 21–23 (2010)
4. J. Gaidhane Prashant, J. Nigam Madhav, A. Kumar, P.P. Mohan, Design of interval type-2 fuzzy precompensated PID controller applied to two-DOF robotic manipulator with variable payload. ISA Trans. (2018)
5. C. Wang, S. Wang, Parameters setting of PID controller based on genetic algorithm. Comput. Simul. **22**(10), 112–115 (2005)
6. R. Dong, H. Wu, Research on performance of vehicle active Suspension based on fuzzy PID composite control. J. Gulin Univ. Aerospace Technol. **25**(02), 194–198 (2020)
7. C. Zhan, C. Cheng, S. Sun, Research on air suspension control strategy based on fuzzy adaptive PID controller. J. Wuhan Univ. Technol. **39**(05), 934–938
8. H. Yang, The design and simulation research on active control system for commercial vehicle cab suspension. Hunan Univ. (2014)

Chapter 54
Health Condition Assessment of Hydraulic System Based on Cloud Model and Dempster–Shafer Evidence Theory

Shuaijie Mei, Mei Yuan, Jin Cui, Shaopeng Dong, and Juanru Zhao

Abstract Hydraulic transmission systems are widely used in industry because of their high output power and compact structure. To cope with the ambiguity and uncertainty in the process of hydraulic system health monitoring, this paper adopts the combination of cloud model and Dempster–Shafer evidence theory for multi-sensor data fusion from three levels: data layer, feature layer, and decision layer, which effectively avoids the problem of high conflict of evidence in Dempster–Shafer theory and completes the assessment of health status of a complex hydraulic system. Firstly, the cloud parameters are calculated to establish the expert knowledge base. Secondly, the membership matrix is used to obtain the basic probability assignments of the evidence. Then, the fusion decision of combining the same type of sensors with evidence iterations is used to improve the efficiency of fusion, and finally, Dempster's rule is performed to obtain the hydraulic system health status assessment results. The feasibility and effectiveness of this method are verified on a real data set.

S. Mei · M. Yuan · S. Dong · J. Zhao
School of Automation Science and Electrical Engineering, Beihang University, Beijing, China
e-mail: shuaijiemei@buaa.edu.cn

M. Yuan
e-mail: yuanm@buaa.edu.cn

S. Dong
e-mail: dspsx@buaa.edu.cn

J. Zhao
e-mail: audrey_zhao@buaa.edu.cn

J. Cui (✉)
Research Institute for Frontier Science, Beihang University, Beijing, China
e-mail: jincui@buaa.edu.cn

M. Yuan · J. Cui · S. Dong
Ningbo Institute of Technology, Beihang University, Ningbo, China

G. Liu and F. Cen (eds.), *Advances in Precision Instruments and Optical Engineering*, Springer Proceedings in Physics 270, https://doi.org/10.1007/978-981-16-7258-3_54

54.1 Introduction

Hydraulic systems have become indispensable transmission systems in many fields under their high stability, high transmission ratio, and adaptability to complex operating conditions [1]. To ensure the reliability of hydraulic systems operating in harsh environments, it is necessary to monitor the condition health of hydraulic system components [2]. The first is a model-based approach based on the modeling of the physical and structural information of the hydraulic system, which usually has poor monitoring results due to the inability to obtain enough detailed information about the complex structure. The second is a statistical approach based on historical measurement data and fault characteristic information [3], including the deep learning approach [4], which has become popular in recent years, the shortcoming of this approach is that it requires a large amount of historical data and has high requirements for data completeness and certainty [5]. Therefore, considering the problems of uncertainty and randomness in the operation of hydraulic systems, the above methods are not better applicable. Dempster–Shafer (D–S) evidence theory is a means of decision-making based on expert experience and has good advantages in the problem of ambiguity [6]. However, in the process of multi-sensor fusion, the evidence theory often suffers from the problem of conflicting or even contradictory evidence from multiple sources [7]. In this paper, we use the improved D-S evidence theory combined with the cloud model, which organically combines the fuzziness and randomness in the concept of uncertainty using cloud model [8], and calculates the cloud parameters of each state parameter as well as the affiliation degree; then, the basic probability assignment matrix of each evidence in D–S theory is obtained from the affiliation degree of each state parameter; further, to solve the problem of high conflict of evidence in D–S evidence theory, the isomorphic sensor evidence is averaged and iterated, and finally, dempster rule evidence synthesis is performed for heterogeneous sensors to obtain the evaluation results of hydraulic system health status.

54.2 Materials and Methods

54.2.1 Cloud Model Characteristics

Let U be a quantitative domain of arbitrary dimension expressed through exact numerical values, C be a qualitative concept within this domain, x be a quantitative value, a random realization of C, $x \in U$, and x be a random number with a stable tendency for the determinacy $\mu(x) \in [0, 1]$ of C.

$$\mu : U \rightarrow [0, 1] \forall x \in U x \rightarrow \mu(x) \tag{54.1}$$

Then the distribution of x over the theoretical domain U is called a cloud, and each x is called a cloud drop. The normal cloud model expresses the numerical characteristics of a qualitative concept in terms of a set of mutually independent parameters that together reflect the uncertainty and wholeness of the concept, thus enabling better quantitative analysis [9]. The numerical characteristics of a cloud usually contain three parameters: expectation E_x, entropy E_n, and superb entropy H_e. λ is a constant value determined according to the ambiguity and randomness of specific different parameters.

$$E_x = \overline{x} = \frac{\sum_{m=1}^{k} x}{k} \tag{54.2}$$

$$E_n = \sigma_x = \sqrt{\frac{1}{k}\sum_{m=1}^{k}(x - \overline{x})^2} \tag{54.3}$$

$$H_e = \lambda \tag{54.3}$$

54.2.2 D–S Evidence Theory

D–S evidence theory is an imprecise inference theory approach that addresses uncertainty due to lack of knowledge, and uses the "identification fram" Θ to represent the set of data to be fused, and gives a function $m : 2^{\Theta} \rightarrow [0,1]$ if it satisfies

$$m(\varnothing) = 0, \sum_{A \subset \Theta} m(A) = 1 \tag{54.5}$$

Then m is called the set of basic credibility of such identification frame Θ, if A is contained in the identification frame Θ, $m(A)$ is called the basic credibility function of A. The basic credibility function $m(A)$ represents the magnitude of the credibility of A itself.

For an arbitrary set, D–S evidence inference gives a notion of credibility function

$$Bel(A) = \sum_{B \subset A} m(B) \tag{54.6}$$

Suppose there exists an A contained in this recognition framework Θ, then give the following definition:

$$pl(A) = 1 - Bel(\overline{A}) Dou(A) = Bel(\overline{A}) \tag{54.7}$$

where pl is called $Bel's$ likelihood function, and Dou is called $Bel's$ doubt function. Then according to this definition, we can call $pl(A)$ the seeming truth of A, and $Dou(A)$ can be called the doubtfulness of A. According to the plausibility synthesis law proposed by Dempster, then we give the synthesis law for two plausibility degrees

$$m(A) = m_1 \oplus m_2 = \frac{\sum_{A_i \bigcap B_j = A} m_1(A_i) m_2(B_j)}{1 - \sum_{A_i \bigcap B_j = \Theta} m_1(A_i) m_2(B_j)} \tag{54.8}$$

Using $m_1, m_2, \ldots, m_n$ to represent the credibility distribution function of n data, and these n data are independent of each other, then multiple credibilities can be written in the following form after fusion:

$$m(A) = m_1 \oplus m_2 \oplus \cdots \oplus m_n = \frac{\sum_{\cap A_i = A} \prod_{i=1}^{m} m_i(A_i)}{1 - \sum_{\cap A_i = \Theta} \prod_{i=1}^{m} m_i(A_i)} \tag{54.9}$$

54.2.3 *Improved D–S Evidence Theory Based on Cloud Model*

The flow chart of improved D–S evidence theory based on the cloud model for hydraulic system health condition assessment is shown in Fig. 54.1. It mainly includes the calculation of cloud model feature parameters, the establishment of cloud model knowledge base, the calculation of membership function and basic probability assignment, the fusion of D–S evidence, and the fusion result by fusion decision.

Suppose that there are n classes of faults in the expert system knowledge base of the system: $F_1, F_2, F_3, \ldots F_n$, each class of faults has m characteristic parameters: $x_{i1}, x_{i2}, \ldots, x_{im}$, where $x_{ij} (j = 1, 2, ..m)$ denotes the jth parameter of the ith class of faults. In industry, there are two main types of parameters for industrial equipment,

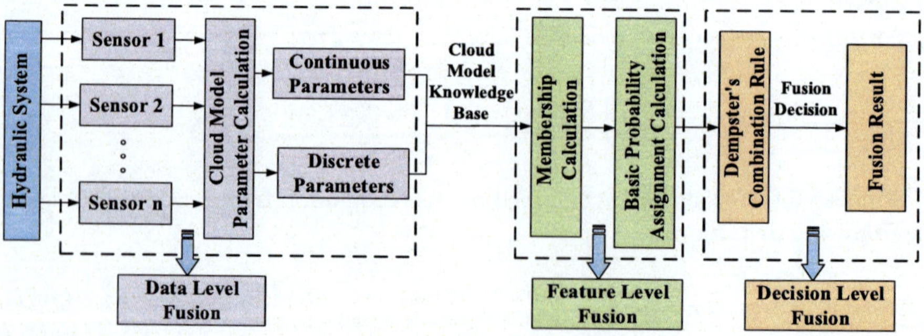

Fig. 54.1 Block diagram of multi-sensor data fusion in hydraulic system

discrete and continuous parameters, so these two can be modeled separately. The required a priori knowledge is obtained through historical information to construct the fault knowledge base.

For the modeling of continuous parameters, since the values of variables are different under different fault conditions, the cloud with the main action region as the bilateral constraint region can be used to approximate the modeling. If the value interval in the fault signal measured under the rth fault mode is $[C_{\min}(r), C_{\max}(r)]$, the median value of the constraint can be adopted as the expected value, and the specific cloud model parameters are calculated as follows:

$$E_{xij}(r) = \frac{(C_{\min}(r) + C_{\max}(r))}{2} \tag{54.10}$$

$$E_{nij}(r) = \frac{(C_{\min}(r) + C_{\max}(r))}{6} \tag{54.11}$$

$$H_{eij} = \lambda \tag{54.12}$$

For the modeling of discrete parameters, the cloud model of the expert system fault knowledge base can be established directly by experimentally measuring the mathematical expectation and standard deviation of the variable parameters (see 54.2–54.4). For the characteristic variables of discrete parameters, the membership degree is calculated as follows, which is the same as continuous parameters.

$$\mu_{ij} = e^{\frac{-\left(x_j - E_{xij}\right)^2}{2\left(E'_{nij}\right)^2}} \tag{54.13}$$

where $\mu_{ij}(k)$ is the membership degree of the jth characteristic parameter of a fault signal obtained by the measurement with respect to the jth characteristic pattern of the i-th class of faults in the expert knowledge base, E_{xij} denotes the expectation value previously obtained in the expert knowledge base, and E'_{nij} is a normal random number generated with the entropy E_{nij} as the expectation and the superentropy H_{eij} as the standard deviation and the normal random number generated. This leads to the affiliation matrix

$$R_{m \times n} = \begin{bmatrix} \mu_{11} & \mu_{12} & \cdots & \mu_{n1} \\ \mu_{12} & \mu_{22} & \cdots & \mu_{n2} \\ \vdots & \vdots & \vdots & \vdots \\ \mu_{1m} & \mu_{2m} & \cdots & \mu_{nm} \end{bmatrix} \tag{54.14}$$

To improve the credibility and accuracy of the fusion results, the membership degree matrix $R_{m \times n}$ is normalized.

$$\gamma_{ij} = \mu_{ij} / \sum_{i=1}^{n} \mu_{ij}, j = 1, 2, \ldots m \tag{54.15}$$

The uncertainty of the actual measurement signal due to the errors caused by the circumstances such as the measurement environment and the measurement method in the actual project is represented by the variable θ. Where $max(\mu_{i1}, \mu_{i2}, \ldots, \mu_{nm})$ denotes the maximum value of each element in each row of the membership degree matrix.

$$\theta_j = 1 - \max(\mu_{i1}, \mu_{i2}, \ldots, \mu_{nm}), j = 1, 2, \ldots m \tag{54.16}$$

Thus, the basic probability assignment function can be computationally determined as

$$\begin{cases} m(\Theta_j) = \theta_j, j = 1, 2, \ldots m \\ m(F_{ij}) = (1 - \theta_j)\gamma_{ij}, i = 1, 2, \ldots n \end{cases} \tag{54.17}$$

where $m(\Theta_j)$ denotes the basic probability assignment of the jth evidence uncertainty in the test sample, and $m(F_{ij})$ denotes the basic probability assignment of the jth characteristic parameter of the fault signal obtained from the measurement compared to the jth characteristic value of the ith fault in the expert knowledge base. The basic probability assignment matrix $M_{m\times(n+1)}$ for m rows and $n + 1$ columns can be obtained after considering both measurement data and uncertainty

$$M_{m\times(n+1)} = \begin{bmatrix} m(R_{11}) & \cdots & m(R_{n1}) & \theta_1 \\ m(R_{12}) & \cdots & m(R_{n2}) & \theta_2 \\ \vdots & \ddots & \vdots & \vdots \\ m(R_{1m}) & \cdots & m(R_{nm}) & \theta_m \end{bmatrix} \tag{54.18}$$

In order to solve the problems of low sensitivity among fault features, high conflict among fused evidence and large uncertainty, this study determines the weights of fused evidence by two aspects and reallocates the weights by the uncertainty coefficient ω_j^{σ} and the overall support coefficient ω_j^{s} of the evidence, respectively, so as to mitigate the conflict problem among the evidence, and after that, use Dempster's rule for evidence fusion. Let ω_j be the weight coefficient of the jth fault feature measured after fusing the evidence, then ω_j should satisfy the condition that

$$\sum_{j=1}^{m} \omega_j = 1, \omega_j \geq 0 \tag{54.19}$$

$$\omega_j = 0.5\omega_j^{\sigma} + 0.5\omega_j^{s}, j = 1, 2, \ldots, m, 0 \leq \omega_j \leq 1 \tag{54.20}$$

In the actual industry, there will be errors when measuring and collecting data due to the layout location of heterogeneous sensors, environmental conditions, and other factors, and the weight coefficient determined by the uncertainty brought by the sensor measurement and the fusion evidence is defined as the uncertainty coefficient. Let the relative measurement error of the sensor be χ_j, then

$$\chi_j = \sqrt{\sigma_j^2 / E(x_j)},\ j = 1, 2, \ldots, m \tag{54.21}$$

$$\omega^{\sigma} = \frac{1}{\chi_j + \theta_j} / \sum_{k=1}^{m} \frac{1}{\chi_j + \theta_j},\ j = 1, 2, \ldots, m \tag{54.22}$$

where $E(x_j)$, σ_j^2 denote the mean and variance of the jth fault characteristic parameter, respectively. The overall support coefficient indicates the mutual support between the evidence and the evidence, and the overall support of the evidence is determined by the distance between the evidence, assuming the existence of two pieces of evidence m_j and m_d, and defining the distance function as

$$d(m_j, m_d) = \frac{\|m_j - m_d\|}{\sqrt{s}} = \sqrt{\frac{1}{s} \sum_{i=1}^{s} (m_{ji} + m_{di})^2 j, d = 1, 2, .., m} \tag{54.23}$$

Then the overall support of the evidence is

$$\eta(m_j) = \sum_{d=1, d \neq j}^{m} (1 - d(m_j, m_d)) j = 1, 2, \ldots, m \tag{54.24}$$

where the larger $\eta(m_j)$ indicates that the higher the support of the evidence in the overall evidence, the less conflict with other evidence, and thus the greater the weight of the evidence in the final fusion. The overall support coefficient of the evidence is calculated as

$$\omega_j^s = \eta(m_j) / \sum_{j=1}^{m} \eta(m_j) j = 1, 2, \ldots, m \tag{54.25}$$

In order to reduce the number of evidence in the final fusion, reduce the running time and improve the fusion efficiency, after obtaining the basic probability assignment matrix and the fusion weight coefficients, the combined evidence iterations are performed on the homogeneous sensor information, and then the final fusion is performed by the Dempster combination rule after the iterations to obtain the decision results. Let there be j pieces of evidence generated by the feature parameters obtained from the homogeneous sensors, the average iterative evidence is calculated as

$$m(\Theta) = \sum_{j=1}^{m} \omega_j \theta_j, m(F) = \sum_{j=1}^{m} \omega_j m(F_{ij}) i = 1, 2, \ldots, n \quad (54.26)$$

54.3 Results and Discussion

This work uses a publicly available real dataset of complex hydraulic systems, which has been publicly released by the UC Irvine Machine Learning Repository [2, 10]. The author has developed a hydraulic test bench to measure the state data of this hydraulic system through multiple real and virtual sensors, from which the characteristics of the hydraulic system under different faults are analyzed. This study is illustrated with one of the cooling state health states, which are divided into three operating states: Close to Total Failure (CTF), Reduced Efficiency (RE), and Full Efficiency (FE). For each condition, 150 sets of data are selected for the calculation to obtain a priori knowledge, and 60 sets of data are selected as tests to verify the results.

54.3.1 Cloud Model Parameters Calculation

The actual industry faces the problem of inconsistent sensor sampling rate, to unify the data length, this paper adopts the unified data length utilizing time–frequency domain feature extraction, 24 common time–frequency domain features are selected in this paper [11], and finally, the cloud model feature matrix data of the cooling condition of the hydraulic system of 3*24 is obtained, and the value of super entropy is taken as a constant value of 0.1, as shown in Table 54.1. Due to space limitation, only the first five parameters of a set of test data are shown in all the following tables.

Table 54.1 Cloud characteristics of partial time–frequency features

Condition type		Feature 1	Feature 2	Feature 3	Feature 4	Feature 5
Cloud expectation	CTF	19.8686	0.2613	19.8678	19.8704	0.5098
	RE	27.7732	0.2889	27.7724	27.7748	0.5411
	FE	47.1203	0.2618	47.1199	47.1210	0.5123
Cloud entropy	CTF	0.1921	0.0637	0.1920	0.1923	0.1151
	RE	0.2339	0.0747	0.2339	0.2340	0.1222
	FE	0.3311	0.0726	0.3311	0.3311	0.1167

Table 54.2 Partial basic probability assignment matrix

Time and frequency characteristics	CTF	RE	FE	Uncertainty
1	0.0000	0.0000	0.8047	0.1953
2	0.3225	0.3301	0.3365	0.0109
3	0.0000	0.0000	0.7231	0.2769
4	0.0000	0.0000	0.7928	0.2075
5	0.3239	0.3288	0.3448	0.0025

Table 54.3 Partial final fusion evidence and results

Evidence	CTF	RE	FE	Uncertainty
1	0.0077	0.0093	0.0117	0.0031
2	0.0083	0.0071	0.0094	0.0062
3	0.0108	0.0060	0.0128	0.0015
4	0.0096	0.0108	0.0105	0.0010
5	0.0064	0.0065	0.0087	0.0183
Θ	0.1750	0.1441	0.6809	0.0000

54.3.2 Calculation of the Basic Probability Assignment Matrix

According to the obtained cloud model parameters, calculate the membership degree of each parameter relative to the corresponding parameter of the cooling state of the hydraulic system, calculate the uncertainty of each piece of evidence, to obtain the basic probability assignment matrix of the hydraulic system as shown in Table 54.2.

54.3.3 Iteration of Homogeneous Sensor Merging Evidence

The same sensors in this dataset are iterated to merge evidence, 8 heterogeneous sensor evidence information are obtained, and finally, these 8 shreds of evidence are fused by applying Dempster's rule to obtain the cooling condition health assessment of this hydraulic system, as shown in Table 54.3, and the final fusion result of this test data is FE.

54.3.4 Experimental Results

The 60 sets of data of the cooling condition of each type of hydraulic system are subjected to the above experimental calculation, and the final operating condition

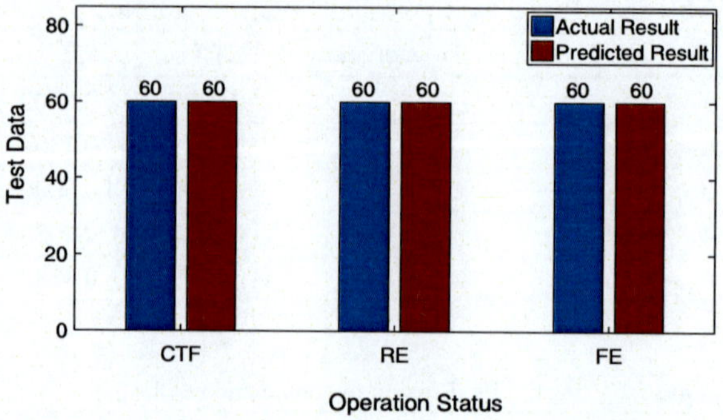

Fig. 54.2 Experimental results

classification results are shown in Fig. 54.2. It can be seen that the classification accuracy of the three operating states of the cooling condition of this hydraulic system reached 100%, and achieved quite good classification results, which proved the feasibility and effectiveness of the improved D–S evidence theory based on the cloud model.

54.4 Conclusions

In this paper, a hydraulic system health assessment method based on the cloud model and D–S evidence theory is proposed to address the problem of ambiguity and randomness of each assessment state quantity in hydraulic system health state assessment. To cope with the problem of the inconsistent sampling rate of hydraulic system acquisition sensors in the industry, a time–frequency domain feature analysis is performed to unify the data length; then, a multi-source information uncertainty fusion method for hydraulic system condition monitoring is constructed using the cloud model from quantitative to qualitative modeling and using D–S evidence theory to obtain the assessment results of hydraulic system health status. The validity and feasibility of the method are verified by real data sets to provide a basis for the condition maintenance of the hydraulic system.

Acknowledgements The research is supported by Beijing Natural Science Foundation (grant number L212033), and the National Key Research and Development Program of China (grant number 2019YFB1705502).

References

1. P. Guo, J. Wu, X. Xu, Y. Cheng, Y. Wang, Health condition monitoring of hydraulic system based on ensemble support vector machine, in *2019 Prognostics and System Health Management*

Conference (PHM-Qingdao) (2019), pp. 1–5. https://doi.org/10.1109/PHM-Qingdao46334.2019.8942981

2. N. Helwig, E. Pignanelli, A. Schuetze, Ieee, *Condition Monitoring of a Complex Hydraulic System using Multivariate Statistics* (2015 IEEE International Instrumentation and Measurement Technology Conference, 2015), pp. 210–215
3. T.M.A. Manghai, R. Jegadeeshwaran, G. Sakthivel, R. Sivakumar, D.S. Kumar, IEEE, Condition monitoring of hydraulic brake system using rough set theory and fuzzy rough nearest neighbor learning algorithms, in *2019 Ieee International Symposium on Smart Electronic Systems* (IEEE Computer Soc, Los Alamitos (in English), 2019), pp. 229–232
4. C. Konig, A.M. Helmi, Sensitivity analysis of sensors in a hydraulic condition monitoring system using CNN models. Sensors **20**(11), 3307. https://doi.org/10.3390/s20113307
5. L. Wang, K. N. Teng, W. M. Lv, System level health condition assessment method of complex equipment under uncertainty based on D-S evidence theory, in *2014 International Conference on Management Science & Engineering*, H. Lan Ed., (International Conference on Management Science and Engineering-Annual Conference Proceedings, IEEE, New York, 2014), pp. 435–441
6. G.Z. Zhao, A.G. Chen, G.X. Lu, W. Liu, Data fusion algorithm based on fuzzy sets and D-S theory of evidence, (in English). Tsinghua Sci. Technol. **25**(1), 12–19. https://doi.org/10.26599/tst.2018.9010138
7. F.Y. Xiao, A new divergence measure for belief functions in D-S evidence theory for multisensor data fusion (in English). Inf. Sci. **514**, 462–483. https://doi.org/10.1016/j.ins.2019.11.022
8. S. Kaparthi, A. Mann, D.J. Power, An overview of cloud-based decision support applications and a reference model (in English). Stud. Inform. Control **30**(1), 5–18. https://doi.org/10.24846/v30i1y202101
9. Y. Li, A. Wang, X. Yi, Based on normal cloud model and D-S evidence theory comprehensive evaluation of operation state of fire control system, in *2019 International Conference on Sensing, Diagnostics, Prognostics, and Control (SDPC)*, (2019), pp. 877–883. https://doi.org/10.1109/SDPC.2019.00167
10. S.S. Chawathe, *Condition Monitoring of Hydraulic Systems by Classifying Sensor Data Streams* (2019 IEEE 9th Annual Computing and Communication Workshop and Conference, 2019), pp. 898–904
11. Y. Lei, Z. He, Y. Zi, Q. Hu, Fault diagnosis of rotating machinery based on multiple ANFIS combination with GAS. Mech. Syst. Signal Proc. **21**(5), 2280–2294 (2007). https://doi.org/10.1016/j.ymssp.2006.11.003

Chapter 55
Design of CAN Communication Network for Tandem Hybrid Tractor

Shao Fengbo and Xu Liyou

Abstract With the rapid development of tractors, traditional tractors are rapidly changing to electrical structures. Traditional wiring methods cannot adapt to the information exchange and sharing between the electronic control units of hybrid tractors. In order to solve these problems, the advantages of four topological structures are analyzed and compared. The two-channel CAN communication network is used to construct the whole machine CAN network architecture. Based on the SAE J1939 protocol, the communication network node is designed, and the whole machine CAN communication network protocol is developed in combination with the tandem hybrid tractor structure and the characteristics of the working conditions. The CANoe software is used to test the network and calculate the network load rate for analysis, which verifies that the communication network system can meet the requirements of real time, accuracy, and stability.

55.1 Introduction

As an indispensable tool for agricultural production, tractors are the main driving force for field operations and play an important role in the development of modern agriculture [1]. In order to cope with energy shortages and greenhouse gas emissions worldwide, energy saving and environmental protection are important themes in the field of industrial technology in the world today [2]. The energy crisis and the increasing environmental pollution have forced traditional fuel tractors to transform into hybrid tractors and pure electric tractors. The tandem hybrid tractor keeps the engine in the best working condition, reduces engine emissions and fuel consumption, and meets the needs of environmental protection and energy saving.

The rapid development of network technology and bus technology has laid a solid foundation for the development of automotive electronic technology [3]. The traditional point-to-point communication method cannot meet the information sharing, real time, and reliability among too many electronic control units [4]. Using CAN

S. Fengbo (✉) · X. Liyou
Henan University of Science and Technology, Luoyang, Henan, China
e-mail: shaofb123@163.com

G. Liu and F. Cen (eds.), *Advances in Precision Instruments and Optical Engineering*, Springer Proceedings in Physics 270, https://doi.org/10.1007/978-981-16-7258-3_55

bus technology, each electronic control unit can be connected to a network, and data and information can be transmitted in a shared manner to realize networked digital communication and control functions.

CAN is a controller area network, and it is a field control bus with a two-layer protocol serial communication structure at the physical layer and the data link layer [5]. CAN bus data communication has outstanding reliability, real time, and flexibility [6]. CAN network technology is widely used in the field of tractors, providing effective solutions for real-time vehicle control and complicated wiring problems [7].

55.2 Design of CAN Communication Network Structure

55.2.1 *Node Definition*

Each node controller of the series hybrid tractor is connected through the CAN bus to form a vehicle control system, which controls the operation of each component separately, so as to realize the function of tractor field operation and transportation [8]. Its network nodes include: engine management system, battery management system, motor control system, ISG motor control system, gearbox control system, braking system, steering system, PTO control system, lighting system, instrument display system, lifting system, farm tools Control system, and fault detection system. As the work items of series hybrid tractors are changeable, corresponding nodes can be reserved for later expansion.

55.2.2 *Analysis and Selection of Network Topology*

There are four common network topologies on vehicles, namely: bus, star, ring, and mesh network topology. The characteristics of the four types of topology are:

- *Bus Structure*

The bus-type structure uses a line as the transmission medium, and all nodes in the network are connected to the bus line through specific connection hardware.

In the bus-type topology, all information in the network is transmitted on the same line, and all nodes can receive the information on the bus. The information sent by any node is based on the node as the center and is sent to both ends along the bus. This propagation mode determines the characteristics of the broadcast transmission of the bus-type network topology. It has the advantages of strong sharing of resources, simple node expansion and operation, simple and flexible structure, high network response speed, low equipment price, and high reliability. Due to the ends of the bus will be a reflection of the signal, need to be in the ends of the bus to terminal impedance with the bus impedance matching, in order to reduce the end signal reflection.

- *Star Structure*

The gateway is the centralized processing center for the information of other nodes in the network, and the information is transmitted to each node after passing through the gateway. The star structure has the characteristics of simple structure and easy construction and management, but the existence of the central point makes each node to pass through the central point before transmission, resulting in a lot of waste of lines. Once the central point fails, the entire network will stop working, which places high demands on the central point.

- *Ring Structure*

Different from the bus network, each node of the ring network is connected in series on a ring medium, and the information is transmitted in sequence, and the information is transmitted in a specific direction. After a week through the network, it returns to the sending node, and the sending node completes the recovery and deletion of the information. The advantages of the ring structure are simple structure, easier implementation, simple information transmission path, and definite time delay. However, since the information must pass through each node in turn to reach the receiving node, if there are too many nodes, the transmission of messages can be affected. And once any node fails, the whole system will be paralyzed, which will cause the problem of high network failure rate.

- *Mesh Structure*

Each node in the network is connected to each other, as long as there is a demand for information transmission, it can be connected with the corresponding line, and the line with the shortest path can be flexibly selected during the process of information transmission. The mesh structure makes maximum use of the resources of the entire network, which not only realizes the sharing of information, but also improves the efficiency of information transmission. The prerequisite for the realization of the function is to build a complex network, and only by adding a high-cost management system, there can be good communication efficiency. If there is a problem with the local area network, it will cause network failure and congestion of other nodes.

The requirements of the vehicle network topology are the characteristics of high node modularization, good expansion performance, reliability, and real-time requirements. Comprehensive comparison of various network topologies, the bus-type topology is more suitable for serial hybrid tractor on-board communication network [9].

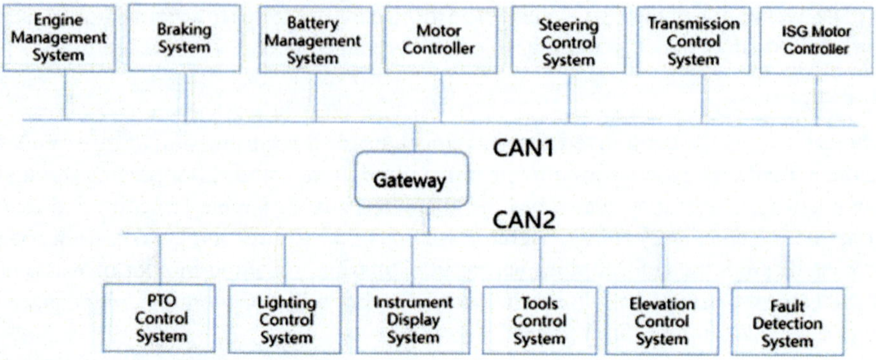

Fig. 55.1 Tandem hybrid tractor network topology

55.2.3 Communication Network Design

Because the bus-type network topology has the advantages of simple structure and high scalability, this article is based on the use of the bus-type network topology, as shown in Fig. 55.1. The communication rate of a single CAN bus is generally 125 kbit/s–1 Mbit/s. In order to ensure the real time and effectiveness of the signal acquisition of the control nodes on each CAN line, comprehensive considerations are adopted for each CAN bus rate to adopt 500 kbit/s [10]. The overall design of the CAN network is mainly to determine the number of CAN nodes and the information content that each node needs to receive and send [11]. The node engine management system, battery management system, motor control system, ISG motor control system, and gearbox The control system, braking system, and steering system are set on CAN1 to meet high real-time requirements. Set the PTO control system, lighting system, instrument display system, lifting system, farm tool control system, and fault detection system on CAN2. In order to ensure the reliability of bus communication, it is necessary to design a 120 Ω terminal resistance on the nearest and farthest controller of the system to eliminate the reflection of the signal in the communication circuit and reduce the risk of signal distortion [12].

55.3 Bus Network Communication Protocol

55.3.1 Message ID Design

Before defining the 29-bit identifier ID of each message, according to the rules of the application layer protocol, the priority, node source address, parameter group code definition is analyzed and formulated, and the communication matrix is designed. There are 8 levels of priority, and the message priority can be set from the highest

Table 55.1 Message ID

Nodes	ID	PGN	P	R	DP	PF	PS	SA
Engine management system	0 × FEDF1	00FEDF	0	0	0	FE	223	01
Battery management system	0 × FEB32	00FEB3	0	0	0	FE	199	02
Motor controller	0 × 4F11F3	00F11F	1	0	0	F1	31	03
ISG motor controller	0 × 4F18D4	00F18D	1	0	0	F1	31	04
Steering control system	0 × 4FEE85	00FEE8	1	0	0	FE	232	05
Transmission control system	0 × 4FEC76	00FEC7	1	0	0	FE	223	06
Braking system	0 × 4FEFA7	00FEFA	1	0	0	FE	250	07
PTO control system	0 × 8FEEF8	00FEEF	2	0	0	FE	240	08
Elevation control system	0 × CFEC39	00FEC3	3	0	0	FE	195	09
Tools control system	0 × CFEC5A	00FEC5	3	0	0	FE	197	10
Instrument display system	0 × 10FE6CB	00FE6C	4	0	0	FE	108	11
Fault detection system	0 × 18FEB1C	00FEB1	6	0	0	FE	177	12
Lighting control system	0 × 1CFE6ED	00FE6E	7	0	0	FE	110	13

0 to the lowest 7. In the SAE J1939 protocol, the default priority of all control messages is 3, and the default priority of all other messages, dedicated, request, and ACK messages are 6 [13]. Considering the safety, real time, and accuracy of the series hybrid tractor, the engine management system and the battery management system have the highest safety requirements, and the two nodes are set to the highest priority 0. The motor controller, ISG motor controller, steering control system, gearbox control system, and braking system nodes are set to priority 1. The priority of PTO control system, lifting system, and farm tool control system is set to 3. The priorities of the instrument display system, light control system, and fault detection system are set to 4, 6, and 7 in sequence.

The source address can be assigned by arranging the numbers in sequence, regardless of the priority, update speed, or importance of the message [14]. Define engine management system, battery management system, motor controller, ISG motor controller, steering control system, gearbox control system, braking system, PTO control system, lifting system, farm tool control system, instrument display system, fault The source addresses of the detection system and the lighting control system are 01, 02, 03, 04, 05, 06, 07, 08, 09, 10, 11, 12, and 13. The ID information of the message is shown in Table 55.1.

55.3.2 *Timing Design of Message*

Due to the difference between the structure and electronic control unit of the tandem hybrid tractor and the traditional tractor, the SAE J1939 protocol was referred to when designing the node message, which was redefined according to the specific

Table 55.2 Message timing

Send node	Name	Receive node	Number	Cycle (ms)	Priority
Engine management system	EI	Correspond node	1	10	0
Battery management system	BMSI	Correspond node	1	50	0
Motor controller	MCUI	Correspond node	1	50	1
ISG motor controller	ISGI	Correspond node	1	50	1
Steering control system	SCSI	Correspond node	1	50	1
Transmission control system	TCSI	Correspond node	1	50	1
Braking system	BCSI	Correspond node	1	50	1
PTO control system	PTOI	Correspond node	1	100	2
Elevation control system	ECSI	Correspond node	1	100	3
Tools control system	ATCSI	Correspond node	1	100	3
Instrument display system	IDSI	Correspond node	1	100	4
Fault detection system	MDSI	All nodes	1	100	6
Lighting control system	LCSI	Correspond node	1	100	7

situation of the tractor. According to the message timing of the tractor, as shown in Table 55.2.

55.4 CAN Bus Communication Network System Test

In order to verify the communication quality and reliability of the CAN bus communication network system, the CANoe network test tool developed by Vector is used to test on this system.

55.4.1 Message Test

According to the message ID and message sequence, a database is established through the CANoe database tool CANdb++, and simulation nodes, messages, signals, and environmental variables are added to the database [15]. Import the database into a network project with two CAN buses to test the CAN bus network. The Trace window of the CANoe software is a data tracking window, which can dynamically record all the information of the CAN network in real time. The number of messages displayed in the window is 13, the time of the test, the selection of the bus channel, the ID of the message, the data type, and the data content of each message collected. The test shows that the message is sent and received normally.

55.4.2 *Communication Quality Test*

The CAN Statistics window of the CANoe software can perform statistics on the messages and get the communication quality information of the network, showing the bus load rate, message frame rate, total message frame, error frame frequency, and total error frame information, as shown in Figs. 55.2 and 55.3. The test results show that the maximum load factor of CAN1 is 6.43%, the average is 6.43%, and the minimum is 6.23%. The maximum value of CAN2 load factor is 1.65%, the average value is 1.64%, and the minimum value is 1.48%. The bus load rate of CAN1 and CAN2 is lower than the standard safety threshold of 30%, no error frame occurred during the test, and the system performance is stable.

Fig. 55.2 CAN1 network load rate

CAN Channel: CAN 1 - CAN1

Statistic	Curr...	Min	Max	Avg
Busload [%]	6.43	6.23	6.43	6.43
Min. Send D...	0.000	n/a	n/a	n/a
Burst Time ...	2.032	2.032	2.032	2.032
Bursts [total]	2086550	n/a	n/a	n/a
Frames per ...	7	7	7	7
Std. Data [...	0	0	0	0
Std. Data [...	0	n/a	n/a	n/a
Ext. Data [...	220	213	220	220
Ext. Data [...	2295...	n/a	n/a	n/a
Std. Remote...	0	0	0	0
Std. Remote...	0	n/a	n/a	n/a
Ext. Remote...	0	0	0	0
Ext. Remote...	0	n/a	n/a	n/a
Errorframes...	0	0	0	0
Errorframes...	0	n/a	n/a	n/a
Chip State	Simu...	n/a	n/a	n/a
Transmit...	0	n/a	0	n/a
Receive ...	0	n/a	0	n/a
Transceiver...	0	n/a	n/a	n/a

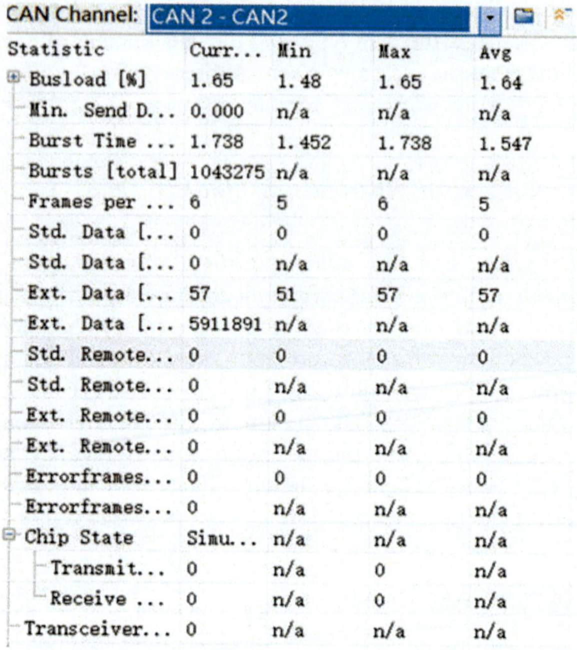

CAN Channel: CAN 2 - CAN2

Statistic	Curr...	Min	Max	Avg
Busload [%]	1.65	1.48	1.65	1.64
Min. Send D...	0.000	n/a	n/a	n/a
Burst Time ...	1.738	1.452	1.738	1.547
Bursts [total]	1043275	n/a	n/a	n/a
Frames per ...	6	5	6	5
Std. Data [...	0	0	0	0
Std. Data [...	0	n/a	n/a	n/a
Ext. Data [...	57	51	57	57
Ext. Data [...	5911891	n/a	n/a	n/a
Std. Remote...	0	0	0	0
Std. Remote...	0	n/a	n/a	n/a
Ext. Remote...	0	0	0	0
Ext. Remote...	0	n/a	n/a	n/a
Errorframes...	0	0	0	0
Errorframes...	0	n/a	n/a	n/a
Chip State	Simu...	n/a	n/a	n/a
Transmit...	0	n/a	0	n/a
Receive ...	0	n/a	0	n/a
Transceiver...	0	n/a	n/a	n/a

Fig. 55.3 CAN2 network load rate

55.4.3 Calculation of Bus Load Rate

This system uses two CAN bus networks, the baud rate is 500 kbps, the total number of binary transmissions within 1 s is 500,000 bit, the maximum number of bits required for a frame of message is 128 bit, the maximum that CAN1 and CAN2 can send within 1 s The number of messages is 500 000/128 = 3906. According to the sending frequency of the messages, the number of messages sent within 1 s can be calculated as.

CAN1: $(1/10 + 1/50 \times 7) \times 1000 = 240$

CAN2: $1/100 \times 5 \times 1000 = 50$

CAN1 network load rate: $240/3906 \times 100\% = 6.14\%$

CAN2 network load rate: $50/3906 \times 100\% = 1.28\%$.

There is a certain error between the calculated bus network load rate and the network load rate tested by the software, but the error is small. Considering that the calculation is the result of an ideal state, there may be queuing and interference in the transmission process when the actual message is sent. They have a good consistency.

55.5 Conclusions

In view of the current situation in my country where there is little research on the communication network of the hybrid tractor vehicle control system, tandem hybrid tractors are taken as the research object, the network topology structure is analyzed, and the CAN bus network structure design is carried out. According to the structure and actual requirements of the series hybrid tractor, the CAN network communication protocol was developed on the basis of the SAE J1939 protocol, and the specific information of the messages and nodes was designed. Using CANoe software to test the Controller Area Network system, it verifies the real-time and accuracy of the communication system. It provides reference for the development of services hybrid tractors in China.

Acknowledgements This article is one of the phased achievements of the Open Project of the State Key Laboratory of Tractor Power Systems (SKT2021002).

References

1. P.J. Wang, J.W. Li, S. Gao, Design and development of tractor control system based on model design. J Guangxi Univ. (Nat. Sci. Edition) **5**(43), 1695–1703 (2018)
2. M.N. Liu, *Design of Drive System of Series Hybrid Tractor* (Henan University of Science and Technology, 2014)
3. Z.B. Xia, *Design and Realization of High-Level Environment of Body Network* (Harbin Institute of Technology, 2009)
4. J. Huang, *Research on Body Control Network Based on CAN Bus* (Harbin Institute of Technology, 2007)
5. M.N. Liu, B. Han, L.Y. Xu, CAN bus network design of bifurcated power electric tractor. Peer-to-Peer Netw. Appl. **14**, 2306–2315
6. C.J. Jin, *Car Body Network Design Based on CAN Bus* (University of Electronic Science and Technology of China, 2009)
7. L.Y. Xu, Z.L. Zhou, B.B. Wang, Study on matching strategies of hydro mechanical continuously variable transmission system of tractor. Int. J. Digital Content Technol. Appl. **7**(4), 843–849 (2013)
8. Y.N. Chen, B. Xie, K. Liu, Design of CAN bus communication network system for electric tractor. J. Agric. Mech. Res. **39**(9), 233–238 (2017)
9. H.L. Xu, *Research on the Communication Network of Dual-Motor Independent Drive Electric Tractor Based on CAN* (Henan University of Science and Technology, 2019)
10. D.L. Wang, H.W. Zhang, Z.X. Gan, Research on the electronic and electrical architecture of plug-in hybrid vehicles based on CAN bus, in *The 14th Henan Automotive Engineering Science and Technology Symposium* (Pu Yang, 2017) pp. 384–386
11. Z.H. Xiao, S. Li, Application of CAN bus technology in hybrid electric vehicles. J. Chongqing Univ. (Nat. Sci. Edition) **6**, 68–70 (2005)
12. L.J. Yang, L.T. Chen, N.N. Lv, Research on the electronic and electrical architecture of the whole vehicle based on CAN bus, in *China Society of Automotive Engineering Annual Meeting* (Shang Hai, 2015), pp. 280–283
13. H.L. Xu, L.Y. Xu, X.H. Liu, CAN communication network design and hardware-in-the-loop test for electric tractors. J. Agric. Equipment Vehicle Eng. **58**(4), 1–5+12 (2020)

14. F. Luo, Z.C. Sun, Principle, Design and Application of Automobile CAN Bus System (Publishing House of Electronics Industry, Beijing, 2010)
15. J. Wu, Y.B. Li, J. Li, et al., CAN bus for automobile driving force control based on SAEJ1939 protocol. J. Jilin Univ. (Eng. Technol. Edition) **39**(04), 855–858 (2009)

Chapter 56
Research on Testability Fault Diagnosis Based on Deep Learning

Yufeng Long, Xianjun Shi, Yufeng Qin, and Jiapeng Lv

Abstract Research on data-driven testability fault diagnosis methods has attracted much attention in recent years. Now more and more scholars use deep learning method for fault diagnosis. Nevertheless, this approach suffers from two challenges, one is that the detection accuracy of ordinary deep learning methods is not good enough, the other is how to use deep learning for testability analysis. To address these two challenges, we propose a multi-stage convolutional neural network for testability fault diagnosis. The proposed method can learn features from raw data. The validation results show that the proposed method maintains a good diagnosis performance under different working conditions and gives higher testability analysis.

56.1 Introduction

Fault diagnosis has attracted continuous attention because of its great importance to industrial engineering, which is one of the most essential procedures in a wide range of machinery, such as helicopter, aeroengine, wind turbine, and high-speed train. Especially when you detect a node, you can know the diagnosis of the current node or the fault of the next node.

Traditional fault diagnosis methods mainly contain feature extraction using signal processing methods [1], wavelet-based methods [2], and fault classification [3] using machine learning approaches. However, facing heterogeneous massive data, feature extraction methods and traditional machine learning could not be suitable for this

Y. Long (✉) · X. Shi · Y. Qin · J. Lv
Naval Aviation University, Yantai, China

X. Shi
e-mail: sxjaa@sina.com

Y. Qin
e-mail: hj_qyf082@163.com

J. Lv
e-mail: lvjiapeng001@163.com

G. Liu and F. Cen (eds.), *Advances in Precision Instruments and Optical Engineering*, Springer Proceedings in Physics 270, https://doi.org/10.1007/978-981-16-7258-3_56

situation. Thus, how to perform diagnosis more precisely and efficiently is still a challenging problem.

Zhibin Zhao [4] performed a comprehensive evaluation of four models, including multi-layer perception (MLP), autoencoder (AE), convolutional neural network (CNN), and recurrent neural network (RNN). Hoang et al. [2] provided a comprehensive review of three popular deep learning (DL) algorithms (AE, DBN, and CNN) for bearing fault diagnosis. Zhang et al. [5] reviewed the machine learning and DL-based algorithms for bearing fault diagnosis systematically and also provided a comparison of the classification accuracy of CWRU with different DL-based methods. Xu proposed a new intelligent diagnosis method based on an elaborately designed deep neural network for failure detection of the wind turbine, which solved the problem of unbalanced distribution with regard to SCADA data [6]. Combined a CNN with a Naive Bayes data fusion proposal, Chen et al. applied DL theory to nuclear power plant inspection [7]. Zhang et al. constructed a new unsupervised learning method called general normalized sparse filtering, which was used for fault diagnosis of rolling bearing and planetary gearbox [8]. Pei Cao [9] proposed transfer learning architecture for Fault Diagnosis, also consists features extraction and a fully-connected stage (Fig. 56.1).

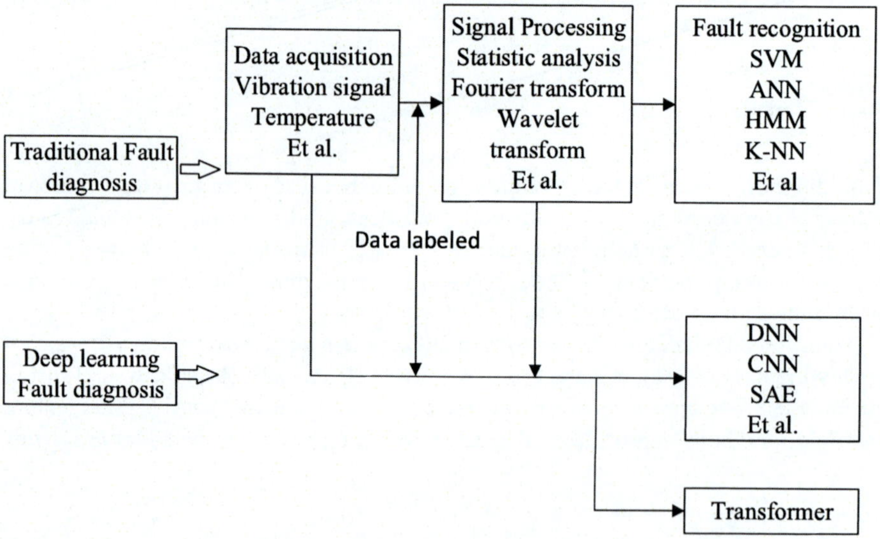

Fig. 56.1 Diagnosis method

56.2 Testability Fault Diagnosis

Testability fault diagnosis state in the following terms. Given a set of measurements taken at preselected test points, the target under test can properly be stimulated at preselected injection points and isolate points [10]. General testability fault analysis is a single fault diagnosis, while the actual testability analysis is multi-fault diagnosis. Deep learning fault analysis could detect multi faults, we use deep learning in testability analysis in this part.

56.2.1 Testability Analysis

Diagnostic strategy is the focus of design for testability, which is the key to improve the fault diagnosis ability, enhancing the diagnostic efficiency, and reduce the full life cycle cost. Dependence matrix (D matrix) is the basis of realizing diagnosis strategy and testability analysis, and it is also the "data bridge" for transforming testability model into diagnosis strategy [11].

The optimal testability problem, known as the test planning problem, is to design a testability strategy that clearly isolates the fault state to meet the requirements with the minimum expected test cost. The OTP parameters are defined as the five-tuple (S, P, T, C, D). $S = \{s_0, s_1, \ldots, s_m\}$ is a set of statistically independent failure states associated with the system. $P = \{p(s_0), p(s_1), \ldots, p(s_m)\}$ is a priori probability vector associated with the set of failure states S. $T = \{t_1, t_2, \ldots, t_n\}$ is a finite set of n reliable binary outcome tests, where each test t_j checks a subset of S. $C = \{c_1, c_2, \ldots, c_n\}$ is a set of test costs measured in terms of time, manpower requirements, or other economic factors. $D = [d_{ij}]$ is a binary matrix of dimension $(m + 1) \times n$ called dependence matrix which represents the relationship between the set of failure states S and the set of tests T, where $d_{ij} = 1$ if the test t_j monitors failure state s_i otherwise $d_{ij} = 0$. The OTP parameters of one small-scale system are shown in Table 56.1 [12].

Table 56.1 OTP parameters for small scale example

S	Accuracy				P
S1	t1	t2	t3	t4	0.08
S2	1	0	0	0	0.14
S3	0	1	1	0	0.26
S4	0	0	1	1	0.22
costC	1.0	2.0	3.0	2.0	

56.2.2 Deep Residual Network Based on Fault Detection and Diagnosis Model

Most deep learning techniques are capable of automatic non-linear feature extraction. Compared to other deep learning methods, such as plain CNN, the advantage of the ResNet includes a higher training speed, easier gradient transmission, and deeper neural network with less gradient vanishing or explosion, and so on [13]. Therefore, in this paper, the ResNet is explored as a multi-stage backbone (multi-stage backbone has been used in many CV problems [14]) to detect and diagnose the normal status and different faults. The architecture of the original ResNet of 34-layers is proposed for the complex image classification, which is composed of some convolution layers, pooling layers, and fully-connected layers (Fig. 56.2).

After the feature extraction and compression by the convolution layers, the linear classifier is used to compute the classification results. In this paper, the linear classifier is composed of a fully-connected (FC) layer and a Softmax function. Specifically, the FC layer connects every neuron from one layer to another, which has the same principle as a multi-layer perceptron. However, the output of the FC is a continuous value that cannot directly represents the discrete classification results. To implement

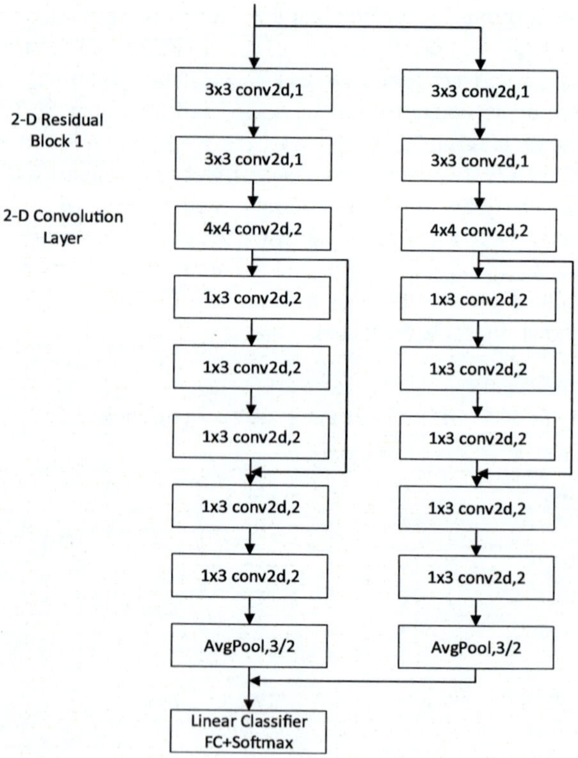

Fig. 56.2 Multi-stage ResNet backbone

classification, the Softmax function is used to normalize the input vector into a vector of the probability belonging to each class, and the class with the maximum probability is chosen as the final prediction output.

$Softmax(x_i) = \frac{e^{x_i}}{\sum_{k=1}^{C_0} e^{x_k}}, i = 1, 2, \ldots, C_0)$ (1).

where the x_i is one of the elements of the output layer and C_0 is the dimension of the output.

56.2.3 Deep Learning in Testability Analysis

Deep learning fault detection can detect many kinds of fault just with one detection. In fact, many nodes detection include many fault types, not only one fault. We combine the discriminant diagram of the fault diagnosis tree of the testable D matrix. The Fig. 56.3 shows the discriminant diagram of the fault diagnosis tree of the testable D matrix, which is a fault diagnosis tree constructed according to Table 56.1. The original test nodes are t1 ~ t4. If the type of each node failure is N_y, and the result of each type of failure is y_i, using Softmax makes the probability of failure in a certain type high, and the probability of not being in a certain type is low. Diagnose the fault according to the fault diagnosis tree in turn (Table 56.2).

In the testable fault D matrix, the probability $P = \{p(s_1), \ldots, p(s_m)\}$ is a priori probability vector associated with the set of fault states S. In deep learning, in testability analysis, P is a priori probability, while P contain N_y faults. We can use many test nodes to get the P, for $p(s_1) = \frac{\sum_{j=N} p(s_1)_j}{N}$, $p(s_1)$ is vector N_y. We can obtain the empirical value P based on the value of multiple measurements.

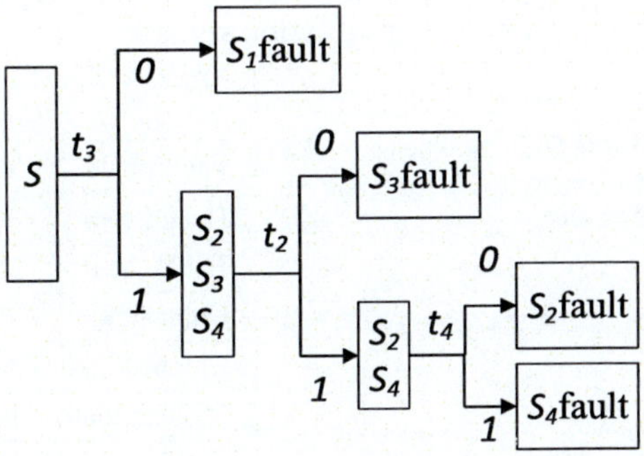

Fig. 56.3 Deep learning in testability analysis

Table 56.2 Original data accuracy

The test process can be:
Require: The fault diagnosis tree of testability D matrix can generally determine the fault type according to data and experience, and the output type of each node is y_{t_j}, Where t_j is t1 ~ t4, y_{t_j} include N_y fault types, T_I is the detection threshold. When it is greater than the threshold, it is considered as the fault. When it is less than the threshold, it is not considered as the fault, T_I can be selected as 0.7
t_3 test if $y_{t_{3i}} > T_I, i \in 1, \ldots, N_y$, to $\{S_2, S_3, S_4\}$
if $y_{t_{2i}} > T_I, i \in 1, \ldots, N_y$, to $\{S_2, S_4\}$
if $y_{t_{4i}} > T_I, i \in 1, \ldots, N_y$, to $\{S_4\}$
else to $\{S_2\}$
else to $\{S_3\}$
else to $\{S_1\}$

56.3 Experiments

56.3.1 Datasets

Southeast University (SEU) gearbox datasets were provided by Southeast University [15, 16]. SEU datasets contained two sub-datasets, including a bearing dataset and a gear dataset, which were both acquired on drivetrain dynamic simulator (DDS). There were two kinds of working conditions with rotating speed-load configuration (RS-LC) set to be 20 Hz–0 V and 30 Hz–2 V shown in Table 56.3. The total number of classes was equal to twenty according to Table 56.3 under different working conditions. Within each file, there were eight rows of vibration signals, and we used the second row of vibration signals.

Table 56.3 Southeast University (SEU) gearbox datasets

Fault mode	Table column	Subhead	Subhead
Health gear	20 Hz–0 V	Health bearing	20 Hz–0 V
Health gear	30 Hz–2 V	Health bearing	30 Hz–2 V
Chipped tooth	20 Hz–0 V	Inner ring	20 Hz–0 V
Chipped tooth	30 Hz–2 V	Inner ring	30 Hz–2 V
Missing tooth	20 Hz–0 V	Outer ring	20 Hz–0 V
Missing ttooth	30 Hz–2 V	Outer ring	30 Hz–2 V
Root fault	20 Hz–0 V	Inner + outer ring	20 Hz-0 V
Root fault	30 Hz–2 V	Inner + outer ring	30 Hz–2 V
Surface fault	20 Hz–0 V	Rolling element	20 Hz–0 V
Surface fault	30 Hz–2 V	Rolling element	30 Hz–2 V

56.3.2 Deep Fault Detection

We use 1D data and 1 * 1024 dimensional input values, convert them into 64 * 512 matrix through up sampling and then send them into the convolution network in the figure above. Finally, the output FC is 64 * 20 dimension, and the probability of belonging to a certain type is obtained through softmax. So the deep learning fault detection flow includes three parts: One part is data acquisition, data acquire from test device's node as the same as t1~t4. Two part is data or data transform, data transform is the main wavelet transform [2]. Three is model train and model test. We test the accuracy when the original data and wavelet transform enters into the network. We compare the methods proposed by SAE (sparse autoencoder), AE(autoencoder), LeNet, AlexNet, and ResNet (Fig. 56.4, Tables 56.4 and 56.5).

Compared with other networks, the accuracy of our proposed method is improved by 1%–3%. When the accuracy reaches 99%, it is difficult to improve the accuracy. As can be seen from Fig. 56.5, the distribution of data after wavelet transform is obvious.

56.3.3 Deep Learning in Testability Analysis

Assuming we have detected each point t1~t4. The type of fault at each point is the type of SEU fault. We can intuitively carry out testable diagnosis and analysis by substituting the value of probability into D matrix. Figure 56.6 is N_{yi} fault type. Assuming (a) is N_{y2} fault type, (a) have not fault, because the numerical value less than T_I detection threshold. Assuming (b) is N_{y10} fault type, so (b) have S_2, S_3, S_4 fault, the fault type is N_{y10}.

When it is necessary to isolate a certain type of fault, we can intuitively judge it through a histogram. N_{yi} is fault type. Form Fig. 56.7, when test point t2~t4, the type of N_{yi} numerical value greater than T_I detection threshold, So the fault type is N_{yi}. The test method is shown in 56.2.

56.4 Conclusion

We propose a multi-stage convolutional neural network for testability fault diagnosis. The proposed method can learn features from raw data. The validation results show that the proposed method maintains a good diagnosis performance under different working conditions and gives higher testability analysis. Compared with other networks, the accuracy of our proposed method is improved by 1%–3%. When the accuracy reaches 99%, it is difficult to improve the accuracy. While in complex occasions, the matrix D is not greatly established. How to use deep learning in testability analysis is also a pivotal problem needed to be thought.

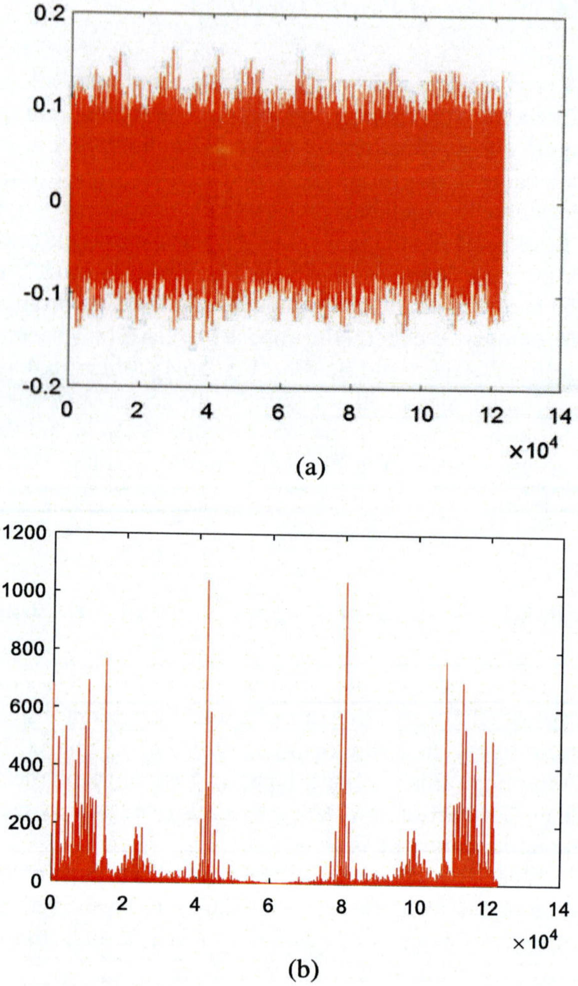

Fig. 56.4 Fault datasets. **a** Is original data, **b** is wavelet transform. x-axis is the number of data, y-axis is amplitude

Table 56.4 Original data accuracy

Architecture	Accuracy
SAE(sparse autoencoder) [2]	0.5714
AE(autoencoder) [2]	0.5429
LeNet	0.6103
AlexNet	0.6801
ResNet	0.7002
Our	0.7306

Table 56.5 Wavelet transform accuracy

Architecture	Accuracy
SAE(sparse autoencoder)	0.8356
AE(autoencoder)	0.8408
LeNet	0.8238
AlexNet	0.9210
ResNet	0.9802
Our	0.9905

1 13%	0	0	0
0	1 12%	1 21%	0
0	0	1 21%	1 15%
0	1 12%	1 21%	1 15%

(a)

1 0.6%	0	0	0
0	1 0.5%	1 78%	0
0	0	1 78%	1 80%
0	1 0.5%	1 78%	1 80%

(b)

Fig. 56.5 N_{yi} fault type. Assuming **a** is N_{y2} fault type and **b** is N_{y10} fault type

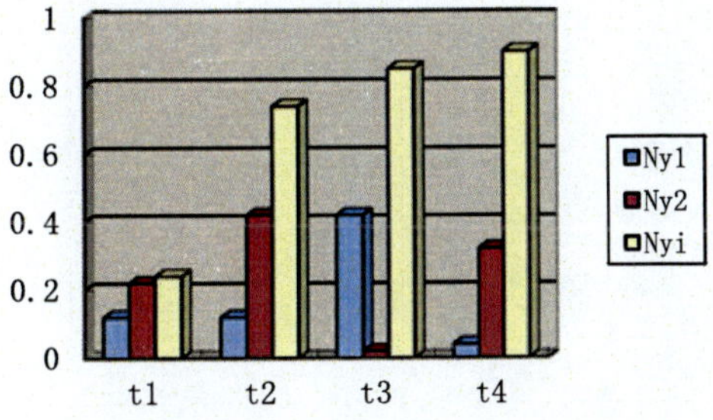

Fig. 56.6 Fault type detection histogram

References

1. Z. Zhao, S. Wu, B. Qiao, S. Wang, X. Chen, Enhanced sparse period-group lasso for bearing fault diagnosis. IEEE Trans. Industr. Electron. **66**, 2143–2153 (2018)
2. D.-T. Hoang, H.-J. Kang, A survey on deep learning based bearing fault diagnosis. Neurocomputing **335**, 327–335 (2019)
3. C. Sun, M. Ma, Z. Zhao, X. Chen, Sparse deep stacking network for fault diagnosis of motor. IEEE Trans. Industr. Inf. **14**, 3261–3270 (2018)
4. Z. Zhao, T. Li, J. Wu, C. Sun, Deep learning algorithms for rotating machinery intelligent diagnosis: an open source Benchmark study. ISA Trans (2020)
5. S. Zhang, S. Zhang, B. Wang, T.G. Habetler, Deep learning algorithms for bearing fault diagnosticsx-a comprehensive review. IEEE Access **8**, 29857–29881 (2020)
6. L. Chen, G. Xu, Q. Zhang, X. Zhang, 'Learning deep representation of imbalanced SCADA data for fault detection of wind turbines.' Measurement **139**, 370–379 (2019)
7. F.-C. Chen, M.R. Jahanshahi, 'NB-CNN: Deep learning-based crack detection using convolutional neural network and Naïve Bayes data fusion.' IEEE Trans. Ind. Electron. **65**(5), 4392–4400 (2018)
8. Z. Zhang, S. Li, J. Wang, Y. Xin, Z. An, 'General normalized sparse filtering: A novel unsupervised learning method for rotating machinery fault diagnosis.' Mech. Syst. Signal Process. **124**, 596–612 (2019)
9. P. Cao, S. Zhang, J. Tang, Pre-processing-free gear fault diagnosis using small datasets with deep convolutional neural network-based transfer learning. IEEE Access 26241–26253, (2017)
10. L. Xu, Z. Cheng, H. Yan, et al., A frame design of helicopter health monitoring and diagnosis system based on testability, in *2017 Prognostics and System Health Management Conference (PHM-Harbin)* (IEEE, 2017)
11. T. Heng, *Reaearch of Fault Diagnosis Strategy Based on Testability D Matrix*. Dalian University of Technology. Ph.D (2019)
12. X.H. Qiu, Y.T. Hu, B. Li, Sequential fault diagnosis using an inertial velocity differential evolution algorithm. Int. J. Autom. Comput. (2019)
13. Z.C. Chen, P. Lin, Deep residual network based fault detection and diagnosis of photovoltaic arrays using current-voltage curves and ambient conditions. Energy Convers. Manage. (2019)
14. S.W. Zamir, A. Arora, S. Khan, *Multi-Stage Progressive Image Restoration* (CVPR, 2021)
15. SEU gearbox datasets, [Online], Available: https://github.com/cathysiyu/Mechanical-datasets. Accessed 2019, September.
16. S. Shao, S. McAleer, R. Yan, P. Baldi, Highly accurate machine fault diagnosis using deep transfer learning. IEEE Trans. Industr. Inf. **15**, 2446–2455 (2018)